Lecture Notes in Control and Information Sciences 416

Andreas Varga, Anders Hansson,
and Guilhem Puyou (Eds.)

Optimization Based Clearance of Flight Control Laws

A Civil Aircraft Application

Springer

Editors

Dr. Andreas Varga
DLR - Oberpfaffenhofen
German Aerospace Center
Institute of Robotics and Mechatronics
82234 Wessling
Germany
E-mail: andreas.varga@dlr.de

Dr. Guilhem Puyou
AIRBUS
Stability and Control Department
316 route de Bayonne
31060 Toulouse Cedex 03
France
E-mail: guilhem.puyou@airbus.com

Prof. Anders Hansson
Linköpings Universitet
Department of Electrical Engineering
Division of Automatic Control
SE-581 83 Linköping
Sweden
E-mail: hansson@isy.liu.se

ISBN 978-3-642-22626-7 e-ISBN 978-3-642-22627-4

DOI 10.1007/978-3-642-22627-4

Lecture Notes in Control and Information Sciences ISSN 0170-8643

Library of Congress Control Number: 2011934146

Typeset & Cover Design: Scientific Publishing Services Pvt. Ltd., Chennai, India.

Printed on acid-free paper

9 8 7 6 5 4 3 2 1

springer.com

Foreword

As far as the laws of mathematics refer to reality, they are not certain; and as far as they are certain, they do not refer to reality.

Albert Einstein

Since the early days of aviation, engineers with inventive talents had to overcome tremendous challenges. From building vehicles that can fly in a controllable manner designed and flown by pioneers like Lilienthal, Wright brothers, Sperry in the beginnings, to developing todays modern comfortable and reliable vehicles for daily routine all-weather operations (Airbus, Boeing and others), there were many creative efforts to improve performance (aircraft size, endurance, speed), minimize structural weight, provide necessary thrust, and guarantee safe flight operations. Nowadays all these inventions assure the high mobility of the modern human society in a global world. The aeronautical challenges were drivers for many new technologies and methodologies that are commonly used in other industries today.

In the last three decades, the requirements for the design of high-performance flight control systems that enhance automation of flight, initiated a number of ingenious technological developments. Flight control law design is one of the areas where aeronautical engineers are pioneering new technologies.

During development (design and test) of flight control laws, engineers rely on mathematical models. Inevitably such models cannot mimic all aspects of a highly complex, physical plant as a modern high-performance jet airplane and its environment (atmosphere, air traffic, etc.) with absolute fidelity. The above quotation from Einstein describes the fundamental difficulty that control engineers are facing, when striving the clearance of flight control laws on basis of mathematical models.

However, engineers (named after the Latin word *ingenium*, meaning innate quality, especially mental power, hence a clever invention) often find solutions even to the most challenging problems. To prove that an aircraft is safe to fly

Contents

Part I: Clearance of Civil Aircraft

Andrea Garulli
Dipartimento di Ingegneria dell'Informazione, Università di Siena,
Via Roma 56, 53100 Siena, Italy
e-mail: garulli@dii.unisi.it

Anders Hansson (Editor)
Division of Automatic Control, Linköping University,
SE-581 83 Linköping, Sweden
e-mail: hansson@isy.liu.se

Simon Hecker
DLR - Oberpfaffenhofen, Institute of Robotics and Mechatronics,
82234 Wessling, Germany
e-mail: simon.hecker@hm.edu

Hans-Dieter Joos
DLR - Oberpfaffenhofen, Institute of Robotics and Mechatronics,
82234 Wessling, Germany
e-mail: dieter.joos@dlr.de

Fabien Lescher
ONERA, Systems Control and Flight Dynamics Department,
2 avenue Edouard Belin, 31055 Toulouse Cedex 4, France
e-mail: fabien.lescher@onera.fr

Johan Löfberg
Division of Automatic Control, Linköping University,
SE-581 83 Linköping, Sweden
e-mail: johanl@isy.liu.se

Yannick Losser
AIRBUS Operation SAS,
316 route de Bayonne, 31060 Toulouse Cedex 03, France
e-mail: yannick.losser@supaero.org

Alfio Masi
Dipartimento di Ingegneria dell'Informazione, Università di Siena,
Via Roma 56, 53100 Siena, Italy
e-mail: masi@dii.unisi.it

Philippe Menard
AIRBUS Operation SAS,
316 route de Bayonne, 31060 Toulouse Cedex 03, France
e-mail: philippe.menard@airbus.com

Rafael Fernandes de Oliveira
AIRBUS Operation SAS,
316 route de Bayonne, 31060 Toulouse Cedex 03, France
e-mail: rafael@rafael.aero

Andrew K. Packard
Department of Mechanical Engineering, University of California,
Berkeley, CA, 94720, USA
e-mail: apackard@berkeley.edu

Sina Khoshfetrat Pakazad
Division of Automatic Control, Linköping University,
SE-581 83 Linköping, Sweden
e-mail: sina.kh.pa@isy.liu.se

Christos Papageorgiou
Division of Automatic Control, Linköping University,
SE-581 83 Linköping, Sweden
e-mail: papageorgiou@isy.liu.se

Simone Paoletti
Dipartimento di Ingegneria dell'Informazione, Università di Siena,
Via Roma 56, 53100 Siena, Italy
e-mail: paoletti@dii.unisi.it

Daniel Petersson
Division of Automatic Control, Linköping University,
SE-581 83 Linköping, Sweden
e-mail: petersson@isy.liu.se

Harald Pfifer
DLR - Oberpfaffenhofen, Institute of Robotics and Mechatronics,
82234 Wessling, Germany
e-mail: harald.pfifer@dlr.de

Guilhem Puyou (Editor)
AIRBUS Operation SAS,
316 route de Bayonne, 31060 Toulouse Cedex 03, France
e-mail: guilhem.puyou@airbus.com

Clément Roos
ONERA, Systems Control and Flight Dynamics Department,
2 avenue Edouard Belin, 31055 Toulouse Cedex 4, France
e-mail: clement.roos@onera.fr

Peter Seiler
Department of Aerospace Engineering and Mechanics,
University of Minnesota, Minneapolis, MN, 55455, USA
e-mail: seiler@aem.umn.edu

Daniel Skoogh
Swedish Research Defence Agency (FOI), System Technology,
SE-16490 Stockholm, Sweden
e-mail: daniel.skoogh@foi.se

Robert Stich
Cassidian, Air Systems, Flight Control,
Rechliner Straße, 85077 Manching, Germany
e-mail: Robert.Stich@cassidian.com

Ercüment Türkoğlu
Dipartimento di Ingegneria dell'Informazione, Università di Siena,
Via Roma 56, 53100 Siena, Italy
e-mail: turkoglu@dii.unisi.it

Andreas Varga (Editor)
DLR - Oberpfaffenhofen, Institute of Robotics and Mechatronics,
82234 Wessling, Germany
e-mail: andreas.varga@dlr.de

Ragnar Wallin
Department of Electrical Engineering, Linköping University,
SE-581 83 Linköping, Sweden
e-mail: ragnarw@isy.liu.se

Acronyms

ADS	Air Data System
AIRSAFE	Aircraft Integrated Resilient Safety Assurance and Failsafe Enhancement
ALN	Alenia Aeronautica
AoA	Angle of Attack
AoS	Angle of Sideslip
AvSP	Aviation Safety Program
BAES	BAE Systems
CAS	Control Augmentation System
CAST	Commercial Aviation Safety Team
CoG	Centre of Gravity
CFCL	Clearance of Flight Control Laws
CT	Center (fuel) Tank
DE	Differential Evolution
DIRECT	Dividing Rectangles Method
DOF	Degrees of Freedom
DLR	Deutsches Zentrum für Luft- und Raumfahrt (German Aerospace Centre)
EADS	European Aeronautic Defence and Space Company
EADS-C	EADS – Casa (Spain)
EADS-D	EADS – Defense & Security, Military Air Systems (Germany)
EPC	Eurofighter Partner Company
ES	Evolutionary Strategy
FAA	Federal Aviation Administration
FAR	Federal Aviation Regulation
FBW	Fly-By-Wire
FCS	Flight Control System
FJT	FCS Joint Team
FOI	Totalförsvarets Forskningsinstitut (Swedish Defence Research Agency)
GA	Genetic Algorithm

Chapter 1
Introduction

Anders Hansson and Andreas Varga

Abstract. We describe the background and motivation of the research carried out within the COFCLUO project in developing and applying optimization techniques to the clearance of flight control laws for civil aircraft.

1.1 Background

To prove to the certification authorities that an aircraft is safe to fly is a long and complicated process. It is the responsibility of the manufacturer to show that the aircraft complies with the certification specifications, and especially the airworthiness requirements specified in the European and US regulations JAR Part 25 [3] and FAR Part 25 [4], respectively. These requirements consists of many different *clearance criteria* that has to be met by the designed flight control system. The main activity in this context is the *clearance of flight control laws* (CFCL). One typical example for a safety critical requirement is the stability of the controlled aircraft in all flight conditions (characterized by different Mach number, altitude, dynamic pressure, or angle of attack), and for all possible values of aircraft parameters (like mass or position of the centre of gravity).

Before the first manned flights can be executed, the CFCL must be performed to prove that the controlled aircraft meets all clearance criteria. For this purpose, extensive computer aided simulations and robustness assessment are performed. Both computations rely on a high fidelity nonlinear

Anders Hansson
Division of Automatic Control, Linköping University, SE-581 83 Linköping, Sweden
e-mail: `hansson@isy.liu.se`

Andreas Varga
Institute of Robotics and Mechatronics, DLR - Oberpfaffenhofen,
82234 Wessling, Germany
e-mail: `andreas.varga@dlr.de`

A. Varga et al. (Eds.): Opti. Based Clearance of Flight Control Laws, LNCIS 416, pp. 3–9.
springerlink.com

dynamics model of the aircraft which explicitly includes the dependencies of all parametric uncertainties (such as mass, centre of gravity, aerodynamics coefficients).

For the assessment of the already mentioned stability criterion, analysis based on linear models are currently performed. Such models are obtained by trimming and linearization of the nonlinear models around steady state flight conditions for fixed values of parameters. The resulting linearized models are therefore only valid for small perturbations around the equilibrium state. By using a sufficiently large number of different linear models obtained for different values of the uncertain parameters and different flight conditions, the stability of the nonlinear model can be investigated by assessing the stability of the linearized models. In this way, any instability detected for the linearized models will disqualify the flight control system design.

Several other clearance criteria for civil aircraft are also formulated in terms of linearized models. The state of the art of industrial certification practice for all these criteria is to investigate a finite (possibly large) number of different linear models, selected from a continuum of models corresponding to each possible value of the parameters and flight conditions. Even if, as today, many thousands of different linear models are investigated on fairly fine grids for uncertain parameters and flight conditions, it is still possible to miss the model corresponding to the most critical parameter combination. Moreover, it is well known that robust control methods based on gridding of the parameter space are highly computationally demanding and suffer from the curse of dimensionality. With an optimization based clearance approach this problem can be overcome in an automated and systematic way.

Even if the whole continuum of linear models is found to meet all the clearance criteria, the underlying nonlinear system may not perform satisfactory, it might actually be unstable. As mentioned above, traditionally this is investigated by both off-line and piloted simulations using the high fidelity aircraft nonlinear model. However, stability for any other manoeuvres than the ones investigated in the simulations cannot be induced from such simulations. Since there are infinitely many different manoeuvres to investigate, it will never be possible to prove stability of an aircraft using only simulations. With optimization based clearance it is believed that this challenge may be overcome.

Impact

Over the next 10 to 15 years vehicle manufacturing is among the groups of technologies expected to hold the most relevant transport application and the most relevant applications to improve vehicle efficiency, [2]. For air transport this will contribute to improve or develop among others sub-sonic civil aircraft with increased capacity, lower emission, noise and fuel consumption. Technologies relevant for vehicle manufacturing are, among others, the *rapid modeling* and *rapid prototyping* for vehicle design. The main drivers for this are competition due to globalization, emerging markets, and flexibility required to meet increased pressure from customers. The efficient and reliable CFCL can be seen as a key technology supporting rapid prototyping of aircraft, with an important impact on increasing flight safety and reducing manufacturing costs.

Previous Research

The shortcomings of the standard industrial clearance process was recognised already in 1999 by the Group for Aeronautical Research and Technology in Europe (GARTEUR), which established the Flight Mechanics Action Group 11 (AG11) to address the CFCL. The AG11 investigated several new techniques and published its findings in [1]. The industrial evaluation pointed out the optimization based CFCL as the most promising approach. The reasons for this are several:

Generality: It can be used for both frequency-domain and time-domain analysis, and for both linear and non-linear models. Moreover, there is no limitation on the number of parametric uncertainties that can be investigated.
Reliability: It will not miss points in parameter space or flight envelope lying in-between a grid.
Non-conservatism: It does not itself add conservatism to the clearance problem as do many other alternative approaches to the traditional griding or Monte-Carlo analysis based CFCL.

It should be mentioned that the GARTEUR group focused its efforts on military applications but that Chapter 24 in the aforementioned reference [1] was devoted to considerations for clearance of civil transport aircraft in which Robert Luckner (formerly at AIRBUS Deutschland GmbH, now at Technical University Berlin) writes: "Therefore, methods and techniques that support an efficient automated flight control laws clearance process – as the optimization-based worst-case method – are needed and should attract researchers' attention. To investigate and demonstrate the benefits of new analysis methods for the civil clearance process, an equivalent civil benchmark problem should be formulated. This benchmark should consist of a civil aircraft model plus controller and a baseline solution. Special emphasis should be placed on the analysis of nonlinear criteria. A combination of these techniques with optimization-based worst-case search should be considered."

An important finding of the GARTEUR AG11 is the complementarity between worst-case search methods and global robustness assessment approaches as μ-analysis. For example, the latter approach can guarantee the stability of a continuum of uncertain parametric models and provide bounds on admissible values of uncertain parameters. The limitations of this approach are related to the significant modelling effort necessary to develop *linear parameter varying* (LPV) models which approximate with a sufficient accuracy the nonlinear aircraft model for all values of uncertain parameters over the full or parts of the flight envelope. Increasing the accuracy of approximations is typically associated with increased modelling and analysis efforts (due to higher dimension of uncertain models or using several uncertain models). Furthermore, these methods generally provide only bounds on the admissible parameter values, but not the worst-case parameter combinations. On the other side, the worst-case search based methods aim to determine worst-case parameter combinations, and thus can easily disqualify a given design by determining just a single value where a clearance criterion is violated. However, when a violation of a criterion is not detected even by using global search based approaches, the worst-case search method cannot guarantee global stability, because the existing remote risk of failure of any global search algorithm. In the light of this discussion it is clear that a combination of the two approaches can provide the best results in terms of the reliability of clearance and in detecting critical parameter combinations.

Enhancement of State-of-the-Art

From what has been said previously it is clear that optimization based clearance is one of the most promising approaches for CFCL for civil aircraft. Despite the fact that much work had been done on optimization based CFCL for military aircraft, there were certain difficulties with the method that needed further attention – it was not just the application of available theory to a civil aircraft that needed to be performed. For civil aircraft, dynamics related to the flexible structure require different, more detailed and thus larger models than what is necessary for military aircraft. Therefore new, integrated models were developed and a special attention was paid to the fast trimming and linearization of these models. Also the question of how to obtain rational approximations of the state space matrices of the linear parameter-varying systems resulting from the linearization was addressed. This was essential in order to build so-called Linear Fractional Transformation based parametric models, which are the state-of-the art model representations used in robustness analysis of control systems. Another main difference in clearance of civil aircraft as compared to military aircraft is the emphasis on protection violation which calls for other types of clearance criteria. There are many more failure cases to be investigated for a civil aircraft. Therefore new optimization based clearance techniques where developed to address this challenge, among others by using global search methods and convex relaxation methods. In addition to this, the optimization algorithms for CFCL often have tuning parameters, which for the ordinary engineer might be difficult to understand. Because of this, easy-to-use graphical user interfaces (GUIs) were developed. Also some optimization problems might have such a large dimension, or the number of problems to be solved might be so large, that answers might not be found in reasonable time with off-the-shelf optimization solvers. Hence parallel computations were employed.

Objectives

It is important to keep in mind that the questions addressed in this project were not purely technical, since industry was already technically able to successfully clear flight control laws. The main industrial benefits of the new methods should be related to reducing the involved effort and cost, while getting sufficiently reliable results, or increasing the reliability of the analysis results with a reasonable amount of effort. Therefore a benchmark problem was defined according to current industrial standards, and the results obtained from optimization-based clearance was compared with a baseline traditional solution based on gridding the parameter space and testing the flight control laws for a finite number of manoeuvres.

1.2 The COFCLUO Project

The results achieved in the COFCLUO project were obtained on the basis of a strong and fruitful cooperation among 6 partner organizations involving a major civil aircraft manufacturer, three research establishments and two universities:

1. Linköping University – Project coordinator (Sweden, Linköping)
2. AIRBUS France SAS (France, Toulouse)
3. Deutsches Zentrum für Luft- und Raumfahrt e.V. (DLR-Oberpfaffenhon, Germany, Wessling)
4. Swedish Defense Research Agency (FOI, Sweden, Stockholm)
5. Office National d'Etudes et de Recherches Aerospatiales (ONERA, France, Toulouse)
6. Universita degli Studi di Siena (Italy, Siena)

The work to be carried out was divided into four main work packages (WPs). Besides the WP0 for project management, there were work packages for aircraft modeling (WP1), optimization techniques for clearance (WP2) and evaluation of results (WP3). The allocation of the work carried out within WP1, WP2 and WP3 approximately corresponds to the successive activities carried out in the project. A prerequisite for optimization based clearance in WP2 was that suitable models to use in the optimization had been developed in WP1, while a prerequisite for the evaluation of the proposed optimization techniques in WP3 was that they had been developed and validated in WP2.

The project work was structured in four logical phases: (1) conception, (2) study, (3) design and (4) implementation. The conception phase was fully carried out while preparing the project-proposal. The study phase was mainly performed during the first 12 months. During this phase surveys of existing technologies were performed and the need for future development assessed. Also initial models were developed. During the design phase the research and development of new technologies for optimization based clearance of flight control laws was carried out. Also a baseline solution to clearance was implemented. Some further modeling issues were also dealt with. The design phase was mainly performed during months 7–24. During the implementation phase software was written, which implemented the technologies developed, and finally the software was evaluated on the benchmark against the baseline solution. This work was mainly carried out during the last 12 months.

Project Achievements

The main achievements of the COFLCUO Project are documented in the *project deliverables*, which are technical reports, of which the public ones can be downloaded from `http://www.cofcluo.isy.liu.se`.

Two international workshops addressing the topic of clearance of flight control laws have been organized within the COFCLUO project. Their web-sites are located at `http://www.unisi.it/eventi/cofcluo/index.htm` and `http://www.cofcluo.isy.liu.se`.

Finally, an invited session was organized at the 6th IFAC Symposium on Robust Control Design in Haifa in 2009. In total, 11 scientific papers have been published within the project. This book summarizes the main achievements of the project.

1.3 Outline of the Book

The book consists of six parts. In part one the background and motivation of the COFCLUO Project is described in Chapter 1 and the clearance benchmark problem for a representative civil aircraft is formulated in Chapter 2.

Part two consists of 4 chapters, which describe the development of LPV uncertainty models and their *linear-fractional representations* (LFRs) expressed via *linear fractional transformations* (LFTs). Chapter 3 describes the generation of LPV models and the corresponding LFRs starting from the high-fidelity nonlinear aircraft dynamics model. Chapter 4 is dedicated to the generation of LFRs from a collection of large order integral linearized models which describe the flexible aircraft dynamics in various load configurations. Chapter 5 addresses the development of uncertainty models for the controller as LFRs. The LFRs for the open-loop aircraft and controller are then feedback connected to obtain the LFRs for the linearized closed-loop system. Chapter 6 describes a new approached to identify LPV uncertainty models by minimizing a suitable input-output error norm.

Part three is the main theoretical part of the book and describes various optimization-based approaches to address different aspects of the clearance problem. Chapter 7 deals with linear robustness analysis methods using enhanced μ-analysis techniques. Chapter 8 describes general worst-case search based methods to solve clearance problems relying on global optimization techniques. The efficiency of these methods can be tremendously improved on parallel architecture machines by exploiting various features of the underlying optimization algorithms. In Chapter 9 clearance methods for stability related criteria relying on Lyapunov-stability-based robustness analysis techniques are described, while Chapter 10 presents *integral quadratic constraints* (IQC) based techniques applicable to linear stability and performance robustness analysis. In Chapter 11 nonlinear programming methods suitable to address worst-case pilot input determination are described.

Part four parallels part three by addressing the applications of various methods to concrete clearance criteria formulated in the benchmark problem. The application of the μ-analysis technique for the analysis of eigenvalue and stability margin criteria is presented in Chapter 12. In Chapter 13 parallel computations based global optimization methods are used for the clearance of various nonlinear simulation based clearance criteria related to nonlinear stability and protection control laws. The Lyapunov-stability-based framework is employed in Chapter 14 to the analysis of the aeroelastic stability criterion and of the un-piloted nonlinear stability criterion. The IQC-based analysis technique is applied in Chapter 15 to both linear and nonlinear stability criteria using both rigid and flexible aircraft models. The nonlinear programming algorithms described in Chapter 11 are applied to compute worst-case pilot inputs in Chapter 16.

The industrial evaluation is the topic of fifth part. The currently employed baseline solution is compared in Chapter 17 with the new optimization based clearance techniques. In Chapter 18 the industrial perspectives for the applicability of the new methods are discussed and final concluding remarks are presented.

Part six includes three surveys of leading experts and practitioners describing alternative approaches and their relevance to the clearance problematic. Chapter 19 presents a nonlinear analysis technique for stability robustness assessment based on the estimation of suitable regions of attraction. Applications to two aircraft models are also presented. Chapter 20 describes NASA's approach to validation and verification of safety critical systems. Finally, in Chapter 21 the clearance approach currently employed in EADS for advanced fighter aircraft is presented.

References

1. Fielding, C., Varga, A., Bennani, S., Selier, M. (eds.): Advanced techniques for clearance of flight control laws. LNCIS, vol. 283. Springer, Heidelberg (2002)
2. Cahill, E., Scapolo, F.: The futures project—Technology map. Technical report, European Commission, Directorate-General, Joint Research Centre, Institute for Technological Studies, Seville (1999)
3. Joint Aviation Requirements (JAR) – 25: 'Large Aeroplanes'. Joint Aviation Authorities, Hoofddorp, The Netherlands
4. Federal Aviation Regulations (FAR) Part 25: Airworthiness Standards Transport Category Airplanes. Federal Aviation Administration (FAA), USA

Chapter 2
Clearance Benchmark for a Civil Aircraft

Guilhem Puyou and Yannick Losser

Abstract. This chapter describes the benchmark problem which will be used in the following chapters to address the clearance of flight control laws. Two sub-problems are considered. The first one, called "nonlinear benchmark", aims at validating the aircraft behavior close to the operating domain limits using a nonlinear rigid body model. The second one, called "integral benchmark", uses large scale linear models including flexible structural modes and represents a challenging problem for robust stability analysis methods. For each benchmark the underlying models and the associated clearance criteria to be validated are presented. The underlying models include detailed models of the flight mechanics or structural mechanics, together with the flight control laws to be assessed, as well as simplified models of the actuators and sensors. The clearance criteria cover a large range of certification requirements, as linear stability and performance analysis, or time domain performance evaluations. The current AIRBUS clearance methodology is also described and realistic expectations are stated regarding potential cost savings (both clearance effort and time) by using enhanced clearance technologies.

2.1 Introduction

The overall flight control law clearance problem that we propose to analyse is divided into two parts: the nonlinear benchmark and the integral benchmark. Both are representative of different issues that occur in the flight control law validation process. After introducing the needs for these two benchmarks and the main drivers that led to this choice, we will give a short overview of the AIRBUS flight control law philosophy by presenting its main objectives and

Guilhem Puyou · Yannick Losser
AIRBUS, Stability and Control Department, Toulouse, France
e-mail: guilhem.puyou@airbus.com, yannick.losser@supaero.org

A. Varga et al. (Eds.): Opti. Based Clearance of Flight Control Laws, LNCIS 416, pp. 11–36.
springerlink.com

key features. Then we will explain in detail, for both benchmarks, the main model components and the criteria to be fulfilled. Lastly we will briefly introduce the current AIRBUS practices for validating these sets of requirements so that we can first understand the main issues and then at the end measure the enhancements provided by the methods developed in this book.

2.1.1 Nonlinear Benchmark

The first benchmark considers the clearance of a flight control law assuming that the aircraft is rigid. It uses a nonlinear model of the closed loop aircraft including a nonlinear model of the control law. The major issue that drives this choice is the ability to perform a validation of requirements on the whole flight domain considering a wide class of pilot inputs and wind perturbations.

In-house control law design methodologies are usually used to address the nonlinearity of the flight mechanics model. Only Monte-Carlo based time domain simulation techniques have so far been used in the AIRBUS validation process to validate the effects of nonlinearities. In the current civil aircraft model many of nonlinearities are introduced by the control laws themselves. Examples of flight control law nonlinearities are: saturated integrators, switches between protection laws (see Sect. 2.2), dead-zones and saturations in control command allocation, and saturations and rate limits on command signals delivered by the control laws to the actuators.

The presence of all these nonlinear elements indicates that flight control law clearance is not an easily tractable problem and only time domain simulations on a nonlinear simulator are suited for clearance purposes. We reach here the second issue. Considering that flight control laws must be tolerant to any pilot action and still perform well for a wide class of wind perturbations, how can it be proved that the whole set of inputs is cleared? Current AIRBUS practices for clearance will be detailed in the last part of this chapter. Regarding requirement assessment in a nonlinear context, the employed techniques are mainly based on flight domain gridding approaches which are known to be highly time consuming. Hence the hope is that new techniques for clearance will significantly reduce validation times.

2.1.2 Integral Benchmark

The second benchmark considers the clearance of a flight control law for a flexible aircraft. Because of highly optimized overall aircraft designs allowing larger aircraft sizes, the bandwidth of the bending modes becomes lower and closer to the rigid body modes. Then two major issues arise: the control laws must not violate the structural load limits, and a proper design of the control laws can help designers in optimizing the aircraft structure by reducing the load limits. However, even if attention is often focused on the flexible body part, the structural behavior is coupled to the rigid aircraft, and hence the models include both rigid and flexible parts. This is the reason why the benchmark is called 'Integral'.

The most interesting features of this benchmark is that the dynamics are of high order. The fact that the models in general would be nonlinear is not so relevant. Hence we will only consider a linear model, but of high order. This will be challenging for two reasons. First, analysing models of high order is difficult per se. Second, the models are only available on a finite and coarse grid of values in the flight envelope and on a finite and course grid of values of uncertain parameters. Hence it will be very challenging to say anything about the points which are not grid points, which is of course highly desirable from a clearance point of view.

2.2 Description of Flight Control Laws

First the flight control law to be cleared will be described. Three different loops can be identified in the overall aircraft control architecture in Fig. 2.1: the control loop, the guidance loop and the navigation loop. Each loop is characterized by different time scales and control objectives. Here we focus on the clearance of the flight control laws acting at the control loop level. In what follows we will describe the main functionalities provided by the flight control laws. Their detailed architecture is presented later in the modeling part of Sect. 2.3.5.

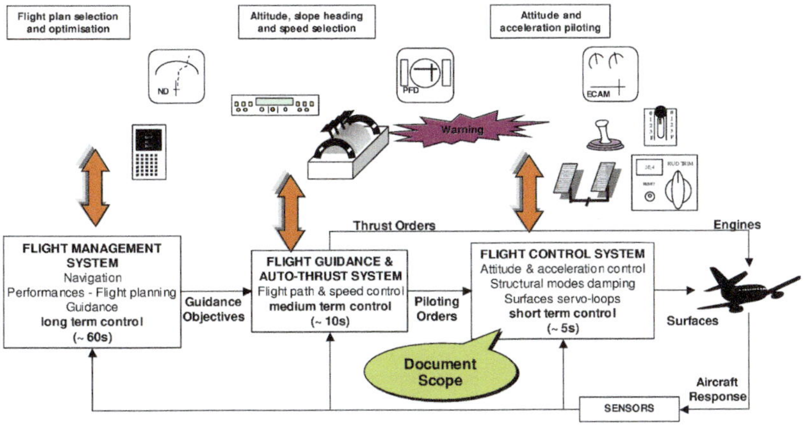

Fig. 2.1 Aircraft global control architecture with control, guidance and navigation loops

2.2.1 Flight Control Laws Philosophy

The design of architecture of the AIRBUS flight control laws is based on fulfilling the following two requirements:

1. In the usual operating domain, the normal flight control laws shall provide an instinctive piloting with the same behavior as a conventional aircraft, as well as an accurate and comfortable control, i.e. stability and manoeuvrability, homogeneous aircraft response with respect to pilot inputs in the whole normal flight envelope, compensation of aircraft configuration changes, minimisation of turbulence effect on the flight path and the bank angle, balanced effort in pitch and roll, and safe behavior in the case of engine asymmetry detection.
2. In extreme situations, flight control laws shall provide protections in order to remain in the safe operating domain and reduce the risk of over-controlling/over-stressing the aircraft while at the same time giving the highest authority to the pilot in order to achieve best possible aircraft performances, e.g. for avoidance manoeuvers.

This leads to the definition of two flight envelopes as illustrated in Fig. 2.2. The normal flight envelope inside which the normal laws are activated and the peripheral flight envelope inside which a set of protection laws ensures both full authority and safety to the pilot. Once in the peripheral envelope, pilots must maintain a permanent stick/pedal deflection to keep the current attitude. From the moment the pilot releases the stick/pedal, the aircraft returns to the normal flight envelope. This is the reason why we talk about "soft limits" for the normal flight envelope, which pilots can decide to override, and "hard limits" for the peripheral envelope.

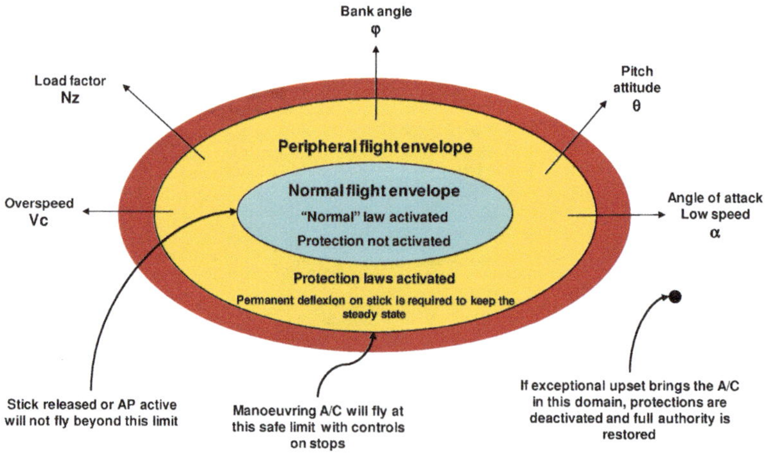

Fig. 2.2 Flight envelopes and protected parameters

2.2.2 Longitudinal Axis

In this section the longitudinal normal law and the associated protections are described.

Pitch Law

The pitch law (called also *Nz normal law*) is used to control the flight path of the aircraft through a load factor demand. An impulse on the stick leads to a flight path angle change. At constant speed, the flight path angle remains roughly constant when the stick is released. In order to stay within the safe flight envelope, the commanded load factor is limited to $[-1g,\ 2.5g]$ in clean configuration and to $[0,\ 2g]$ in high lift configurations.

Angle of Attack Protection

The angle of attack protection law (also called *low speed protection*) ensures static stability at low speed and protects the aircraft against stall, while providing the best possible manoeuvrability when necessary (typically avoidance manoeuvres). The angle of attack limits are "α prot" (α_{prot}) with neutral stick position and "α max" (α_{max}) with full back-stick as can be seen in Fig. 2.3. The control must also provide adequate roll manoeuvrability and compensate pitch-up aerodynamic effects.

Low Energy Awareness and Alpha-Floor

The objective of these functions is to enhance pilot awareness of a low energy situation by providing the necessary level of energy depending on the flight phase. Low energy awareness is the first level of warning which tells the pilot to increase thrust. Alpha-floor is the second level of warning, and it automatically applies full thrust.

High Speed Protection

The objective of the high speed protection law is to limit the possible speed/Mach exceeding maximum speed (Vmax) and Mach number (Mmax). The Vmax/Mmax target is V_{MO} (maximum operating speed)/M_{MO} (maximum operating Mach number) stick free and roughly V_{MO}/M_{MO}+15kts full forward stick. The goal is to protect the aircraft from the ultimate value that corresponds to the structural limit V_D (diving speed)/M_D (diving Mach number).

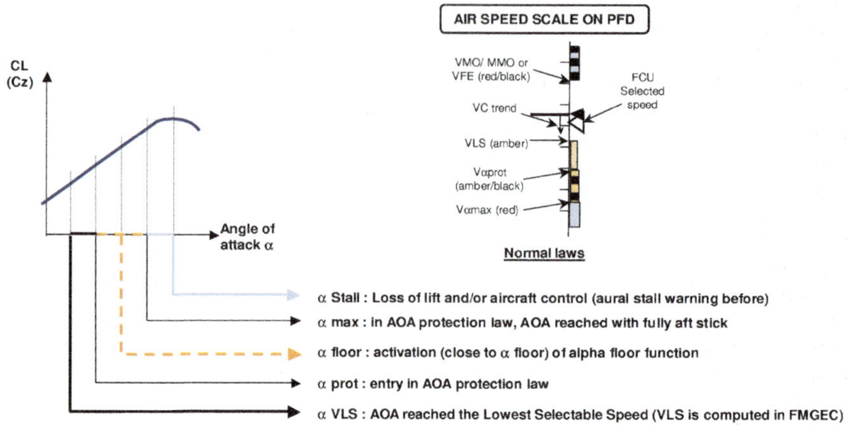

Fig. 2.3 AOA and Speed scales

Pitch Attitude Protection

The objective of the pitch attitude protection law is to enhance the effectiveness of the angle of attack and high speed protections in extreme conditions, by limiting the aircraft pitch dynamic close to the angle of attack and speed limits.

2.2.3 Lateral Axis

Similarly to the longitudinal axis, the lateral axis control laws and protections are briefly introduced.

Lateral Law

The lateral law (called also *lateral normal law*) controls the roll and yaw axes of the aircraft, through roll rate and sideslip demands. The roll stick deflection is translated into a roll rate demand at zero sideslip. An impulse on the stick leads to a bank angle change. When the stick is released, the bank angle is stabilized in the neutral stability domain. The pedal input commands a combination of sideslip and roll angle. The maximum roll rate demand must not exceed 15°/s. In the case of one external engine failure at low speed, the roll rate authority must be limited to 7.5°/s.

Roll Attitude Protection

A neutral spiral stability must be achieved up to 33° bank angle for constant bank angle with stick at neutral, while spiral stability, i.e. real part of the mode strictly negative, must be restored above 33° bank angle, i.e. bank angle comes back to 33° with stick at neutral. The objective is to limit the maximum bank angle to 66°.

2.3 The Nonlinear Benchmark Model

The closed loop nonlinear aircraft model is shown in the Simulink representation in Fig. 2.4. It includes the pilot inputs, flight control laws, actuators, flight mechanics and sensors. A simple wind model providing only ramps or constant inputs has been used. In what follows we highlight the main features of the model to emphasize its complexity and representativeness. A detailed description of all closed-loop model components is available in [4].

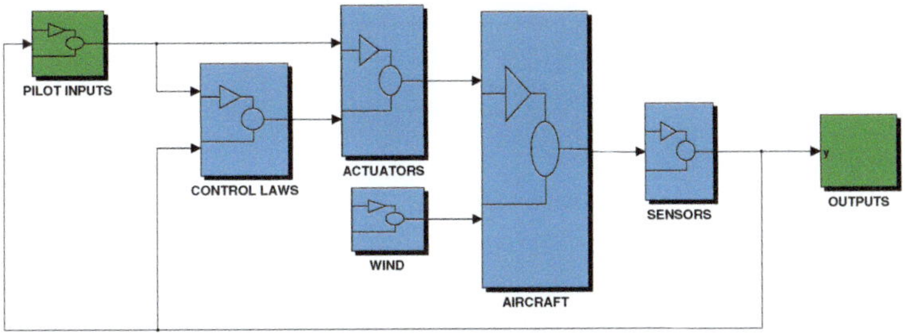

Fig. 2.4 Simulink closed-loop aircraft model overview

2.3.1 Flight Envelopes

The aircraft benchmark model aims to be representative of a civil aircraft augmented with flight control laws in the normal and peripheral flight envelopes shown in Fig. 2.2. The limiting values of the main flight mechanics variables are detailed in Table 2.1. The mass and center of gravity location are varying parameters, which are connected in accordance with the mass and balance diagram shown in Fig. 2.5.

Table 2.1 Flight envelopes definition

Variable	Normal envelope	Peripheral envelope
Load factor	$-1g \leq n_z \leq 2.5g$	
Pitch attitude	$-15° \leq \theta \leq 30°$	
Angle of attack	$-5° \leq \alpha \leq \alpha_{prot}$	$-5° \leq \alpha \leq \alpha_{max}$
Altitude	$0 \leq h \leq 41000ft$	
Speed (in flight)	$V_{\alpha_{prot}} \leq V_{cas} \leq V_{MO}$	$V_{\alpha_{max}} \leq V_{cas} \leq V_D$
Mach (in flight)	$0.2 \leq M \leq M_{MO}$	$0.2 \leq M \leq M_D$
Roll angle	$-33° \leq \phi \leq 33°$	$-66° \leq \phi \leq 66°$

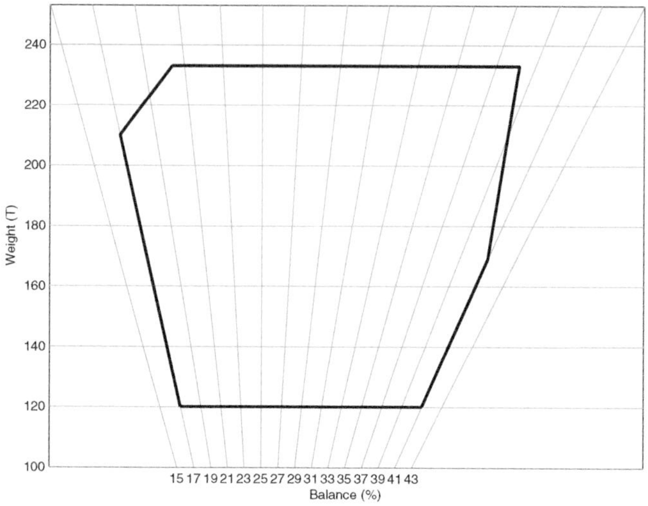

Fig. 2.5 Mass (tonnes) and balance diagram (percentage of the aerodynamic mean chord)

2.3.2 Pilot Inputs and Pilot Model

The aircraft is controlled through the pilot inputs described in Table 2.2: the side stick with both longitudinal and lateral inputs, the pedals, the high lift configuration lever, the airbrakes lever and the throttle lever.

Table 2.2 Pilot inputs

Variable	Meaning	Values	Conventions
δp_m	Lateral stick input	$-16° \leq \delta p_m \leq 16°$	Left turn: $\delta p_m \geq 0$
δq_m	Longitudinal stick input	$-16° \leq \delta q_m \leq 16°$	Nose down: $\delta q_m \geq 0$
δr	Pedal input	$-21° \leq \delta r \leq 21°$	
CONF	High lift configuration	0, 1, 2, 3, 4, 5	Clean:0, Full:5
AF	Airbrakes lever command	$0 \leq \text{AF} \leq 1$	Full airbrakes:1
Π	Thrust on mass input		

For simulation purposes, the values for the inputs variable δp_m, δq_m, δr, and Π can either be defined by the user or independently generated by a pilot model which roughly corresponds to an enhanced[1] set of auto-pilot hold modes. We define a flight path angle hold mode to manage the pitch axis and

[1] Enhanced means that it includes some nonlinearities such as dead-zones and saturations to be representative for a human pilot.

control δq_m, a heading hold mode to manage the roll axis and control δp_m, a sideslip angle hold mode to manage the yaw axis and control δr, and a speed hold mode, named auto-thrust, to manage the thrust and control Π. Each mode can be independently activated so that users can mix inputs provided by them and the pilot model.

2.3.3 Actuators and Sensors

It is important to emphasize that the focus of the nonlinear benchmark model is on evaluating handling qualities. Therefore only simple models for actuators, sensors and onboard computers are considered. It aims to be representative of the main dynamics, to the first order at least, and the saturation effects when relevant. A typical model for an actuator is based on three elements as shown in Fig. 2.6. These are a first or second order transfer function which represents actuator position dynamic, a position saturation that could be asymmetric, and a rate limiter. The input is the demanded actuator position and the output is the real actuator position.

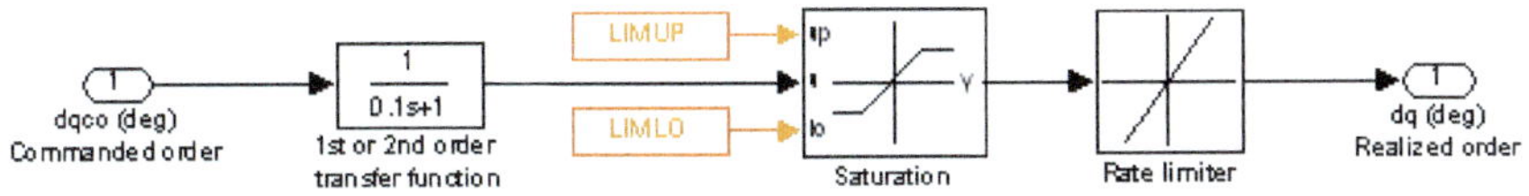

Fig. 2.6 Simplified actuator modelling

2.3.4 Flight Mechanics

The mathematical description of aircraft flight mechanics is based on the fundamental principles of dynamics. In the nonlinear benchmark we use a quaternion based model to describe the aircraft motion in the form (see [2] for details)

$$
\begin{aligned}
m\dot{V_{tas}} &= F_a + F_g + F_p \\
\dot{\bar{q}} &= \frac{1}{2}\bar{q} \otimes \Omega \\
I_T \dot{\Omega} &= M_a + M_p - \Omega \times I_T \Omega
\end{aligned}
$$

where the intervening variables are described in Table 2.3. In the above equations "$\otimes$" and "$\times$" are standard notations for quaternion and cross products, respectively. The main forces and moments acting on the aircraft are due to the aerodynamic effects (F_a and M_a), gravity (F_g) and engine thrust (F_p and M_p). The aim of this part is to describe how these various forces and moments are modeled in order to give an overall view of the behavior of the aircraft and of the model which is associated with it.

Table 2.3 Interpretation of model variables and parameters.

Notation	Meaning
m	mass
F_a	sum of the aerodynamic forces
F_g	sum of the gravity forces
F_p	sum of the engines forces
$\bar{q}$	quaternion vector describing attitude
Ω	angular velocity vector in the aircraft coordinate system
I_T	inertial tensor
M_a	aerodynamic forces induced momentum at center of gravity
M_p	engine forces induced momentum at center of gravity

Aerodynamic Loads

The main loads used in the flight mechanics equations are the aerodynamic loads. Normally, these loads are represented macroscopically by a set of moments and forces applied to the center of gravity and projected either onto the aircraft coordinate system or onto the aerodynamic coordinate system. These aerodynamic loads and moments depend on the state of the system, i.e. velocities, attitude, altitude, external conditions such as velocity and direction of the wind, etc., the configuration of the aircraft such as slats, flaps, spoilers, etc., and the position of the aerodynamic control surfaces such as ailerons, rudder, elevators, etc. The determination of these loads is most often based on the identification of a set of aerodynamic coefficients $(C_x, C_y, C_z, C_l, C_m, C_n)$, which enters in the expressions of the components of the forces $(F_{a_x}, F_{a_y}, F_{a_z})$ and moments $(M_{a_x}, M_{a_y}, M_{a_z})$ along the axes (x, y, z) as

$$\begin{aligned} F_{a_x} &= -\tfrac{1}{2}\rho S V_{tas}^2 C_x, & F_{a_y} &= -\tfrac{1}{2}\rho S V_{tas}^2 C_y, & F_{a_z} &= -\tfrac{1}{2}\rho S V_{tas}^2 C_z, \\ M_{a_x} &= \tfrac{1}{2}\rho S L V_{tas}^2 C_l, & M_{a_y} &= \tfrac{1}{2}\rho S L V_{tas}^2 C_m, & M_{a_z} &= \tfrac{1}{2}\rho S L V_{tas}^2 C_n, \end{aligned}$$

where ρ is the air density, L is the wing span, and S is the reference area. The global aerodynamic coefficients $(C_x, C_y, C_z, C_l, C_m, C_n)$ in the nonlinear benchmark are expressed as nonlinear functions of the previously mentioned variables through a neural network. For simplicity, these loads are expressed in the aircraft coordinate system according to Fig. 2.7.

Gravity

The vertically oriented gravity force $F_g = mg$ is considered to be applied in the center of gravity of the aircraft. The components $(F_{g_x}, F_{g_y}, F_{g_z})$ of this force along the axes (x, y, z) in the aircraft coordinate system of Fig. 2.8 can be expressed as

$$F_{g_x} = mg\sin\theta, \quad F_{g_y} = mg\cos\theta\,\sin\phi, \quad F_{g_z} = mg\cos\theta\,\cos\phi$$

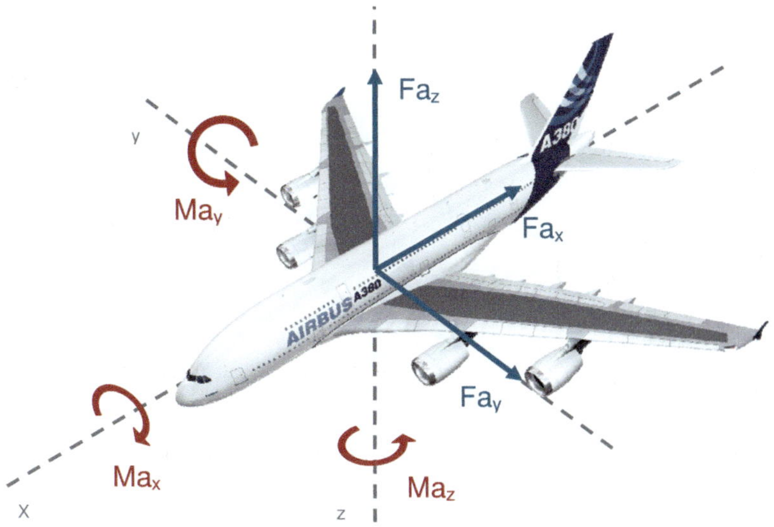

Fig. 2.7 Aerodynamic loads and moments in the aircraft coordinate system

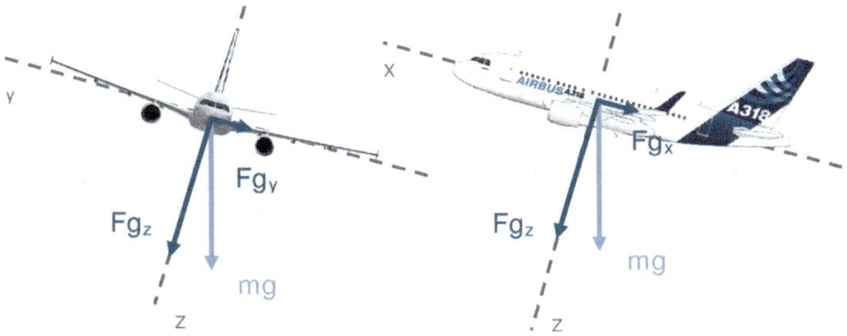

Fig. 2.8 Projection of gravitational force in aircraft coordinate system

Forces and Moments Generated by the Engines

The force generated by the engine thrusts allows to achieve and maintain a desired longitudinal velocity of the aircraft. The total force is $F_p = F_{p_R} + F_{p_L}$, representing the sum of forces F_{p_R} and F_{p_L} generated by the right and left engines, respectively. The force generated by each engine is directed along the engine's longitudinal axis, which is in general different from that of the aircraft coordinate system shown in Fig. 2.9. Dissimilar thrusts of the left and right engines can help in the lateral control of the aircraft by generating a torque M_p in the yaw axis. The total force and moment have respectively the components $(F_{p_x}, F_{p_y}, F_{p_z})$ and $(M_{p_x}, M_{p_y}, M_{p_z})$ in the aircraft coordinate system.

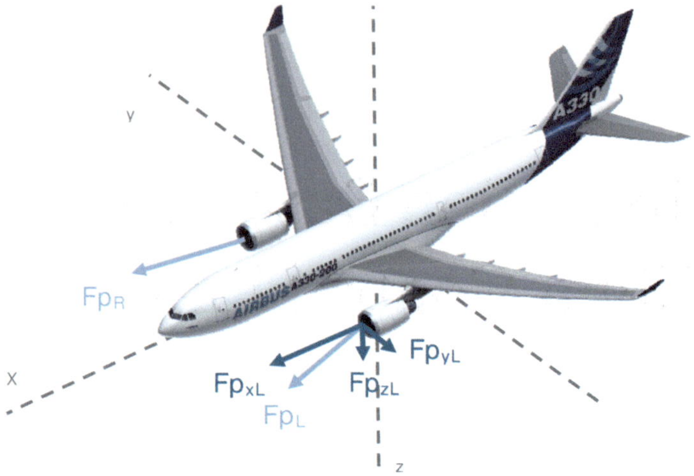

Fig. 2.9 Representation of engine thrust forces

2.3.5 Control Laws

As already mentioned in Sect. 2.2, the nonlinear benchmark considers manual control only, i.e. the autopilot and guidance laws are not included in the flight controls to be validated. However, for better manoeuvre management, the autothrust and a pilot model are included in the model to maintain constant states (speed, flight path during lateral manoeuvre ...). Moreover, since failure reconfiguration is not part of the study, these elements are not included in the flight control laws. The different embedded laws and their functionalities have been described in Sect. 2.2 in detail. In Fig. 2.10 and Fig. 2.11, simplified views of each control axis are presented including the "normal" laws, protections and control command allocation named "kinematics". The longitudinal control gets longitudinal stick pilot inputs (DQM) and delivers symmetric control demands to the elevators (ODQ*) and horizontal trim (OIH), while the lateral control gets lateral stick pilot inputs (DPM) and provides antisymmetric control demands to the inner ailerons (OAILI), outer ailerons (OAILE), spoilers (OSPi), and rudder (OLDEGRYD).

The normal control laws (longitudinal and lateral) are exclusively employed for linear system analysis based clearance. Figs 2.12 and 2.13 provide some extra information on these laws. They mainly consist of proportional output feedback with an integral control error feedback to ensure zero steady-state tracking error. Their gains are scheduled to cover the whole operating domain. Notations are the following: V_c is the conventional speed (V_{cas}), CONF is the slat/flap configuration and X_G is the center of gravity position. More details about AIRBUS flight control law design are given in [1].

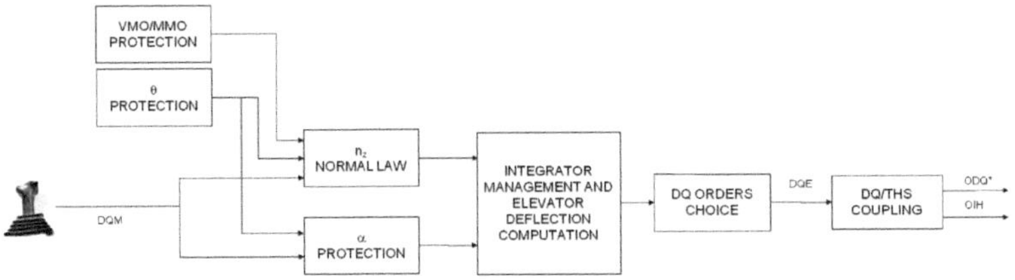

Fig. 2.10 Longitudinal axis control

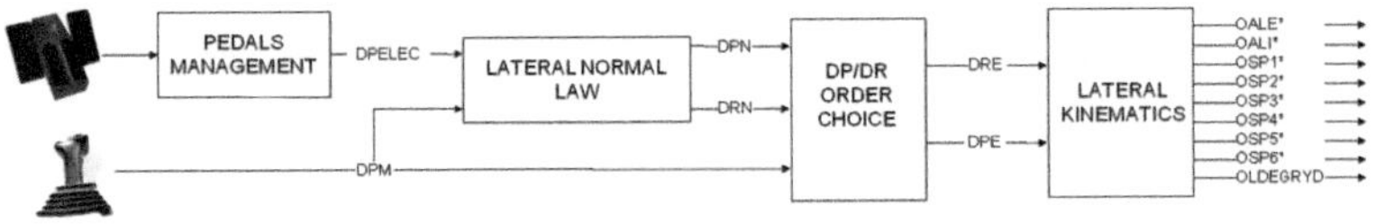

Fig. 2.11 Lateral axis control

2.4 Clearance Criteria for the Nonlinear Benchmark

In this section we define a set of clearance requirements for the nonlinear benchmark. The description of each requirement involves the specification of some *initial conditions*, as for example, the relevant mass and balance values, the range of flight points and/or special settings for control device deflections. The definition of the corresponding clearance criteria involves the specification of the manoeuvers to be simulated for their evaluation or the necessary indicators to be computed (e.g., eigenvalues for stability analysis). The purpose of the clearance is to show that the clearance criteria exhibit robustness with respect to various uncertainties and variabilities, as for example, unknown external inputs such as special wind profiles, varying parameters of aircraft or of various manoeuvers, changes in the aerodynamics data, variations of flight points, failure cases, etc. The list of considered clearance problems is not exhaustive. It represents merely a selection of clearance problems which would probably most benefit from enhanced computer-aided clearance methodologies in terms of reduction of analysis efforts, especially, by substantially reduced analysis times.

2.4.1 Un-piloted Aircraft Stability

Criterion Description

In any trimmed point in the overall flight domain, the closed-loop aircraft must remain stable in the absence of pilot actions (i.e., for an un-piloted

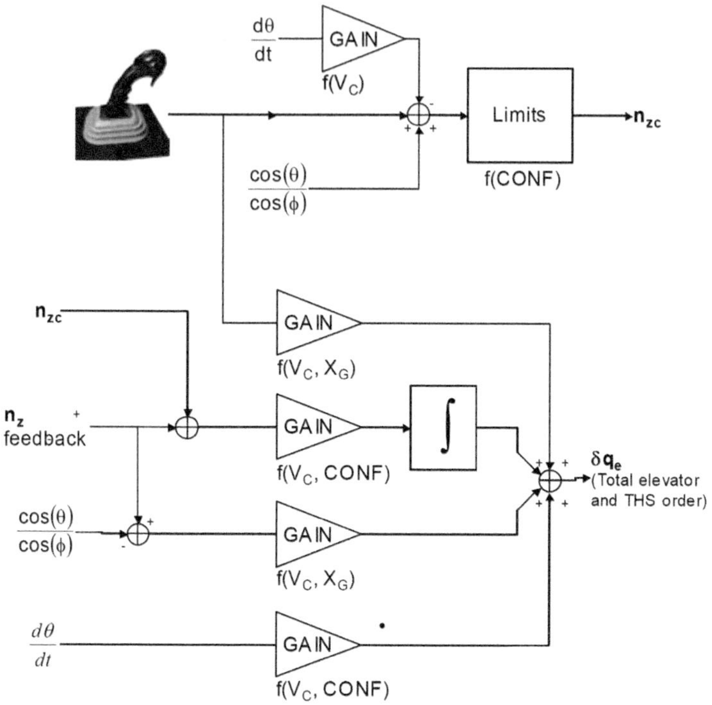

Fig. 2.12 Simplified view of longitudinal normal law

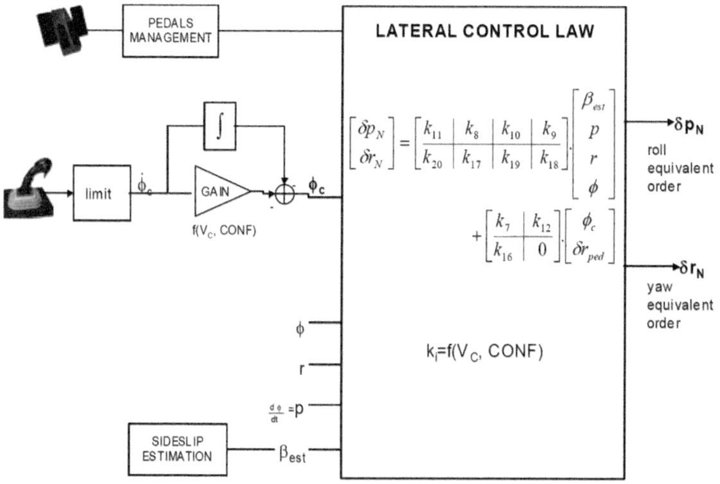

Fig. 2.13 Simplified view of lateral normal law

aircraft[2]). Because the pilot is nevertheless still in the loop, stability can be relaxed to the existence of slowly divergent modes, provided that the time of doubling of any divergent state variable is above 6s.

Uncertainties

Besides the usual variabilities regarding mass and flight points, the stability must be assessed also in the case of erroneous CG data used for the scheduling of the control laws. In particular, the robustness to CG data must be guaranteed to cover pilot errors when entering payload data into the system before take-off. Even when wrong CG values are used in the control laws (between the minimum and maximum CG values), the closed-loop stability must be guaranteed. Therefore the CG scheduling value must be considered as an uncertain parameter in the robustness analysis.

2.4.2 Manoeuvrability Requirements for the Longitudinal Axis

Criterion Description

In any trimmed point in the peripheral flight domain, the flight domain protections should not limit the aircraft capability for avoidance manoeuvres in the absence of pilot actions. The following two cases are considered:

- the low speed case: the pilot must be able to modify the short term trajectory within the flight envelope. When pulling the stick for $\Delta t = 5s$, the short term response of $C_z(t)$ should be compared to the natural aircraft capability by computing the ratio

$$r = \frac{\displaystyle\int_0^{\Delta t} C_z(t)dt}{C_{z,\alpha_{max}} \Delta t}$$

  The ratio r should be greater than $\frac{1}{\sqrt{2}}$. The coefficient $C_{z,\alpha_{max}}$ is the maximum lift coefficient value in the current aircraft configuration at the current flight point.
- the high speed case: between V_{MO} and V_{MO}+15kts, the nose-down authority, measured by the induced variation of vertical load factor n_z, must remain greater than $0.3g$.

Manoeuvre Description

Based on past experiences, the following manoeuvres are known to be the most demanding ones:

[2] Of course, the aircraft should remain stable even with a pilot in the loop, but this is more difficult to assess, and is considered to be a distinct clearance criterion.

- the low speed case: longitudinal side-stick is constantly forward deflected within the capabilities of the device ($-16° \leq \delta q_m \leq 0$). Other control devices are kept at their trim positions.

Device inputs	Objective / Manoeuvre
δq_m	constant nose up demand
δp_m	no input
δr	no input
Π	constant thrust level (auto-thrust off)

- the high speed case: longitudinal side-stick is constantly upward deflected within the capabilities of the device ($0 \leq \delta q_m \leq 16°$).

Device input	Objective / Manoeuvre
δq_m	constant nose down demand
δp_m	no input
δr	no input
Π	constant thrust level (auto-thrust off)

Uncertainties

Global aerodynamic coefficient uncertainties of $\pm 10\%$ on the values of C_x, C_y, C_z, C_l, C_m and C_n are assumed.

2.4.3 Flight Domain Protection

Criteria Description

In any trimmed point in the normal flight domain, for any pilot inputs and wind perturbations, the aircraft response must remain in steady state within the peripheral envelope defined in Table 2.1. Accordingly, flight domain protection clearance criteria have been defined in [3] for load factor, pitch attitude, angle of attack, altitude, Mach number, and roll angle protections. For the pitch attitude θ and angle of attack α protections, some transient overshoot of the bounds can be tolerated, where the overshoot of bounds should be less than 1° at high Mach numbers ($M > 0.5$) and less than 2° at low Mach numbers ($M \leq 0.5$).

By "any pilot inputs" we mean that the control device inputs can be arbitrary time functions provided that they fulfill the following constraints on the steering device capabilities:

- longitudinal stick: $-16° \leq \delta q_m \leq 16°$, $\left|\frac{d\delta q_m}{dt}\right| < 30°/s$
- lateral stick: $-16° \leq \delta p_m \leq 16°$, $\left|\frac{d\delta p_m}{dt}\right| < 30°/s$
- pedals: $-21° \leq \delta r \leq 21°$, $\left|\frac{d\delta r}{dt}\right| < 20°/s$

For clearance purposes, only violations of the flight envelope boundaries are of interest. Nevertheless, for analysis and insight, the worst-case parameter combinations are of interest, even if the boundaries are not violated.

Manoeuvre Description

The flight domain protection should be assessed for any pilot input. This implies that the set of manoeuvres contains infinitely many elements. Nevertheless, from previous experience, and in order to decrease the number of varying parameters, we may restrict the attention to some kind of "sizing manoeuvre" as described in [3]. Unless otherwise specified, the auto-thrust will be considered on.

Uncertainties

The following uncertainties must be considered:

- global aerodynamic coefficients uncertainties: ±10% on the values of C_x, C_y, C_z, C_l, C_m and C_n;
- wind gradient occurrence: gradient between [1, 5kts/s], amplitude <20kts, any orientation, see Fig. 2.14;
- failure cases: one engine out.

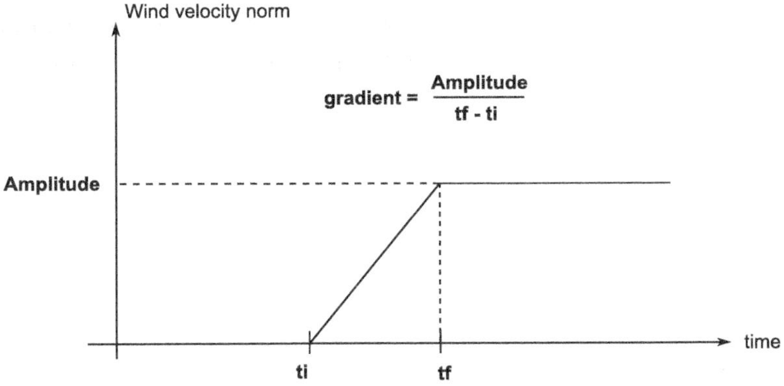

Fig. 2.14 Wind gradient

2.5 The Integral Benchmark Model

In this section the integral benchmark model is described. The model components are the same as for the nonlinear benchmark model shown in Fig. 2.4. The only difference is within the aircraft model itself, where a description of the flexible modes is added to the rigid body dynamics.

2.5.1 Flexible Aircraft Model

Structural Dynamics Equation of Motion

The integral model is obtained from the following Laplace-transformed second order differential equation which describes the structural dynamics with generalized aerodynamic loads

$$\mu s^2 q(s) + \beta s q(s) + \gamma q(s) = F_m(s) + F_t(s) + F_g(s) \tag{2.1}$$

where q is the truncated modal coordinates vector (including the rigid motion modes and flexible motion modes truncated to the most relevant ones), (μ, β, γ) represents the generalized mass, dissipation and stiffness matrices, respectively, F_m are the aerodynamic forces, F_t are the surfaces deflection induced forces, and F_g are the turbulence forces. Using the expression of the dynamic pressure $q_{dyn} := \frac{1}{2}\rho V_{tas}^2$, these forces can be modeled as

$$\begin{aligned} F_m(s) &= q_{dyn} FGM(s) q(s), \\ F_t(s) &= q_{dyn} FGT(s) \delta(s), \\ F_g(s) &= q_{dyn} FGM_g(s) w(s), \end{aligned}$$

where $FGM(s)$, $FGT(s)$ and $FGM_g(s)$ are large scale improper transfer-function matrices of the particular forms

$$FGM(s) = F_0 + F_1 s + F_2 s^2 + \sum_{i=1}^{N_{ret}} F_{i+2} \frac{s}{s+\gamma_i} \tag{2.2}$$

$$FGT(s) = F_{t0} + F_{t1} s + F_{t2} s^2 + \sum_{i=1}^{N_{ret}^t} F_{t(i+2)} \frac{s}{s+\gamma_{ti}} \tag{2.3}$$

$$FGM_g(s) = F_{g0} + F_{g1} s + F_{g2} s^2 + \sum_{i=1}^{N_{ret}^g} F_{g(i+2)} \frac{s}{s+\gamma_{gi}} \tag{2.4}$$

The intervening coefficient matrices F_* and filter poles γ_* result from a FEM analysis.

State-Space Model

The transfer-function matrix corresponding to an input-out description of the model (2.1) from the inputs δ and w to some outputs y and l (defined as selected components of q) is guaranteed to be proper. Therefore, it can be realized as a first order state-space model of the form

$$\begin{aligned} \dot{X} &= AX + B\begin{pmatrix} \delta \\ w \end{pmatrix} \\ \begin{pmatrix} y \\ l \end{pmatrix} &= CX + D\begin{pmatrix} \delta \\ w \end{pmatrix} \end{aligned} \tag{2.5}$$

where X is the state vector (including q and $\dot{q}$, but also delays introduced by the rational terms $\frac{s}{s+\gamma_\star}$), δ is the vector of control surface inputs, w is the vector of wind disturbance inputs, y is the vector of measured outputs (e.g., attitude or attitude rate), and l is a vector of load outputs at selected locations on the aircraft structure. The state space model (A, B, C, D) corresponds to a single flight condition and a fixed mass configuration.

Several linearized models are usually provided to cover a representative set of flight conditions and mass configurations. All matrices of such state space models are sparse, but for all flight conditions and mass configurations they have the same sparsity pattern of zero and nonzero elements as shown in Fig. 2.15 for the matrices A and B. The large state vector dimensions of about 250 (may vary among the models) is one of the main computational challenges for the analysis methods.

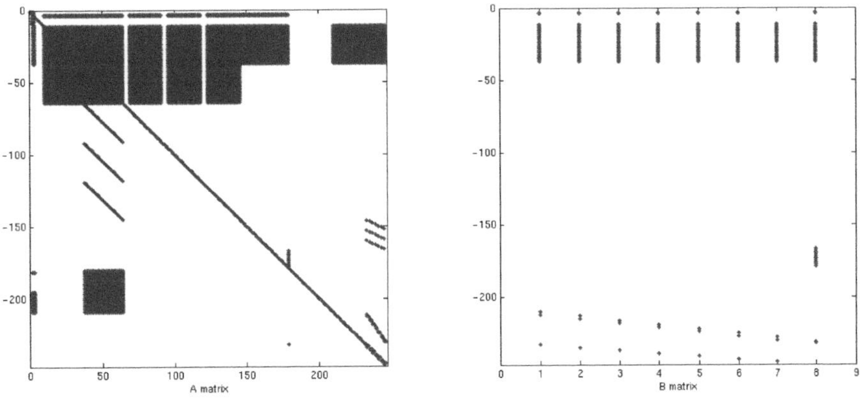

Fig. 2.15 Sparsity patterns of A and B matrices

2.5.2 Mass Configurations and Flight Points

The mass configurations are described in the terms of percentages of the Pay Load (PL) and of three fuel tank loads: the Trim Tank (TT) which is located in the horizontal tail plane, the Center Tank (CT) which is located in the center wing box, and the Outer Tank (OT) which is symmetrically located in the wing part. We propose the following reduced number of load configurations that can be independently combined:

- 2 PL configurations: 0% and 100%
- 1 TT configuration: 0%
- 3 CT configurations: 0%, 50% and 100%
- 3 OT configurations: 0%, 50% and 100%

The above selection gives 18 mass configurations for each flight point defined by a combination of values of the Mach number M and speed V_{cas}. In the benchmark definition, 9 combinations are considered for three values of M and three values of V_{cas}. The corresponding values of altitude h, V_{tas} and air density ρ are given in Table 2.4.

Table 2.4 Flight points

M	V_{cas} (kts)	Flight point name	h (ft)	ρ $(kg \cdot m^{-3})$	V_{tas} $(m \cdot s^{-1})$
0.70	205	M70_V1	40825.50	0.290	206.55
0.70	285	M70_V2	26233.57	0.525	215.66
0.70	365	M70_V3	13732.22	0.803	226.68
0.86	255	M86_V1	41294.11	0.283	253.76
0.86	310	M86_V2	32751.14	0.414	257.61
0.86	365	M86_V3	25046.82	0.548	266.27
0.91	275	M91_V1	40819.91	0.290	268.52
0.91	320	M91_V2	34182.28	0.392	270.85
0.91	365	M91_V3	28082.68	0.492	278.17

By counting both longitudinal and lateral models for all mass configurations and all flight points, we have altogether 324 models of which 162 are longitudinal models and 162 are lateral models.

2.6 Clearance Criteria for the Integral Benchmark

2.6.1 Aeroelastic Stability

Criterion Definition

Assume that the linearized closed-loop system in Figure 2.4 has the state-space realization $(A_{cl}, B_{cl}, C_{cl}, D_{cl})$. To asses the aeroelastic stability of the closed-loop linear system, all eigenvalues λ_i of A_{cl} must have negative real parts. The clearance of the *aeroelastic stability criterion* (also known as *eigenvalue criterion*) must cover all operating points within the flight envelope and all mass configurations.

2.6.2 Stability Margins

Criterion Description

Depending on the development phase, strict stability (as defined in Sect. 2.6.1) is sometime not enough to cover model uncertainties. Then SISO phase and gain stability margins are required. A usual requirement is to provide

- absolute phase margin $\geq 90°$, and
- gain margin ≥ 6 dB.

The margins are computed by considering the open-loop transfer function obtained when breaking the loop at the input of one actuator while leaving the other loops closed. In the case of multiple inputs system, margins should be computed for each actuator. On the Nichols plot of the open-loop frequency response, the gain margin is defined by the gain value (in dB) when the phase is equal to -180° +/- 360°, while the stability margin is defined by the phase value when the gain is equal to 0dB.

Varying Parameters

Those margins have to be satisfied whatever the fuel tank load ratios, which can vary between empty (0%) and full (100%), and for any flight point (i.e. speed/Mach value).

2.6.3 Comfort with Respect to Turbulence

Description

The comfort criterion uses specific acceleration outputs γ that represent vertical or lateral load factors in different parts of the aircraft. Comfort filters representing the passenger seat transfer function are added after these accelerations to compute evaluation outputs γ_{cf}, see [5] for further details. We use a von Kármán filter with white noise input $\widetilde{w}$ to describe the wind, as required by the Military Specification MIL-F-8785C. No pilot input is used in this case (u_c=0). The closed-loop model used for analysis is presented in Fig. 2.16.

Criterion Description

Let $T_{\widetilde{w}\to\gamma_{cf}}(s)$ be the stable transfer function between the wind input $\widetilde{w}$ and the evaluated output γ_{cf} in Fig.2.16. Since $\widetilde{w}$ is assumed to be a white noise input, we can compute the variance σ_c of γ_{cf} using the relation

$$\sigma_c^2 = \frac{1}{2\pi}\int_{-\infty}^{+\infty} \left|T_{\widetilde{w}\to\gamma_{cf}}(j\omega)\right|^2 \mathrm{d}\omega = \left|\left|T_{\widetilde{w}\to\gamma_{cf}}(s)\right|\right|_2^2 \tag{2.6}$$

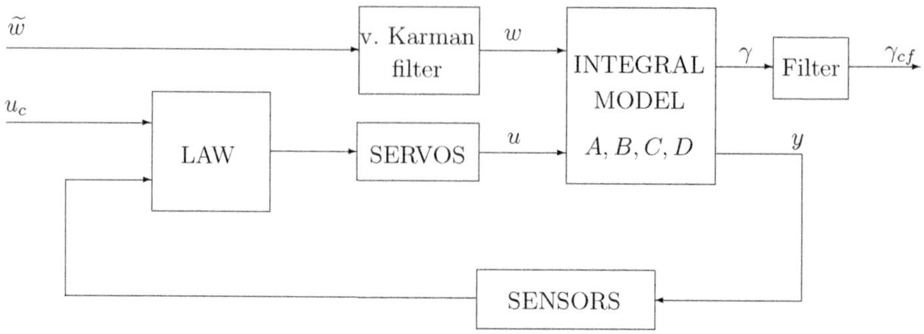

Fig. 2.16 Closed loop with aircraft integral model, control law, white noise input and comfort filter

Let $\sigma_{c_{CL}}$ and $\sigma_{c_{OL}}$ be the evaluations of σ_c with and without the control law, respectively. The fulfillment of the comfort requirements is achieved if

$$\frac{\sigma_{c_{CL}}}{\sigma_{c_{OL}}} \leq k,$$

where k is the comfort improvement factor. A basic requirement is obtained with $k = 1$, but smaller values can be used for k if the control laws aim at improving the natural aircraft comfort level.

Varying Parameters

The comfort requirement has to be satisfied whatever the fuel tank load ratios, that can vary between empty (0%) and full (100%), and for any flight point, i.e. speed/Mach value. It has also to be evaluated for different points along the fuselage to be representative for different seats in the aircraft.

2.7 Current AIRBUS Practices

In this section we describe the current AIRBUS practices for validation. Starting from current standards, we will highlight where major enhancements can be expected by using new clearance methods.

2.7.1 Validation Methods

Validation history is closely linked to design history. Figure 2.17 describes both design methods used along the program life from feasibility studies to series support and validation methods used to verify the designed laws. The starting point of the design process is the mechanics of flight and load models delivery. These models are the reference behavior models used everywhere

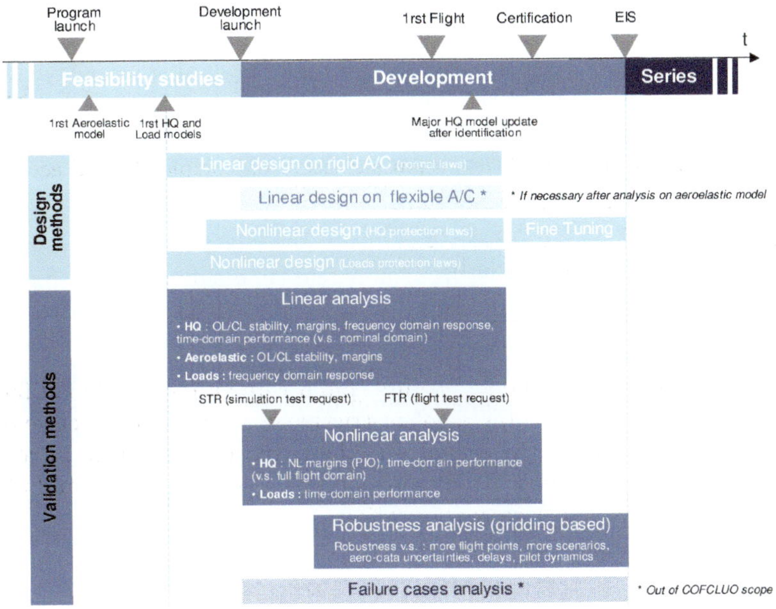

Fig. 2.17 Validation methods

in AIRBUS. We usually start with a linear design of the normal law inside the nominal flight envelope for the rigid body aircraft model. Then nonlinear design is performed to tune the protection laws which keep the aircraft inside the peripheral envelope. Finally load protections are designed either in a linear context when dealing with turbulence or in a nonlinear context when dealing with gust and manoeuver. A possible fourth linear design step might be required after evaluation of the aeroelastic margins. There are many iterative loops during the development phase due to model updates. The last one follows at the end of the identification phase. After that we can assume that only fine tuning is performed to adjust the local behavior of the closed loop.

Based on the design cycle, we first use linear analysis methods to validate the linear design. This includes: closed-loop eigenvalue study, i.e. stability and handling qualities related to pole placement, gain and phase margins, time-domain response to basic analytical pilot inputs such as steps, and frequency domain response. Then nonlinear design is validated using nonlinear analysis methods such as: nonlinear stability margins, i.e. admissible phase and gain perturbations for a nonlinear model until instability is reached, nonlinear time domain response to basic analytical inputs but also to more complex scenarios involving protections. After analysis of basic design points, robustness is tested on a finer grid including: more flight points, more flight scenarios, and additional uncertainties such as data processing delays, pilot dynamics etc.

The failure case analysis is also a highly time consuming activity which is, however, out of the scope of this project.

2.7.2 Validation Means

To perform the different types of analysis described above, a wide set of validation means are available, from non real-time simulators to flight test aircraft. All these simulators use the same flight mechanics and load models when available, but representativeness of the system parts such as actuators, sensors and computers are different. At start, the validation process uses a non-real-time fully simulated aircraft model, i.e. non-real-time desktop simulators. A typical example is the MATLAB-Simulink model provided to support the benchmark definition. Other model building and simulation environments exist within AIRBUS. ATOSMA is dedicated to handling quality studies and manual laws, SIMPA is dedicated to auto-pilot law validation and ATLAS to load computation. Only analytical inputs are available. To perform piloting tasks we must introduce a pilot model that will control the different axes for given objectives on speed, sideslip angle, heading, flight path angle, etc. Later in the validation process a real-time desktop simulator is used. This consists of an interactive fully simulated aircraft model. It provides simplified control devices so that design engineers can interact with the simulation and perform some basic scenarios which are more complex than single objective control. We then switch to advanced simulators. This simulators contain a true representation of the cockpit and the piloting devices. We can here begin to involve flight test pilots in the validation process. Those simulators also progressively introduce real systems in the loop, i.e.

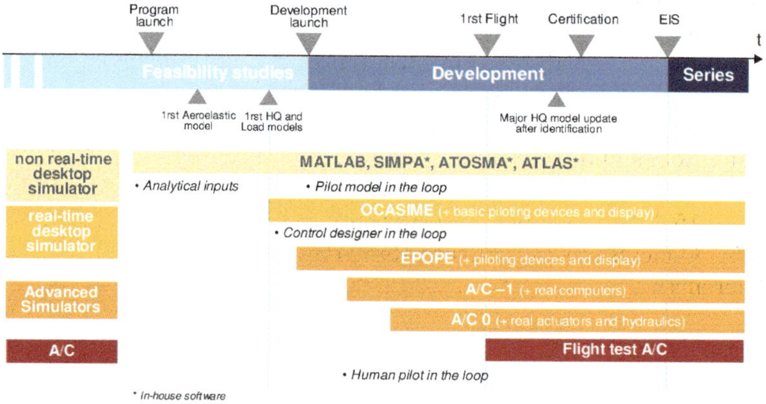

Fig. 2.18 Validation means

on-board computers, hydraulic systems, electrical systems, actuators, etc. The last validation means is of course the most representative: the real aircraft.

2.7.3 Validation Coverage

The different validation means are used to assess clearance criteria fulfillment on the specified flight domain. Correlation between the flight domain validation coverage and the use of validation means is shown in Figure 2.19.

The depicted ratios are not necessarily representative of the current situation. We only want to highlight the main features. The very first part of validation is performed on the non-real-time simulators. This includes linear model analysis such as stability, frequency domain response, etc. and time domain simulation on the nonlinear model using basic scenarios: analytical inputs such as steps, pulses, etc. or one piloting task using a pilot model. Today, within the context of manual laws clearance, non-real-time simulations are only used to perform analysis on fine grids. Monte-Carlo based statistical validation methods are solely implemented for the auto-pilot validation, where the set of inputs is more limited. Introduction of the statistical validation process to the manual flight control laws is an ongoing process. It shall be noted that the latter problem is more complex because of the wide class of inputs, i.e. pilot and wind inputs combinations.

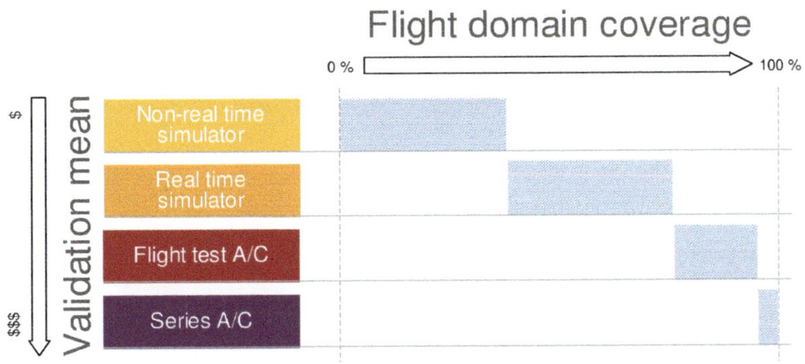

Fig. 2.19 Current situation

For the time being, fixed grid validations are used to pre-determinate worst cases used for real-time simulations. Desktop simulations are widely used by designers to perform full flight domain validation and to help select worst case scenarios that need to be tested during advanced simulator sessions with a test pilot. When the simulation provides satisfactory results, flight tests can begin. After an identification campaign and a model update only a few worst case scenarios will be flight tested.

These scenarios mainly include certification requirements, simulation worst-cases where pilot judgment is mandatory, and some pilot scenarios coming from previous aircraft development knowledge.

2.8 AIRBUS Expectations

Our expectations for the future are detailed in Figure 2.20. The main goal we want to achieve is to widen the use of un-piloted desktop simulators in order to reduce the piloted simulation activities. This will reduce the cost. From our point of view the major task will be the automation of finding worst case combinations so that we can reduce our piloted simulations and flight tests. Nevertheless it is important to note that flight tests cannot be drastically reduced, mainly because we always need both to validate the model accuracy and to fulfill the certification requirements.

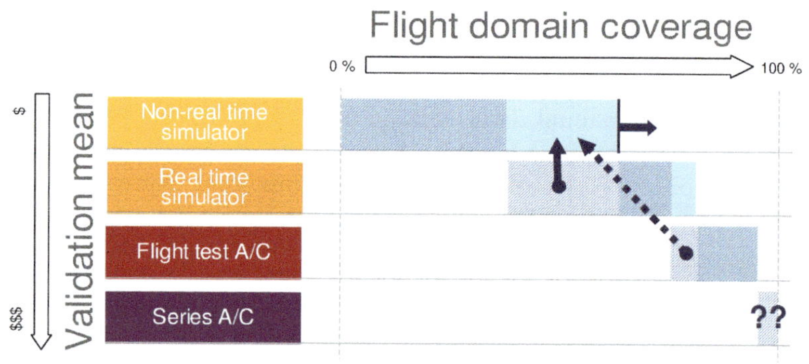

Fig. 2.20 Trend

References

1. Jeanneau, M.: Unsteady character of fly-by-wire flight controls: roundup and research orientations at Airbus. In: 6th ASCC, Bali (July 2006)
2. Phillips, W.F.: Mechanics of Flight. John Wiley and Sons, Chichester (2004)
3. Puyou, G.: D1.1.1 Clearance problem definition, Part 1 - Nonlinear model. COFCLUO Report
4. Puyou, G.: D1.1.2 Models delivery, Part 1 - Nonlinear model. COFCLUO Report
5. Kubica, F., Madelaine, B.: Passenger comfort improvement by integrated control law design. RTO-MP-036 (1999)

Part II
Generation of Linear Uncertain Models

Chapter 3
Generation of LPV Models and LFRs for a Nonlinear Aircraft Model

Simon Hecker and Harald Pfifer

Abstract. In this chapter we present a general approach to generate a linear parameter varying (LPV) state-space model, which approximates a nonlinear system with high accuracy and is well suited for LFT-based robustness analysis. A Jacobian-based linearisation of the nonlinear parametric aircraft model is performed first to generate a set of linearised state-space models describing the local behaviour of the nonlinear aircraft for a representative set of parameter values and flight conditions. These models are then approximated by a unique LPV model, by using multivariable polynomial fitting techniques in combination with global optimisation. The objective is to find an LPV model which guarantees a specified approximation accuracy and simultaneously leads to a linear fractional representation (LFR) of least possible order. For this, a gap metric constraint on the input-output transfer-function error is included in the optimisation problem. The effectiveness of the proposed method is demonstrated by generating high accuracy LPV models and the corresponding LFRs for the COFCLUO nonlinear aircraft model.

3.1 Introduction

Various dynamic systems can be described by nonlinear differential equations

$$\begin{aligned} \dot{x} &= f(x, u, p) \\ y &= g(x, u, p) \end{aligned} \tag{3.1}$$

with state vector $x(t)$ confined to some operating region $X \subset \mathbb{R}^n$, input vector u and output vector y. These systems may depend on a parameter vector

Simon Hecker · Harald Pfifer
Institute of Robotics and Mechatronics, DLR - Oberpfaffenhofen,
82234 Wessling, Germany
e-mail: simon.hecker@hm.edu, harald.pfifer@dlr.de

A. Varga et al. (Eds.): Opti. Based Clearance of Flight Control Laws, LNCIS 416, pp. 39–57.
springerlink.com

p which is either not exactly known (i.e., uncertain) or is time-varying and belongs to an admissible parameter value set Π, i.e., $p \in \Pi$. The analysis or control design for such systems ensuring the stability and performance requirements for all allowable parameter variations and over the whole range of operating conditions is a highly complex task and can be addressed only by employing advanced techniques like μ-analysis/synthesis [1] or Linear Parameter Varying (LPV) control [3].

Therefore (3.1) is usually approximated by an LPV model of the form

$$\begin{aligned} \dot{x} &= A(\delta)x + B(\delta)u \\ y &= C(\delta)x + D(\delta)u, \end{aligned} \tag{3.2}$$

where the matrices $A(\delta)$, $B(\delta)$, $C(\delta)$, $D(\delta)$ depend rationally on δ [1]. Note that besides the parameter vector p, the vector δ may also include components of the state vector x, thus allowing to cover state dependent nonlinearities in the representation given by (3.2) [4]. Note that depending on the LPV generation method the vectors x, u and y in (3.2) may differ from the corresponding vectors in (3.1). As an example, (3.2) may be generated by interpolating a set of LTI models, which are generated by linearising (3.1) around a certain set of equilibrium points. In this case x, u and y in (3.2) will describe only small variations around equilibrium points of (3.1).

Finally, (3.2) is transformed into a Linear Fractional Representation (LFR) [1], which is a standard form to apply modern robust control techniques like μ-analysis/synthesis [1]. These techniques are usually computationally demanding and one may obtain more accurate results (e.g. tighter bounds for μ) when using least order LFRs [5]. Therefore LFRs of high accuracy and low complexity (order) are required. Once (3.2) is available, very powerful and efficiently implemented techniques [6, 7] exist for the transformation of (3.2) into an almost least order LFR. However, the minimal achievable order of the resulting LFR mainly depends on the complexity (order of rational or polynomial approximations) and structure of (3.2). Thus the main emphasis must be put on an optimal generation of the LPV model (3.2) such that it is of high accuracy and optimally fits for low order LFR generation.

For this purpose we present a general procedure, starting with the generation of a set of linear time-invariant (LTI) state-space systems obtained by linearisation of (3.1) at certain equilibrium points (trim points). Least-squares multivariable polynomial fitting is used to approximate the single elements of the state-space matrices and to find a single LPV model (3.2) covering the whole set of LTI equilibrium models. Within the fitting process, sophisticated methods are employed to reduce the complexity of the polynomials by eliminating monomials with negligible influence on the accuracy of the polynomial approximations. Furthermore, the element-wise polynomial fitting is part of a global optimisation loop, where a genetic algorithm tunes the polynomial structure of each state-space matrix element with the objective to minimise the order of the related LFR model, which will be generated from the LPV model. To guarantee the accuracy of the transfer function of the LPV model, the maximum ν-gap error between the set of equilibrium models and the LPV model is calculated and included as a constraint for the optimisation.

It is important to emphasize that we directly choose to minimise the overall achievable LFR order for the optimisation of the LPV model instead of reducing the polynomial order/complexity of its single elements. This may allow to increase the accuracy, while keeping the same order for the resulting LFR. To see this, consider the matrix with two parameters

$$A(\delta) = \begin{bmatrix} \delta_1^2 & a_{12}(\delta_1, \delta_2) \\ 0 & \delta_2^2 \end{bmatrix} \tag{3.3}$$

where the entry a_{12} may only have small variations within a given parameter value set. In order to reduce the complexity of $A(\delta)$ one may therefore decide to choose a_{12} to be constant. However, in terms of the LFR order, one may also choose a_{12} as a second order polynomial (e.g., $a_{12} = \delta_1 + \delta_1^2 + \delta_2^2$) without increasing the resulting LFR order for $A(\delta)$.

Note, that in several cases one may directly derive (3.2) from (3.1) via symbolic calculations. However, especially in aeronautical applications the nonlinear models usually include highly nonlinear functions (neural networks, tables) or may only be given for a discrete set of conditions (linear aeroelastic models) such that the generation and approximation of a set of LTI state-space models is the only way to apply LPV and LFT-based robust control techniques.

In sections 3.2 and 3.3 we will describe the overall procedure for the generation of "optimal" LPV models. In section 3.4 the efficiency of the procedure is demonstrated by the generation of highly accurate LFRs for the COFCLUO nonlinear aircraft benchmark model.

3.2 Basic Procedure for the Generation of LPV Models

The starting point for the generation of LPV models is a nonlinear parametric model as given in (3.1). For this model, a grid of m linear, time-invariant state-space systems for a pre-specified set of flight conditions with parameter values $\delta^{(k)}$, $k = 1, ..., m$ is generated. Note that it must be guaranteed that all elements of the grid have consistent state, input and output vectors, i.e. the vectors have the same dimensions and physical interpretation. In the following the grid-point state-space matrices with transfer matrix $G_k(s) = D_k + C_k(sI - A_k)^{-1}B_k$ are alternatively represented in concatenated form as

$$S_k = \begin{bmatrix} A_k & B_k \\ C_k & D_k \end{bmatrix}, \quad k = 1 \ldots m.$$

The goal is to calculate a parameter dependent matrix $S(\delta)$, which best approximates all matrices S_k as a function of the parameter vector δ. The basic fitting procedure is divided into four steps:

A Separate the known physical relations and the unknown parts of S_k, that shall be approximated.
B Check if the variation of an element $s_{i,j}$ of S_k, $k = 1, \ldots, m$ within the grid points has only a negligible influence on the accuracy of the transfer function of $S(\delta)$ given as $G(s, \delta) = D(\delta) + C(\delta)(sI - A(\delta)^{-1})B(\delta)$, such that it can be approximated by a constant value or a low order polynomial.
C If the variation of an element $s_{i,j}$ has a significant influence on $G(s, \delta)$, then find a polynomial approximation for the element $s_{i,j}$ as a function of the parameter vector δ. This step includes a so-called rank deficient polynomial basis reduction based on QR decomposition with column pivoting.
D Finally, perform a full rank basis reduction of the polynomial approximations using suitable measures for the significance of individual monomials.

These four steps are embedded in an optimisation loop, which will be described later on.

3.2.1 Specification of Known Relations

For many dynamical systems physical insight into the system may be available, so that the dependency of $S(\delta)$ on the parameter vector δ may be partly known. Consider for instance a matrix as in (3.3) with two parameters, where the polynomial dependencies of all elements except a_{12} are known. Hence, the generation process has to approximate only the element a_{12}.

3.2.2 Element-Wise Significance Check

During the element-wise significance check, for each element $s_{i,j}$ of the set of matrices S_k a so-called influence coefficient $IC_{i,j}$ is determined. An element has a low influence coefficient if its variation among the set of grid point LTI models does not significantly influence the transfer function of the models (in terms of the ν-gap metric [8]) and if it is sufficiently accurate to approximate an element with its mean value of all the grid point models. Therefore for each $s_{i,j}$ a set of concatenated state-space matrices $S_{k_{i,j}}$, $k = 1, \ldots, m$ with transfer matrix $G_{k_{i,j}}$ is generated, where all entries are equal to the entries of the set S_k except the entry $s_{i,j}$, which is chosen as the mean value of the m grid point values $s_{i,j}$. Finally, the influence coefficient $IC_{i,j}$ of $s_{i,j}$ is defined as

$$IC_{i,j} = \max_k(\delta_\nu(G_k, G_{k_{i,j}})), \; k = 1, \ldots, m,$$

where δ_ν denotes the ν-gap metric, which is introduced in [8] as a measurement of the distance between two systems. A ν-gap metric value of one states that two systems are far apart, whereas zero means that they are identical.

During the polynomial approximation, which is described in the next section, the significance $IC_{i,j}$ is used to adapt the accuracy of the polynomial approximation, which may further help to reduce the order of the resulting LFR. If $IC_{i,j}$ is smaller than a pre-specified threshold ρ_1, the constant mean value of the grid points is used to approximate the element $s_{i,j}$. If $IC_{i,j}$ is larger than ρ_2, then $s_{i,j}$ will be approximated with a polynomial of high accuracy, which is defined as a given maximum root mean square error (RMSE). If $IC_{i,j}$ is between ρ_1 and ρ_2, then the desired RMSE for the polynomial approximation will be increased/multiplied by the factor q, which is illustrated in Fig. 3.1. Hence, if an element $s_{i,j}$ has a small significance $IC_{i,j}$ close to ρ_1, the RMSE for its polynomial approximation will be doubled, which will lead to a simpler and lower order polynomial approximation. Note, the thresholds ρ_1 and ρ_2 will be free parameters for the optimisation described in section 3.3.

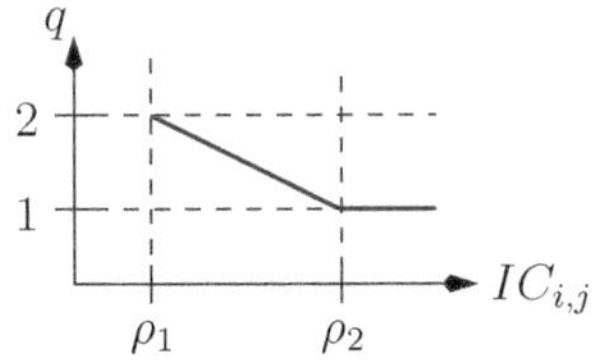

Fig. 3.1 Relation between $IC_{i,j}$ and q

3.2.3 Multivariable Polynomial Fitting

The algorithm for finding polynomial approximations of the single matrix elements is based on a least squares fitting by employing monomial polynomial bases. In addition, sophisticated methods for the reduction of monomial bases [9] are used to obtain the simplest possible approximation.

In the following $\delta_i^{(k)}$ denotes the numerical value of the parameter δ_i at the kth grid point, y is a vector including the m grid point values of an element $s_{i,j}$ and b is a vector including the polynomial coefficients. In a first step, a matrix X will be built, which considers all possible monomials for a multivariable polynomial of a given order. As an example, the matrix X for a two parametric, second order polynomial $p(\delta) = [1\ \delta_1\ \delta_2\ \delta_1^2\ \delta_1\delta_2\ \delta_2^2]\,b$ is given by

$$X = \begin{bmatrix} 1 & \delta_1^{(1)} & \delta_2^{(1)} & {\delta_1^{(1)}}^2 & \delta_1^{(1)}\delta_2^{(1)} & {\delta_2^{(1)}}^2 \\ \vdots & \vdots & \vdots & \vdots & \vdots & \vdots \\ 1 & \delta_1^{(m)} & \delta_2^{(m)} & {\delta_1^{(m)}}^2 & \delta_1^{(m)}\delta_2^{(m)} & {\delta_2^{(m)}}^2 \end{bmatrix} \tag{3.4}$$

With the above defined matrix X the following least squares problem is solved.

$$\min_b \frac{1}{2}\|y - Xb\|_2^2 \tag{3.5}$$

$$\frac{1}{2}\frac{\partial\|y - Xb\|_2^2}{\partial b} = X^T Xb - X^T y = 0 \tag{3.6}$$

Since the assumption of X having full column rank may not hold especially for higher order polynomial approximations, (3.6) cannot be directly solved for b by inverting $X^T X$. Instead, the optimal coefficients b^* of (3.5) are found by means of singular value decomposition (SVD) [2] of X, which is defined in the following way. For an arbitrary matrix $X \in \mathbb{R}^{m\times n}$, there exist unitary, orthogonal matrices $U \in \mathbb{R}^{m\times m}$ and $V \in \mathbb{R}^{n\times n}$, such that

$$X = U\Sigma V^T.$$

The matrix Σ is a diagonal matrix containing the singular values of X in descending order. The rank r of the matrix X corresponds to the non-zero singular values, so that a diagonal matrix $\Sigma_r \in \mathbb{R}^{r\times r}$ can be defined which only contains the non-zero singular values. Further, the first r columns of U and V can be written in $U_r \in \mathbb{R}^{m\times r}$ and $V_r \in \mathbb{R}^{n\times r}$ respectively, since the neglected columns do not contribute to the result of the SVD [10].

With the so defined matrices Σ_r, U_r and V_r, (3.5) can finally be solved for the optimal coefficients b^* as

$$b^* = V_r \Sigma_r^{-1} U_r^T y. \tag{3.7}$$

In case of a nearly rank deficient X, some of its singular values are almost zero, which may drastically reduce the accuracy of the solution of (3.7). A way to deal with nearly rank deficient matrices in case of a least squares problem is to use a truncated SVD solution [11]. Instead of the real rank a numerical rank is used by setting all singular values below a specified threshold to zero.

The algorithm developed in the present work iteratively increases the polynomial order of the approximation and computes the new polynomial coefficients by employing (3.7) until either the relative root mean square error RMSE defined in (3.8) drops below a specified maximum ρ_3 or the improvement in the relative RMSE becomes insignificant. Those two values (i.e. the maximum allowed error ρ_3 and the minimum error decrement ρ_4) are used as additional tuning parameters in the optimal generation process as described in section 3.3.

$$RMSE = \frac{\|e\|_2}{\|y\|_2} \qquad \text{with} \qquad e = y - Xb^* \tag{3.8}$$

Remark: Recall, that in order to distinguish between elements with significant and negligible variations, the RMSE and the RMSE decrement will both be multiplied by the factor q, which was defined in the last section. If the corresponding $IC_{i,j}$ is below ρ_1 then no polynomial fitting will be performed and the constant mean value will be used to approximate the element $s_{i,j}$.

In addition to the already described polynomial fitting based on a singular value decomposition, a sophisticated monomial basis reduction is integrated, in order to use as few monomials for the approximation as possible. The data matrix X includes all possible monomials and some of them might be redundant and can be omitted. For this purpose a so-called rank deficient or nearly rank deficient basis reduction is performed.

In case of nearly rank deficient basis reduction, a subset of columns of X, which is numerically most linear independent shall be found and used for solving the least square problem [9, 11]. It is therefore possible to eliminate the nearly redundant (nearly linear dependent) columns of X which reduces the number of monomials needed for the approximation. The basis reduction is based on a QR decomposition with column pivoting and is described in [9]. Let $V_r \in \mathbb{R}^{n \times r}$ be the matrix formed from the first r columns of V. Define $\overline{V}$ as

$$\overline{V} = V_r^T \mathrm{diag}(b_1^*, \ldots, b_n^*)$$

Then, the QR decomposition with column pivoting of $\overline{V}$ has the form

$$\overline{V} = Q \begin{bmatrix} R_{11} & R_{12} \end{bmatrix} P^T$$

With the permutation matrix P a new data matrix Z is calculated according to (3.9) and the first r columns of Z are used to refit the polynomial approximation [9], where

$$Z = XP. \tag{3.9}$$

3.2.4 Full Rank Basis Reduction

The full rank basis reduction is conducted as a final step of the fitting algorithm after a feasible approximation has been found. Some monomials of this polynomial approximation may not have a significant influence so that they can be eliminated. A good measure of the significance of certain monomials is the magnitude of the whole term, i.e. the product of the coefficient with its corresponding column of X. For this reason a utility factor u for a term j is defined according to (3.10) [9].

$$u_j = \frac{||b_j^* X_j||_2}{||y||_2} \tag{3.10}$$

During the full rank basis reduction, the column corresponding to the lowest utility factor is omitted from X and a new set of polynomial coefficients is calculated with the reduced X. This procedure is repeated until the RMSE of the new approximation has significantly increased with respect to the original approximation. The threshold ρ_5 is specified as the relative increase with respect to the original RMSE in percentage.

The presented approach via singular value decomposition and basis reduction introduced by [9] and implemented in this work has been compared to classical linear regression methods as provided in the Matlab Statistics Toolbox. Both algorithms yield similar results but the SVD-based one has performed faster and numerically more robust in case of rank deficient and nearly rank deficient problems in several test cases.

3.3 Optimisation of the Linear Parameter Varying Model

The structured singular value computation and similar LFT based stability analysis are computational demanding. Thus for an efficient analysis the availability of low order LFRs is vital, which will also yield more accurate analysis results [5, 12]. Such an optimal, low order LFR, which still possesses a sufficient accuracy can be obtained by solving an optimisation problem. Instead of minimising the LFR order directly, it is approximated by a lower bound as described in [6], which reduces the computational effort drastically.

For a given LPV model $S(\delta)$ with $\delta \in \mathbb{R}^l$ the lower bound can be calculated as follows: Substitute all but one parameter δ_i with random values and compute a one parametric LFR with minimal order r_i. Note, that for single parametric systems one can always calculate a minimal order LFR. Repeat this procedure for all parameters. Finally, the lower bound is given by $r = \sum_{i=1}^{l} r_i$.

The above defined lower bound can then be minimised over the following optimisation parameters:

- Thresholds ρ_1 and ρ_2 from the significance check of a single element
- Maximum RMSE for the polynomial approximation ρ_3
- Minimum decrement ρ_4 of the RMSE when increasing the order of the polynomial approximation
- Maximum allowed increase ρ_5 of the RMSE during the full rank basis reduction

These five parameters are concatenated in the vector ρ. For given numerical values of the optimisation parameters in ρ, it is first decided which elements are significant and therefore are considered for the polynomial approximation. For the elements with a sufficiently high influence coefficient, multivariable polynomials are fitted by means of the algorithms described in the previous section.

3.3.1 Optimisation with ν-Gap Metric Constraint

In addition to the minimisation of the LFR order via its lower bound, an easily verifiable criteria for the accuracy of the LPV model is required, in the form of the distance between the LPV model $S(\delta)$ and the grid-point LTI models. For this, the ν-gap metric δ_ν already introduced at the element-wise significance check is used, which has the advantage that it can also be used for unstable systems.

By introducing the maximum ν-gap metric between the LPV model and the set of LTI models as an additional optimisation constraint $\delta_{\nu,max} = \max \delta_\nu(G_k(s), G(s,\delta^{(k)}))$, $k = 1, \ldots, m$, the optimisation problem can be written in the following way:

$$\min_{\rho} \left(r\left(\rho\right)\right) \;\; \text{subject to} \;\; \delta_{\nu,max}(\rho) < \nu_{max}. \tag{3.11}$$

$G(s,\delta^{(k)})$ denotes the transfer function of the system corresponding to $S(\delta)$ for δ evaluated at the kth entry, i.e. $G(s,\delta^{(k)}) = D(\delta^{(k)}) + C(\delta^{(k)})(sI - A(\delta^{(k)}))^{-1}B(\delta^{(k)})$. The value ν_{max} is chosen depending on the desired approximation accuracy of the LPV model.

Due to the fact that neither $r(\rho)$ nor $\delta_{\nu,max}(\rho)$ are continuously differentiable in ρ, a gradient based optimisation algorithm is unsuitable and instead a global search algorithm is used. In the present work the differential evolution algorithm proposed by [13] and implemented in [14] is applied. For the COFLCUO models this optimisation typically took around 20 minutes with Matlab 2007b under Windows XP on an Intel Dual Core T2500 (2GHz, 1GB RAM) computer.

Concerning the initial choice of ρ, we usually started with very low values yielding very accurate but high order models and then the optimiser increased the values to reduce the complexity until the maximum allowed model error was reached.

3.3.2 Optimisation of the Polynomial Coefficients

Once the optimisation of (3.11) is finished, the polynomial structure and also the resulting LFR order are fixed. However, a further reduction of the maximum ν-gap between the LPV model $S(\delta)$ and the LTI models may be achieved by additional minimisation of $\delta_{\nu,max}$ over all the polynomial coefficients. So far, aside from the influence coefficient computation, each element has been treated individually in the algorithm. In the end, however, the system's behaviour reflects the quality of the solution, not the best individual approximation. Therefore, the following optimisation problem is proposed

$$\min_{\text{Coeffs}} \delta_{\nu,max},$$

where Coeffs denotes a vector including the polynomial coefficients of all elements $s_{i,j}(\delta)$ of $S(\delta)$. In our procedure, the polynomial coefficients are constrained within a twenty percent band of the original solution, in order to stay close to the solution obtained from the previous optimisation.

Note, that during this second optimisation the structure and order of the polynomial fittings within the LPV model are not changed and only the coefficients are (slightly) tuned in order to improve the accuracy of the model. Hence at least for the models under consideration the cost function of the optimisation was smooth and fast gradient based algorithms could be used for this second optimisation. For the COFLCUO models this second optimisation typically took around five minutes with Matlab 2007b under Windows XP on an Intel Dual Core T2500 (2GHz, 1GB RAM) computer.

3.4 Application to the COFLCUO Nonlinear Aircraft Model

The aircraft model includes many nonlinearities, e.g. the aerodynamic equations are represented as neural networks, and an exact LPV representation based on nonlinear symbolic equations would be very complex. Therefore we decided to apply the procedure presented in sections 3.2 - 3.3 to generate an accurate model approximation in LPV form which can be transformed into an LFR of low complexity.

3.4.1 The Aircraft Model

The nonlinear aircraft model is implemented in Simulink and consists of an actuator block, an aircraft dynamics block and a sensor block. The inputs, outputs and states (without sensor and actuator states) of the aircraft model are given in Tables 3.1 - 3.3. As no guidance loops have been considered in the project, the states describing the aircraft position were removed and the application of numerical order reduction also removed one unobservable state of the quaternions from the linearised state-space models, such that the final LPV models and LFRs only have nine states.

The uncertain and varying parameters of the aircraft model are given by the aircraft mass m, the x-position of the center of gravity X_{cg}, the calibrated airspeed V_{cas} and the Mach number M. These parameters are allowed to vary within some given domains, the weight/balance domain and the $V_{cas} - M$ flight envelope. In order to obtain LPV models and LFRs of low complexity, we splitted the overall parameter domain into 16 subdomains (yielding 16 LPV models and LFRs). The subdomains are shown in Figures 3.2-3.4. Depending on the mass of the aircraft (light or heavy cases) we used different $V_{cas} - M$ domains, which can be seen in Fig. 3.3 and Fig. 3.4.

Note that in order to allow an independent variation of the uncertain parameters m and X_{cg} in the LPV models - which will simplify the application

Table 3.1 Inputs to aircraft model

Input	Description
1	Elevator
2	Horizontal Stabilizer
3	Rudder
4	Inner left aileron
5	Inner right aileron
6	Outer left aileron
7	Outer right aileron
8	Left spoilers 2 and 3
9	Right spoilers 2 and 3
10	Left spoilers 4 and 5
11	Right spoilers 4 and 5

Table 3.2 Outputs of aircraft model

Output	Description
1	Pitch angle θ
2	Roll angle ϕ
3	Roll rate p
4	Pitch rate q
5	Yaw rate r
6	Lateral load factor n_y
7	Vertical load factor n_z
8	Angle of sideslip β

of μ-analysis techniques - it was initially decided to choose rectangular domains for these two parameters. With the actual set of models a small region of the weight and balance domain is not covered, however, on the one hand it would be no problem to generate additional models that cover the remaining regions and on the other hand the actual set of models was considered to be sufficient to demonstrate in principle the applicability of the clearance methods and tools developed within the project.

3.4.2 Trimming and Linearisation

In order to apply the method described in section 3.2 we adapted the very accurate and efficient tools for trimming (`trimex.m`) and linearisation (`linmod3.m`) from DLR to the nonlinear aircraft model in Simulink. The function `trimex.m` relies on efficient nonlinear system solvers available via the *mex*-function interfaces to nonlinear system solvers and least-squares routines from the subroutine libraries MINPACK [15] and PORT [16]. A

Table 3.3 States of aircraft model

State	Description
1	Quaternion 1
2	Quaternion 2
3	Quaternion 3
4	Quaternion 4
5	Roll rate p
6	Pitch rate q
7	Yaw rate r
8	Ground speed V_x
9	Ground speed V_y
10	Ground speed V_z
11	x-Position
12	y-Position
13	z-Position

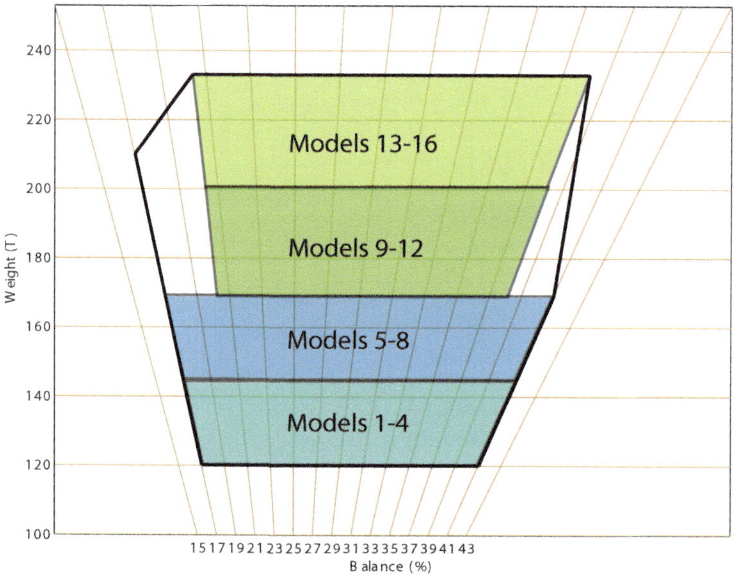

Fig. 3.2 Weight and balance domains

very useful feature implemented in `trimex.m` is the optional trimming with simple bounds on the trim variables. The superiority of the new trimming tool `trimex.m` over the standard Matlab tool `trim.m` in what concerns speed (factor of 10 faster) and reliability (accuracy and feasibility) of the results has been demonstrated in many trimmability studies within the project. These

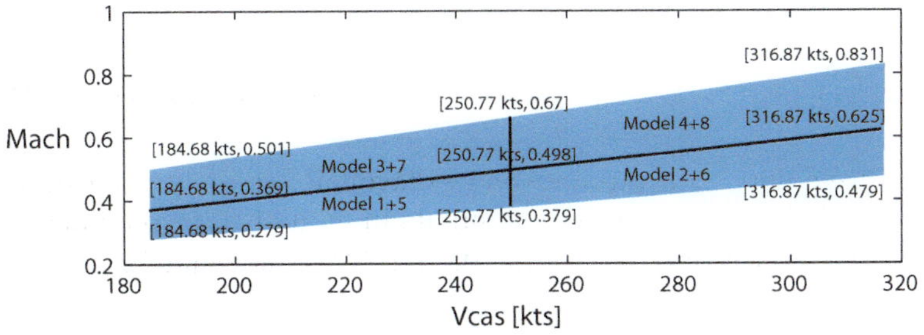

Fig. 3.3 $V_{cas} - M$ domain for low masses (models 1-8)

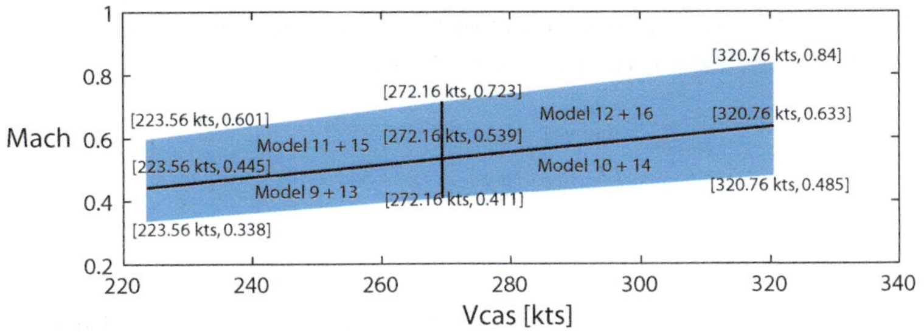

Fig. 3.4 $V_{cas} - M$ domain for high masses (models 9-16)

tools allow to easily generate a huge grid of linear state-space systems within the admissible flight envelope of the aircraft.

For the generation of the LPV models, in each of the 16 parameter domains we used 3 gridpoints for each parameter resulting in 81 (3^4) gridpoint LTI-models per subdomain and an overall set of $16 \times 81 = 1296$ models. In addition we generated a finer grid of LTI-models consisting of 625 (5^4) gridpoints per subdomain (overall set of $16 \times 625 = 10000$ LTI-models) that were used for the validation of the LPV-models.

3.4.3 Generation of an LFR for the Actuator Model and the Sensor Model

In the aircraft model the actuator dynamics are independent from the parameters m, X_{cg}, M and V_{cas}. Each actuator is described by a first order system including a rate limiter and a deflection limiter (see Fig. 3.5).

To simplify the robust stability analysis of the aircraft model, the limiters are substituted by dead-zones (see Fig. 3.6), where in the LFR the dead-zones are included as artificial parameters.

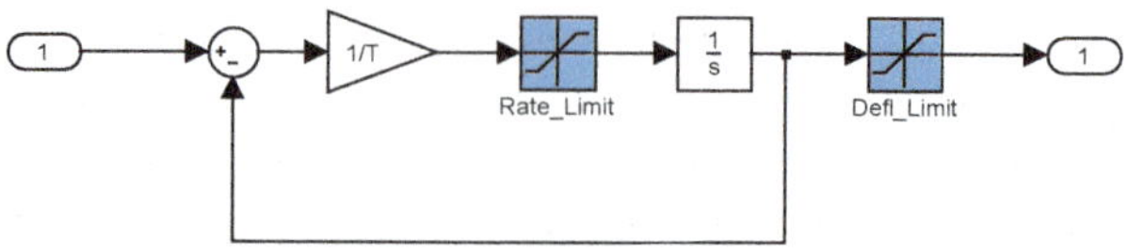

Fig. 3.5 Actuator model

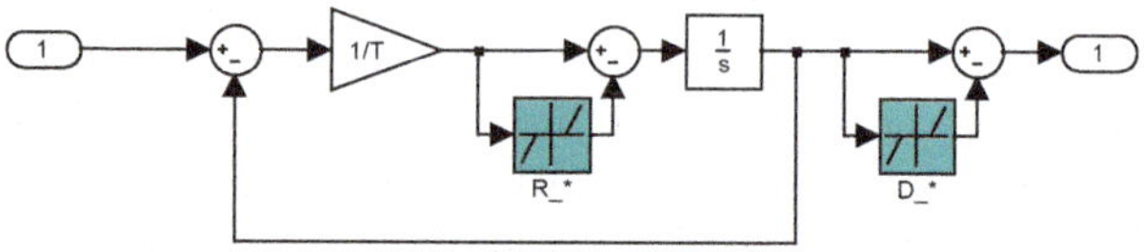

Fig. 3.6 Actuator model including dead-zones

The LFR (M, Δ) of a single actuator is given as

$$M = \left[\begin{array}{ccc|c} -1/T & 0 & 1 & 1/T \\ -1 & 0 & 0 & 0 \\ 1/T & 0 & 0 & -1/T \\ \hline 1 & 1 & 0 & 0 \end{array}\right], \; \Delta = \mathrm{diag}(1/s, D_*, R_*).$$

where T is the time constant, D_* is the dead-zone parameter for the deflection limiter and R_* is the dead-zone parameter for the rate limiter of the actuator. In the LFRs the $*$ is substituted by the name of the corresponding control surface.

The sensor model has 17 inputs that are used to calculate the 8 outputs that are used as controller inputs and performance outputs. In order to reduce the complexity of the resulting LFR model the algebraic and parameter dependent part (only affine dependence on X_{cg}) of the sensor model was considered to be part of the flight dynamics model, thus reducing the outputs of the flight dynamics model from 17 to 8. Therefore the remaining part of the sensor model only consists of 8 first order linear filters representing the dynamics of the sensors, which can directly be transformed into an LFR.

3.4.4 Generation of LPV Models and LFRs for the Flight Dynamics Model

The method described in sections 3.2 - 3.3 was applied to generate optimised LPV-models for the nonlinear aircraft dynamics. Due to the small number of grid-points (81 per subdomain), the 16 LPV models had a very similar polynomial structure resulting in LFRs of orders around 270 for all the models.

Note, that the LTI grid point models have been numerically reduced, diagonalised and balanced in order to obtain minimal order state-space representations that can be interpolated. Therefore the knowledge about the parametric dependence of the states was lost and the first step of the LPV model generation process (see section 3.2) was omitted.

The accuracy and structure of the 16 LFRs is listed in Table 3.4, where the maximum ν-gap error and the mean ν-gap error between the LFR and the grid-point models is shown. Furthermore, the structure of the Δ matrix, with $\Delta = \text{diag}(X_{cg}I_{n_1}, mI_{n_2}, M\, I_{n_3}, V_{cas}I_{n_4})$ is given. For the generation of the LFRs from the optimised LPV models the symbolic pre-processing tools described in [6] were used. It is important to emphasize that a direct application of the standard object-oriented LFR generation method [6] without any pre-processing or numerical order reduction would result in LFRs of orders around 24000. However, as there is a lot of exploitable structure in the LPV models (common factors in the rows and columns of the LPV models), symbolic pre-processing methods allowed to reduce this incredible large order down to orders around 270, which is sufficiently small for the application of the efficient μ-analysis and Lyapunov function based analysis techniques as proposed in the COFCLUO project.

Note, that the lower bound LFR orders obtained from the LPV model optimisation were ranging around 190 and the orders of 270 obtained by symbolic pre-processing are comparatively large. However, there is no way to evaluate the accuracy of this lower bound and the orders around 270 could even be minimal.

Note, that the dynamics of the LPV models and LFRs are exactly the same as no numerical reduction/approximation was performed during the generation of the LFRs.

Table 3.4 Accuracies and orders of LFRs

Model No.	$\max(\delta_\nu)$	$\text{mean}(\delta_\nu)$	(n_1, n_2, n_3, n_4)	$\sum n_i$
1	0.013	0.0028	(81,49,51,85)	266
2	0.044	0.0051	(61,50,85,51)	247
3	0.044	0.0033	(84,56,51,85)	276
4	0.029	0.0025	(66,53,85,51)	255
5	0.023	0.0026	(62,71,85,51)	269
6	0.025	0.0030	(81,50,85,51)	267
7	0.023	0.0049	(78,55,85,51)	269
8	0.024	0.0029	(75,55,51,85)	266
9	0.041	0.0053	(85,73,85,51)	294
10	0.022	0.0020	(69,62,85,51)	267
11	0.050	0.0031	(81,56,68,68)	273
12	0.024	0.0032	(71,47,51,85)	254
13	0.040	0.0090	(81,55,85,51)	272
14	0.040	0.0055	(79,56,85,51)	271
15	0.043	0.0032	(78,53,51,85)	267
16	0.047	0.0047	(84,86,85,51)	306

3.4.5 Validation of the LPV Models and LFRs of the Flight Dynamics Model

In a first step it was very important to validate the set of linearised grid point models. Therefore a comparison between step responses of the nonlinear simulation and the grid point LTI models was performed. In several cases it was necessary to adapt the perturbations, which are used to compute the central difference in the Matlab linearisation function *linmod.m*. In Fig. 3.7 an example of this comparison for a rudder step response with outputs p, q, r are shown. Here the dominating dynamics (in roll and yaw rate) are matched very good. However, there will always be small discrepancies (even locally), which can be seen in this case when after some seconds the nonlinear simulation shows some pitch movement, which is not excited in the LTI model.

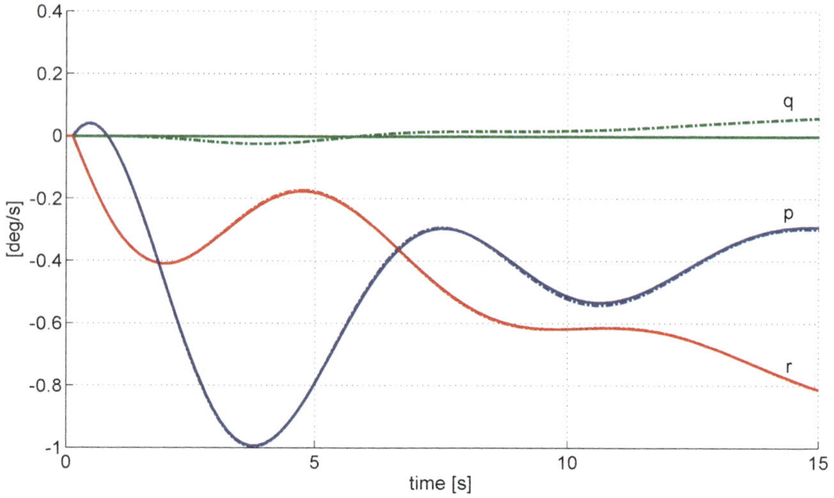

Fig. 3.7 Angular rates for rudder step input of nonlinear (dash-dotted line) and LTI (solid line) model

After validation of the LTI grid-point models the next step is to compare the LTI models with the LPV model. For this the ν-gap metric between the LPV model and the fine LTI-model validation grid (625 grid-points per parameter subdomain) was calculated, which only resulted in a negligible increase of the ν-gap values as given in Table 3.4. In addition, Bode magnitude plots of the LPV and LTI models were compared (see an example in Fig. 3.8). These usually showed almost no difference, which was already indicated by the small values of ν-gap metric errors (< 0.05 in average).

One of the main achievements of the fitting algorithm of section 3.2 was to eliminate all unnecessary monomials from the polynomial approximations of

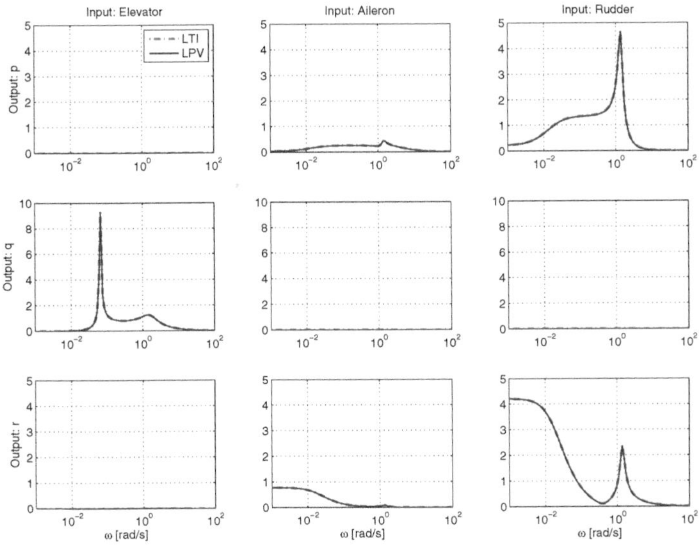

Fig. 3.8 Comparison of LTI and LPV model bode magnitude plots

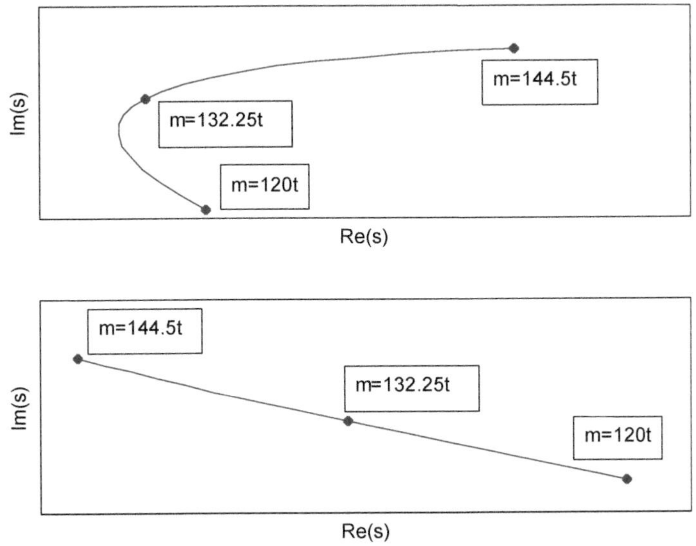

Fig. 3.9 Example of the movement of two complex eigenvalues

the state-space matrix elements, which resulted in a very smooth transition of the LPV model eigenvalues between the grid-points.

As an example the transitions of two complex eigenvalues of the aircraft model valid in subdomain 1 are shown in Fig. 3.9 for a variation of the mass m between $120t$ and $144.5t$. The grid for the generation of the LPV model included the values at $120t$, $132.25t$ and $144.5t$ and the transition of the eigenvalues between these grid points is very smooth.

3.5 Conclusion

A general algorithm for generating linear parameter varying (LPV) models has been developed, which can be applied to arbitrary nonlinear systems, as long as the system behaviour can be accurately described/approximated with polynomial or rational parametric state-space systems. The bases for efficient LFT based methods like μ-analysis are low order and accurate models in LFT form. Hence, an optimisation problem is derived to find an accurate parametric approximation of the nonlinear system, which provides an optimal structure in terms of least order LFR generation. Additionally, state-of-the-art algorithms as proposed in [6] are used for the transformation of the LPV model into an LFR.

In the present work, these methods have been successfully applied to generate LFRs of high accuracy and reasonable order for the COFCLUO nonlinear benchmark model. These LFRs will be combined with the nonlinear controller in LFR form (see chapter 5) to allow the application of LFT based clearance methods.

References

1. Zhou, K., Doyle, J.C.: Essentials of Robust Control. Prentice-Hall, Englewood Cliffs (1998)
2. Golub, G.H., Van Loan, C.F.: Matrix Computations, 3rd edn. Johns Hopkins University Press, Baltimore (1996)
3. Packard, A.: Gain scheduling via linear fractional transformations. System and Control Letters 22, 79–92 (1994)
4. Leith, D.J., Leithead, W.E.: Survey of gain-scheduling analysis and design. International Journal of Control 73(11), 1001–1025 (2000)
5. Hecker, S.: Improved mu-analysis results by using low order uncertainty modelling techniques. AIAA Journal of Guidance, Control and Dynamics 31(4), 962–969 (2008)
6. Hecker, S.: Generation of low order LFT Representations for Robust Control Applications. VDI, Verlag (2007)
7. Hecker, S., Varga, A., Magni, J.: Enhanced LFR-toolbox for Matlab. Aerospace Science and Technology 9(2), 173–180 (2005)
8. Vinnicombe, G.: Frequency domain uncertainty and the graph topology. IEEE Transactions on Automatic Control 38(9), 1371–1383 (1993)

9. Lin, J.G.: Modelling test responses by multivariable polynomials of higher degrees. Siam J. Sci. Comput. 28(3), 832–867 (2006)
10. Skogestad, S., Postlethwaite, I.: Multivariable Feedback Control. John Wiley and Sons, Chichester (2005)
11. Björk, Å.: Numerical Methods for Least Square Problems. SIAM, Philadelphia (1996)
12. Herrnberger, M., Heller, M., Paul, R., Hecker, S., Sachs, G.: Advanced uncertainty modeling and robustness analysis for the basic control system of a modern jet trainer. In: Proceedings of AIAA Conference on Guidance, Navigation and Control (2007)
13. Storn, R., Price, K.: Differential evolution - a simple and efficient adaptive scheme for global optimization over continuous spaces. Technical report TR-95-012, ICSI (1995)
14. Joos, H.-D.: MOPS - Multi-Objective Parameter Synthesis. User guide, DLR (2007)
15. Moré, J.J.: User's Guide for MINPACK-1. Applied Mathematics Division Report ANL-80-74, Argonne National Laboratory, Argone, IL (1980)
16. Fox, P.A., Hall, A.P., Schryer, N.L.: The PORT mathematical subroutine library. ACM Trans. Math. Softw. 4(2), 104–126 (1978)

Chapter 4
Generation of LFRs for a Flexible Aircraft Model

Clément Roos

Abstract. The flexible aircraft described in Chapter 2 is modelled as a collection of linear time-invariant models which correspond to different combinations of aircraft payload configurations and flight conditions. In this chapter we propose a methodology to convert such a set of models into a linear fractional representation (LFR), to be used in several robustness analysis methods. The modelling challenge consists in building a unique LFR starting from a set of large order state-space models, having different sizes and different physical meanings of state vectors. The resulting LFR has a moderate size and satisfactorily approximates the aircraft behaviour, by providing almost exact match of the system eigenvalues and frequency responses. This model serves as basis of the clearance techniques relying on μ-analysis, Lyapunov-based analysis and IQC-based analysis.

4.1 Introduction

The current industrial approach to clearance of flight control laws is thoroughly described in Chapter 2. Basically, it consists in gridding the considered parametric domain and checking a set of criteria at each point of the grid. The main drawback of such a strategy is that clearance is restricted only to the considered grid points and nothing can be assessed for the remaining points in the parametric domain. Moreover, significant costs are involved to perform this task. Several techniques, such as μ-analysis (see Chapter 7 and [1,2,3]), Lyapunov-based analysis (see Chapter 9 and [4]) and IQC-based analysis (see Chapter 10 and [5]), could be efficient alternatives. Indeed, they allow to determine quickly whether the clearance requirements are fulfilled on a continuous parametric domain. Unfortunately, most of these techniques

Clément Roos
ONERA, Systems Control and Flight Dynamics Department,
2 avenue Edouard Belin, 31055 Toulouse Cedex 4, France
e-mail: clement.roos@onera.fr

A. Varga et al. (Eds.): Opti. Based Clearance of Flight Control Laws, LNCIS 416, pp. 59–77.
springerlink.com

require the aircraft models to be expressed as linear fractional representations (LFRs), which usually are not available in the aeronautical industry.

Moreover, because of a better overall aircraft design optimisation and an increase in aircraft sizes, the bandwidth of the bending modes tends to be larger and now often comes very close to the one of the rigid body modes. It must then be checked that control laws do not alter neither the stability properties of the first flexible modes nor the over-shoot loads objectives assigned by structural aircraft design [6]. To address these issues, the flexible aircraft behaviour need to be modeled with high fidelity. This is usually done by using a collection of high order linear time-invariant state-space models (so-called reference models), which correspond to different combinations of aircraft payload configurations and flight conditions. The building of LFRs using the linear fractional transformation (LFT)-based modelling process becomes very challenging, since both the size and the meaning of state variables of the reference models can vary according to the considered point in the parametric domain. In this context, the main contribution of the present chapter is to propose a new methodology to convert a set of reference models describing the aeroelastic dynamics of the aircraft for various payload configurations and flight conditions into a suitable LFR, with a special emphasis on LFT reduction techniques. Note that another approach to the same issue is proposed in [7].

The outline of the chapter is as follows. Sect. 4.2 states the problem, while Sect. 4.3 describes the LFT modelling and reduction strategy. Numerical results are then presented in Sect. 4.4 to demonstrate that the resulting LFRs are highly representative of the reference models and that their complexity is compatible with the use of robustness analysis tools.

4.2 Problem Statement

4.2.1 Description of the Reference Models

Let us consider the reference set of $N = 162$ open-loop longitudinal (or equivalently lateral) models $(G_i(s))_{i\in[1,\,N]} \leftrightarrow (A_i, B_i, C_i, D_i)_{i\in[1,\,N]}$ introduced in Chapter 2. These models describe both the rigid and the flexible dynamics of a civil passenger aircraft on a 5-D grid corresponding to the mass configuration (amount of fuel in the center/outer tanks δ_{CT}/δ_{OT}, embarked payload δ_{PL}) and the flight point (Mach number M, calibrated air speed V_{cas}). More precisely, the grid is composed of

- 2 payload configurations: 0% and 100%,
- 3 outer tank configurations: 0%, 50% and 100%,
- 3 center tank configurations: 0%, 50% and 100%,
- 3 values of the Mach number: 0.70, 0.86 and 0.91,
- 3 values of the calibrated air speed for each value of M, as shown in Table 4.1.

Table 4.1 Description of the flight domain

Mach number	0.70	0.86	0.91
	205	255	275
Calibrated air speed (kts)	285	310	320
	365	365	365

4.2.2 LFT Modelling Objective

Let $\delta = (\delta_{CT}, \delta_{OT}, \delta_{PL}, M, V_{cas})$. Let $\overline{\delta}$ denote the normalised version of vector δ, *i.e.* $\overline{\delta} \in [-1, 1]^5$ on the grid. With reference to Fig. 4.1(a), the LFT modelling task considered in this chapter is to convert the aforementioned set of aeroelastic models into a suitable open-loop LFR $\mathcal{F}_u(M(s), \Delta)$, where y_r is composed of the control surfaces deflections and possibly the wind velocity, while y consists of the flight parameters measured by the sensors and possibly a load output or an acceleration signal. In the present context, Δ is a block-diagonal matrix

$$\Delta = \mathrm{diag}\,(\overline{\delta}_{CT} I_{n_1}, \overline{\delta}_{OT} I_{n_2}, \overline{\delta}_{PL} I_{n_3}, \overline{M} I_{n_4}, \overline{V}_{cas} I_{n_5}) \qquad (4.1)$$

where I_{n_j} stands for the n_j-by-n_j identity matrix and $n_\Delta = \sum n_j$ is the total size of Δ.

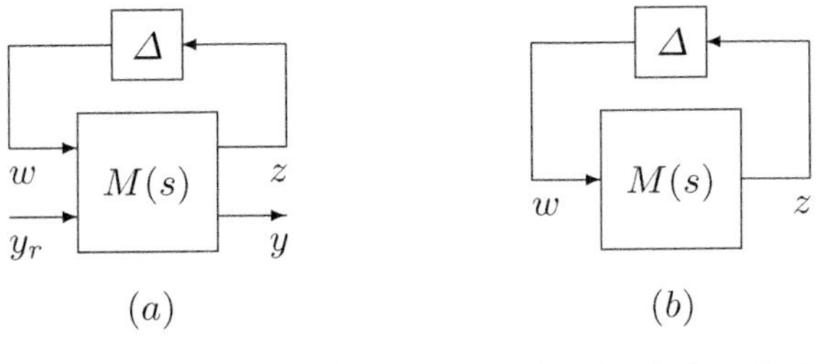

Fig. 4.1 Structure of the LFRs used to assess either loads/comfort (a) or stability (b) criteria

The open-loop LFR should be highly representative of the initial models, in the sense that its eigenvalues and frequency responses should almost exactly match those of the initial models on the 5-D grid. A special attention should also be paid to the trajectories of its eigenvalues and the continuum of its frequency responses to ensure that they are as regular as possible on the whole continuous parametric domain.

Moreover, its complexity should remain compatible with the use of robustness analysis tools. Indeed, the open-loop LFR will be used in Chapter 5 to build different kinds of closed-loop LFRs, so as to assess several clearance criteria. As an example, two such criteria are described below

- *Turbulence loads or comfort criterion.* Loads level or vertical acceleration induced by turbulence must not overshoot a given value. The associated closed-loop LFR is depicted in Fig. 4.1(a), where y_r is the wind velocity and y denotes either a load output or an acceleration signal at a specific point of the aircraft fuselage.
- *Un-piloted aircraft stability criterion.* The closed-loop plant must remain stable on the whole operating domain. The associated closed-loop LFR is depicted in Fig. 4.1(b).

4.2.3 Challenging Issues

Generating a suitable open-loop LFR from the set of aeroelastic models introduced in Chapter 2 has several challenging aspects:

- Due to the presence of numerous flexible modes, the reference models $(G_i(s))_{i\in[1,N]}$ are of relatively high orders, with state dimensions of order 300. Moreover, their state matrices are ill-conditioned, which makes their reduction tedious.
- The dimensions of the state variables may vary from one grid point to another. Some flexible modes are present in one model and missing from another model for different mass configurations and/or flight conditions. This is inherent to the industrial modelling process.
- The grid is very coarse. Therefore, achieving a good fit over the whole domain of variation of the mass and flight parameters is a demanding interpolation task.
- The parametric structure of the models is unknown.

Each of these aspects may prevent the direct use of existing methods to obtain a suitable LFR. Indeed, LFT modelling for a flexible plant is notably investigated in [8, 9], but under the assumption that the structure of the initial models is available.

The methodology presented in Sect. 4.3 allows to overcome the aforementioned limitations. Moreover, it is worth pointing out that it can be applied to dynamic models for which the analytical structure is unknown and for different use (e.g., evaluating the robustness properties of a plant or designing self-scheduled controllers). This can be achieved by tuning the LFT reduction algorithm described in Sect. 4.3.4, so as to favor either high reliability or low complexity.

4.3 Description of the Method

In this section a complete methodology is described to convert a set of reference aeroelastic models $(G_i(s))_{i\in[1,N]}$ defined on a parameter grid into an LFR. A preprocessing step described in Sect. 4.3.1 first generates reduced models with consistent state space matrices (same order and same physical meaning of states) and accurate modal content. These models are then interpolated and converted into an LFR in Sect. 4.3.2, and the special case of a coarse grid is considered in Sect. 4.3.3. An algorithm is finally proposed in Sect. 4.3.4 to reduce the LFR complexity while maintaining a satisfactory approximation accuracy.

4.3.1 Generation of Reduced and Consistent Models

A modal truncation is first performed on each reference model: the flexible modes whose frequency is higher than a given bound $\overline{\omega}$ are eliminated, as well as all other unwanted modes, such as the delay modes introduced in Chapter 2 to model the generalised aerodynamic loads. This allows to preserve the exact values of the remaining modes. A new set of state space models $(G_i^{red}(s))_{i\in[1,\,N]}$ is thus obtained with a block-diagonal structure of the state matrix. These models remain highly representative of the initial ones on a sufficiently large frequency interval, but whose size makes them compatible with the development of reasonably low order LFRs.

The next step consists in reordering as much as possible the state vectors of these reduced models in order to ensure modal consistency (i.e., the same physical meaning of state components), which is crucial in the perspective of interpolating their state space matrices in Sect. 4.3.2. Two strategies can be adopted:

- If the state vectors of the reference models are consistent, the states of the reduced models can be reordered by comparing the eigenvectors contained in the modal shape matrices used for reduction. The similarity between two eigenvectors v_1 and v_2 is intuitively measured by the angle σ between them, which makes $\cos^2(\sigma) = \frac{(v_1^T v_2)^2}{|v_1|^2|v_2|^2}$ a good indicator.
- Otherwise, a cruder approach can be implemented, which consists in comparing the distances between the eigenvalues of different reduced models, as well as their trajectories in the complex plane.

In both cases, a recursive algorithm can be implemented to investigate step by step all reduced models. In addition, the following strategy is proposed in case some flexible modes are missing for certain grid points:

- If the influence of such a mode on the frequency responses of the models in which it is present is not significant, the mode is eliminated.
- Otherwise, it is artificially added to all the other models (the associated elementary A matrix is simply obtained by linear interpolation).

A change of state coordinates is then performed to scale the reduced models $G_i^{red}(s) \leftrightarrow (A_i^{red}, B_i^{red}, C_i^{red}, D_i^{red})$. The idea is to impose the same structure to each A_i^{red} and to the first column of each B_i^{red} matrix. For example, the submatrices $A_{i,j}^{red}$ and $B_{i,j}^{red}$ associated to the jth mode of the ith model are expressed as follows

$$\left\{\begin{array}{l} A_{i,j}^{red} = \begin{pmatrix} 0 & 1 \\ -|\lambda_{i,j}|^2 & 2\,\Re(\lambda_{i,j}) \end{pmatrix} \\ B_{i,j}^{red} = \begin{pmatrix} 1 & B_{i,j}^1 \\ 0 & B_{i,j}^2 \end{pmatrix} \end{array}\right. \quad \text{if } \lambda_{i,j} \text{ is a complex eigenvalue} \tag{4.2}$$

$$\left\{\begin{array}{l} A_{i,j}^{red} = \lambda_{i,j} \\ B_{i,j}^{red} = \begin{pmatrix} 1 & B_{i,j}^1 \end{pmatrix} \end{array}\right. \quad \text{if } \lambda_{i,j} \text{ is a real eigenvalue}$$

The computation of A_i^{red} and B_i^{red} is then straightforward

$$\left\{\begin{array}{l} A_i^{red} = \operatorname{diag}(A_{i,j}^{red}) = \begin{pmatrix} \ddots & & \\ & A_{i,j}^{red} & \\ & & \ddots \end{pmatrix} \\ B_i^{red} = \begin{pmatrix} \vdots \\ B_{i,j}^{red} \\ \vdots \end{pmatrix} \end{array}\right. \tag{4.3}$$

This allows to avoid an imbalance between the models which would prove harmful during the interpolation performed in Sect. 4.3.2. Moreover, it is desirable to have as many constant terms as possible in the matrices to be interpolated, so as to reduce the complexity of the resulting LFR. Note also that the use of a companion form for $A_{i,j}^{red}$ is not a trivial choice: numerous tests have indeed revealed that this structure is a good choice in terms of LFR complexity.

The reference full-size models are ill-conditioned and numerical problems can alter the aforementioned reduction step. The frequency responses of some reduced models thus have to be improved, so that they remain as close as possible to the ones of the reference models on the frequency interval $[0, \overline{\omega}]$. The frequency error between a reference model $G_i(s)$ and its reduced counterpart $G_i^{red}(s)$ is defined by

$$H_i(\omega) = xF_i(\omega) - G_i(\omega) \tag{4.4}$$

where $x = \begin{pmatrix} C_i^{red} & D_i^{red} \end{pmatrix}$ contains the optimisation parameters and

$$F_i(\omega) = \begin{pmatrix} (j\omega I - A_i^{red})^{-1} B_i^{red} \\ I \end{pmatrix} \tag{4.5}$$

$$G_i(\omega) = C_i(j\omega I - A_i)^{-1} B_i + D_i \tag{4.6}$$

The idea is to minimise the following quantity

$$\sqrt{\frac{1}{2\pi}\sum_{j} \operatorname{trace}\left(H_i^*(\omega_j)H_i(\omega_j)\right)(\omega_{j+1}-\omega_j)} \tag{4.7}$$

It is similar to the $\mathcal{H}_2$-norm except that the integral term has been replaced by a sum on a frequency grid, which must be fine enough to ensure that the approximation is reasonable. The next proposition shows that minimising (4.7) *w.r.t.* C_i^{red} and D_i^{red} is equivalent to minimising a quadratic criterion, which can be performed easily. It is assumed in the sequel that the considered frequency grid is regular: the term $\omega_{j+1}-\omega_j$ is constant for all j and can thus be ignored.

Proposition 4.1. *[7] Let $u_1,\dots,u_n$ and $l_1(\omega_j),\dots,l_n(\omega_j)$ denote the rows of x and the columns of $F_i(\omega_j)G_i^*(\omega_j)$ respectively. Then*

$$\sum_{j} \operatorname{trace}\left(H_i^*(\omega_j)H_i(\omega_j)\right) = \tilde{x}^T Q\tilde{x} - 2\tilde{x}^T f + c \tag{4.8}$$

where

$$\tilde{x} = \begin{pmatrix} u_1 & \dots & u_n \end{pmatrix}^T \tag{4.9}$$

and

$$Q = \sum_{j} \operatorname{diag}\left(\underbrace{F_i(\omega_j)F_i^*(\omega_j),\dots,F_i(\omega_j)F_i^*(\omega_j)}_{n\ repetitions}\right) \tag{4.10}$$

$$f = \sum_{j} \begin{pmatrix} \Re(l_1(\omega_j)) \\ \vdots \\ \Re(l_n(\omega_j)) \end{pmatrix} \tag{4.11}$$

$$c = \sum_{j} \operatorname{trace}\left(G_i(\omega_j)G_i^*(\omega_j)\right) \tag{4.12}$$

Minimising (4.7) is thus equivalent to computing $\tilde{x}$ such that $Q\tilde{x} = f$.

In practice, Q is almost always invertible and thus $\tilde{x} = Q^{-1}f$. Moreover, an obvious advantage of (4.7) is that it is possible to consider only a finite frequency interval $[0,\overline{\omega}]$. Following the same lines, it is also possible to optimise B_i^{red} and D_i^{red}. A biconvex optimisation procedure can thus be implemented to optimise alternatively C_i^{red}, D_i^{red} and B_i^{red}, D_i^{red}.

At the end of this preprocessing step, a set of reduced models with consistent state space matrices and accurate modal/frequential content is available.

4.3.2 Polynomial Interpolation and LFT Modelling

The elements of the state space matrices $(A_i^{red}, B_i^{red}, C_i^{red}, D_i^{red})_{i\in[1,\,N]}$ can now be interpolated. Let $(z_i)_{i\in[1,\,N]}$ be the values taken by one of them on the grid. In the perspective of building an LFR, either a polynomial or a rational interpolation must be performed. The former prevents the appearance of discontinuities and is preferred here, although a higher degree is sometimes necessary to model accurately a large variation in the element. The following expression is thus assumed

$$z(\overline{\delta}) = \sum_{k=1}^{n_p} \gamma_k p_k(\overline{\delta}) \tag{4.13}$$

where $(p_k)_{k\in[1,n_p]}$ is a set of multivariate polynomials and $(\gamma_k)_{k\in[1,n_p]}$ are parameters to be determined. Let

$$P = \begin{pmatrix} p_1(\overline{\delta}^{(1)}) & \dots & p_{n_p}(\overline{\delta}^{(1)}) \\ \vdots & \ddots & \vdots \\ p_1(\overline{\delta}^{(N)}) & \dots & p_{n_p}(\overline{\delta}^{(N)}) \end{pmatrix} = \begin{pmatrix} P_1 & \dots & P_{n_p} \end{pmatrix} \tag{4.14}$$

$$\Gamma^T = \begin{pmatrix} \gamma_1 & \dots & \gamma_{n_p} \end{pmatrix} \;,\; Z^T = \begin{pmatrix} z_1 & \dots & z_N \end{pmatrix} \tag{4.15}$$

where $\overline{\delta}^{(i)}$ is the value of $\overline{\delta}$ at the ith point of the grid and $(P_k)_{k\in[1,n_p]}$ are called the modelling functions. The objective is to minimise the quadratic error between $z(\overline{\delta})$ and $(z_i)_{i\in[1,N]}$ on the grid, *i.e.* to compute

$$\Gamma_{opt} = \arg \min_{\Gamma\in\mathbb{R}^{n_p}} J(\Gamma) \tag{4.16}$$

where

$$J(\Gamma) = (Z - P\Gamma)^T(Z - P\Gamma) \tag{4.17}$$

An intuitive choice for $(p_k)_{k\in[1,n_p]}$ is

$$\left\{ \begin{array}{ccc} [-1,\,1]^m & \rightarrow & \mathbb{R} \\ (\overline{\delta}_1, ..., \overline{\delta}_m) & \rightarrow & \overline{\delta}_1^{i_1} ... \overline{\delta}_m^{i_m} \end{array}, i_1 \leq d_1, ..., i_m \leq d_m \right\} \tag{4.18}$$

where $\overline{\delta}_j$ is the jth element of $\overline{\delta}$, m is the length of $\overline{\delta}$ and $d_1, \dots, d_m$ are user-defined integers. In this context, the solution of (4.16) is

$$\Gamma_{opt} = (P^T P)^{-1} P^T Z \tag{4.19}$$

An alternative consists in choosing orthogonal modelling functions such that $P_k^T P_l = 0\ \forall k \neq l$, as proposed in [10]. Thus

$$\gamma_{k,opt} = \frac{P_k^T Z}{P_k^T P_k} \quad \forall k \in [1,\, n_p] \tag{4.20}$$

and the corresponding value J_{opt} of $J(\Gamma)$ is given by

$$J_{opt} = Z^T Z - \sum_{k=1}^{n_p} \frac{(P_k^T Z)^2}{P_k^T P_k} \tag{4.21}$$

Using orthogonal modelling functions is relevant. Indeed, $P^T P$ becomes diagonal and can be trivially inverted. Moreover, it can be observed in (4.21) that the reduction in the least squares criterion $J(\Gamma)$ resulting from the inclusion of P_k does not depend on P_j for all $j \neq k$. This allows to evaluate each orthogonal modelling function in terms of its ability to reduce $J(\Gamma)$, regardless of which other functions are selected. This property is exploited in Sect. 4.3.4 to reduce the LFR complexity.

In the present section, a classical basis (4.18) is defined, from which standard modelling functions are obtained. The solution of (4.16) is thus given by (4.19). Once polynomial approximations $A(\overline{\delta}), B(\overline{\delta}), C(\overline{\delta}), D(\overline{\delta})$ of $(A_i^{red},\, B_i^{red},\, C_i^{red},\, D_i^{red})_{i\in[1,\,N]}$ are available, the structured tree decomposition algorithm of [11] is applied to get an LFR. This algorithm is implemented in the function `symtreed.m` of the LFR Toolbox for Matlab [12].

4.3.3 Special Case of a Coarse Grid

The direct application of the aforementioned interpolation technique often gives satisfactory results. Nevertheless, if the considered grid is too coarse, the eigenvalues of the resulting LFR sometimes follow inconsistent trajectories outside the grid. In this context, the following algorithm is introduced, which exploits the block-diagonal structure of the state matrices of the models $(A_i^{red}, B_i^{red}, C_i^{red},\, D_i^{red})_{i\in[1,\,N]}$: it allows to refine the grid only for the rigid or flexible modes that cannot be interpolated satisfactorily over the initial grid.

Algorithm 4.1 (interpolation over a coarse grid). *Let $\mathcal{L} = 0$. For each rigid or flexible mode of the models to be interpolated:*

1. *Build an LFR $\mathcal{L}_j$ from the submodels $(A_{i,j}^{red}, B_{i,j}^{red}, C_{i,j}^{red}, D_{i,j}^{red})_{i\in[1,\,N]}$ by interpolation over the initial grid $\mathcal{G}_1$.*
2. *Compute the eigenvalues of $\mathcal{L}_j$ on a finer grid $\mathcal{G}_2$. If the values obtained on $\mathcal{G}_2$ are included (to a given tolerance) in the convex envelopes of those computed on $\mathcal{G}_1$, set $\mathcal{L} \leftarrow \mathcal{L} + \mathcal{L}_j$ and stop. Otherwise, go to step 3.*
3. *Identify the points of $\mathcal{G}_2$, for which the distances between the eigenvalues of $\mathcal{L}_j$ and the associated convex envelopes are the largest, and add them to $\mathcal{G}_1$. Generate fictitious submodels associated to these new points by linear interpolation of the existing ones. Using an augmented basis of multivariate*

polynomials, build a new LFR $\mathcal{L}_j$ by interpolation over the augmented grid $\mathcal{G}_1$, and go back to step 2.

The choice of a block-diagonal (modal) form for the reduced models $(A_i^{red}, B_i^{red}, C_i^{red}, D_i^{red})_{i\in[1,\,N]}$ proves relevant in the perspective of applying Algorithm 4.1. Indeed, it allows to increase the size of the grid and the order of the polynomials used for interpolation only for the eigenvalues that cannot be interpolated correctly over the initial grid. To some extent, this amounts to minimising the increase of complexity of the LFR. The procedure described in Algorithm 4.1 is illustrated in Sect. 4.4.3.

4.3.4 Low Order LFR Generation Procedure

It is generally desirable to have an LFR with as low complexity as possible. A three step procedure is thus introduced to reduce the size of Δ while keeping a reasonable accuracy.

First step. The influence of the variation of each element of the matrices A_i^{red}, B_i^{red}, C_i^{red} and D_i^{red} on the frequency responses of G_i^{red} is evaluated in terms of $\mathcal{H}_2$-norm. More precisely, let $G_{i,c}^{red}$ be obtained from G_i^{red} by replacing a given element of $A_i^{red}, B_i^{red}, C_i^{red}$ or D_i^{red} by its mean value for all $i \in [1,\,N]$. Let $\Delta G_i^{red}(\omega_j) = G_i^{red}(\omega_j) - G_{i,c}^{red}(\omega_j)$. If

$$\max_{i\in[1,\,N]} \sqrt{\frac{\sum_j \operatorname{trace}\left(\Delta G_i^{red*}(\omega_j)\Delta G_i^{red}(\omega_j)\right)}{\sum_j \operatorname{trace}\left(G_i^{red*}(\omega_j)G_i^{red}(\omega_j)\right)}}$$

is below a pre-specified threshold ϵ_1, then the mean value of this element is used for all models and no polynomial interpolation is performed in the next step.

Second step. Each varying element of the state space models $(A_i^{red}, B_i^{red}, C_i^{red}, D_i^{red})_{i\in[1,\,N]}$ is interpolated. For this, the following algorithm can be used to identify the most relevant monomials of (4.18) which allow to obtain an interpolation error lower than a given value ϵ_2. All the other monomials are eliminated, which amounts to a certain extent to minimising the complexity of the interpolation formula.

Algorithm 4.2 (Identification of the Most Relevant Monomials).

1. *Let Z be the data to interpolate as defined in (4.15). Let τ_1 and τ_2 be two small predefined tolerances (e.g., $\tau_1 = 10^{-6}$ and $\tau_2 = 10^{-6}$). Choose the maximum allowable interpolation error ϵ_2, and build a basis of multivariate monomials $(\xi_k)_{k\in[1,\,n_p]}$ as defined in (4.18), where d_i is the maximum degree in the ith variable.*

2. *Let* $(\Xi_k)_k$ *be the modelling functions associated to* $(\xi_k)_k$ *as defined in (4.14). Build an orthogonal basis* $(P_k)_k$ *from* $(\Xi_k)_k$ *using a Gram-Schmidt procedure. Compute the value (4.20) of* Γ_{opt} *which solves the optimisation problem (4.16). Let* $Z_{opt} = P\Gamma_{opt}$ *and* $\mathcal{S}_{test} = \left\{i, \frac{(P_i^T Z)^2}{P_i^T P_i} \leq \tau_1 Z^T Z\right\}$.
3. *Let* $\mathcal{S}_{elim} = \varnothing$. *For each element* i *of* $\mathcal{S}_{test}$*:*
   a. *let* $(\tilde{\Xi}_k)_k = (\Xi_k)_k \setminus \left\{\Xi_j, j \in \tilde{\mathcal{S}}_{elim}\right\}$, *where* $\tilde{\mathcal{S}}_{elim} = \mathcal{S}_{elim} \cup i$, *and build an orthogonal basis* $(\tilde{P}_k)_k$ *from* $(\tilde{\Xi}_k)_k$,
   b. *compute* $\tilde{Z} = \tilde{P}\tilde{\Gamma}$, *where* $\tilde{\Gamma}$ *solves the optimisation problem (4.16),*
   c. *if* $\|Z - \tilde{Z}\|_2 > \epsilon_2 \|Z\|_2$, *interrupt step (3) and go to step (4) ; otherwise, if* $\|Z_{opt} - \tilde{Z}\|_2 \leq \tau_2 \|Z_{opt}\|_2$, *let* $\mathcal{S}_{elim} = \tilde{\mathcal{S}}_{elim}$ *and* $Z_{opt} = \tilde{Z}$.
4. *Let* $(\xi_k)_k \leftarrow (\xi_k)_k \setminus \{\xi_j, j \in \mathcal{S}_{elim}\}$. *If step 3 has been interrupted, stop. Otherwise, let* $\tau_1 \leftarrow 10\tau_1$ *and* $\tau_2 \leftarrow 10\tau_2$ *if* $\mathcal{S}_{elim} = \varnothing$ *and then go back to step 2.*

Step 2 determines a set of candidate monomials that are likely to be eliminated. Step 3 then decides whether these monomials are to be eliminated or not. The gradual increase in the tolerances τ_1 and τ_2 allows to eliminate progressively some of the monomials until the maximum allowable interpolation error ϵ_2 is reached. The remaining ones can then be used to interpolate the considered element. Once all the elements are interpolated, an LFR is generated as detailed at the end of Sect. 4.3.2.

Third step. The frequency error between the LFR $\mathcal{F}_u(M(s), \Delta)$ and the reference models $(G_i(s))_{i\in[1,\,N]}$ is minimised. The following quantity is considered

$$\sqrt{\frac{1}{2\pi}\sum_{i,j} \text{trace}\,(H_i^*(\omega_j)H_i(\omega_j))(\omega_{j+1} - \omega_j)} \tag{4.22}$$

where $H_i(\omega_j) = \mathcal{F}_u(M(\omega_j), \Delta^i) - G_i(\omega_j)$. The matrices C, D and B, D of the LFR are then optimised alternatively using the method presented in Sect. 4.3.1, which can be easily adapted to the LFR framework [7].

The choice of ϵ_1, ϵ_2 in the aforementioned procedure allows to handle the trade-off between accuracy and complexity.

4.4 Numerical Results

The methodology proposed in Sect. 4.3 is now applied to the set of longitudinal aeroelastic models described in Sect. 4.2 (see also Chapter 2). Note that similar results can be obtained for the set of lateral models but they are not detailed here for the sake of conciseness.

4.4.1 Construction of the LFR

The first step consists in choosing the interpolation parameters:

- The aircraft mass is obviously described by the payload $\overline{\delta}_{PL}$, as well as the center and outer tanks configurations $\overline{\delta}_{CT}$ and $\overline{\delta}_{OT}$.
- The flight domain can be characterised by any two of the following parameters: Mach number, true air speed $\overline{V}_{tas}$, calibrated air speed, altitude and air density. The choice can be made by observing the shape of the domain for several couples of parameters. This is illustrated in Fig. 4.2, where the domain is plotted in both the $(\overline{M}, \overline{V}_{tas})$ and $(\overline{M}, \overline{V}_{cas})$ planes. In the first case, the domain is far from being square and the nine grid points are placed very irregularly. In particular, some of them are very close, which tends to generate an LFR with inconsistent behaviour between the points. On the contrary, a much more regular grid is obtained in the $(\overline{M}, \overline{V}_{cas})$ plane, which corresponds to the retained configuration.

Remark 4.1. As already mentioned, the flight domain is not square whatever combination of two parameters is used to describe it. Thus, a preprocessing step described in Chapter 12 is sometimes necessary before some analysis techniques such as μ-analysis can be applied. Note that such a transformation results in more complex LFRs than those presented in Tables 4.2 and 4.3.

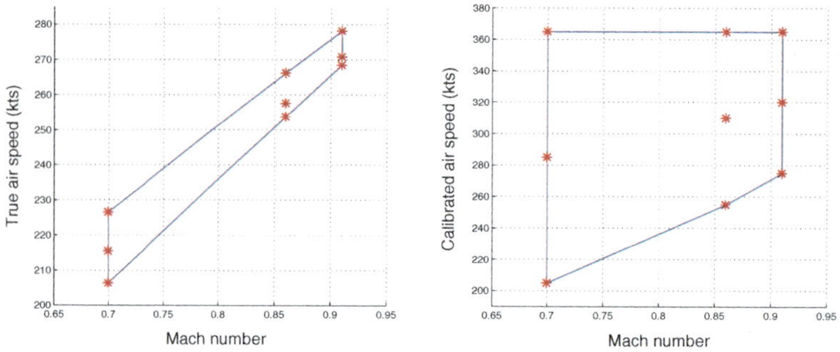

Fig. 4.2 Representation of the flight domain in the (M, V_{tas}) and (M, V_{cas}) planes

The interpolation method described in Sections 4.3.1 to 4.3.3 is then applied. The grid is composed of 3 values for $\overline{\delta}_{CT}$, $\overline{\delta}_{OT}$, $\overline{M}$, $\overline{V}_{cas}$ and only 2 for $\overline{\delta}_{PL}$. A basis of multivariate monomials of degree 2 in $\overline{\delta}_{CT}$, $\overline{\delta}_{OT}$, $\overline{M}$, $\overline{V}_{cas}$ and 1 in $\overline{\delta}_{PL}$ thus allows to perform an exact interpolation in the sense that the eigenvalues and the frequency responses obtained while evaluating the resulting LFR on the initial grid exactly match those of the reduced models $(A_i^{red}, B_i^{red}, C_i^{red}, D_i^{red})_{i\in[1,N]}$.

Several open-loop LFRs corresponding to different parametric domains and values of $\overline{\omega}$ are generated. Their structure is as depicted in Fig. 4.1(a):

- y_r is composed of the elevator deflection and possibly the wind velocity,
- y consists of the flight parameters measured by the sensors (pitch rate, vertical load factor) and possibly a load output or an acceleration signal.

Their dimensions are specified in Tables 4.2 and 4.3, where n_s, n_{io} and n_Δ denote the number of states, the number of inputs/outputs and the size of Δ respectively. Note that the order of the LFRs dedicated to the evaluation of loads and comfort criteria is slightly higher. Indeed, several turbulence delays (see Chapter 2) have been kept to model the wind influence more accurately.

Table 4.2 Open-loop LFRs dedicated to the evaluation of stability criteria

Parameters in the Δ block	$\overline{\omega} = 15\,rad/s$			$\overline{\omega} = 50\,rad/s$		
	n_s	n_{io}	n_Δ	n_s	n_{io}	n_Δ
$\overline{M}, \overline{V}_{cas}$	10	1/2	**48**	22	1/2	84
$\overline{\delta}_{CT}, \overline{M}, \overline{V}_{cas}$	10	1/2	96	22	1/2	175
$\overline{\delta}_{OT}, \overline{\delta}_{CT}, \overline{M}, \overline{V}_{cas}$	10	1/2	193	22	1/2	344
$\overline{\delta}_{PL}, \overline{\delta}_{OT}, \overline{\delta}_{CT}, \overline{M}, \overline{V}_{cas}$	10	1/2	277	28	1/2	610
$\overline{\delta}_{CT}$	10	1/2	14	22	1/2	26
$\overline{\delta}_{OT}, \overline{\delta}_{CT}$	10	1/2	46	22	1/2	82
$\overline{\delta}_{PL}, \overline{\delta}_{OT}, \overline{\delta}_{CT}$	10	1/2	72	28	1/2	166

Table 4.3 Open-loop LFRs dedicated to the evaluation of loads and comfort criteria

Parameters in the Δ block	$\overline{\omega} = 15\,rad/s$			$\overline{\omega} = 50\,rad/s$		
	n_s	n_{io}	n_Δ	n_s	n_{io}	n_Δ
$\overline{M}, \overline{V}_{cas}$	12	2/3	60	30	2/3	100
$\overline{\delta}_{CT}, \overline{M}, \overline{V}_{cas}$	12	2/3	120	30	2/3	231
$\overline{\delta}_{OT}, \overline{\delta}_{CT}, \overline{M}, \overline{V}_{cas}$	12	2/3	241	30	2/3	436
$\overline{\delta}_{PL}, \overline{\delta}_{OT}, \overline{\delta}_{CT}, \overline{M}, \overline{V}_{cas}$	12	2/3	346	36	2/3	790
$\overline{\delta}_{CT}$	12	2/3	18	30	2/3	30
$\overline{\delta}_{OT}, \overline{\delta}_{CT}$	12	2/3	58	30	2/3	98
$\overline{\delta}_{PL}, \overline{\delta}_{OT}, \overline{\delta}_{CT}$	12	2/3	91	36	2/3	**207**

Sections 4.4.2 and 4.4.3 focus on the LFR whose size is printed in bold type in Table 4.3, while Sect. 4.4.4 considers the one whose size is printed in bold type in Table 4.2.

4.4.2 Validation on the Grid

Several frequency and modal criteria are defined to evaluate the similarity between the LFR and the reference models on the grid. Let $H_i(\omega) = \mathcal{F}_u(M(\omega), \Delta^i) - G_i(\omega)$.

$$\mathcal{H}_2 \text{ criterion} : \delta_2 = \max_{i \in [1,N]} \sqrt{\frac{\sum_j \text{trace}\,(H_i^*(\omega_j)H_i(\omega_j))}{\sum_j \text{trace}\,(G_i^*(\omega_j)G_i(\omega_j))}} \tag{4.23}$$

$$\mathcal{H}_\infty \text{ criterion} : \delta_\infty = \max_{i \in [1,N]} \frac{\overline{\sigma}(H_i(\omega))}{\overline{\sigma}(G_i(\omega))} \tag{4.24}$$

$$\text{modal criterion} : \delta_\lambda = \max_{i \in [1,N]} \max_j \frac{|\lambda_i^j - \lambda_{ref,i}^j|}{|\lambda_{ref,i}^j|} \tag{4.25}$$

where $\lambda_{ref,i}^j$ and λ_i^j denote the jth eigenvalue of the ith reference model (including all the lag states) and of the LFR evaluated at the ith grid point respectively. Here, $\delta_2 = 5\%$, $\delta_\infty = 6\%$ and $\delta_\lambda = 10^{-5}\%$. All errors are very low, thus proving the relevance of the method, and particularly of the reduction step and the biconvex optimisation performed in Sect. 4.3.1. Fig. 4.3 shows the results obtained at one of the worst-case grid points. The frequency response on the right plot is associated to the transfer function between the vertical wind velocity and the vertical acceleration at the central point of the aircraft fuselage. A square without a cross on the left plot corresponds to an eigenvalue that has been eliminated at the very beginning of the modelling process. A cross without a square means that a flexible mode is missing in the reference model and has been added by linear interpolation (see Sect. 4.3.1). Note also that the rigid behaviour of the aircraft is represented by four states in the reference model: the altitude z, the pitch angle θ, and their derivatives. It can be equivalently modelled by only two states, namely the angle of attack α and the pitch rate q. This explains why the LFR does not exhibit any pole at the origin.

Remark 4.2. The interpolation error is always low on the grid, which can lead to the computation of an over-optimistic modelling error. A more realistic approach would be to generate the LFR using a coarse grid, and then to compute the modelling error on a denser grid. Unfortunately, such a strategy cannot be applied in the present context, since only a very coarse grid is available.

4.4.3 Validation on the Whole Continuous Domain

It is certainly important to ensure that the eigenvalues and the frequency responses obtained when the LFR is evaluated on the initial grid precisely match those of the reference models. But a special attention should also be paid to the eigenvalues of the LFR in order to ensure that they are as realistic

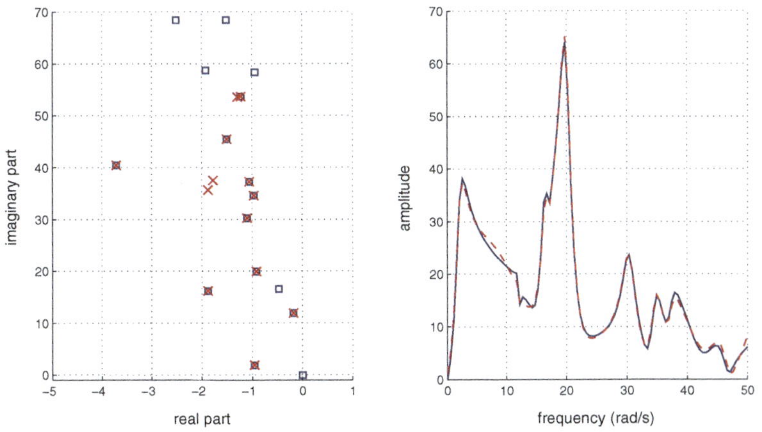

Fig. 4.3 Comparison of eigenvalues (left) and frequency responses (right) on the grid (squares and solid lines: reference models - crosses and dashed lines: LFR)

as possible between the grid points. It is notably desirable that they cannot become too different from the values they take on the grid, which motivated the introduction of the convex envelopes at step 2 of Algorithm 4.1. Fig. 4.4 illustrates, for a given eigenvalue λ, the relevance of the interpolation technique introduced in Algorithm 4.1. The convex envelope corresponding to the values of λ on the initial grid $\mathcal{G}_1$ is represented by dashed lines. If the LFR computed at step 1 is evaluated on a finer grid $\mathcal{G}_2$ (crosses), it can be observed on Fig. 4.4(a) that several configurations exist for which the value of λ is unrealistic, and sometimes even unstable. The worst-case configuration, *i.e.* the configuration for which the value of λ is the farthest from the convex envelope, is then identified (circle) and added to $\mathcal{G}_1$. A fictitious submodel is then generated and a new LFR is computed. It can be observed on Fig. 4.4(b) that the resulting cloud of points is significantly smaller (there is no configuration anymore, for which this mode is unstable). This whole procedure is applied once again, and it can finally be observed on Fig. 4.4(c) that all crosses are now included in or located very near the convex envelope. The interpolation is thus satisfactory and the algorithm is terminated.

It should also be emphasised that the increase in complexity due to the refinement of the grid at step 3 of Algorithm 4.1 is negligible. For instance, if the initial grid is used to interpolate the whole reduced models, *i.e.* if only step 1 is performed:

- $n_\Delta = 195$, which is only little lower than the value of 207 obtained with the whole algorithm,
- the resulting LFR exhibits an unrealistic behaviour between the grid points, as shown on Fig. 4.4(a).

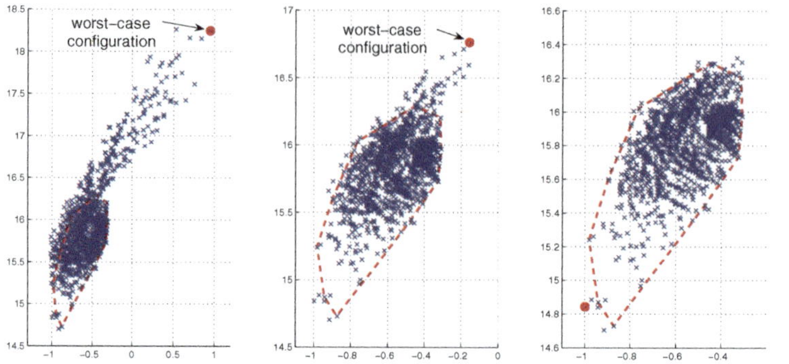

(a) interpolation on the initial grid (b) first refinement of the grid (c) second refinement of the grid

Fig. 4.4 Refinement of the initial grid

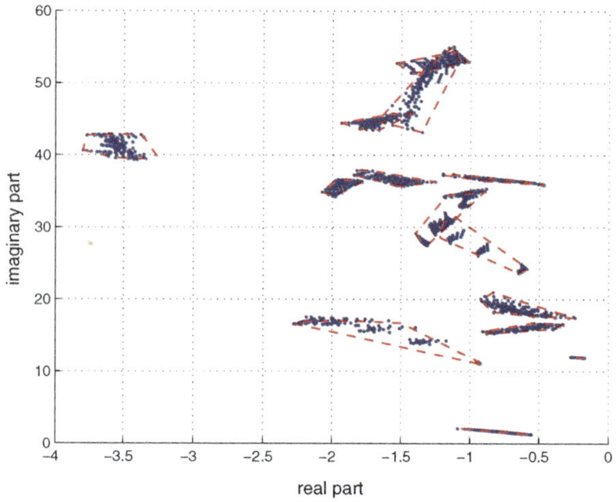

Fig. 4.5 Eigenvalues of the LFR computed on a fine grid (dots) and convex envelopes of the eigenvalues computed on the initial grid (dashed lines)

Fig. 4.5 shows the eigenvalues of the LFR computed on a very fine grid. It appears that Algorithm 4.1 is efficient, since the clouds of points associated to the various eigenvalues (dots) are almost included in the corresponding convex envelopes (dashed lines).

Although no comparison can be made outside the grid points, another strong requirement is that the LFR behaviour remains realistic, *i.e.* that its eigenvalues and frequency responses vary as linearly as possible. The whole domain cannot be checked exhaustively, but investigating the principal directions provides a good indication of the validity of the LFR. Here, both the modal trajectories and the frequency responses continuum are quite regular, as can be seen in Fig. 4.6 and Fig. 4.7.

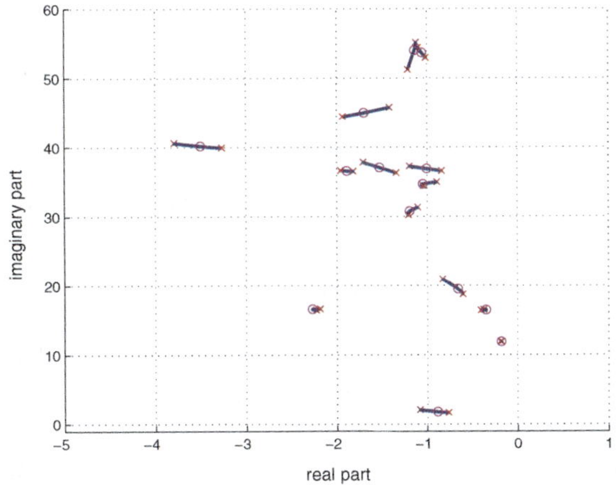

Fig. 4.6 Modal trajectories w.r.t. $\overline{\delta}_{OT}$ (crosses and circles: reference models)

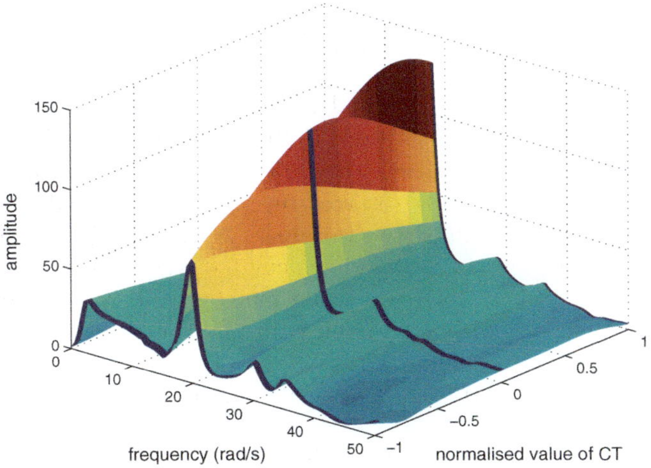

Fig. 4.7 Frequency response continuum w.r.t. $\overline{\delta}_{CT}$ (solid lines: reference models)

4.4.4 Evaluation of the Low Order LFR Generation Procedure

Designing a self-scheduled controller can become a computational burden if the size of the considered open-loop LFR is too high, which motivates an efficient low order LFR generation procedure. In this context, the method of Sect. 4.3 is used to build an open-loop LFR with two varying parameters ($\overline{\delta}_{CT}$ and $\overline{\delta}_{OT}$) as depicted in Fig. 4.1(a), where y_r is the elevator deflection and

Table 4.4 Impact of the LFT reduction algorithm on complexity and accuracy

Type of LFR	n_Δ	δ_2	δ_∞	δ_λ
initial	46	2%	8%	10^{-5}%
reduced	22	10%	13%	8%

y denotes the flight parameters outputs (pitch rate and vertical load factor). Note that setting $\overline{\omega} = 15\, rad/s$ is not restrictive in a design perspective due to the limited bandwidth of the actuators. The reduction algorithm introduced in Sect. 4.3.4 is then applied. Results are summarised in Table 4.4 and plotted in Fig. 4.8 for the worst-case configuration.

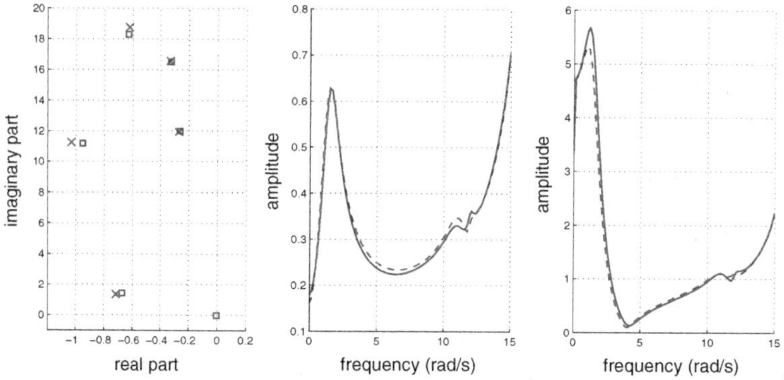

Fig. 4.8 Comparison of eigenvalues (left) and frequency responses (right) on the grid (squares and solid lines: reference models - crosses and dashed lines: reduced LFR)

The size of Δ is halved and accuracy remains quite sufficient for design purposes. The LFT reduction algorithm thus proves conclusive.

4.5 Conclusion

A whole methodology is described in this chapter to convert a set of aeroelastic models into an LFR. It can be applied to any kind of purely numerical models for which the analytical structure is unknown. A preprocessing step first generates reduced order models with consistent state space matrices and accurate modal/frequential content. A basis of multivariate modelling functions is then defined and a polynomial interpolation is performed using the method of [10]. The structured tree decomposition of [11] is finally applied to build an LFR. An efficient algorithm is also proposed, which allows to select only the most relevant modelling functions during the interpolation step and thus to prevent data overfitting. This significantly reduces the

LFR complexity while maintaining a satisfactory accuracy. The application of this methodology to a flexible aircraft proves conclusive. The resulting LFRs are indeed highly representative of the plant behaviour, in the sense that their eigenvalues and frequency responses almost exactly match those of the reference models at the considered grid points. Moreover, the trajectories of their eigenvalues and the continuum of their frequency responses remain quite regular on the whole continuous parametric domain. It should finally be underlined that their reasonable complexity makes them exploitable both to evaluate the robustness properties of the plant and to design self-scheduled controllers, as in demonstrated in Chapters 7, 9 and 10.

References

1. Roos, C., Biannic, J.-M.: Efficient computation of a guaranteed stability domain for a high-order parameter dependent plant. In: Proceedings to the American Control Conference, Baltimore, Maryland (2010)
2. Biannic, J.M., Ferreres, G.: Efficient computation of a guaranteed robustness margin. In: Proceedings of the 16th IFAC World Congress, Prague, Czech Republic (2005)
3. Bates, D.G., Kureemun, R., Mannchen, T.: Improved clearance of a flight control law using μ-analysis techniques. Journal of Guidance, Control and Dynamics 26(6), 869–884 (2003)
4. Garulli, A., Masi, A., Paoletti, S., Turkoglu, E.: Clearance of flight control laws via parameter-dependent Lyapunov functions. In: Proceedings of the 6th IFAC Symposium on Robust Control Design, Haifa, Israel (2009)
5. Papageorgiou, C., Falkeborn, R., Hansson, A.: Formulation of the stability margins clearance criterion as a convex optimization problem. In: Proceedings of the 6th IFAC Symposium on Robust Control Design, Haifa, Israel (2009)
6. Puyou, G., Losser, Y.: Formulation of the clearance benchmark problem. Technical Report COFCLUO D3.1.1, AIRBUS France SAS (August 2007)
7. Torralba, J., Demourant, F., Puyou, G., Ferreres, G.: A method for flexible aircraft LFT modelling. In: Proceedings of the European Control Conference, Budapest, Hungary (2009)
8. Bennani, S., Beuker, B., Willem van Staveren, J., Dijkgraaf, J.P.: Numeric LFT generation for robust aircraft flutter analysis. In: Proceedings of the IEEE Conference on Computer Aided Control System Design, Glasgow, Scotland, pp. 260–265 (2002)
9. Cumer, C., Ferreres, G., Humbert, M.: Flutter detection with robustness tools. In: Proceedings of the International Forum on Aeroelasticity and Structural Dynamics, Amsterdam, Netherlands (2003)
10. Morelli, E.A., DeLoach, R.: Wind tunnel database development using modern experiment design and multivariate orthogonal functions. In: Proceedings of the 41st AIAA Aerospace Sciences Meeting and Exhibit, Reno, NV (2003)
11. Cockburn, J.C., Morton, B.G.: Linear fractional representations of uncertain systems. Automatica 33(7), 1263–1271 (1997)
12. Magni, J.-F.: User manual of the Linear Fractional Representation Toolbox (version 2.0). Technical Report 5/10403.01F, ONERA/DCSD (2006), `http://www.onera.fr/staff-en/jean-marc-biannic/docs/lfrtv20s.zip`

Chapter 5
Generation of LFRs for a Nonlinear Controller and Closed-Loop Aircraft Models

Carsten Döll, Fabien Lescher, and Clément Roos

Abstract. In this chapter the block-diagram based generation approach of low order Linear Fractional Representations (LFRs) is applied to the nonlinear controller described in Chapter 2. For this purpose, all individual parameter dependent blocks as well as all nonlinear blocks like saturations or rate limiters are replaced by their LFR counterparts in the block-diagram structure of the controller. Look-up tables are approximated by rational expressions before their replacements. The overall controller LFR is extracted from the resulting block-diagram by a standard linearization technique. The resulting controller LFR is interconnected with the LFRs of the actuators, sensors and of the nonlinear rigid aircraft or linear flexible aircraft generated in Chapters 3 or 4 to obtain the LFRs of the corresponding closed-loop systems. These closed-loop LFRs are used for stability and performance analysis in Chapters 12–16.

5.1 Introduction

The generation of *linear fractional representations* (LFRs) for the nonlinear controller introduced in Chapter 2 can be performed employing the methodology of Chapter 3 applied to the nonlinear aircraft model. Thus, a set of linearised controller models can be generated first via successive trimming and linearisations of the nonlinear controller for a given set of flight conditions. Then, parameter fitting has to be performed to obtain LFRs which interpolate these models. Alternative interpolation techniques are also described in Chapter 4. From an user's point of view, the resulting controller LFR would

Carsten Döll · Fabien Lescher · Clément Roos
ONERA, Systems Control and Flight Dynamics Department,
2 avenue Edouard Belin, 31055 Toulouse Cedex 4, France
e-mail: `carsten.doll@onera.fr`, `fabien.lescher@onera.fr`,
`clement.roos@onera.fr`

A. Varga et al. (Eds.): Opti. Based Clearance of Flight Control Laws, LNCIS 416, pp. 79–109.
springerlink.com

be a black-box input/output behaviour. Therefore, the main drawback of this approach is that the knowledge of the readable and comprehensive controller structure is entirely lost.

A second drawback is that, if for example, one component of the controller (e.g., a certain gain) is changed during the overall aircraft design process to cope with new stability or performance constraints or newly acquired ground or flight test results, it would generally not be possible to simply adjust only this element and its corresponding LFR block to update the existing LFRs. Instead, the complete trimming, linearisation and interpolation process has to be repeated, which can be very time consuming.

To overcome both above drawbacks by maintaining the readability of the controller structure in the final LFR and allowing an easy modification of the controller LFR after a change in one of its components, an alternative LFR generation procedure will be employed based on the use of block-diagrams under MATLAB-Simulink[1] [1]. In essence, each uncertain parametric block in the controller block-diagram is replaced by its LFR counterpart. The block-diagram is hence not changed, it is still readable. If some elements of the controller are later changed, it is sufficient to change only these blocks and the corresponding LFR counterparts. The overall LFR of the controller is finally computed by employing a standard linearization technique. This procedure applies the LFR objects block-set for use with MATLAB-Simulink, which has been developed by ONERA in the past in order to simplify the manipulation of interconnected LFRs [2].

This chapter is organized as follows. The nonlinear controller structure is first explained in Sect. 5.2. The generation of LFRs for the nonlinear controller is then detailed in Sect. 5.3. In Sect. 5.4, the controller LFRs are combined with either the LFRs for the nonlinear rigid aircraft of Chapter 3 or the LFRs for the linear flexible aircraft of Chapter 4 as well as the LFRs for the actuators and the sensors in order to create the LFR for the closed-loop system. The resulting closed-loop LFRs are used by the analysis teams for stability and performance analysis, see Chapters 12–16. In Sect. 5.5, the controller and the closed-loop LFRs are evaluated with respect to linearised models obtained from the initial nonlinear system.

5.2 Description of the Nonlinear Controller

The nonlinear controller is split into a longitudinal controller and a lateral controller. Both controllers contain several component blocks whose dynamics contain gains which are scheduled with respect to various flight parameters. Most of the gains are also scheduled with respect to the aircraft configuration. This dependency is however not taken into account in the following, as just the clean configuration is considered.

[1] MATLAB and Simulink are registered trademarks of *The Mathworks Inc.*

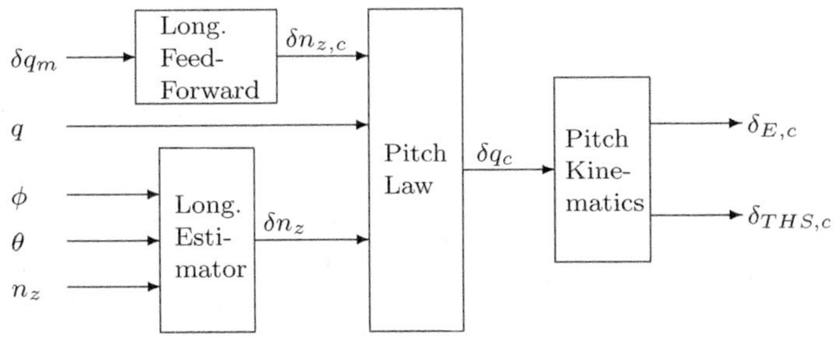

Fig. 5.1 The structure of the longitudinal controller

The longitudinal controller structure is depicted in Fig. 5.1. The pilot pitch input δq_m is first translated into an equivalent demanded vertical load factor $\delta n_{z,c}$ by a nonlinear relation including a dead-zone and a deflection saturation within the longitudinal feedforward block without any dynamics. The measurements are the pitch rate q, the bank angle ϕ, the pitch angle θ and the vertical load factor n_z. ϕ, θ and n_z are translated within the longitudinal estimator into a correction signal δn_z using the following relation

$$\delta n_z = n_z - \frac{\cos\theta}{\cos\phi} \tag{5.1}$$

For example, during horizontal flight, $n_z = 1$, $\theta \approx 0$ and $\phi = 0$, thus the resulting correction $\delta n_z \approx 0$. However, during a standard turn manoeuvre with $\phi = 30^o$, the resulting correction δn_z is $1 - \frac{2}{\sqrt{3}} \approx -0.15$. The aircraft must pitch up in order to hold the altitude during the turn. This relation is therefore called the *turn coordination.* There are no dynamics in the longitudinal estimator.

In the pitch law block, an equivalent pitch deflection demand δq_c is computed. The control contains two proportional terms and one integral term. In total, five gains are used which are scheduled with respect to the calibrated airspeed V_{cas}, Mach number M and center of gravity location $X_{cg,p}$, which the pilot has chosen in function of the loading of the aircraft. $X_{cg,p}$ can therefore be different from the aircraft's true X_{cg}.

The controller outputs are the demanded elevator deflection $\delta_{E,c}$ and the demanded horizontal tailplane deflection $\delta_{THS,c}$, which are computed within the pitch kinematics block. The equivalent pitch deflection δq_c is first multiplied with the tailplane efficiency and then filtered by a first order *linear time invariant* (LTI) low pass filter and limited in rate and position in order to determine the slow tailplane deflection. On the other hand, the fast elevator deflection is determined from δq_c by subtracting a slow deflection which is equivalent to the tailplane deflection divided by its control surface efficiency. The resulting signal is then filtered by an LTI second order low pass filter whose cut-off frequency is 10^3 times higher than the one of the tailplane. It

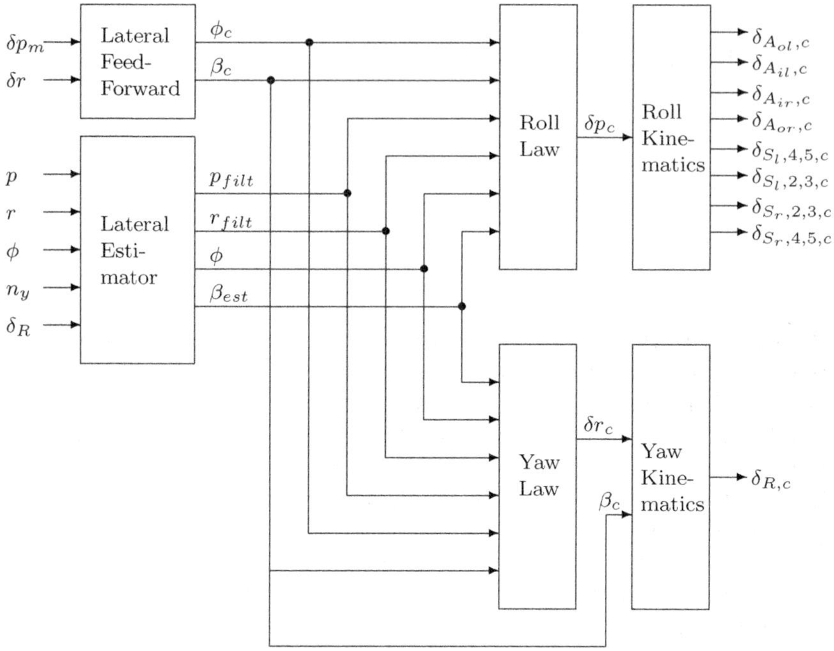

Fig. 5.2 The structure of the lateral controller

is finally limited in rate and position. The tailplane efficiency is given as a look-up table in function of M.

The lateral controller is depicted in Fig. 5.2. Within the lateral feedforward block, the pilot roll input δp_m is first translated into an equivalent internal signal $\delta p_{m,c}$ by a nonlinear relation including a dead-zone and a deflection saturation. On the one hand, $\delta p_{m,c}$ is directly multiplied with a gain, which is scheduled with respect to V_{cas}. On the other hand, it is also fed through an integrator with a constant gain. Both signals are then summed to get the demanded bank angle ϕ_c. The pilot pedal input δr is directly transmitted as the demanded sideslip angle β_c. The lateral feedforward block is hence a system of order one.

The lateral estimator processes the measurements of the roll rate p, the yaw rate r, the bank angle ϕ, the lateral load factor n_y and the deflection of the rudder δ_R. p and r are filtered within the lateral estimator by two LTI first order low pass filters, while ϕ is directly transmitted. An estimated sideslip β_{est} is calculated from the measurements using the lateral side force equation

$$\beta_{est} = -\frac{1}{C_{y\beta}(M,\alpha)} \Big[C_{yr}(M,\alpha)\, r\, \frac{c}{V_{tas}} + C_{yp}(M,\alpha)\, p\, \frac{c}{V_{tas}} \cdots + C_{y\delta_R}(M,\alpha)\, \delta_R + K_{n_y}(V_{cas}, m_p)\, n_y \Big] \quad (5.2)$$

where the four aerodynamic coefficients are given as look-up tables in function of M and the angle of attack α, the gain K_{n_y} is a function of V_{cas} and the mass of the aircraft m_p which is entered by the pilot. Note that m_p can be different from the real aircraft mass m. c is the aerodynamic chord and V_{tas} the true airspeed which is itself a function of M and V_{cas}. Two LTI first order low pass filters are used during the estimation of β_{est}. The lateral estimator is hence a system of order four.

Within the roll law block, an equivalent roll demand δp_c is computed. The control law contains five proportional terms and one term with first order dynamics. In total, six gains are used, which are scheduled with respect to V_{cas}.

The roll outputs are the demanded deflections of the four inner and outer, left and right ailerons $\delta_{A,c}$ and the four groups of left and right spoilers $\delta_{S,c}$. They are computed within the roll kinematics block. δp_c is first filtered by an LTI second order low pass filter before being limited in position. This signal is then split into the various aileron and spoiler deflections by using their corresponding surface efficiencies and the corresponding lever arms. In our case, LTI gains and nonlinear input/output relations are involved, but no dynamics are added.

Table 5.1 The input-output description of the controller

Inputs	Outputs	States x_c	Scheduling Parameters
δp_m	$\delta_{E,c}$	1 integrator in the pitch law	M
δq_m	$\delta_{THS,c}$	1 integrator in the roll law	V_{cas}
δr	$\delta_{R,c}$	1 integrator in the yaw law	V_{tas}
δ_R	$\delta_{A_{il,c}}$	filter order 1 on $\delta_{THS,c}$	$X_{cg,p}$
θ	$\delta_{A_{ir,c}}$	filter order 2 on $\delta_{E,c}$	α
ϕ	$\delta_{A_{ol,c}}$	filter order 2 on δp_c	m_p
p	$\delta_{A_{or,c}}$	filter order 2 on δr_c	
q	$\delta_{S_{l,2,3,c}}$	1 integrator in the lat. feedforward	
r	$\delta_{S_{r,2,3,c}}$	filter order 1 on p	
n_y	$\delta_{S_{l,4,5,c}}$	filter order 1 on r	
n_z	$\delta_{S_{r,4,5,c}}$	2 filters order 1 for β_{est}	

Within the yaw law block, an equivalent yaw demand δr_c is computed. The control law contains four proportional terms and one first order term. In total, five gains are used which are scheduled with respect to V_{cas}.

The yaw output is the demanded deflection of the rudder $\delta_{R,c}$ and is computed within the yaw kinematics block. δr_c is first filtered by an LTI second order low pass filter before being limited in rate and position. The resulting signal is added to β_c and a second time saturated in position in order to get finally $\delta_{R,c}$. The position saturations are function of V_{cas}. No dynamics are added.

Table 5.1 summarises the inputs, outputs, states and scheduling parameters of the nonlinear controller. Thus, the controller is a system with 11 inputs, 11 outputs, and 15 states. The internal gains are scheduled via 6 scheduling parameters. For the generation of the controller LFR, three additional scheduling variables, namely δq_m, θ and ϕ, will also be included.

5.3 Generation of the Controller LFRs

In the following subsections, it is illustrated step by step how the controller LFR is built by preserving the longitudinal and lateral block structures given in Fig. 5.1 and Fig. 5.2, respectively. The controller LFR is expressed as an upper *linear fractional transformation* (LFT) $\mathcal{F}_u(M_c, \Delta_c)$ (see Chapter 4), where all known and fixed dynamics are put together in the (nominal) LTI controller with transfer-function matrix $M_c(s)$, while the scheduling parameters, uncertainties and the nonlinearities (like saturations) are parts of the controller perturbation matrix Δ_c. The $M_c(s)$ and Δ_c blocks are interconnected via artificial inputs w_c and outputs z_c as shown in Fig. 5.3. The controller inputs are the pilot inputs δp_m, δq_m, and δr, the rudder deflection δ_R, and the sensor measurements (see Table 5.1 for details), while the outputs are the demanded control surface deflections $\delta_{*,c}$.

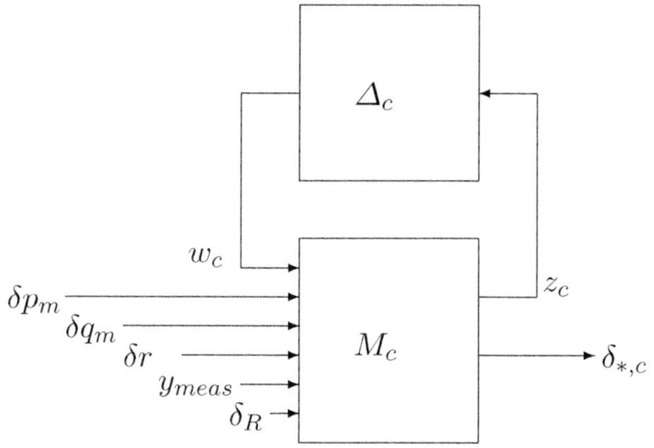

Fig. 5.3 Upper LFT form of the controller

In contrast to Chapters 3 and 4, the LFR generation process exploits and preserves the original block-diagram structure of the controller. The first step is to identify the individual blocks which have to be transformed into LFRs. Parameter varying gains like scheduled controller gains or tabulated aerodynamic coefficients, both given as look-up tables, have first to be replaced by rational expressions and then transformed into equivalent LFRs. The employed approach is illustrated in Sect. 5.3.1. All kind of nonlinearities as rate limiters, saturations, trigonometric or irrational functions have also to be replaced by appropriate LFRs. Possible approaches for this are described in Sections 5.3.2-5.3.3. LTI systems with constant parameters enter without modification in the final LFR of the controller and are therefore ignored at this stage. In the controller structure, 52 blocks which need to be handled have been identified. For the moment being, the identification of these blocks has been done manually, however it is possible to completely automate this step. This automation becomes mandatory for complicated block-diagrams with a huge number of blocks. After having built the LFRs of all blocks, the overall LFR corresponding to the overall block-diagram of the controller is determined as described in Sect. 5.3.4.

5.3.1 LFRs for Parameter Varying Gains

35 blocks have been identified with parameter varying gains, including dependencies like $V_{tas}(M, V_{cas})$, the gain $K_{n_y}(V_{cas}, m_p)$, four aerodynamic coefficients $C_{y*}(M, \alpha)$ and 29 scheduled gains, which are functions of M, V_{cas} and/or $X_{cg,p}$. All these gains are given as look-up tables and we can assume for them a functional dependence of the form $\zeta(\rho)$, where ρ is one (or contains several) scheduling variable(s).

To build an LFR of each $\zeta(\rho)$, the corresponding look-up tables have to be replaced by explicit polynomial or rational expressions in the parameters ρ. Typically, approximations based on polynomial or rational interpolation are used. The first one prevents the appearance of discontinuities and is preferred here, although a higher degree is sometimes necessary to model accurately the table. In what follows, we describe shortly the interpolation approach using multivariate polynomials.

Let $\overline{\zeta}(\rho)$ be a polynomial approximation of $\zeta(\rho)$ of the form

$$\overline{\zeta}(\rho) = \sum_{k=1}^{n_p} \gamma_k p_k(\rho), \tag{5.3}$$

where $p_k(\rho)$, for $k = 1, \ldots, n_p$ is a set of multivariate polynomials and γ_k, for $k = 1, \ldots, n_p$, are constant parameters to be determined.

For example, we can determine γ_k which minimise the quadratic error $\sum_{i=1}^{N}(\zeta(\rho_i) - \overline{\zeta}(\rho_i))^2$, where ρ_i, $i = 1, \ldots, N$ are suitably chosen values of ρ on a grid. The solution of this least squares problem can be done similarly as described in Sect. 3.2. The approximation problem can also be solved by choosing orthogonal modelling functions [3] applying (4.20) and (4.21), as shown in Sect. 4.3.2.

To illustrate the polynomial interpolation in the mono-variable case, let us consider the lateral gain $LSG_{07}(V_{cas})$ which is scheduled with respect to the calibrated airspeed V_{cas}. It is a 1-D look-up table and its tabulated values are linearly interpolated as shown in Fig. 5.4(a) (continuous line). It is a

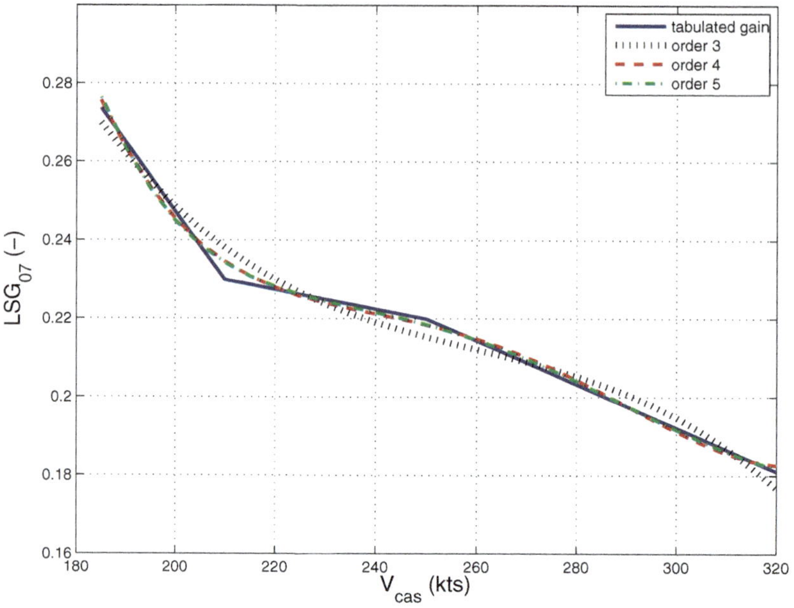

(a) Polynomial approximations

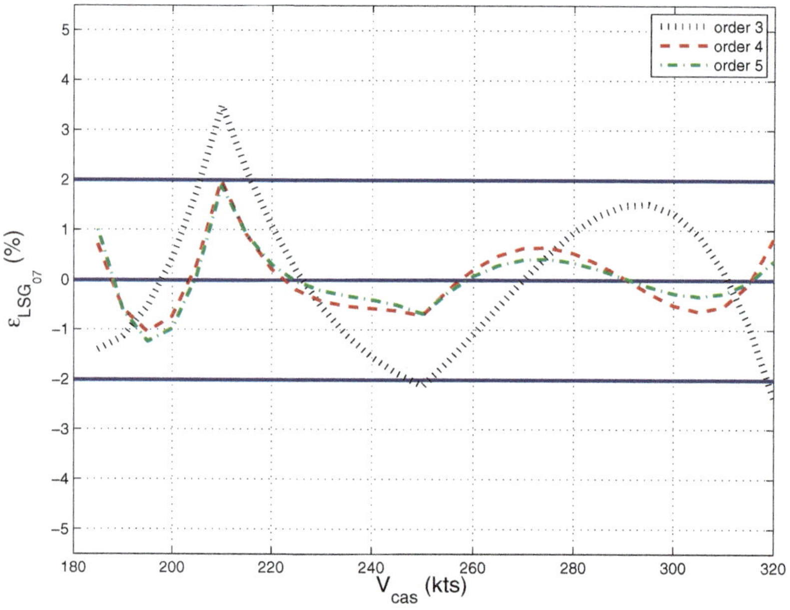

(b) Approximation errors

Fig. 5.4 Polynomial approximation of the scheduled gain $LSG_{07}(V_{cas})$ on the whole range $V_{cas} \in [185,\ 320]$ kts monotonically decreasing function with three slopes, a central flat and two steep ones.

Fig. 5.4(b) illustrates that at least a 4th-order polynomial is required in order to satisfy $\epsilon_{max} \leq 2\%$ on the whole V_{cas}-range [185, 320] kts. The resulting polynomial is hence of the following structure

$$\overline{LSG}_{07}(V_{cas}) = \gamma_1 + \gamma_2\, V_{cas} + \gamma_3\, V_{cas}^2 + \gamma_4\, V_{cas}^3 + \gamma_5\, V_{cas}^4 \tag{5.4}$$

which, with the help of a Horner form based representation, can be realized as a fourth order LFR of the form $\mathcal{F}_u(M, V_{cas} I_4)$. Note that a direct realization based on the power representation of each term would lead to an LFR of order 10.

It is possible to further reduce the order of the resulting LFRs by splitting the parameter range into sub-domains. The initial range $\Gamma = [185, 320]$ kts of V_{cas} can be split into two smaller sub-domains defined as $\Gamma_1 = [185, 250]$ kts and $\Gamma_2 = [250, 320]$ kts. For a given precision level $\epsilon_{max} \leq 2\%$, polynomials of lower order can be used to approximate $LSG_{07}(V_{cas})$ within the smaller ranges. For example, in the sub-domain Γ_1 LSG_{07} can be approximated by a third order polynomial, while in the sub-domain Γ_2 LSG_{07} can be approximated by a linear function. Actually, the linear approximation can be used even on a larger domain $\Gamma_3 = [225, 320]$ kts (which includes Γ_2). The LFRs corresponding to Γ_1 and Γ_2 (also Γ_3) have orders 3 and 1, respectively.

The polynomial interpolation using orthogonal modelling functions has been performed with the new function `data2poly.m` included in the LFR Toolbox for MATLAB [4]. This function automatically selects the degree of polynomials to guarantee a desired approximation accuracy. The creation of the elementary LFRs, the series and parallel couplings, as well as the Horner factorisation are part of the symbolic pre-processing tools implemented in the functions `sym2lfr.m` and `symtreed.m`. The polynomial interpolation and the creation of the corresponding LFR can be fully automated. This represents a substantial gain in time because of the important number of parameter varying blocks, which have usually to be treated.

Additional symbolic manipulations, as extracting common factors, employing special evaluation sequences, or separation of variables can help to reduce the overall order of the controller LFR. For six blocks out of 35, symbolic pre-processing allowed to reduce the global LFR order with at least one for each block.

5.3.2 LFRs for Saturations and Rate Limiters

Three rate limiters and five saturations can been found in the initial block-diagrams of the controller. Fig. 5.5(a) shows how rate limiters and saturations are modelled in the initial controller block-diagram. A rate limitation block is added after a first order transfer function with a certain time constant T

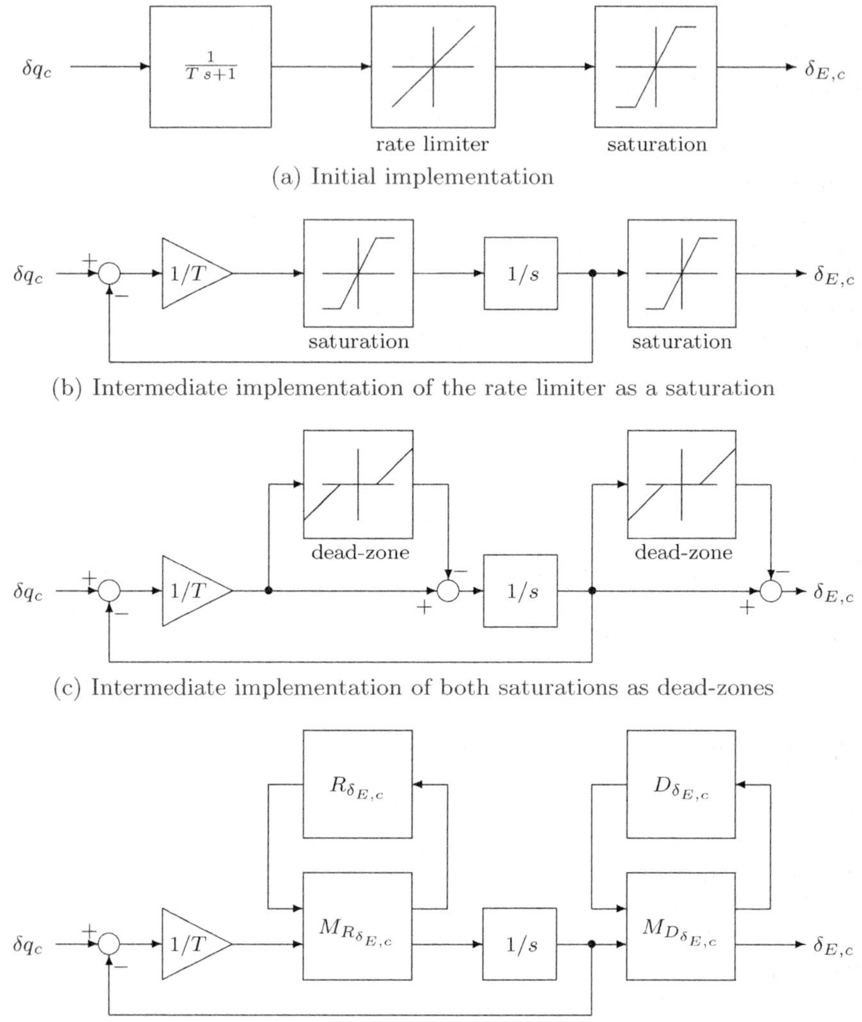

(d) LFR implementation

Fig. 5.5 The implementation of rate limiters and saturations

and before the saturation. This type of modelling would introduce additional dynamics during linearisation.

To reduce the state dimension of the overall resulting controller LFR, the rate limiters can also be modelled as depicted in Fig. 5.5(b). In this case, the first order transfer function with rate limiter and saturation is modelled as a feedback connection containing an integrator, saturation and a gain $1/T$, with the output of the integrator fed back. The rate signal is hence the input of the integrator. A rate limitation can be naturally modelled by a saturation of this signal. Therefore, rate limitations and position saturations are both treated as saturations.

A saturation can be replaced by a dead-zone as illustrated in Fig. 5.5(c). Such a dead-zone can be described as an LFR of the form

$$\mathcal{F}_u\left(\left[\begin{array}{c|c} 0 & 1 \\ \hline -1 & 1 \end{array}\right], \Delta_{DZ}\right),$$

where Δ_{DZ} is the uncertainty associated to the dead-zone operator. The LFRs for rate limiters and saturations are depicted in Fig. 5.5(d). Separate dead-zones are used for a deflection saturation and for a rate limiter. Dead-zones and hence saturations belong to the family of so-called *piece-wise affine nonlinearities* (PWANL). A differentiated dead-zone corresponds to a *piece-wise constant slope*, where the slope is equal to zero inside the dead-zone and to one outside the dead-zone. Anti-windup methods [5, 6] are able to cope with this kind of LFRs.

5.3.3 LFRs for Nonlinear Input/Output Relations

Eight blocks describing static nonlinear input/output relations are present in the controller block structure. To illustrate the treatment of these blocks, let us consider the nonlinear relation

$$\delta p_{m,c} = f(\delta p_m) \tag{5.5}$$

in the lateral feedforward between the pilot roll input δp_m and the equivalent control input $\delta p_{m,c}$. The continuous line in Fig. 5.6 shows $\delta p_{m,c}$ as a function of δp_m. The function graph is symmetric with respect to the origin, but for clarity, it is shown only in the first quadrant of the $(\delta p_m, \delta p_{m,c})$ plane. As it can be observed, the values of δp_m are limited to the range $[-15^o, +15^o]$, and there is also a dead-zone for values of the input $\delta p_{m,c}$ within $[-0.5^o, +0.5^o]$. In the rest, the input-output dependence is almost linear with a slope of about 0.77.

This nonlinearity can be handled in several ways to obtain the corresponding LFRs. The function $f(\delta p_m)$ is a PWANL and can be expressed exactly using a combination of a dead-zone and a saturation as shown in Sect. 5.3.2. The main advantage of this type of modelling is that the PWANL can be transcribed as a standard LFR in the form

$$\mathcal{F}_u\left(\left[\begin{array}{cc|c} 0 & 0 & 2 \\ 0 & 0 & 1 \\ \hline 0.38 & -0.77 & 0 \end{array}\right], \begin{bmatrix} \Delta_{DZ_1} & 0 \\ 0 & \Delta_{DZ_2} \end{bmatrix}\right),$$

where Δ_{DZ_1} and Δ_{DZ_2} are the uncertainties associated to the dead-zone operators.

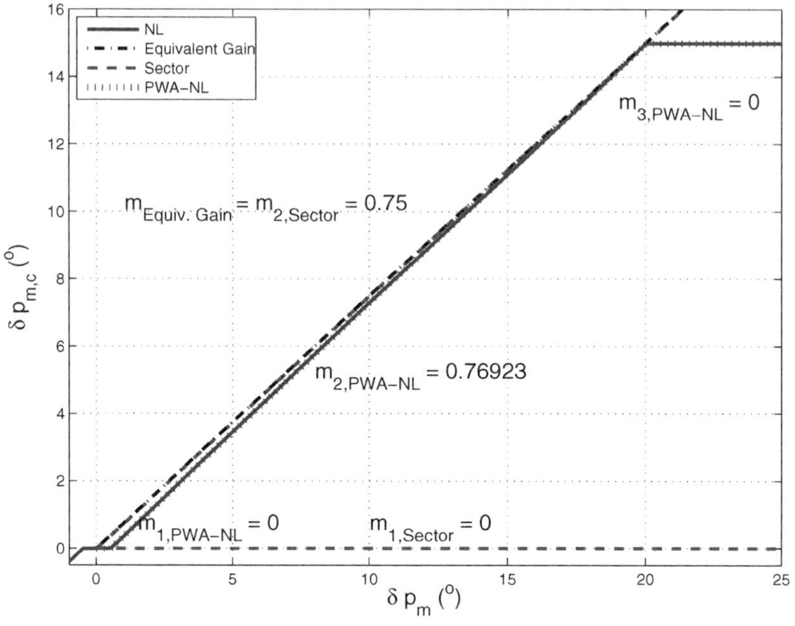

Fig. 5.6 Nonlinear input/output relation between δp_m and $\delta p_{m,c}$

A less accurate modelling uses the approximation with a sector function between the horizontal axis (with zero slope) and the line between the origin and the value of $\delta p_{m,c} = 15^o$ for $\delta p_m = 20^o$ with a slope of 0.75. The border of the sector is indicated by the dashed lines in Fig. 5.6. The corresponding LFR can be simply expressed as

$$\mathcal{F}_u\left(\left[\begin{array}{c|c} 0 & 1 \\ \hline 1 & 0 \end{array}\right], \Delta_{DS}\right),$$

where Δ_{DS} is the operator corresponding to the sector defined by the slope range [0, 0.75]. This modelling is however conservative, because it does not take into account the limitation of $\delta p_{m,c}$ at 15°. Also, for analysis special algorithms like algorithms based on Integral Quadratic Constraints (IQC) must be used [7].

Frequently, an *equivalent gain* m_e can be used to approximate $f(p_m)$ as $m_e p_m$, which represents a linear approximation of f over the whole domain. For our example, this is the dash-dotted line plotted in Fig. 5.6 through the origin and the point $(20^o, 15^o)$, with a slope of 0.75. No LFR modelling is necessary in this case. Although simple, the equivalent gain based approximation ignores both the saturation of $\delta p_{m,c}$ at 15° as well as the presence of the dead-zone.

The use of an equivalent gain is well suited for differentiable nonlinearities. Assuming f in (5.5) would be differentiable, the corresponding equivalent gain is $m_e = \partial f / \partial(\delta p_m)$ evaluated in the corresponding trim condition.

In the controller block, seven nonlinear input/output relations were implemented as equivalent gains for the purpose of the LFR generation.

The equivalent gain based approximation can be extended to approximate nonlinear input/output relations via parameter varying gains. This approach has been chosen for the nonlinear input/output relation between δq_m and δn_{z_c} in the feedforward block of the longitudinal controller. In this case, the gain explicitly depends on the trim value of δq_m. Using a polynomial approximation of the gain, a third order *quasi*-LFR can be determined, where the trim value of δq_m appears as an uncertain parameter.

This technique has been also applied to generate the LFR for the turn coordination relation (5.1), which can be linearized as

$$\Delta \delta n_z = \begin{bmatrix} 1 & \dfrac{\sin\theta}{\cos\phi} & -\dfrac{\cos\theta}{\cos\phi}\tan\phi \end{bmatrix} \begin{bmatrix} \Delta n_z \\ \Delta\theta \\ \Delta\phi \end{bmatrix} \tag{5.6}$$

where θ and ϕ represent the corresponding trim values. The sine-, cosine- and tangent-functions can be very accurately approximated for small angle values by truncated Taylor-series, as for example,

$$\begin{aligned} \cos(\theta) &\cong 1 - \frac{\theta^2}{2} \\ \sin(\phi) &\cong \phi\left(1 - \frac{\phi^2}{6}\right) \\ \tan(\phi) &\cong \phi\left(1 + \frac{\phi^2}{3}\right) \end{aligned} \tag{5.7}$$

The resulting LFR has a Δ-matrix of the form

$$\Delta_{Est,lon} = \operatorname{diag}\left(\theta\, I_5, \phi\, I_7\right) \tag{5.8}$$

5.3.4 LFRs for the Overall Nonlinear Controller

The result of the analysis of the previous sections is a collection of 52 LFRs for all parameter dependent and nonlinear blocks identified in the initial controller block-structure. Replacing these blocks with their LFR counterparts fully preserves the block-structure. For example, the resulting LFR of the longitudinal controller block-diagram is depicted in Fig. 5.7. The longitudinal feedforward, estimator and the pitch law blocks are shown in the upper part, whilst the longitudinal kinematics blocks are plotted in the lower part.

If one or several of these elements change during the control law design process, it is easy to replace it or them with new LFR(s). This also holds

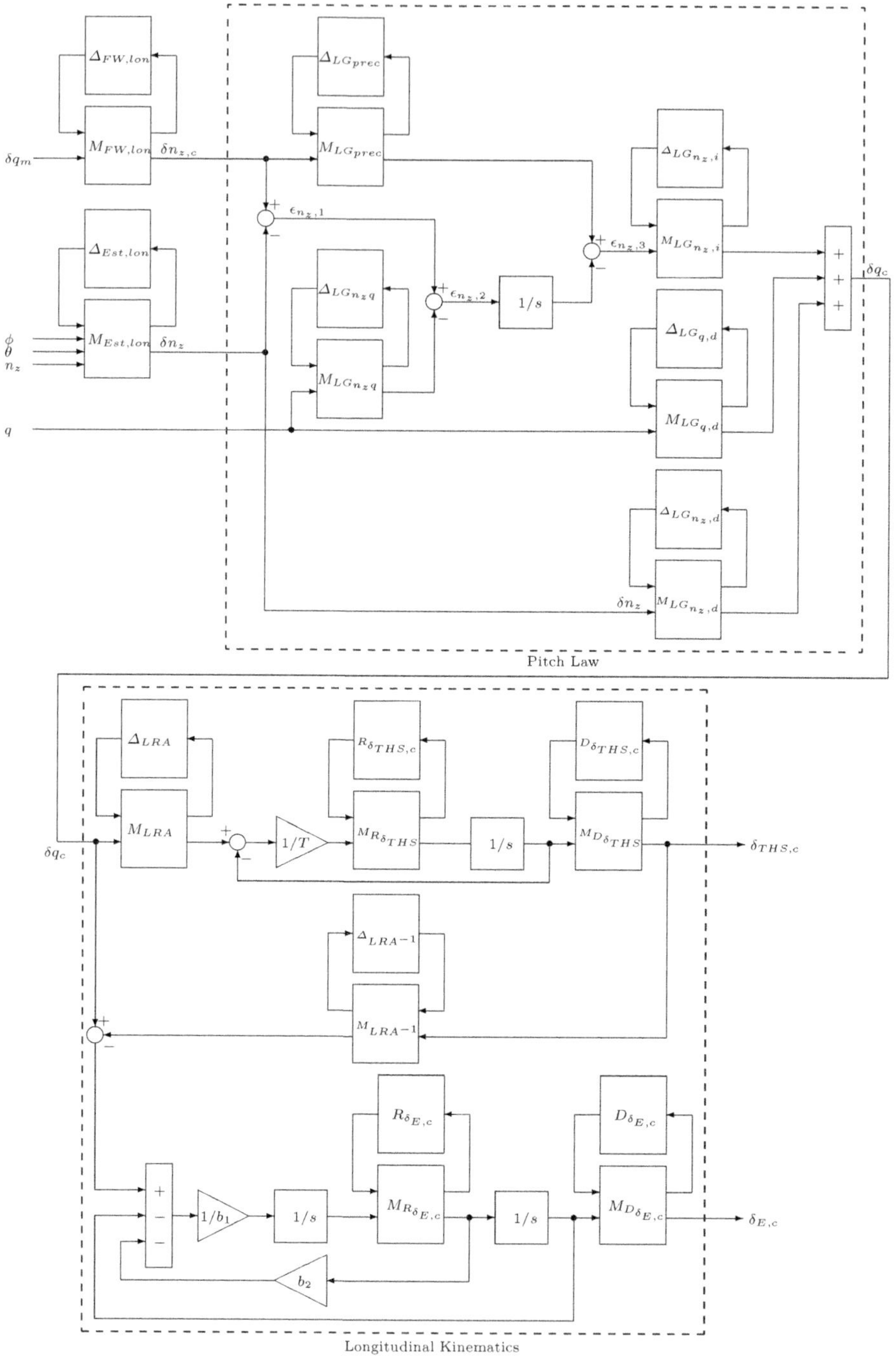

Fig. 5.7 The resulting longitudinal controller LFR block-diagram

for blocks where different LFRs exist. All gains depending on V_{cas} possess four LFRs corresponding to four different V_{cas}-ranges (Γ, Γ_1, Γ_2, Γ_3) defined in Subsection 5.3.1). All nonlinear input/output relations are usually replaced by equivalent gains, but could also be replaced by LFRs for sectors or PWANLs. The same holds for blocks whose LFRs have to be changed, as the underlying polynomial interpolation or the modelling as equivalent gain are not precise enough. In all these cases, it is not necessary to recompute the whole controller LFR, which could be a very time-consuming task.

The LFR of the global controller of Fig. 5.3 can be extracted from the block-diagram of Fig. 5.7 as follows. The i-th LFR block of the controller ($i = 1, \ldots, 52$) has input $u_{c,i}$ and output $y_{c,i}$ and is represented with the help of the LFT-representation as

$$y_{c,i} = \mathcal{F}_u(M_{c,i}, \Delta_{c,i})u_{c,i},$$

where $\Delta_{c,i}$ is a diagonal matrix and $M_{c,i}$ is a constant matrix partitioned as

$$M_{c,i} = \begin{bmatrix} M_{11,i} & M_{12,i} \\ M_{21,i} & M_{22,i} \end{bmatrix}$$

such that the square matrices $M_{11,i}$ and $\Delta_{c,i}$ have the same order. The LFT-based representation corresponds to a feedback connection (see the blocks in Fig. 5.7), which involves the additional inputs $w_{c,i}$ and outputs $z_{c,i}$ connected by an internal feedback connection

$$w_{c,i} = \Delta_{c,i} z_{c,i}$$

The interconnection of all LFRs and LTI components of the controller results in an LFR which can be described by a system

$$\begin{aligned} \dot{x}_c &= A_c x_c + \begin{bmatrix} B_{c,1} & B_{c,2} \end{bmatrix} \begin{bmatrix} w_c \\ u_c \end{bmatrix} \\ \begin{bmatrix} z_c \\ y_c \end{bmatrix} &= \begin{bmatrix} C_{c,1} \\ C_{c,2} \end{bmatrix} x_c + \begin{bmatrix} D_{c,11} & D_{c,12} \\ D_{c,21} & D_{c,22} \end{bmatrix} \begin{bmatrix} w_c \\ u_c \end{bmatrix} \end{aligned} \tag{5.9}$$

and a feedback connection

$$w_c = \widetilde{\Delta}_c z_c$$

To perform the coupling, first all LFRs are prepared for coupling by cutting out all Δ_i-matrices and introducing the artificial inputs w_i and outputs z_i. What remains are the static transfers M_i and dynamic and static LTI blocks. To obtain the system matrices in (5.9), a standard linearization tool can be employed (e.g., the `linmod` function of MATLAB), using the trivial equilibrium point $x_{c,0} = 0$, $u_{c,0} = 0$, $w_{c,0} = 0$. A similar coupling takes place to obtain Δ_c by stacking all inputs $w_{c,i}$ and all outputs $z_{c,i}$ in the corresponding vectors w_c and z_c, respectively. The matrix $\widetilde{\Delta}_c$ results diagonal, with the scheduling and uncertain variables on its diagonal. For analysis purposes with standard tools (e.g., μ-analysis), the inputs in w_c and corresponding outputs in z_c have to be permuted to obtain $\widetilde{\Delta}_c$ in a block diagonal form

$$\widetilde{\Delta}_c = \text{diag}\left(\delta_{c,1} I_{n_1}, \cdots, \delta_{c,\ell} I_{n_\ell}\right),$$

where each parameter $\delta_{c,i}$ appears only in a single block. The permutation performed on $\widetilde{\Delta}_c$ must be also performed on the columns of $B_{c,1}$ and rows of $C_{c,1}$. The overall LFR representation of the controller in Fig. 5.3 corresponds to the following definition of matrices M_c and Δ_c

$$M_c = \left[\begin{array}{cc|c} A_c & B_{c,1} & B_{c,2} \\ C_{c,1} & D_{c,11} & D_{c,12} \\ \hline C_{c,2} & D_{c,21} & D_{c,22} \end{array}\right], \quad \Delta_c = \begin{bmatrix} \frac{1}{s} I_{n_c} & 0 \\ 0 & \widetilde{\Delta}_c \end{bmatrix},$$

where n_c is the dimension of the linearized controller state vector.

The block-diagram based LFR generation process can be performed automatically using a new function `slk2lfr`, from ONERA's *Simulink handling of LFR objects* extension [2] to the LFR toolbox [4]. Table 5.2 shows the number of inputs, outputs and states of all resulting controller LFRs. The longitudinal (Lon.) and lateral (Lat.) controllers are decoupled in terms of states and outputs. The numbers of outputs and states of the global controller are equal to the sum of the outputs or the states of the lateral and longitudinal controllers, respectively. ϕ is the only input which is used in both the lateral and longitudinal controllers. Hence, the global controller uses just 11 inputs. For the list of inputs, outputs and states, refer to Table 5.1.

Table 5.2 Inputs, outputs and states of the controller LFRs

Controller	Lon.	Lat.	Lon. & Lat.
Number of inputs	5	7	11
Number of outputs	2	9	11
Number of states	4	11	15

In Table 5.3 the corresponding Δ_c-blocks are compared. The varying parameters are M, V_{cas}, $X_{cg,p}$, α and m_p (see Sect. 5.3.1). θ and ϕ stem from the turn coordination (see Sect. 5.3.3) and δq_m from the longitudinal feedforward (see Sect. 5.3.3). The four nonlinearities in the longitudinal controller LFR concern the rate and position saturations of $\delta_{E,c}$ and $\delta_{THS,c}$

$$\Delta_{NL,c,lon} = \text{diag}\left(D_{\delta_{E,c}}, R_{\delta_{E,c}}, D_{\delta_{THS,c}}, R_{\delta_{THS,c}}\right) \tag{5.10}$$

whilst the four nonlinearities in the lateral controller concern the rate and position saturations of δr_c, the position saturation of $\delta_{R,c}$ and the position saturation of δp_c

$$\Delta_{NL,c,lat} = \mathrm{diag}\left(D_{\delta r_c}, R_{\delta r_c}, D_{\delta_{R,c}}, D_{\delta p_c}\right) \tag{5.11}$$

where D and R are the notations for the dead-zones corresponding to position saturations and rate limiters introduced in Sect. 5.3.2. For the global controller holds

$$\Delta_{NL,c,lon+lat} = \mathrm{diag}\left(\Delta_{NL,c,lon}, \Delta_{NL,c,lat}\right) \tag{5.12}$$

resulting in eight nonlinearities.

Table 5.3 The Δ_c-block structure of the resulting controller LFRs

Controller/	n_i									$\sum n_i$
V_{cas} Range	M	V_{cas}	$X_{cg,p}$	δq_m	θ	ϕ	α	m_p	Nonlinearities	
Lat. on Γ	13	28	-	-	-	-	6	2	4	53
Lat. on Γ_1	13	15	-	-	-	-	6	2	4	40
Lat. on Γ_2	13	11	-	-	-	-	6	2	4	36
Lat. on Γ_3	13	16	-	-	-	-	6	2	4	41
Lon. on Γ	8	13	3	3	3	5	-	-	4	39
Lon. on Γ_1	8	13	3	3	3	5	-	-	4	39
Lon. on Γ_2 & Γ_3	8	8	2	3	3	5	-	-	4	33
Lon. & Lat. on Γ	21	41	3	3	3	5	6	2	8	92
Lon. & Lat. on Γ_1	21	28	3	3	3	5	6	2	8	79
Lon. & Lat. on Γ_2	21	19	2	3	3	5	6	2	8	69
Lon. & Lat. on Γ_3	21	24	2	3	3	5	6	2	8	74

The size of the Δ_c-block of the global controller is the sum of the sizes of the individual Δ_c-blocks of the lateral and longitudinal controllers. The size of the global Δ_c-block of the coupled longitudinal/lateral nonlinear controller is 92. The size of the LFR for the longitudinal controller on its own is 39 and the one for the lateral controller is 53. Thanks to the split of the initial V_{cas}-range into 3 sub-domains Γ_1, Γ_2 and Γ_3, the size of the global Δ_c-block can be significantly reduced from 92 to 79 for the $V_{cas} \in \Gamma_1$ or even to 69 for $V_{cas} \in \Gamma_2$. The center of gravity $X_{cg,p}$ and the mass m_p vary within the whole weight-balance domain given in Fig. 3.2. The calibrated air speed V_{cas} and the Mach number M vary within the $V_{cas} - M$ domains for low and high masses given in Fig. 3.3 and 3.4.

A final normalization of the ranges of variables in the Δ_c-matrix is required to complete the generation of the controller LFRs. This means that all δ_i are shifted and/or scaled to lie in the interval $[-1, 1]$. For this purpose, we employ the same ranges for $X_{cg,p}$, m_p, V_{cas} and M as those employed for the generation of LFRs for the nonlinear aircraft model in Chapter 3 and for the LFRs of the linear flexible aircraft models in Chapter 4. For the nonlinear aircraft model 1, $X_{cg,p}$ is normalized on $X_{cg,p} \in [17\%, 41\%]$. The normalization of V_{cas} is performed with respect to the different ranges employed in

Subsection 5.3.1. For example, for aircraft model 1, V_{cas} lies in $V_{cas} \in \Gamma_1$, the normalization range [185, 250] kts is hence used. The corresponding ranges for M and m_p are $M \in [0.28,\ 0.50]$ and $m_p \in [120t,\ 144.5t]$. The corresponding normalized controller LFR will be named for further references as *controller LFR 1*.

5.4 Generation of the Closed-Loop LFR

In order to generate the closed-loop LFRs for stability and performance analysis, it is now sufficient to interconnect the aircraft LFRs with the controller LFRs via the LFRs for the sensors and actuators.

The LFR of the sensors is treated during the modelling of the nonlinear rigid aircraft. The actuators do not have parameter varying elements. Their deflections are just limited by rate limitations and position saturations which are different from those in the controller. However, the saturations can depend on V_{cas}. $\Delta_{NL,act}$ is expressed as

$$\begin{aligned}\Delta_{NL,act} = \text{diag}\Big(&D_{\delta_{A_{il}}}, R_{\delta_{A_{il}}}, D_{\delta_{A_{ir}}}, R_{\delta_{A_{ir}}}, D_{\delta_{S_{l,2,3}}}, R_{\delta_{S_{l,2,3}}}, \dots \\ &D_{\delta_{S_{l,4,5}}}, R_{\delta_{S_{l,4,5}}}, D_{\delta_{S_{r,2,3}}}, R_{\delta_{S_{r,2,3}}}, D_{\delta_{S_{r,4,5}}}, R_{\delta_{S_{r,4,5}}}, \dots \\ &D_{\delta_R}, R_{\delta_R}, D_{\delta_E}, R_{\delta_E}, D_{\delta_{THS}}, R_{\delta_{THS}}\Big)\end{aligned} \tag{5.13}$$

See Sect. 3.4.3 for more details on Δ_{sens}, Δ_{act} as well as D and R. Various Δ_c are given in Table 5.3. Sixteen $\Delta_{A/C}$ for the nonlinear rigid aircraft model are given in Table 3.4 and fourteen $\Delta_{A/C}$ for the linear flexible aircraft model are given in Table 4.2.

5.4.1 Closed-Loop LFRs for the Nonlinear Model Performance Analysis

Fig. 5.8 illustrates for example the interconnection structure for the turn coordination performance analysis between the pilot roll input δp_m and the bank angle ϕ and sideslip angle β.

Using the same procedure as in Sect. 5.3.4, the LFR of the closed-loop system can be derived. It corresponds to Fig. 5.9 with one input δp_m and two outputs ϕ and β. The total number of states is always 40 where nine states stem from the aircraft, nine states from the actuators, seven states from the sensors and 15 states from the controller. The nominal transfer matrix M_{cl} is fed-back with the perturbation matrix Δ_{cl} involving M, V_{cas}, m, X_{cg}, $X_{cg,p}$, α, θ, ϕ, m_p and nonlinearities corresponding to

$$\Delta_{NL,cl} = \text{diag}\left(\Delta_{NL,c},\ \Delta_{NL,act}\right) \tag{5.14}$$

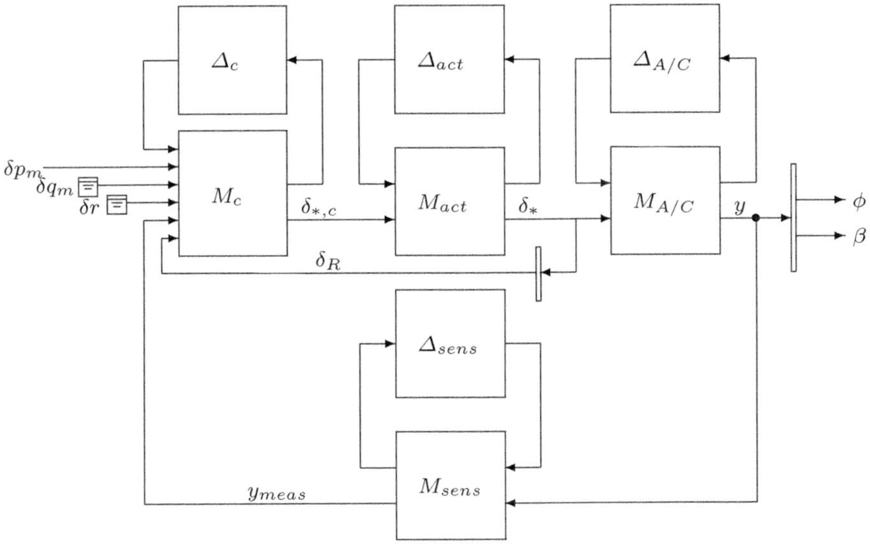

Fig. 5.8 The closed-loop for turn coordination performance analysis

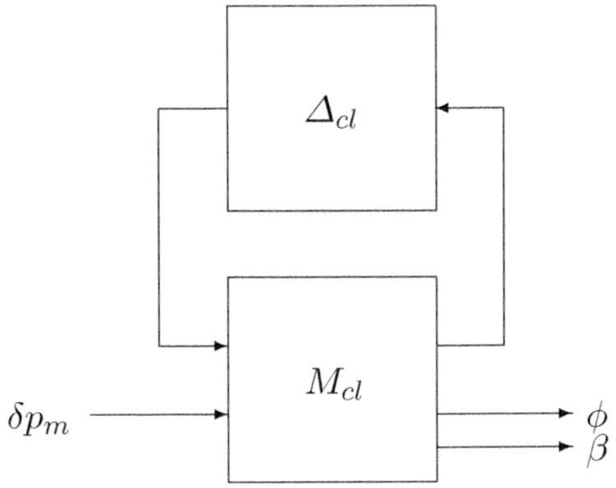

Fig. 5.9 The closed-loop LFR for turn coordination performance analysis

with $\Delta_{NL,c}$ given by (5.12) with a size of 8 and $\Delta_{NL,act}$ given by (5.13) with a size of 18.

Several closed-loop LFRs using the interconnection structure of Fig. 5.8 have been generated. In all cases, the LFR for the nonlinear aircraft model 1 is used. Hence, the *controller LFR 1* has to be used.

As the turn coordination analysis criteria do not depend on the pilot's pitch input δq_m, δq_m can be set to zero in the Δ_c-matrix in all cases. It will hence disappear from the resulting overall Δ_{cl}-matrix as shown in Table 5.4. θ is defined as

$$\theta = \alpha + \gamma \tag{5.15}$$

Table 5.4 Δ_{cl}-block of the resulting closed-loop LFRs

Cases	n_i								$\sum n_i$
	M	V_{cas}	m	X_{cg}	$X_{cg,p}$	ϕ	m_p	Nonlinearities	
1	70	113	46	76	2	3	2	26	338
2	70	113	48	78	-	3	-	26	338
3	70	113	48	78	-	-	-	26	335
4	85	131	122	64	-	-	-	26	428
5	85	147	122	64	-	-	-	26	444
6	70	113	48	78	-	-	-	26	335

For horizontal flight, γ is always equal to zero, so that θ equals to α. It is replaced by α in the Δ_{cl}-matrix. In the cases 1, 2, 3 and 6, α is set to the nominal trim value $\alpha_0 = 3^o$ for the aircraft model 1. θ and α will hence simply disappear from the resulting overall Δ_{cl}-matrix. The size of the Δ_{cl}-matrix in case 1 is 338. It corresponds to the sum of the sizes of the Δ_i-matrices. In cases 4 and 5, α is replaced by the trim value which is itself a function of M, V_{cas}, X_{cg} and m. The corresponding LFR is computed using the approach of Subsection 3.4.4. Its Δ_α-matrix is

$$\Delta_\alpha = \text{diag}(M\, I_2, V_{cas}\, I_2, X_{cg}\, I_5, m\, I_2) \tag{5.16}$$

The size of the Δ_{cl}-matrix increases significantly from 338 to 428 in case 4. This is a drawback for the analysis-algorithms. If the analysis-algorithms are able to cope with such a Δ_{cl}-matrix, the analysis result will however be less conservative, as θ and α are adapted in the controller to the real flight condition, whereas in cases 1, 2, 3 and 6, θ and α correspond just to a mean value.

For cases 2-6, it is supposed that the controller $X_{cg,p}$ and m_p are set by the pilot corresponding to the real aircraft X_{cg} and m. In these cases, $X_{cg,p}$ and m_p disappear from the overall Δ_{cl}-matrix, however the corresponding X_{cg} and m entries increase by two. The overall size in case 2 does not change with respect to case 1. The LFR in case 1 can be used to analyse the impact of a wrong controller $X_{cg,p}$ and m_p setting. The LFRs of cases 2-6 are used to analyse nominal settings.

In cases 3-5, ϕ is fixed to a mean value of -5°. This reduces the size of the resulting LFR by three. ϕ disappears from the Δ_{cl}-matrix. The analysis result will however be more conservative, as ϕ is not adapted in the controller during the turn.

In case 5, the actuator saturations depend on V_{cas}. The size of the resulting Δ_{cl}-matrix increases by 16 due to additional V_{cas} terms with respect to case 4.

In the following, let us consider the case 3, i.e. the controller trim conditions are set to $\delta q_m = 0$, $\alpha = \theta = \alpha_0$, $\phi = -5^o$, $X_{cg,p} = X_{cg}$ and $m_p = m$. The simplified actuator model is used, where the saturations do not depend on

Table 5.5 Δ_{cl}-block of the resulting closed-loop LFRs for turn coordination performance analysis with $\delta q_m = 0$, $\alpha = \theta = \alpha_0$, $\phi = -5^o$, $X_{cg,p} = X_{cg}$ and $m_p = m$, the simplified actuators and the 16 aircraft models

Model	n_i					$\sum n_i$
	M	V_{cas}	X_{cg}	m	nonlinearities	
1	70	113	78	48	26	335
2	104	68	58	50	26	306
3	70	111	81	55	26	343
4	104	68	63	52	26	313
5	104	79	60	68	26	337
6	104	68	76	48	26	322
7	104	77	75	53	26	335
8	70	102	71	54	26	323
9	104	73	69	63	26	335
10	104	73	66	60	26	329
11	87	90	77	54	26	334
12	70	107	68	46	26	317
13	104	73	77	53	26	333
14	104	73	75	54	26	332
15	70	107	75	52	26	330
16	104	73	69	63	26	335

V_{cas}. The loop is now closed for all 16 aircraft LFRs with the corresponding 16 controller LFRs. The sizes of the resulting Δ_{cl}-matrices are listed in Table 5.5. The smallest size of 306 is obtained for model 2. The corresponding aircraft LFR is the smallest of all aircraft LFRs. The controller LFR for model 2 is built with the LFR for $V_{cas} \in \Gamma_2$, which is also the smallest controller LFR. The largest size of 343 is obtained for model 3. The controller LFR for model 3 is built with the LFR for $V_{cas} \in \Gamma_1$, which is the second largest controller LFR. All or some of the position saturations and rate limitations can finally be suppressed by the analysis teams by setting the corresponding D and R values in the Δ_{cl}-matrix to zero, which sets the saturation to the zone where it behaves in a linear way.

5.4.2 Closed-Loop LFRs for the Nonlinear Model Stability Analysis

For stability analysis of the nonlinear model, it is sufficient to delete the pilot's input and the outputs in Fig. 5.9. The resulting LFR of the closed-loop system corresponds to Fig. 5.10. The resulting Δ_{cl}-matrix corresponds to case 6 in Table 5.4, which is equal to the one of case 3.

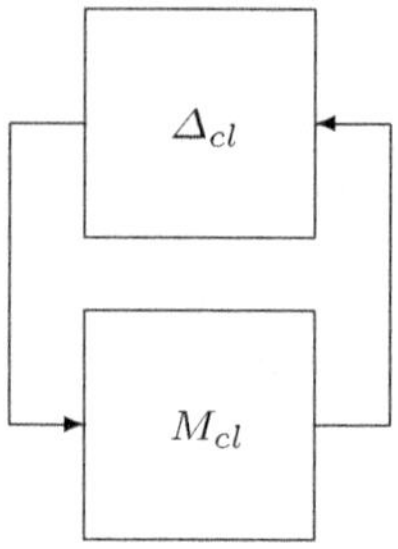

Fig. 5.10 The closed-loop LFR for the stability analysis

In Chapter 3 it has already been explained that the longitudinal and lateral motion of the nonlinear aircraft can be extracted from the overall aircraft LFR. In Table 5.3 it was shown that this also holds for the controller. The closed-loop LFR of Fig. 5.10 can hence be used for the stability analysis of either the combined nonlinear longitudinal/lateral motion or the nonlinear longitudinal or the nonlinear lateral motion.

5.4.3 Closed-Loop LFRs for the Flexible Model Stability and Performance Analysis

Interconnecting the longitudinal controller LFR given in Table 5.3 with one of the fourteen linear flexible aircraft LFRs given in Table 4.2 like in the previous subsections leads to the closed-loop LFRs for stability or performance analysis of the flexible aircraft model. Table 5.6 illustrates for example five stability analysis LFRs with either ten or four flexible modes n_f and 32 or 20 states n_s. The parameters in the Δ_{cl}-block correspond to the flight condition expressed by the Mach number M, the calibrated air speed V_{cas} and the payload PL as well as the filling level in the center (CT) and outer tank (OT).

Table 5.6 Δ_{cl}-block of the resulting closed-loop LFRs for the stability analysis of the flexible aircraft model

LFR number	Parameters in Δ_{cl}	n_f	n_s	Size of Δ_{cl}
1	$\overline{M}, \overline{V}_{cas}$	10	32	87×87
2	$\overline{M}, \overline{V}_{cas}$	4	20	51×51
3	$\overline{M}, \overline{V}_{cas}, \overline{\delta}_{OT}$	4	20	104×104
4	$\overline{M}, \overline{V}_{cas}, \overline{\delta}_{OT}, \overline{\delta}_{CT}$	4	20	203×203
5	$\overline{M}, \overline{V}_{cas}, \overline{\delta}_{OT}, \overline{\delta}_{CT}, \overline{\delta}_{PL}$	4	20	290×290

5.5 Evaluation of the LFR Generation Process

The evaluation of the generation process of the controller and the closed-loop LFRs is performed in order to:

- verify the accuracy of the employed LFR modelling process,
- demonstrate the ability of the LFR framework to represent the nonlinear behaviour of the system due to the presence of elements such as dead-zones and saturations.

For both the LFR of the controller and the LFR of the closed-loop, the evaluation consists in comparing:

- the dynamic properties of the generated LFRs with those of the linearised models of the nonlinear system,
- the time domain simulations of the LFRs and of the nonlinear models.

5.5.1 Evaluation of the LFR of the Nonlinear Controller

In order to analyse its dynamic properties, the nonlinear controller is firstly linearised over a tight grid of its scheduling parameters covering the whole flight envelope. Linearisations are performed by applying the `linmod` function to the Simulink-schemes of the controller. Special care is taken to the linearisation process of the nonlinear elements. In particular, each controller dead-zone is removed from the Simulink-schemes.

These linearised models are compared to the linear models generated from the controller LFR. These linear models are generated by setting the Δ-matrix parameters:

- to the flight condition values for the scheduling parameters V_{cas}, M, $m_p = m$ and $X_{cg,p} = X_{cg}$,
- to the corresponding equilibrium value for the trimmed angle of attack α_0,
- to α_0 for the trimmed pitch angle θ_0, considering that the flight path angle $\gamma = 0$,
- to zero for the trimmed demanded pitch angle $\delta q_{m,0}$,
- arbitrarily to the negative value -5° for the trimmed bank angle ϕ_0,
- to zero for the parameters corresponding to nonlinear elements.

Such values of trimmed variables correspond to a realistic initial configuration for turn coordination analysis. Controller input/output transfers of the two families of linear models are then analysed in order to evaluate the accuracy of the LFR modelling process.

Input/output transfer functions of both linear models extracted from the LFR and linearised models of the longitudinal controller are computed over a tight grid of the three scheduling parameters $X_{cg,p} = X_{cg}$, M and V_{cas} covering the whole flight domain (the longitudinal controller is only scheduled by these three variables). For the transfer function between the j-th input and the i-th output of the controller, the relative error $\epsilon_{i,j}(\rho_k)$ at the flight point ρ_k of the grid is defined as

$$\epsilon_{i,j}(\rho_k) = \max_{\omega \in \mathbb{R}_+} \frac{\left|K_{i,j}(j\omega, \rho_k) - \tilde{K}_{i,j}(j\omega, \rho_k)\right|}{|K_{i,j}(j\omega, \rho_k)|}$$

where $K_{i,j}(j\omega, \rho_k)$ and $\tilde{K}_{i,j}(j\omega, \rho_k)$ correspond respectively to the frequency responses of the linearised model and of the linear model extracted from the LFR. The maximum relative error over the grid $\epsilon_{i,j} = \max_k \epsilon_{i,j}(\rho_k)$ is presented in Table 5.7. Results are satisfactory, except for the column corresponding to the δq_m input. For this input, the amplitudes of errors of about 20% can be explained by the choice of a mean equivalent gain for the representation of the longitudinal feedforward nonlinearities. The mean equivalent gain is calculated by considering δq_m variations of arbitrarily large amplitudes, whereas the linearised model is just valid for very small variations in δq_m. Note however that this gap concerns an external input of the system and does not affect closed-loop stability analysis (but could significantly affect the performance analysis).

Table 5.7 Maximum relative error between transfer functions of the linear models extracted from the LFR and the linearised model of the longitudinal controller

	δq_m	q	ϕ	θ	n_z
$\delta_{E,c}$	22.5%	3.83%	5.33%	5.33%	5.33%
$\delta_{THS,c}$	21.8%	4.70%	6.21%	6.21%	6.21%

Table 5.8 presents maximum relative errors of the input/output transfer functions between linear models generated from the LFR and linearised models of the lateral controller. These transfer functions are calculated over a tight grid of the four parameters $m_p = m$, α, M and V_{cas} that schedule the lateral controller. The whole flight domain is covered. For this comparison, the LFR of the lateral controller is generated by selecting mean equivalent gains adapted to small variations for the description of the nonlinearities of the ailerons and spoilers kinematics functions. For each input/output transfer, a maximum relative error less than 12% is considered acceptable. This is achieved excepting the spoiler input/output transfer functions, for which the maximum error adds up to 16.5% for the input δp_m. This particular high error can be explained by the employed mean equivalent gain approximation within the lateral feedforward (see Sect. 5.3.3).

Nonlinear simulations are also performed to compare time domain responses of the LFRs and nonlinear models of the controller. Simulations of LFRs including nonlinear elements are performed using [2]. For these

Table 5.8 Maximum relative error between transfer functions of the linear models extracted from the LFR and the linearised model of the lateral controller

	δp_m	δr	p	r	ϕ	n_y	r
$\delta_{A_{il,c}}$ and $\delta_{A_{ir,c}}$	3.43%	-	0.63%	-	0.42%	-	-
$\delta_{S_{l,2,3}}$ and $\delta_{S_{r,2,3}}$	16.5%	-	14.2%	-	13.9%	-	-
$\delta_{S_{l,4,5}}$ and $\delta_{S_{r,4,5}}$	16.5%	-	14.2%	-	13.9%	-	-
$\delta_{R,c}$	11.75%	0.0%	9.8%	9.5%	10.1%	11.1%	9.95%

simulations, the LFR of the lateral controller is generated selecting piece-wise affine functions to describe spoilers and ailerons kinematics.

Fig. 5.11 presents the time responses of both the LFR and the nonlinear controllers due to the pilot roll input δp_m, which covers the range $[-20^o, +20^o]$. The LFR controller responses (continuous lines) match almost exactly the nonlinear controller ones (dash-dotted lines), in spite of the presence of dead-zones, rate and position saturations in the nonlinear controller. The generated LFR is able to correctly represent the controller nonlinearities, whose effects are sensitive to inputs of realistic amplitude. The controller LFR seems therefore to be well suited for the response to the δp_m input within the turn coordination performance context. The controller LFR response is closer to the nonlinear system response than to the response of the linearised system. This explains the relatively large error in Table 5.8. The responses of the controller LFR due to the pilot pitch input δq_m match correctly the nonlinear controller ones. There is a small error of about 0.5^o, especially on the horizontal tailplane deflection $\delta_{THS,c}$, which is much better than the relatively large error of Table 5.7. The behaviour of the controller LFR is considered as satisfactory with respect to the nonlinear one.

5.5.2 Evaluation of the Closed-Loop LFR

The dynamic properties of the generated LFR of the closed-loop system have to be compared to those of the nonlinear model. A first step consists in generating linearised models of the nonlinear closed-loop around some operating points. However, the `linmod` function cannot be directly applied to the Simulink-scheme of the closed-loop nonlinear aircraft because of some difficulties which have been identified during the aircraft LFR generation process, see Chapter 3 for more details. Therefore, the linearised models of the aircraft which have been generated during the aircraft LFR generation process (see again Chapter 3) are employed. They are connected to the linearised models of the controllers, actuators and sensors.

To cover the whole flight envelope, linearisations are performed for each one of the 16 flight sub-domains defined during the aircraft LFR generation process in Chapter 3. For each flight sub-domain, 81 flight points are considered, corresponding to a grid on V_{cas}, M, m and X_{cg}. These models

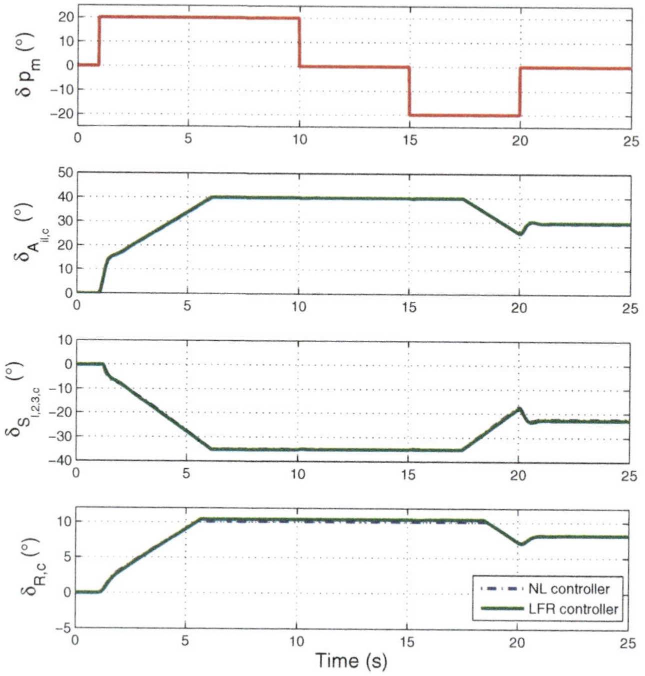

Fig. 5.11 Comparison of controller time responses to a δp_m input at the flight condition: $[V_{cas};\ M;\ m;\ X_{cg}] = [217.7\,kts;\ 0.43;\ 144.5\,t;\ 29\%]$. Dash-dotted: response of the nonlinear controller; solid: response of the LFR of the controller.

are compared with linear models generated from the closed-loop LFR, which are obtained by setting to zero the Δ-matrix parameters corresponding to nonlinear elements.

Fig. 5.12 illustrates the pole chart for the sub-domain 8 (see Table 3.4 and Fig. 3.3 for more details). It corresponds to high V_{cas} and has been identified, amongst all sub-domains, as the worst-case sub-domain in terms of pole matching. The poles of the LFR (markers $\times$) and of the linearised models (markers $+$) are spread relatively far from each other, but the correspondence is still satisfactory.

The pole matching between the linear model $\tilde{M}_{cl}(s, \rho_k)$ generated from the LFR at the flight point ρ_k and the linearised model $M_{cl}(s, \rho_k)$ is evaluated by computing the distance $d(\rho_k)$ between the vectors containing the eigenvalues $\lambda(\tilde{M}_{cl}(s, \rho_k))$ and $\lambda(M_{cl}(s, \rho_k))$

$$d(\rho_k) = \sum_{j=1}^{n_s} \left| \lambda_j(\tilde{M}_{cl}(s, \rho_k)) - \lambda_j(M_{cl}(s, \rho_k)) \right|$$

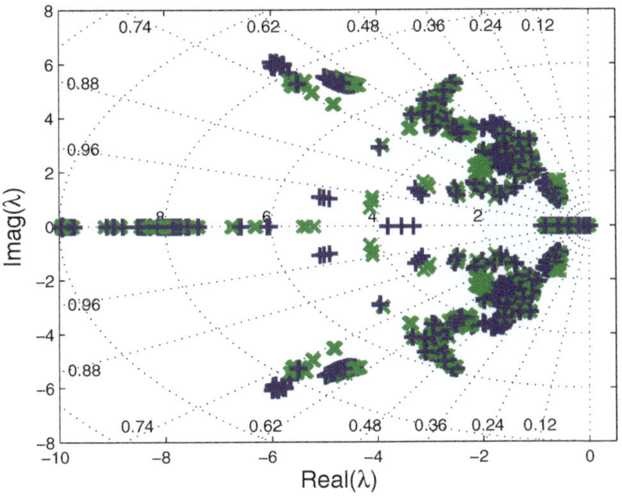

Fig. 5.12 Pole charts of the linear models extracted from the LFR of the closed-loop (Marker ×) and of the linearised model (Marker +) for several flight conditions corresponding to sub-domain 8 (high V_{cas}).

with $\lambda_j(.)$ denoting the j-th eigenvalue and n_s the number of states. Note that the computation of $d(\rho_k)$ requires that the poles of $\tilde{M}_{cl}(s,\rho_k)$ and $M_{cl}(s,\rho_k)$ are correctly paired. For this purpose, the poles of both models are associated, considering their locations, in such a way that the overall distance between the paired poles is minimized.

The worst-matching flight conditions are specifically identified for the transfer functions $\delta p_m \rightarrow \phi$ and $\delta q_m \rightarrow \theta$ by evaluating the matching of the poles of these specified transfer functions over a grid of the variables V_{cas}, M, X_{cg} and m. The worst-matching flight conditions correspond to:

- Worst case 1 for the $\delta p_m \rightarrow \phi$ transfer:
  $[V_{cas}; M; m; X_{cg}] = [316.9\,kts; 0.83; 169\,t; 17\%]$
- Worst case 2 for the $\delta q_m \rightarrow \theta$ transfer:
  $[V_{cas}; M; m; X_{cg}] = [316.9\,kts; 0.63; 144\,t; 17\%]$

For these two cases, frequency and time domain responses of both linear models are relatively close, and the main dynamic properties are reproduced in a satisfactory way by the models generated from the closed-loop LFR. Fig. 5.13 illustrates for instance the dynamic behaviours of the linear model extracted from the LFR and of the linearised models on the worst-matching flight condition 2.

The LFR of the closed-loop is finally compared to the nonlinear closed-loop model by performing some time domain nonlinear simulations. The time responses of the closed-loop LFR and the nonlinear model due to a pilot roll input δp_m are relatively close, except for the response of the pitch angle θ.

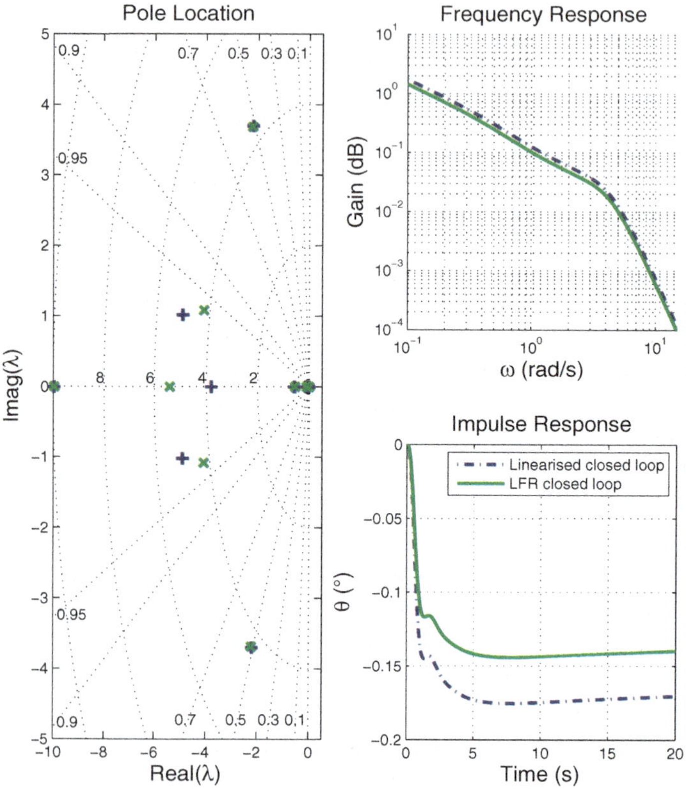

Fig. 5.13 Dynamic closed-loop properties of the $\delta q_m \rightarrow \theta$ transfer for the worst-matching flight condition 2. Marker ×, solid: LFR of the closed-loop; Marker +, dash-dotted: linearised closed-loop.

Only the nonlinear system has an expected behaviour concerning this output: the pitch angle θ of the nonlinear system remains stable after the δp_m input, whereas θ diverges in the case of the closed-loop LFR and the linearised model. Consequently, this phenomenon seems to be due to the linearisation operation, and not to the LFR generation process. The correspondence of the time responses of the different models due to the pitch input δq_m is satisfactory, and the main characteristics of the time responses of the nonlinear model are correctly reproduced by the LFR. Note that, particularly in the case of the response to the pedal input δr, the response of the closed-loop LFR matches more accurately the response of the nonlinear model than the linearised one. This property can be explained by the presence of the nonlinear elements in the closed-loop LFR, which permit to represent some nonlinear behaviour induced notably by the presence of dead-zones and saturations in the controller and actuators LFRs. Hence, these nonlinear simulations also

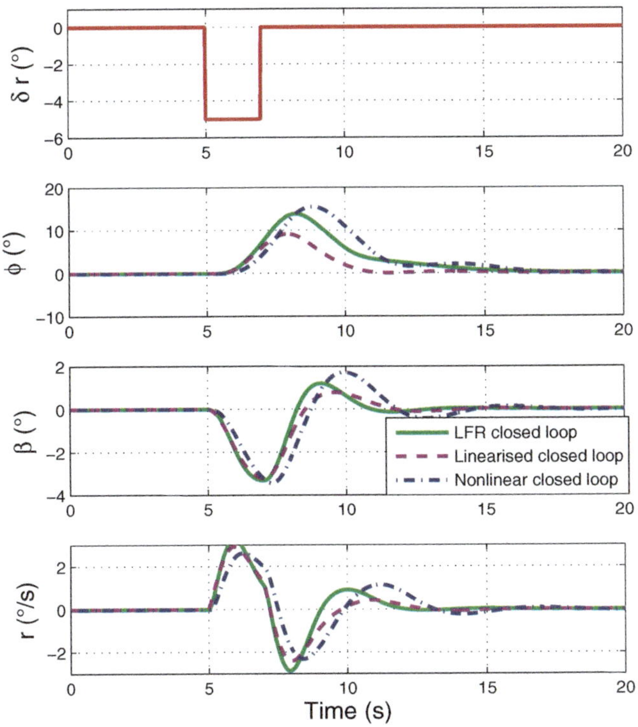

Fig. 5.14 Comparison of closed-loop time responses to a δr input at the flight point: $[V_{cas};\ M;\ m;\ X_{cg}] = [272.2\,kts;\ 0.72;\ 217\,t;\ 29\%]$. Solid: response of the LFR of the closed-loop; dashed: response of the linearised closed-loop; dash-dotted: response of the nonlinear closed-loop.

demonstrate the ability of the LFR framework to correctly represent the dynamic behaviour of such a nonlinear system. Fig. 5.14 presents for example the responses to a pulse on the pedal input δr of the LFR, of the linearised model and of the nonlinear model of the closed-loop.

5.6 Conclusions

An alternative LFR generation procedure based on the block-diagram structure of a given dynamic parameter-varying nonlinear system was introduced in this chapter. It is complementary to the procedures introduced in Chapters 3 and 4, as it keeps by its nature the block-diagram structure and hence the access to all block elements. All varying blocks and all nonlinearities are one by one replaced by their LFR counterparts before generating the overall LFR. The procedure is based on the Simulink extension [2] of the LFR toolbox [4] for use with MATLAB. This is particularly interesting for the generation of controller LFRs. If one or several blocks of the controller change during the industrial design process, it is not necessary to repeat the whole LFR generation procedure, but just to replace the LFRs of the new blocks. That is why this alternative LFR generation procedure has been highlighted on the LFR generation for the controller given in Chapter 2.

Parameter varying blocks are often given as look-up tables, which have to be expressed as rational or polynomial functions before linear fractional transformations can be applied. A polynomial interpolation method was proposed in order to come up with a polynomial with a reduced number of monomials ensuring a chosen precision level. This is very important for low size and precise LFR generation. The polynomial interpolation has already been used during the generation of aircraft LFRs [8, 9].

Finally, the closed-loop LFRs have been built using again the Simulink extension of the LFR toolbox. The LFRs for the nonlinear rigid aircraft of Chapter 3 or the LFRs for the linear flexible aircraft of Chapter 4 have been connected to the actuators, the sensors and the developed controller LFRs. The interconnection of several LFRs using the LFR objects blocks becomes straightforward. Furthermore, the interconnection structure can easily be modified for the various stability and performance analysis applications in Chapters 12–16.

In the near future, the Simulink extension and the LFR toolbox will both be further enhanced. In addition to the polynomial interpolation tools, some rational interpolation tools will be developed, always with the objective to reduce as much as possible the size of the resulting LFR.

References

1. Döll, C.: La Robustesse de Lois de Commande pour des Structures Flexibles. PhD Thesis, Ecole Nationale Supérieure de l'Aéronautique et de l'Espace, SUPAERO, BP 4032, F–31055 Toulouse, France (July 2001)
2. Biannic, J.M., Döll, C., Magni, J.F.: Simulink-based tools for creating and simulating interconnected LFRs. In: IEEE Symposium on Computer–Aided Control System Design, Munich, Germany (October 2006)
3. Morelli, E.A., De Loach, R.: Wind tunnel database development using modern experiment design and multivariate orthogonal functions. In: Proc. 41st AIAA Aerospace Sciences Meeting and Exhibit, Reno, Nevada, pp. 2653–2664 (January 2003)
4. Magni, J.F.: Linear Fractional Representation Toolbox Version 2.0. Technical Report DCSD–T n° 7/04, ONERA-DCSD, ONERA Centre de Toulouse, BP 4025, F–31055 Toulouse, France (January 2004)
5. Biannic, J.M., Tarbouriech, S., Farret, D.: A practical approach to performance analysis of saturated systems with application to fighter aircraft flight controllers. In: Proc. 5th IFAC Symposium on Robust Control Design ROCOND 2006, Toulouse, France (July 2006)

6. Roos, C., Biannic, J.M.: Aircraft–on–ground lateral control by an adaptive LFT–based anti–windup approach. In: IEEE Symposium on Computer–Aided Control System Design, Munich, Germany (October 2006)
7. Megretski, A., Rantzer, A.: System analysis via Integral Quadratic Constraints. IEEE Transactions on Automatic Control 42(6), 819–829 (1997)
8. Döll, C., Bérard, C., Knauf, A., Biannic, J.M.: LFT modelling of the 2–DOF longitudinal nonlinear aircraft behaviour. In: IEEE Symposium on Computer–Aided Control System Design, San Antonio, Texas (September 2008)
9. Döll, C., Bérard, C.: Modelling of the 3–DOF nonlinear aircraft behaviour via small standard forms. In: Proc. 26th ICAS Congress, Anchorage, Alaska, USA (September 2008)

Chapter 6
Identification of LPV State-Space Models Using $\mathcal{H}_2$-Minimisation

Daniel Petersson and Johan Löfberg

Abstract. Advanced robustness analysis methods employed in flight control law clearance such as IQC-analysis and μ-analysis, rely on linear fractional representations (LFRs). These models are usually obtained from linear parameter varying (LPV)-models which approximate the behaviour of the underlying parameter uncertain nonlinear aircraft model. The generation of LPV-models is usually done starting from a collection of linearised state-space models describing the local behaviour of the nonlinear aircraft for a representative set of parameter values and flight conditions. In this chapter we propose an optimisation-based generation method to convert these linear models into an LPV-model, by minimising a suitable $\mathcal{H}_2$ error norm. Although computationally more demanding than the alternative element-wise approximation approach, the new method often produces LPV-models with lower complexity.

6.1 Introduction

The behaviour of a linear parameter varying (LPV)-model can be described by

$$\begin{aligned}\dot{x}(t) &= \mathbf{A}(\delta(t))x(t) + \mathbf{B}(\delta(t))u(t),\\ y(t) &= \mathbf{C}(\delta(t))x(t) + \mathbf{D}(\delta(t))u(t)\end{aligned}$$

where $x(t)$ are the states, $u(t)$ and $y(t)$ are the input and output signals and $\delta(t)$ is the vector of model parameters. In flight control applications, the components of $\delta(t)$ are typically mass, position of centre of gravity and various aerodynamic coefficients, but can also include state dependent parameters

Daniel Petersson · Johan Löfberg
Division of Automatic Control, Linköping University, SE-581 83 Linköping, Sweden
e-mail: petersson@isy.liu.se, johanl@isy.liu.se

A. Varga et al. (Eds.): Opti. Based Clearance of Flight Control Laws, LNCIS 416, pp. 111–128.
springerlink.com

such as altitude and velocity, specifying current flight conditions. In this chapter we will study the case when the parameters vary slowly and not take time dependence of the parameters into account.

Some advanced robustness analysis methods such as IQC-analysis and μ-analysis, see, *e.g.*, [1] and Chapters 7, 9 and 10, require a conversion of the LPV-model into a linear fractional representation (LFR). For this purpose, it is necessary that the parametric matrices $\mathbf{A}(\delta)$, $\mathbf{B}(\delta)$, $\mathbf{C}(\delta)$, $\mathbf{D}(\delta)$ of the LPV-model are rational in δ. This requirement is often violated in LPV-models generated directly from a nonlinear model description, either due to presence of nonlinear parametric expressions or tabulated data in the model. In both cases, rational approximations must be used to obtain a suitable model.

The LPV-models in this chapter are generated by starting from a multi-model system in state-space form

$$G(\delta^{(i)}) = \left[\begin{array}{c|c} \mathbf{A}(\delta^{(i)}) & \mathbf{B}(\delta^{(i)}) \\ \hline \mathbf{C}(\delta^{(i)}) & \mathbf{D}(\delta^{(i)}) \end{array}\right]$$

where each model, $G(\delta^{(i)})$, corresponds to a model sampled in the point $\delta^{(i)}$, for $i = 1, \ldots, N$. The goal is to approximate this multi-model system with a single LPV-model

$$\hat{G}(\delta) = \left[\begin{array}{c|c} \hat{\mathbf{A}}(\delta) & \hat{\mathbf{B}}(\delta) \\ \hline \hat{\mathbf{C}}(\delta) & \hat{\mathbf{D}}(\delta) \end{array}\right]$$

whose state-space realisation depends polynomially on δ, and $G(\delta^{(i)})$ and $\hat{G}(\delta)$ are transfer-function matrices. A frequently used method today is element-wise approximation see, *e.g.*, [2] and Chapter 3. This method interpolates the elements in the system matrices individually with rational or polynomial functions. A possible drawback with this approach is that it fails to take system properties into account, such as input-output relations or the frequency response. Additionally, a prerequisite for the application of this method is that the number of states is the same in all the models and the matrices correspond to the same ordering of states. Other methods that also use interpolation are, *e.g.*, [3], [4], but they transform the models into canonical state-space forms before doing the interpolation. There are also methods that address the input-output relation and try to identify an LPV-model using, *e.g.*, linear regression [5] or nonlinear optimisation [6]. An excellent survey over existing methods can be found in [7].

In this chapter we first formulate an optimisation problem to find an LPV-model that approximates the multi-model system and captures the input-output behaviour as measured by the $\mathcal{H}_2$-norm, [1]. This method was first introduced in [8] and is now extended here. We then present two approaches to solve the optimisation problem that arises. The first of these two methods is then used to create low-order LPV-models of an aircraft, which ultimately are used to generate low-order LFRs.

Common measures to characterise a dynamical system, such as the $\mathcal{H}_2$- or $\mathcal{H}_\infty$-norm, are candidates to use when formulating the problem. One reason for choosing the $\mathcal{H}_2$-norm and not the $\mathcal{H}_\infty$-norm is that the cost function in the optimisation problem becomes differentiable in the optimisation variables and any standard gradient based solvers can be used to perform the optimisation.

6.2 $\mathcal{H}_2$-Minimisation

In this section we formulate an optimisation problem that arise when trying to approximate a multi-model system, consisting of N local models, with an LPV-model. The optimisation problem is formulated such that the sought model should capture input-output behaviour of the multi-model system. The objective is to minimise the error between the true models and the sought LPV-model in the sampled points, in the $\mathcal{H}_2$-norm, *i.e.*, we formulate the optimisation problem

$$\min_{\hat{\mathbf{A}},\hat{\mathbf{B}},\hat{\mathbf{C}},\hat{\mathbf{D}}} \sum_{i=1}^{N} \left|\left| G_i - \hat{G}(\delta^{(i)}) \right|\right|_{\mathcal{H}_2}^2 = \min_{\hat{\mathbf{A}},\hat{\mathbf{B}},\hat{\mathbf{C}},\hat{\mathbf{D}}} V \tag{6.1}$$

where

$$G_i = \left[\begin{array}{c|c} \mathbf{A}_i & \mathbf{B}_i \\ \hline \mathbf{C}_i & \mathbf{D}_i \end{array}\right]$$

are the sampled (given) models and

$$\hat{G}(\delta) = \left[\begin{array}{c|c} \hat{\mathbf{A}}(\delta) & \hat{\mathbf{B}}(\delta) \\ \hline \hat{\mathbf{C}}(\delta) & \hat{\mathbf{D}}(\delta) \end{array}\right]$$

is the LPV-model depending on the parameter δ. We will in this chapter assume that the system matrices in the LPV-model depends polynomially on the parameters, *e.g.*, when we have one parameter

$$\hat{\mathbf{A}}(\delta) = \hat{\mathbf{A}}_{\delta^0} + \hat{\mathbf{A}}_{\delta^1}\delta + \cdots + \hat{\mathbf{A}}_{\delta^{k_A}}\delta^{k_A} \tag{6.2a}$$

$$\hat{\mathbf{B}}(\delta) = \hat{\mathbf{B}}_{\delta^0} + \hat{\mathbf{B}}_{\delta^1}\delta + \cdots + \hat{\mathbf{B}}_{\delta^{k_B}}\delta^{k_B} \tag{6.2b}$$

$$\hat{\mathbf{C}}(\delta) = \hat{\mathbf{C}}_{\delta^0} + \hat{\mathbf{C}}_{\delta^1}\delta + \cdots + \hat{\mathbf{C}}_{\delta^{k_C}}\delta^{k_C}. \tag{6.2c}$$

where $\hat{\mathbf{A}}_{\delta^j}$ are coefficient matrices and δ^j are monomials of δ. We start by looking at the models in one sample point and omit the index i. Later this will be generalised to the case where we have multiple models.

Define the error systems as

$$E = G - \hat{G},$$

this system can be realised in state-space form as

$$E = \left[\begin{array}{c|c} \mathbf{A}_e & \mathbf{B}_e \\ \hline \mathbf{C}_e & \mathbf{D}_e \end{array}\right] = \left[\begin{array}{c|c} \begin{pmatrix} \mathbf{A} & \mathbf{0} \\ \mathbf{0} & \hat{\mathbf{A}} \end{pmatrix} & \begin{pmatrix} \mathbf{B} \\ \hat{\mathbf{B}} \end{pmatrix} \\ \hline \begin{pmatrix} \mathbf{C} & -\hat{\mathbf{C}} \end{pmatrix} & \mathbf{D} - \hat{\mathbf{D}} \end{array}\right]. \tag{6.3}$$

This realisation of the error system will later prove beneficial in rewriting the optimisation problem. Notice that for a continuous time model the $\mathcal{H}_2$-norm is unbounded if the model is not strictly proper, *i.e.*, we need $\mathbf{D} = \hat{\mathbf{D}}$ for all models or that both $\mathbf{D} = \mathbf{0}$ and $\hat{\mathbf{D}} = \mathbf{0}$. We can thus see the problem of finding an approximation $\hat{\mathbf{D}}$ as a separate problem which we do not address in this chapter. We will also throughout the chapter assume that the models we are given are stable, otherwise the $\mathcal{H}_2$-norm will not be defined.

6.2.1 Important Property of $\mathcal{H}_2$-Minimisation

The idea of the method is to capture the input-output behaviour of the model. A benefit with the method, is that even though some elements can depend non-polynomially or as a high order polynomial on the parameters, it can exploit the fact that the realisation is non-unique. The model $G(\delta) = \left[\begin{array}{c|c} \mathbf{A}(\delta) & \mathbf{B}(\delta) \\ \hline \mathbf{C}(\delta) & \mathbf{D}(\delta) \end{array}\right]$ has the same transfer function and input-output behaviour as $G_{\mathbf{T}}(\delta)$, with

$$G_{\mathbf{T}}(\delta) = \left[\begin{array}{c|c} \mathbf{A}_{\mathbf{T}}(\delta) & \mathbf{B}_{\mathbf{T}}(\delta) \\ \hline \mathbf{C}_{\mathbf{T}}(\delta) & \mathbf{D}(\delta) \end{array}\right] = \left[\begin{array}{c|c} \mathbf{T}(\delta)\mathbf{A}(\delta)\mathbf{T}(\delta)^{-1} & \mathbf{T}(\delta)\mathbf{B}(\delta) \\ \hline \mathbf{C}(\delta)\mathbf{T}(\delta)^{-1} & \mathbf{D}(\delta) \end{array}\right].$$

where $\mathbf{T}(\delta)$ is a non-singular transformation matrix that can depend on δ. This means that for every model G_i we are not only limited to find the best approximation between $G_i = \left[\begin{array}{c|c} \mathbf{A}_i & \mathbf{B}_i \\ \hline \mathbf{C}_i & \mathbf{D}_i \end{array}\right]$ and $\hat{G}(\delta^{(i)}) = \left[\begin{array}{c|c} \hat{\mathbf{A}}(\delta^{(i)}) & \hat{\mathbf{B}}(\delta^{(i)}) \\ \hline \hat{\mathbf{C}}(\delta^{(i)}) & \hat{\mathbf{D}}(\delta^{(i)}) \end{array}\right]$, but to find the match between $G_i = \left[\begin{array}{c|c} \mathbf{T}_i\mathbf{A}_i\mathbf{T}_i^{-1} & \mathbf{T}_i\mathbf{B}_i \\ \hline \mathbf{C}_i\mathbf{T}_i^{-1} & \mathbf{D}_i \end{array}\right]$ and $\hat{G}(\delta^{(i)}) = \left[\begin{array}{c|c} \hat{\mathbf{A}}(\delta^{(i)}) & \hat{\mathbf{B}}(\delta^{(i)}) \\ \hline \hat{\mathbf{C}}(\delta^{(i)}) & \hat{\mathbf{D}}(\delta^{(i)}) \end{array}\right]$. We illustrate this by an example.

Example 6.1. Assume we are given samples from the LPV-model

$$\mathbf{A}(\delta) = \begin{pmatrix} 0.40\delta^2 + 3\delta - 3.6 & -\frac{0.40(\delta^3 - 24\delta - 40)}{\delta} & \frac{0.20(27\delta^3 + 55\delta^2 + 37\delta - 160.)}{\delta} \\ 0.40\delta^2 + 3.6\delta - 3.2 & -\frac{0.20(2\delta^3 + 3\delta^2 - 46\delta - 10)}{\delta} & \frac{0.20(27\delta^3 + 23\delta^2 - 96\delta - 20)}{\delta} \\ 1.6\delta - 1.6 & -\frac{0.20(8\delta^2 - 33\delta - 5)}{\delta} & \frac{0.20(23\delta^2 - 68\delta - 10)}{\delta} \end{pmatrix}$$

$$\mathbf{B}(\delta) = \begin{pmatrix} 8 + 7\delta + \delta^2 \\ 6 + 2\delta + \delta^2 \\ 3 \end{pmatrix}$$

$$\mathbf{C}(\delta) = \begin{pmatrix} 0.20 + 0.20\delta & -\frac{0.20(-9\delta + \delta^2 - 10)}{\delta} & -\frac{0.80(-1\delta + 4\delta^2 - 5)}{\delta} \end{pmatrix}$$

$$\mathbf{D}(\delta) = 0$$

It would be difficult to use an element-wise method with low order polynomials to identify this LPV-model due to the rational functions. However, a different realisation of this model is given by

$$\mathbf{A_T}(\delta) = \mathbf{T}(\delta)\mathbf{A}(\delta)\mathbf{T}^{-1}(\delta) = \begin{pmatrix} -2 + \delta & 3 + \delta & 5 + 2\delta \\ 2 + 2\delta & -4 + 3\delta & 1 + 5\delta \\ -8 + 8\delta & 1 + 5\delta & -2 + 3\delta \end{pmatrix}$$

$$\mathbf{B_T}(\delta) = \mathbf{T}(\delta)\mathbf{B}(\delta) = \begin{pmatrix} 1 + \delta \\ 2 + \delta \\ 3 \end{pmatrix}$$

$$\mathbf{C_T}(\delta) = \mathbf{C}(\delta)\mathbf{T}^{-1}(\delta) = \begin{pmatrix} 1 + \delta & 2 + 2\delta & 3 + 3\delta \end{pmatrix}$$

$$\mathbf{D_T}(\delta) = \mathbf{D}(\delta) = 0$$

$$\mathbf{T}(\delta) = \begin{pmatrix} 0.2 & -0.2 & 0.2 \\ 0 & \frac{1}{\delta} & -\frac{2}{\delta} \\ 0 & 0 & 1 \end{pmatrix}$$

obviously, this model is affine in δ.

The example illustrates the important property that when searching in the class of LPV-models with low dependence on δ, it is sometimes possible to find an equivalent model with respect to the input-output relation, even though the given model depends non-polynomially on δ. One important thing to note is that the **D**-matrix is not affected by a state transformation, thus state transformations do not help to simplify the dependence of δ in **D**.

6.2.2 Rewriting the $\mathcal{H}_2$-Norm of the Error System

By definition the $\mathcal{H}_2$-norm of a continuous LTI system, see [1], is

$$||E||^2_{\mathcal{H}_2} = \text{tr}\left[\frac{1}{2\pi}\int_{-\infty}^{\infty} E(j\omega)E(j\omega)^H d\omega\right]. \tag{6.4}$$

To calculate the cost function efficiently we rewrite (6.4) to a numerically more suitable form. This can be done using the Gramians for the system, see [1]. The observability and controllability Gramians, $\mathbf{Q}_e$ and $\mathbf{P}_e$ respectively, for the error system, E, are defined as

$$\mathbf{P}_e = \int_0^\infty e^{\mathbf{A}_e t}\mathbf{B}_e (e^{\mathbf{A}_e t}\mathbf{B}_e)^T dt, \tag{6.5a}$$

$$\mathbf{Q}_e = \int_0^\infty (\mathbf{C}_e e^{\mathbf{A}_e t})^T \mathbf{C}_e e^{\mathbf{A}_e t} dt \tag{6.5b}$$

where $\mathbf{P}_e$ and $\mathbf{Q}_e$ satisfy the Lyapunov equations, see [1],

$$\mathbf{A}_e\mathbf{P}_e + \mathbf{P}_e\mathbf{A}_e^T + \mathbf{B}_e\mathbf{B}_e^T = \mathbf{0}, \tag{6.6a}$$

$$\mathbf{A}_e^T\mathbf{Q}_e + \mathbf{Q}_e\mathbf{A}_e + \mathbf{C}_e^T\mathbf{C}_e = \mathbf{0}. \tag{6.6b}$$

Now we use Parseval's identity to rewrite (6.4) as

$$\begin{aligned} ||E||_{\mathcal{H}_2}^2 &= \operatorname{tr}\left[\int_0^\infty \left(\mathbf{C}_e e^{\mathbf{A}_e t}\mathbf{B}_e\right)\left(\mathbf{C}_e e^{\mathbf{A}_e t}\mathbf{B}_e\right)^T dt\right] = \\ &= \operatorname{tr}\left[\int_0^\infty \left(\mathbf{C}_e e^{\mathbf{A}_e t}\mathbf{B}_e\right)^T\left(\mathbf{C}_e e^{\mathbf{A}_e t}\mathbf{B}_e\right) dt\right]. \end{aligned}$$

By using (6.5) and (6.6) it is possible to rewrite the cost function as

$$||E||_{\mathcal{H}_2}^2 = \operatorname{tr}\mathbf{B}_e^T\mathbf{Q}_e\mathbf{B}_e = \operatorname{tr}\mathbf{C}_e\mathbf{P}_e\mathbf{C}_e^T. \tag{6.7}$$

6.3 Method 1: General Nonlinear Optimisation

In this section, we try to solve the optimisation problem as a general nonlinear optimisation problem, for which we derive expressions for the cost function and the gradient which can be efficiently evaluated.

For the method proposed in this section, we have derive efficiently computable expressions for both the cost function and its gradient. To solve the optimisation problem any gradient based optimisation scheme can be used to actually perform the numerical search for a local optimum [12]. Hence, to test the efficiency of the proposed algorithm, essentially any available commercial or open-source solver can be used. It is thus beyond the scope of this chapter to give any details on how a complete solver is implemented.

6.3.1 Evaluation of the Cost Function

With the realisation (6.3) of E and equations (6.6), if we partition the Gramians $\mathbf{P}_e$ and $\mathbf{Q}_e$ as

$$\mathbf{P}_e = \begin{pmatrix} \mathbf{P} & \mathbf{X} \\ \mathbf{X}^T & \hat{\mathbf{P}} \end{pmatrix}, \quad \mathbf{Q}_e = \begin{pmatrix} \mathbf{Q} & \mathbf{Y} \\ \mathbf{Y}^T & \hat{\mathbf{Q}} \end{pmatrix}$$

we obtain six Sylvester and Lyapunov equations

$$\mathbf{AP} + \mathbf{PA}^T + \mathbf{BB}^T = \mathbf{0}, \tag{6.8a}$$
$$\mathbf{AX} + \mathbf{X}\hat{\mathbf{A}}^T + \mathbf{B}\hat{\mathbf{B}}^T = \mathbf{0}, \tag{6.8b}$$
$$\hat{\mathbf{A}}\hat{\mathbf{P}} + \hat{\mathbf{P}}\hat{\mathbf{A}}^T + \hat{\mathbf{B}}\hat{\mathbf{B}}^T = \mathbf{0}, \tag{6.8c}$$
$$\mathbf{A}^T\mathbf{Q} + \mathbf{QA} + \mathbf{C}^T\mathbf{C} = \mathbf{0}, \tag{6.8d}$$
$$\mathbf{A}^T\mathbf{Y} + \mathbf{Y}\hat{\mathbf{A}} - \mathbf{C}^T\hat{\mathbf{C}} = \mathbf{0}, \tag{6.8e}$$
$$\hat{\mathbf{A}}^T\hat{\mathbf{Q}} + \hat{\mathbf{Q}}\hat{\mathbf{A}} + \hat{\mathbf{C}}^T\hat{\mathbf{C}} = \mathbf{0}. \tag{6.8f}$$

We note that $\mathbf{P}$ and $\mathbf{Q}$ satisfy the Lyapunov equations for the controllability and the observability Gramians for the given system, while $\hat{\mathbf{P}}$ and $\hat{\mathbf{Q}}$ satisfy the Lyapunov equations for the controllability and the observability Gramians for the sought system. With the partitioning of $\mathbf{P}_e$ and $\mathbf{Q}_e$ it is possible to rewrite the cost function, (6.7), as

$$||E||^2_{\mathcal{H}_2} = \mathrm{tr}\left(\mathbf{B}^T\mathbf{QB} + 2\mathbf{B}^T\mathbf{Y}\hat{\mathbf{B}} + \hat{\mathbf{B}}^T\hat{\mathbf{Q}}\hat{\mathbf{B}}\right) \tag{6.9a}$$

$$||E||^2_{\mathcal{H}_2} = \mathrm{tr}\left(\mathbf{CPC}^T - 2\mathbf{CX}\hat{\mathbf{C}}^T + \hat{\mathbf{C}}\hat{\mathbf{P}}\hat{\mathbf{C}}^T\right). \tag{6.9b}$$

The two equations in (6.9) are equivalent and are both useful to simplify the derivations for the gradients later. However, at this point we will only use (6.9a) to calculate the cost function. It is now straightforward to express the cost function for the more general case when we have multiple models, *i.e.*, rewrite the cost function V in (6.1) with the new partitioning

$$V = \sum_i ||E_i||^2_{\mathcal{H}_2} = \sum_i \mathrm{tr}\Big(\mathbf{B}_i^T\mathbf{Q}_i\mathbf{B}_i + 2\mathbf{B}_i^T\mathbf{Y}_i\hat{\mathbf{B}}(\delta^{(i)}) + \\ + \hat{\mathbf{B}}(\delta^{(i)})^T\hat{\mathbf{Q}}_i\hat{\mathbf{B}}(\delta^{(i)})\Big). \tag{6.10}$$

The optimisation problem (6.1) can now be written as

$$\min_{\hat{\mathbf{A}}_{\delta k}, \hat{\mathbf{B}}_{\delta k}, \hat{\mathbf{C}}_{\delta k}} \sum_i \mathrm{tr}\left(\mathbf{B}_i^T\mathbf{Q}_i\mathbf{B}_i + 2\mathbf{B}_i^T\mathbf{Y}_i\hat{\mathbf{B}}(\delta^{(i)}) + \hat{\mathbf{B}}(\delta^{(i)})^T\hat{\mathbf{Q}}_i\hat{\mathbf{B}}(\delta^{(i)})\right). \tag{6.11}$$

Keep in mind the parametrisation of the system matrices introduced in (6.2). Additionally, $\mathbf{P}_i, \mathbf{Q}_i, \hat{\mathbf{P}}_i, \hat{\mathbf{Q}}_i, \mathbf{X}_i$ and $\mathbf{Y}_i$ satisfy the equations

$$\mathbf{A}_i\mathbf{P}_i + \mathbf{P}_i\mathbf{A}_i^T + \mathbf{B}_i\mathbf{B}_i^T = \mathbf{0}, \tag{6.12a}$$

$$\mathbf{A}_i\mathbf{X}_i + \mathbf{X}_i\hat{\mathbf{A}}(\delta^{(i)})^T + \mathbf{B}_i\hat{\mathbf{B}}(\delta^{(i)})^T = \mathbf{0}, \tag{6.12b}$$

$$\hat{\mathbf{A}}(\delta^{(i)})\hat{\mathbf{P}}_i + \hat{\mathbf{P}}_i\hat{\mathbf{A}}(\delta^{(i)})^T + \hat{\mathbf{B}}(\delta^{(i)})\hat{\mathbf{B}}(\delta^{(i)})^T = \mathbf{0}, \tag{6.12c}$$

$$\mathbf{A}_i^T\mathbf{Q}_i + \mathbf{Q}_i\mathbf{A}_i + \mathbf{C}_i^T\mathbf{C}_i = \mathbf{0}, \tag{6.12d}$$

$$\mathbf{A}_i^T\mathbf{Y}_i + \mathbf{Y}_i\hat{\mathbf{A}}(\delta^{(i)}) - \mathbf{C}_i^T\hat{\mathbf{C}}(\delta^{(i)}) = \mathbf{0}, \tag{6.12e}$$

$$\hat{\mathbf{A}}(\delta^{(i)})^T\hat{\mathbf{Q}}_i + \hat{\mathbf{Q}}_i\hat{\mathbf{A}}(\delta^{(i)}) + \hat{\mathbf{C}}(\delta^{(i)})^T\hat{\mathbf{C}}(\delta^{(i)}) = \mathbf{0}. \tag{6.12f}$$

The cost function to the optimisation problem (6.11) is now expressed in the sought variables $\hat{\mathbf{A}}, \hat{\mathbf{B}}, \hat{\mathbf{C}}$, the given data $\mathbf{A}_i, \mathbf{B}_i, \mathbf{C}_i$ and in the different partitions of the Gramians for the error system, *i.e.*, the solutions to the equations in (6.12) that can easily be calculated.

6.3.2 Evaluation of the Gradient

An appealing feature of the proposed nonlinear optimisation approach to solve the problem is that the equations (6.9) are differentiable in the system matrices, $\hat{\mathbf{A}}, \hat{\mathbf{B}}$ and $\hat{\mathbf{C}}$ (see [9, 10]). In addition, the closed form expression you obtain when differentiating the cost function is expressed in the given data ($\mathbf{A}, \mathbf{B}$ and $\mathbf{C}$), the optimisation variables ($\hat{\mathbf{A}}, \hat{\mathbf{B}}$ and $\hat{\mathbf{C}}$) and solutions to equations (6.12), some of them already calculated when calculating the cost function, so you get them for free. To be more precise, the computational effort of computing the derivative is only an additional constant cost once the cost function is calculated.

We start by calculating the gradient with respect to $\hat{\mathbf{A}}$, *i.e.*, $\frac{\partial ||E||^2_{\mathcal{H}_2}}{\partial \hat{\mathbf{A}}}$ where $\left[\frac{\partial ||E||^2_{\mathcal{H}_2}}{\partial \hat{\mathbf{A}}}\right]_{ij} = \frac{\partial ||E||^2_{\mathcal{H}_2}}{\partial \hat{a}_{ij}}$ and $\hat{a}_{ij}$ are the individual elements in $\hat{\mathbf{A}}$. Again we start by looking at the case when we only have one given system and then later extend the result to multiple models and an LPV-model with polynomial dependence in the parameters. $\hat{\mathbf{Q}}$ and $\mathbf{Y}$ depend on $\hat{\mathbf{A}}$ which we need to keep in mind when differentiating (6.9a) with respect to $\hat{\mathbf{A}}$. Hence, $\left[\frac{\partial ||E||^2_{\mathcal{H}_2}}{\partial \hat{\mathbf{A}}}\right]_{ij}$ becomes

$$\left[\frac{\partial \, ||E||^2_{\mathcal{H}_2}}{\partial \hat{\mathbf{A}}}\right]_{ij} = \operatorname{tr}\left(2\hat{\mathbf{B}}\mathbf{B}^T\frac{\partial \mathbf{Y}}{\partial \hat{a}_{ij}} + \hat{\mathbf{B}}\hat{\mathbf{B}}^T\frac{\partial \hat{\mathbf{Q}}}{\partial \hat{a}_{ij}}\right) \tag{6.13}$$

where $\frac{\partial \mathbf{Y}}{\partial \hat{a}_{ij}}$ and $\frac{\partial \hat{\mathbf{Q}}}{\partial \hat{a}_{ij}}$ depend on $\hat{\mathbf{A}}$ via the differentiated versions of equations (6.8d) and (6.8e)

$$\mathbf{A}^T \frac{\partial \mathbf{Y}}{\partial \hat{a}_{ij}} + \frac{\partial \mathbf{Y}}{\partial \hat{a}_{ij}} \hat{\mathbf{A}} + \mathbf{Y} \frac{\partial \hat{\mathbf{A}}}{\partial \hat{a}_{ij}} = \mathbf{0}, \tag{6.14a}$$

$$\hat{\mathbf{A}}^T \frac{\partial \hat{\mathbf{Q}}}{\partial \hat{a}_{ij}} + \frac{\partial \hat{\mathbf{Q}}}{\partial \hat{a}_{ij}} \hat{\mathbf{A}} + \frac{\partial \hat{\mathbf{A}}^T}{\partial \hat{a}_{ij}} \hat{\mathbf{Q}} + \hat{\mathbf{Q}} \frac{\partial \hat{\mathbf{A}}}{\partial \hat{a}_{ij}} = \mathbf{0}. \tag{6.14b}$$

To simplify the calculations of the gradient we need the following Lemma (see [11]).

Lemma 6.1. *If* $\mathbf{M}$ *and* $\mathbf{N}$ *satisfies the following Sylvester equations*

$$\mathbf{AM} + \mathbf{MB} + \mathbf{C} = \mathbf{0}, \quad \mathbf{NA} + \mathbf{BN} + \mathbf{D} = \mathbf{0}$$

then $\operatorname{tr} \mathbf{CN} = \operatorname{tr} \mathbf{DM}$.

Now applying Lemma 6.1 on (6.8) and (6.14) yields

$$\operatorname{tr} \hat{\mathbf{B}} \mathbf{B}^T \frac{\partial \mathbf{Y}}{\partial \hat{a}_{ij}} = \operatorname{tr} \mathbf{X}^T \mathbf{Y} \frac{\partial \hat{\mathbf{A}}}{\partial \hat{a}_{ij}},$$

$$\operatorname{tr} \hat{\mathbf{B}} \hat{\mathbf{B}}^T \frac{\partial \hat{\mathbf{Q}}}{\partial \hat{a}_{ij}} = \operatorname{tr} \hat{\mathbf{P}} \left(\frac{\partial \hat{\mathbf{A}}^T}{\partial \hat{a}_{ij}} \hat{\mathbf{Q}} + \hat{\mathbf{Q}} \frac{\partial \hat{\mathbf{A}}}{\partial \hat{a}_{ij}} \right).$$

Inserting this in (6.13) yields

$$\left[\frac{\partial \, ||E||^2_{\mathcal{H}_2}}{\partial \hat{\mathbf{A}}} \right]_{ij} = \operatorname{tr} \left(2\mathbf{X}^T \mathbf{Y} \frac{\partial \hat{\mathbf{A}}}{\partial \hat{a}_{ij}} + \hat{\mathbf{P}} \left(\frac{\partial \hat{\mathbf{A}}^T}{\partial \hat{a}_{ij}} \hat{\mathbf{Q}} + \hat{\mathbf{Q}} \frac{\partial \hat{\mathbf{A}}}{\partial \hat{a}_{ij}} \right) \right) =$$
$$= 2 \operatorname{tr} \left(\frac{\partial \hat{\mathbf{A}}^T}{\partial \hat{a}_{ij}} \left(\hat{\mathbf{Q}} \hat{\mathbf{P}} + \mathbf{Y}^T \mathbf{X} \right) \right).$$

It follows that $\frac{\partial ||E||^2_{\mathcal{H}_2}}{\partial \hat{\mathbf{A}}} = 2 \left(\hat{\mathbf{Q}} \hat{\mathbf{P}} + \mathbf{Y}^T \mathbf{X} \right)$. Analogously we can calculate the gradients with respect to $\hat{\mathbf{B}}$ and $\hat{\mathbf{C}}$.

$$\frac{\partial \, ||E||^2_{\mathcal{H}_2}}{\partial \hat{\mathbf{A}}} = 2 \left(\hat{\mathbf{Q}} \hat{\mathbf{P}} + \mathbf{Y}^T \mathbf{X} \right) \tag{6.15a}$$

$$\frac{\partial \, ||E||^2_{\mathcal{H}_2}}{\partial \hat{\mathbf{B}}} = 2 \left(\hat{\mathbf{Q}} \hat{\mathbf{B}} + \mathbf{Y}^T \mathbf{B} \right) \tag{6.15b}$$

$$\frac{\partial \, ||E||^2_{\mathcal{H}_2}}{\partial \hat{\mathbf{C}}} = 2 \left(\hat{\mathbf{C}} \hat{\mathbf{P}} - \mathbf{C} \mathbf{X} \right) \tag{6.15c}$$

If we now extend this to the more general form when given multiple models and when the LPV-model has polynomial dependence in the parameters, *e.g.* $\hat{\mathbf{A}}(\delta) = \hat{\mathbf{A}}_{\delta^0} + \hat{\mathbf{A}}_{\delta^1} \delta + \hat{\mathbf{A}}_{\delta^2} \delta^2 + \cdots + \hat{\mathbf{A}}_{\delta^k} \delta^k$, then the gradient of (6.10) with respect to the coefficient matrices $\hat{\mathbf{A}}_{\delta^j}, \hat{\mathbf{B}}_{\delta^j}, \hat{\mathbf{C}}_{\delta^j}$ becomes

$$\frac{\partial V}{\partial \hat{\mathbf{A}}_{\delta^j}} = \frac{\partial\left(\sum_i ||E_i||^2_{\mathcal{H}_2}\right)}{\partial \hat{\mathbf{A}}_{\delta^j}} = 2\sum_i \left(\delta^{(i)}\right)^j \left(\hat{\mathbf{Q}}_i\hat{\mathbf{P}}_i + \mathbf{Y}_i^T\mathbf{X}_i\right)$$

$$\frac{\partial V}{\partial \hat{\mathbf{B}}_{\delta^j}} = \frac{\partial\left(\sum_i ||E_i||^2_{\mathcal{H}_2}\right)}{\partial \hat{\mathbf{B}}_{\delta^j}} = 2\sum_i \left(\delta^{(i)}\right)^j \left(\hat{\mathbf{Q}}_i\hat{\mathbf{B}}_i + \mathbf{Y}_i^T\mathbf{B}_i\right)$$

$$\frac{\partial V}{\partial \hat{\mathbf{C}}_{\delta^j}} = \frac{\partial\left(\sum_i ||E_i||^2_{\mathcal{H}_2}\right)}{\partial \hat{\mathbf{C}}_{\delta^j}} = 2\sum_i \left(\delta^{(i)}\right)^j \left(\hat{\mathbf{C}}_i\hat{\mathbf{P}}_i - \mathbf{C}_i\mathbf{X}_i\right).$$

Remark: To calculate the cost function (6.10), three Lyapunov/Sylvester equations (6.12d,6.12e,6.12f) need to be solved for every i and iteration. Notice that $\mathbf{Q}_i$ in (6.12d) can be precomputed before the algorithm starts. Crucial to notice is that the extra cost to compute the gradient is merely to solve two additional Lyapunov/Sylvester equations (6.12b,6.12c).

In addition, looking at equations (6.12) we see that they all have $\mathbf{A}_i$ and $\hat{\mathbf{A}}_i$ as factors in them, which can be exploited by the Lyapunov solver to solve all of these equations efficiently.

6.4 Method 2: Semidefinite Programming

In this section we try to minimise the error system using a semidefinite programming approach. With the realisation (6.3) and equations (6.6) and (6.7) we rewrite the $\mathcal{H}_2$-norm for a system as a minimisation problem

$$\begin{aligned} &\min_{\mathbf{Q}} \operatorname{tr} \mathbf{B}_e^T\mathbf{Q}\mathbf{B}_e \\ &\text{s.t.} \quad \mathbf{A}_e^T\mathbf{Q} + \mathbf{Q}\mathbf{A}_e + \mathbf{C}_e^T\mathbf{C}_e = \mathbf{0}, \quad \mathbf{Q} \succ 0 \end{aligned}$$

This can be rewritten, using a Schur complement (see [13]), to get an equivalent problem

$$\begin{aligned} &\min_{\gamma,\mathbf{Q}} \gamma \\ &\text{s.t.} \quad \begin{pmatrix} \mathbf{Q}\mathbf{A}_e + \mathbf{A}_e^T\mathbf{Q} & \mathbf{C}_e^T \\ \mathbf{C}_e & -\mathrm{I} \end{pmatrix} \prec 0, \quad \begin{pmatrix} \gamma\mathrm{I} & \mathbf{B}_e^T\mathbf{Q} \\ \mathbf{Q}\mathbf{B}_e & \mathbf{Q} \end{pmatrix} \succ 0. \end{aligned} \tag{6.16}$$

Our objective is to find the system $\hat{G} = \left[\begin{array}{c|c} \hat{\mathbf{A}} & \hat{\mathbf{B}} \\ \hline \hat{\mathbf{C}} & \mathbf{0} \end{array}\right]$ that approximates the system $G = \left[\begin{array}{c|c} \mathbf{A} & \mathbf{B} \\ \hline \mathbf{C} & \mathbf{0} \end{array}\right]$ well in the $\mathcal{H}_2$-norm, *i.e.*, we also want to minimise over the matrices $\hat{\mathbf{A}}, \hat{\mathbf{B}}, \hat{\mathbf{C}}$.

Since the system E is the augmented system

$$E = \left[\begin{array}{c|c} \mathbf{A}_e & \mathbf{B}_e \\ \hline \mathbf{C}_e & \mathbf{0} \end{array}\right] = \left[\begin{array}{c|c} \begin{pmatrix} \mathbf{A} & \mathbf{0} \\ \mathbf{0} & \hat{\mathbf{A}} \end{pmatrix} & \begin{pmatrix} \mathbf{B} \\ \hat{\mathbf{B}} \end{pmatrix} \\ \hline \begin{pmatrix} \mathbf{C} & -\hat{\mathbf{C}} \end{pmatrix} & \mathbf{0} \end{array}\right]$$

and if we partition $\mathbf{Q}$ as

$$\mathbf{Q} = \begin{pmatrix} \mathbf{Q}_{11} & \mathbf{Q}_{12} \\ \mathbf{Q}_{12}^T & \mathbf{Q}_{22} \end{pmatrix}$$

we can rewrite the minimisation problem (6.16) as a new problem, now also optimising over the sought system matrices, as

$$\begin{aligned} &\min_{\gamma,\mathbf{Q},\hat{\mathbf{A}},\hat{\mathbf{B}},\hat{\mathbf{C}}} \quad \gamma \\ &\text{s.t.} \quad \begin{pmatrix} \mathbf{Q}_{11}\mathbf{A} + \mathbf{A}^T\mathbf{Q}_{11} & \mathbf{Q}_{12}\hat{\mathbf{A}} + \mathbf{A}^T\mathbf{Q}_{12} & \mathbf{C}^T \\ \mathbf{Q}_{12}^T\mathbf{A} + \hat{\mathbf{A}}^T\mathbf{Q}_{12} & \mathbf{Q}_{22}\hat{\mathbf{A}} + \hat{\mathbf{A}}^T\mathbf{Q}_{22} & -\hat{\mathbf{C}}^T \\ \mathbf{C} & -\hat{\mathbf{C}} & -\mathrm{I} \end{pmatrix} \prec 0, \\ &\begin{pmatrix} \gamma\mathrm{I} & \mathbf{B}^T\mathbf{Q}_{11} + \hat{\mathbf{B}}^T\mathbf{Q}_{12}^T & \mathbf{B}^T\mathbf{Q}_{12} + \hat{\mathbf{B}}^T\mathbf{Q}_{22} \\ \mathbf{Q}_{11}\mathbf{B} + \mathbf{Q}_{12}\hat{\mathbf{B}} & \mathbf{Q}_{11} & \mathbf{Q}_{12} \\ \mathbf{Q}_{12}^T\mathbf{B} + \mathbf{Q}_{22}\hat{\mathbf{B}} & \mathbf{Q}_{12}^T & \mathbf{Q}_{22} \end{pmatrix} \succ 0. \end{aligned} \tag{6.17}$$

Generalising this to the case when we have a true model sampled at different p-values and want to find an LPV-approximation is straightforward. Rewriting (6.17) again for this problem we get

$$\begin{aligned} &\min_{\gamma_i,\mathbf{Q}_i,\hat{\mathbf{A}}_{\delta^k},\hat{\mathbf{B}}_{\delta^k},\hat{\mathbf{C}}_{\delta^k}} \quad \sum_{i=1}^{N} \gamma_i, \quad k = 1,\ldots,L \\ &\text{s.t.} \quad \begin{pmatrix} \mathbf{Q}_{11,i}\mathbf{A}_i + \mathbf{A}_i^T\mathbf{Q}_{11,i} & \mathbf{Q}_{12,i}\hat{\mathbf{A}}(\delta^{(i)}) + \mathbf{A}_i^T\mathbf{Q}_{12,i} & \mathbf{C}_i^T \\ \mathbf{Q}_{12,i}^T\mathbf{A}_i + \hat{\mathbf{A}}(\delta^{(i)})^T\mathbf{Q}_{12,i} & \mathbf{Q}_{22,i}\hat{\mathbf{A}}(\delta^{(i)}) + \hat{\mathbf{A}}(\delta^{(i)})^T\mathbf{Q}_{22,i} & -\hat{\mathbf{C}}(\delta^{(i)})^T \\ \mathbf{C}_i & -\hat{\mathbf{C}}(\delta^{(i)}) & -\mathrm{I} \end{pmatrix} \prec 0, \\ &\begin{pmatrix} \gamma_i\mathrm{I} & \mathbf{B}_i^T\mathbf{Q}_{11,i} + \hat{\mathbf{B}}(\delta^{(i)})^T\mathbf{Q}_{12,i}^T & \mathbf{B}_i^T\mathbf{Q}_{12,i} + \hat{\mathbf{B}}(\delta^{(i)})^T\mathbf{Q}_{22,i} \\ \mathbf{Q}_{11,i}\mathbf{B}_i + \mathbf{Q}_{12,i}\hat{\mathbf{B}}(\delta^{(i)}) & \mathbf{Q}_{11,i} & \mathbf{Q}_{12,i} \\ \mathbf{Q}_{12,i}^T\mathbf{B}_i + \mathbf{Q}_{22,i}\hat{\mathbf{B}}(\delta^{(i)}) & \mathbf{Q}_{12,i}^T & \mathbf{Q}_{22,i} \end{pmatrix} \succ 0 \\ &i = 1,\ldots,N \end{aligned} \tag{6.18}$$

where $\hat{\mathbf{A}}_{\delta^k}, \hat{\mathbf{B}}_{\delta^k}$ and $\hat{\mathbf{C}}_{\delta^k}$ are the coefficient matrices in (6.2) and L is the highest degree of δ in $\hat{\mathbf{A}}, \hat{\mathbf{B}}$ or $\hat{\mathbf{C}}$. Looking at (6.18) we see that it is bilinear in the variables. To try to solve this a local iterative two-step algorithm can be used, [14]. Start by keeping $\hat{\mathbf{A}}_{\delta^k}, \hat{\mathbf{B}}_{\delta^k}$ constant, then solve (6.18) for $\mathbf{Q}_i, \hat{\mathbf{C}}_{\delta^k}$. Then keep $\mathbf{Q}_{12,i}, \mathbf{Q}_{22,i}$ constant and solve (6.18) for $\mathbf{Q}_{11,i}, \hat{\mathbf{A}}_{\delta^k}, \hat{\mathbf{B}}_{\delta^k}, \hat{\mathbf{C}}_{\delta^k}$. Continue doing this until convergence. Generally bilinear semidefinite programs are very hard to solve [15], and note also that in this case all the $\mathbf{Q}_i$ are optimisation variables and the semidefinite program grows rapidly with the dimension of the models and soon becomes intractable. The semidefinite programming approach described in this section, although having a straightforward format and derivation, was early on discovered to be far too computationally expensive, and was thus omitted from the evaluations. These are the reasons that the rest of the chapter is devoted to the more general nonlinear programming approach described in Sect. 6.3.

6.5 Regularisation of the Optimisation Problem

In the previous section we have tacitly assumed that the given data, (*i.e.*, the state-space matrices in different parameters points) are exact. In a more realistic setting we can assume the presence of errors (*e.g.*, truncation or round-off) in these data. The question is how to cope with these errors and take them into account. The method we propose is to use a problem-specific regularisation, which we will show can be interpreted as a worst-case optimization approach.

To reduce the influence of errors in data, we regularise the original cost function by adding three new terms to the cost function. These are the Frobenius norm of the derivative of the cost function with respect to the given data, $\mathbf{A}, \mathbf{B}$ and $\mathbf{C}$, *i.e.*,

$$\min_{\hat{\mathbf{A}},\hat{\mathbf{B}},\hat{\mathbf{C}}} V + \epsilon_{\mathbf{A}} \left\| \frac{\partial V}{\partial \mathbf{A}} \right\|_F + \epsilon_{\mathbf{B}} \left\| \frac{\partial V}{\partial \mathbf{B}} \right\|_F + \epsilon_{\mathbf{C}} \left\| \frac{\partial V}{\partial \mathbf{C}} \right\|_F . \tag{6.19}$$

As in the case for the gradient of the cost function, the cost function is also differentiable in the matrices $\mathbf{A}, \mathbf{B}, \mathbf{C}$. Using also here Lemma 6.1 and the equations in (6.8) we can calculate the derivative, analogously with the calculations in Sect. 6.3.2.

$$\frac{\partial V}{\partial \mathbf{A}} = 2\left(\mathbf{QP} + \mathbf{YX}^T\right) \tag{6.20}$$

$$\frac{\partial V}{\partial \mathbf{B}} = 2\left(\mathbf{QB} + \mathbf{Y}\hat{\mathbf{B}}\right) \tag{6.21}$$

$$\frac{\partial V}{\partial \mathbf{C}} = 2\left(\mathbf{CP} - \hat{\mathbf{C}}\mathbf{X}^T\right) \tag{6.22}$$

Finding an explicit expression of the derivative of the regularised cost function with these terms inserted can be done using the same methodology as in Sect. 6.3.2, but the details are omitted for brevity.

To motivate the choice of the Frobenius norm in (6.19), and not, *e.g.*, the squared Frobenius norm, we look at the case where we have an unstructured error in the $\mathbf{B}$ matrix.

$$V_\Delta = \mathrm{tr}\left((\mathbf{B}+\Delta)^T\mathbf{Q}(\mathbf{B}+\Delta) + 2(\mathbf{B}+\Delta)^T\mathbf{Y}\hat{\mathbf{B}} + \hat{\mathbf{B}}^T\hat{\mathbf{Q}}\hat{\mathbf{B}}\right) = \tag{6.23}$$

$$= V + \mathrm{tr}\left(2\Delta^T\left(\mathbf{QB} + \mathbf{Y}\hat{\mathbf{B}}\right) + \Delta^T\mathbf{Q}\Delta\right) \tag{6.24}$$

Now maximise this expression with respect to Δ, under the assumption that the error is small,

$$\max_{||\Delta||_F \leq \epsilon} V_\Delta = \operatorname{tr} V + \max_{||\Delta||_F \leq \epsilon} \operatorname{tr}\left(2\Delta^T\left(\mathbf{QB} + \mathbf{Y}\hat{\mathbf{B}}\right) + \Delta^T \mathbf{Q} \Delta\right) = \tag{6.25}$$

$$= V + 2\epsilon \left|\left|\mathbf{QB} + \mathbf{Y}\hat{\mathbf{B}}\right|\right|_F + O(\epsilon^2) \tag{6.26}$$

Here we identify the second term as the Frobenius norm of the derivative of the cost function with respect to $\mathbf{B}$. Analogous calculation can be done when we have a small unstructured error in $\mathbf{C}$. This shows that the approach has clear connections to recently popularised worst-case approaches, [16]. Regarding $\mathbf{A}$, the interpretation is not as clear and is currently an open question.

6.6 Examples

In this section we start with an academic example to shed light to some properties of the proposed method, and then address a practical example illustrating the applicability of the proposed method to realistically sized problems.

When solving the examples, the function `fminunc` in MATLAB was used as the quasi-Newton solver framework. To generate a starting point for the solver, which is an extremely important problem with much research left to do, a balanced realization of the linear model given in the mid-point of the parameter-space was used to initialize $(\hat{\mathbf{A}}_{\delta^0}, \hat{\mathbf{B}}_{\delta^0}, \hat{\mathbf{C}}_{\delta^0})$. All other parameters were initialised to zero. The examples was performed on a computer with 2GB memory, Intel Core2Duo (2.66 GHz) CPU running under CentOS 5.4 with MATLAB version 7.9 (R2009b).

6.6.1 Academic Example

Here a small academic example is presented to show the potential of the new method and to show the importance of addressing system properties.

The system in this example is $G = G_1 G_2$ where $G_1 = \frac{1}{s^2+2\zeta_1 s+1}$ and $G_2 = \frac{9}{s^2+6\zeta_2 s+9}$ with $\zeta_1 = 0.1 + 0.9\delta$ and $\zeta_2 = 0.1 + 0.9(1-\delta)$ and $\delta \in [0, 1]$. The system was sampled in 30 points equidistant in $[0, 1]$, *i.e.* we are given 30 linear models with four states.

The data is given in a state basis where all the elements in the system matrices happen to depend nonlinearly on the parameter δ, see Fig. 6.1. In this basis it will undoubtedly be hard to find a good low-order approximation with an element-wise approach with polynomial dependence of δ. The interesting property of this example is that there exists a state basis where the model has linear dependence on δ, in fact only two elements of the

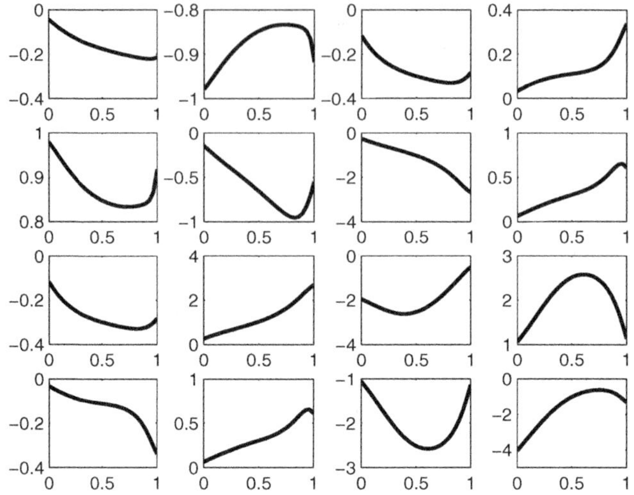

Fig. 6.1 The elements in the **A**-matrix depending on δ. δ varies between 0 and 1 on the x-axis.

system matrix **A** are linear in δ and all other matrix elements in $\mathbf{A}, \mathbf{B}, \mathbf{C}$ are constants.

To test the algorithm, 15 validation points were generated. From the result in Table 6.1 we see that when the proposed method is used, a high accuracy low-order (indeed linear) LPV-model of the system can be found. If we try to obtain a model using an element-wise method with first order polynomials we, of course, obtain a much worse model. Achieving comparable results using an element-wise strategy requires polynomials of order 9. To further illustrate the accuracy in the validation points, root loci are given in Fig. 6.2 and the $\mathcal{H}_2$-norm for the error model in the 15 validation points is shown in Fig. 6.3.

Table 6.1 Numerical results for the academic example

Method	$\sum_i \|\|E_i\|\|_{\mathcal{H}_2}^2$	Degree	Computational Time
Element-wise	0.514	1	0.013 s
Element-wise	$3.79 \cdot 10^{-5}$	9	0.049 s
Method 1	$5.37 \cdot 10^{-7}$	1	179 s

This example illustrates what was said in Sect. 6.2.1. Even though the realisation of the model given is non-polynomial in the parameters, we are able to find the underlying model with only linear dependence. It thus illustrates the importance to look at the input-output relation and not only the individual elements.

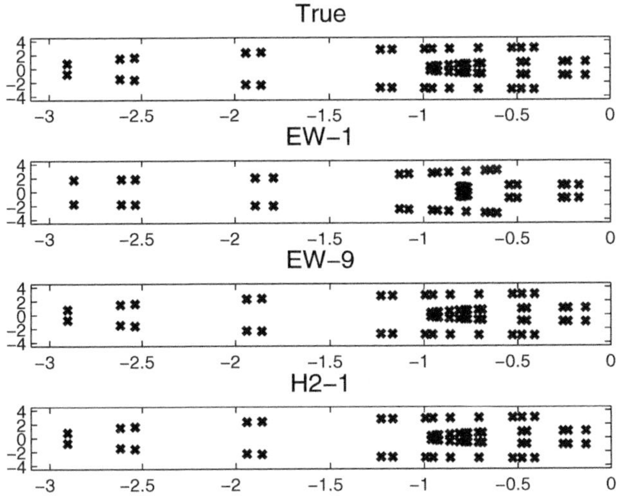

Fig. 6.2 Root loci for the true model, the model approximated with the element-wise method using 1st order and 9th order polynomials and the model approximated with the new method using 1st order polynomials, in the 15 validation points.

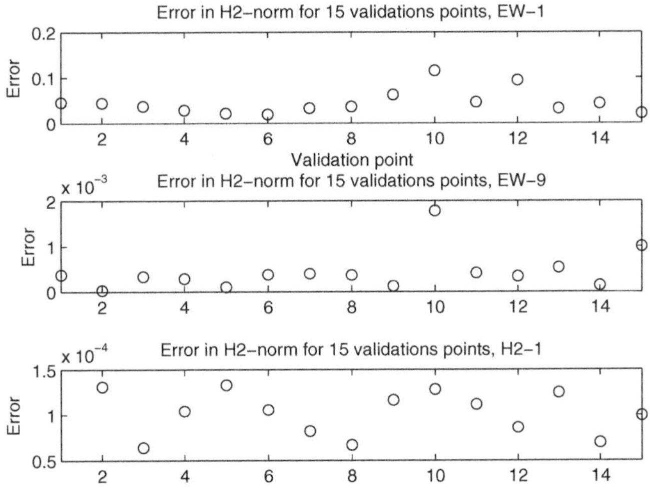

Fig. 6.3 $\mathcal{H}_2$-norm for the true model, the model approximated with the element-wise method using 1st order and 9th polynomials and the model approximated with the new method using 1st order polynomials, in the 15 validation points.

6.6.2 Application Example

The proposed method in Sect. 6.3 is now applied to three sets of longitudinal aeroelastic models. Each of these models have been created from an LFR with the purpose of trying to identify an LPV-model that has a corresponding LFR with a smaller Δ-block than the original model, but still is sufficiently accurate for the kind of analysis described in Chapters 7,9 and 10.

The models depend on three parameters that describe the aircraft mass. These parameters are the payload δ_{PL}, and the centre and outer tanks configurations δ_{CT} and δ_{OT}. The dimensions of the original LFRs can be seen in Table 6.2, where n_s is the number of states in the models, n_{io} is the number of input/outputs and n_Δ is the size of the Δ-block.

Since our goal is to find models which yield LFRs with smaller Δ-block, we try with simple affine LPV-models. The size of the resulting Δ-blocks can be seen in Table 6.3 under $\overline{n}_\Delta$. We validate the models in 100 randomly chosen points in the parameter region using the relative error in $\mathcal{H}_2$- and $\mathcal{H}_\infty$-norm, the results are given in Table 6.3. In Fig. 6.4 the amplitude plot of a frequency response for the LFR and the resulting model for a validation point is plotted, and as we can see they almost completely agree.

Table 6.2 Data for aeroelastic models

Model no.	Parameters in Δ-block	n_s	n_{io}	n_Δ
1	δ_{CT}	22	1/1	20
2	δ_{CT}, δ_{OT}	22	1/1	62
3	$\delta_{CT}, \delta_{OT}, \delta_{PL}$	22	1/1	98

Table 6.3 Resulting models

Model no.	Computational time	mean $\left(\frac{\|\|E_i\|\|^2_{\mathcal{H}_2}}{\|\|G_i\|\|^2_{\mathcal{H}_2}}\right)$	max $\left(\frac{\|\|E_i\|\|^2_{\mathcal{H}_2}}{\|\|G_i\|\|^2_{\mathcal{H}_2}}\right)$	$\overline{n}_\Delta$
1	5h 09m 47s	$3.6 \cdot 10^{-5}$	$7.4 \cdot 10^{-5}$	9
2	7h 50m 46s	$1.5 \cdot 10^{-4}$	$2.2 \cdot 10^{-4}$	18
3	11h 02m 41s	$1.9 \cdot 10^{-4}$	$3.5 \cdot 10^{-4}$	27

Remark: One important thing to note is that we do not take the **D**-matrix into account in our optimisation algorithms. In the original size of the Δ-block of the LFRs, n_Δ, this is included but not in the resulting model. This is because the **D**-matrix is not important for the stability analysis for the closed loop system.

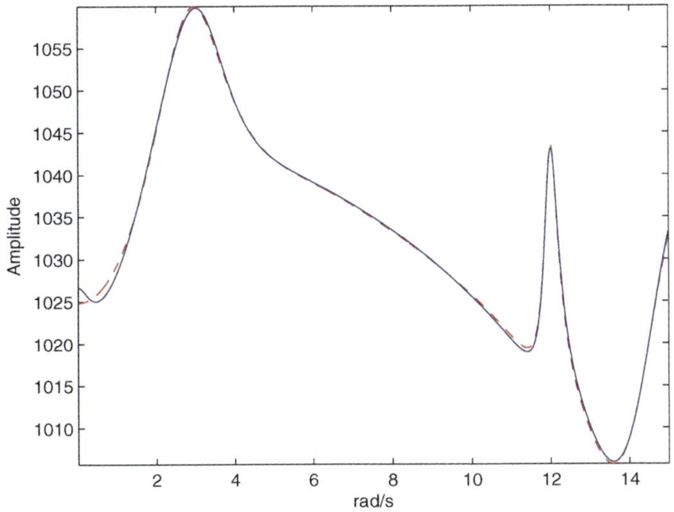

Fig. 6.4 This figure shows an amplitude curve of a frequency response in a validation point for model no. 2. The LFT-model is the solid line and the resulting model from method 1 is the dashed line.

6.7 Conclusions

In this chapter we have proposed two new methods for generating LPV-models. The core concept in both approaches is to preserve input-output relations in the approximation, and not strive to match the actual numbers in the given state-space models. But, because of this, the resulting LPV-models might not be suitable when you want to assess the stability with respect to a boundary of a sector. On the other hand, the frequency responses of LPV-models obtained by an element-wise approach sometimes exhibit an oscillatory behaviour between the points used for interpolation, which this method has not shown to do. One of these methods, the one using standard nonlinear optimisation, has shown good properties on both academic examples and more realistic problems. The method was applied on complex aircraft models, and gave us an approach to generate LFRs with reduced complexity. The method described in this chapter have been implemented in Matlab and can found at `http://www.cofcluo.isy.liu.se`. The method is slower than typical methods used today, *e.g.*, the methods in Chapters 3 and 4, but if we are able to create less complex models, then there is much time to save in the following analysis steps of the model. The semidefinite programming approach, although having a straightforward format and derivation, was early on discovered to be far too computationally expensive, and was thus omitted from the evaluations.

References

1. Zhou, K., Doyle, J.C., Glover, K.: Robust and Optimal Control. Prentice-Hall, Inc., Upper Saddle River (1996)
2. Varga, A., Looye, G., Moormann, D., Grübel, G.: Automated generation of LFT-based parametric uncertainty descriptions from generic aircraft models. Mathematical and Computer Modelling of Dynamical Systems 4, 249–274 (1998)
3. Steinbuch, M., van de Molengraft, R., van der Voort, A.: Experimental modelling and LPV control of a motion system. In: Proc. of the American Control Conference 2003, pp. 1374–1379 (2003)
4. Wassink, M.G., van de Wal, M., Scherer, C., Bosgra, O.: LPV control for a wafer stage: beyond the theoretical solution. Control Engineering Practice 13(2), 231–245 (2005)
5. Wei, X., Del Re, L.: On persistent excitation for parameter estimation of quasi-LPV systems and its application in modeling of diesel engine torque. In: Proc. of the 14th IFAC Symposium on System Identification, pp. 517–522 (2006)
6. Previdi, F., Lovera, M.: Identification of a class of non-linear parametrically varying models. International Journal of Adaptive Control and Signal Processing 17(1) (2003)
7. Tóth, R.: Modeling and Identification of Linear Parameter-Varying Systems, an Orthonormal Basis Function Approach. PhD thesis, Delft University of Technology (2008)
8. Petersson, D., Löfberg, J.: Optimization based LPV-approximation of multi-model systems. In: Proc. of the European Control Conference 2009, pp. 3172–3177 (2009)
9. Van Dooren, P., Gallivan, K.A., Absil, P.A.: $\mathcal{H}_2$-optimal model reduction of MIMO systems. Appl. Math. Lett. 21(12), 1267–1273 (2008)
10. Wilson, D.A.: Optimum solution of model reduction problem. Proc. IEEE 117, 1161–1165 (1970)
11. Yan, W.Y., Lam, J.: An approximate approach to $\mathcal{H}_2$ optimal model reduction. IEEE Transactions on Automatic Control 44(7), 1341–1358 (1999)
12. Nocedal, J., Wright, S.J.: Numerical Optimization. Springer, New York (1999)
13. Boyd, S., El Ghaoui, L., Feron, E., Balakrishnan, V.: Linear Matrix Inequalities in System and Control Theory. Studies in Applied Mathematics, vol. 15. SIAM, Philadelphia (1994)
14. Helmersson, A.: Model reduction using LMIs. In: Proc. of the 33rd IEEE Conference on Decision and Control, pp. 3217–3222 (December 1994)
15. Mesbahi, M., Papavassilopoulos, G.P., Safonov, M.G.: Matrix cones, complementarity problems, and the bilinear matrix inequality. In: Proc. of the 34th IEEE Conference on Decision and Control, pp. 3102–3107 (December 1995)
16. Ben-Tal, A., Nemirovski, A.: Robust optimization - Methodology and applications. Mathematical Programming (Series B) 92, 453–480 (2002)

Part III
Analysis Techniques and Tools

Chapter 7
Enhanced μ-Analysis Techniques for Clearance

Jean-Marc Biannic and Clément Roos

Abstract. A practical method based on μ-analysis is proposed in this chapter to compute a stability robustness margin for high-order LTI plants with real parametric uncertainties. In contrast to grid-based approaches, the validity of this margin is guaranteed on a continuous frequency interval, without any risk of missing critical frequency values. The algorithm to compute the stability margin underlies a recursive procedure to determine a guaranteed stability domain for possibly uncertain parameter dependent plants. Extensions to non-standard robustness problems as well as algorithmic variants to handle the trade-off between conservatism and computational time are also discussed. The application of the proposed method to the clearance of flight control laws is finally addressed for two stability related clearance criteria.

7.1 Introduction

Consider the issue of computing an estimate of the robustness margin for an uncertain linear time-invariant (LTI) system, *i.e.* the maximal size of model uncertainties for which closed-loop stability or performance is still guaranteed. Such a margin can be obtained in terms of the inverse of the maximal structured singular value μ over the frequency range $\mathbb{R}_+$ [1]. Computing the exact value of μ is known to be NP hard [2]. Thus, a guaranteed robustness margin is usually obtained by computing an upper bound of μ on $\mathbb{R}_+$ using the classical polynomial-time algorithm proposed in [3,4], which relies on the determination of suitable scaling matrices.

Nevertheless, this is a difficult optimisation problem with an infinite number of both frequency-domain constraints and optimisation variables. A standard solution consists in restricting the computation of the μ upper bound

Jean-Marc Biannic · Clément Roos
ONERA/DCSD, Toulouse, France
e-mail: `jean-marc.biannic@onera.fr`, `clement.roos@onera.fr`

A. Varga et al. (Eds.): Opti. Based Clearance of Flight Control Laws, LNCIS 416, pp. 131–147.
springerlink.com

to a finite frequency grid. But such a solution can be unreliable, especially in the case of flexible systems, since it is impossible to guarantee that the critical μ peak is not missed between two points of the grid. Moreover, computing a μ upper bound on a fine grid, so as to reduce the probability to miss such a peak, can be time-consuming. It thus appears that there is a real need to develop more reliable and computationally attractive tools, particularly when high-order flexible plants have to be analysed.

The idea to compute a guaranteed μ upper bound on a whole frequency segment is not new. For example, the frequency is condidered as an uncertain parameter in [5], which is repeated as many times as the plant order. A state-space LMI formulation is also proposed in [6]. Nevertheless, these methods usually lead to numerically intractable problems when high-order systems are considered. A computationally very efficient approach is presented in [7,8,9], where strategies for computing a guaranteed μ upper bound over a whole frequency interval and for eliminating some frequency intervals inside which μ is less than a given threshold are merged into a single algorithm. These contributions serve as a basis for the present work, where a more powerful technique is introduced to validate the scaling matrices. As a consequence, the convergence properties of the algorithm are significantly improved and high-order plants with numerous real uncertainties can now be analysed, as demonstrated in Chapter 12. The resulting algorithm is then used to compute a guaranteed stability domain for a possibly uncertain parameter dependent system.

This chapter is organised as follows. The analysis problem is first stated in Sect. 7.2, where all the theoretical background is presented. An efficient algorithm is then introduced in Sect. 7.3 to compute a guaranteed robustness margin for an uncertain LTI plant. It is encompassed in a recursive procedure in Sect. 7.4, which allows to compute a guaranteed stability domain for a possibly uncertain parameter dependent plant. Some extensions and variations are also discussed. The connection with the clearance of flight control laws is finally established in Sect. 7.5.

7.2 Problem Statement and Preliminary Results

7.2.1 Introduction to μ-Analysis

Let us consider the standard interconnection of Fig. 7.1. In the context of μ-analysis, the transfer function matrix $M(s)$ is a stable real-valued continuous-time LTI plant representing the nominal closed-loop system. $\Delta(s)$ is a m-by-m block-diagonal LTI operator, which gathers all model uncertainties. It is composed of real scalar blocks of the form $\delta_r I$, where $\delta_r \in \mathbb{R}$, corresponding to parametric uncertainties, as well as complex scalar blocks of the form $\delta_c I$, where $\delta_c \in \mathbb{C}$, and unstructured transfer matrices representing neglected dynamics. Let $\mathbf{\Delta}$ be the set of matrices with the same structure and nature (real or complex) as $\Delta(j\omega)$.

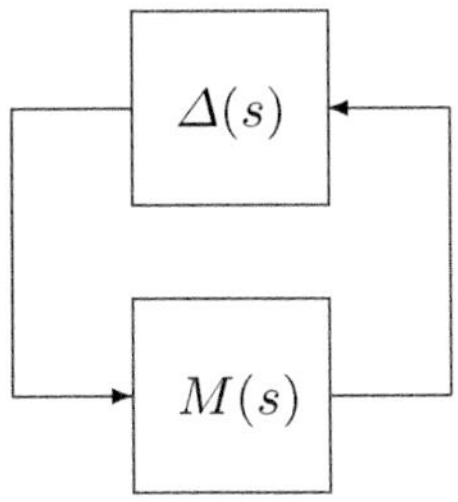

Fig. 7.1 Standard interconnection for robust stability analysis

For a given frequency ω, a singularity appears in the interconnection of Fig. 7.1 along the imaginary axis if $\det(I - M(j\omega)\Delta(j\omega)) = 0$. As a result, the inverse of the structured singular value $\mu_{\mathbf{\Delta}}$, whose definition is recalled below, provides the distance to instability at frequency ω.

Definition 7.1. (structured singular value) If no matrix $\Delta \in \mathbf{\Delta}$ makes $I - M(j\omega)\Delta$ singular, then the structured singular value $\mu_{\mathbf{\Delta}}(M(j\omega))$ is equal to 0. Otherwise, it is defined as the inverse of the size of the smallest matrix $\Delta \in \mathbf{\Delta}$ satisfying $\det(I - M(j\omega)\Delta) = 0$

$$\mu_{\mathbf{\Delta}}(M(j\omega)) = \frac{1}{\min\{k \in \mathbb{R}_+ : \exists \Delta \in k\mathcal{B}_{\mathbf{\Delta}}\,,\, \det(I - M(j\omega)\Delta) = 0\}} \tag{7.1}$$

where $\mathcal{B}_{\mathbf{\Delta}} = \{\Delta \in \mathbf{\Delta} \ : \ \overline{\sigma}(\Delta) < 1\}$ denotes the unit ball of matrices with admissible structure.

The robustness margin k_{max} is then obtained as the inverse of the maximal value of $\mu_{\mathbf{\Delta}}(M(j\omega))$ over the frequency range $\mathbb{R}_+$

$$\frac{1}{k_{max}} = \max_{\omega \in \mathbb{R}_+} \mu_{\mathbf{\Delta}}(M(j\omega)) \tag{7.2}$$

and stability of $M(s)$ is thus guaranteed for all $\Delta(s)$ such that $\Delta(j\omega) \in k_{max}\mathcal{B}_{\mathbf{\Delta}}$ for all $\omega \geq 0$.

Let us now present the classical formulation introduced in [3,4] to compute an upper bound of $\mu_{\mathbf{\Delta}}$. It is essentially based on the use of two frequency dependent scaling matrices $D(\omega)$ and $G(\omega)$ which belong to specific sets $\mathcal{D}$ and $\mathcal{G}$ reflecting the block-diagonal structure and the real or complex nature of Δ [4].

Proposition 7.1. *($\mu_{\mathbf{\Delta}}$ upper bound) Let β be a positive scalar and $\omega > 0$ be a given frequency. If there exist scaling matrices $D(\omega) \in \mathcal{D}$ and $G(\omega) \in \mathcal{G}$ which satisfy*

$$\overline{\sigma}\left(F(\omega)^{-\frac{1}{4}}\left(\frac{D(\omega)M(j\omega)D(\omega)^{-1}}{\beta} - jG(\omega)\right)F(\omega)^{-\frac{1}{4}}\right) \leq 1 \tag{7.3}$$

where $\overline{\sigma}(.)$ *denotes the largest singular value and*

$$\begin{aligned} F(\omega) &= I + G(\omega)^2 \\ \mathcal{D} &= \{D \in \mathbb{C}^{m\times m}, D = D^* > 0 : \forall \Delta \in \mathbf{\Delta},\ D\Delta = \Delta D\} \\ \mathcal{G} &= \{G \in \mathbb{C}^{m\times m}, G = G^* : \forall \Delta \in \mathbf{\Delta},\ G\Delta = \Delta^* G\} \end{aligned}$$

then $\mu_{\mathbf{\Delta}}(M(j\omega)) \leq \beta$.

In the general case, the computation of the robustness margin is a challenging problem with an infinite number of both frequency-domain constraints and optimisation variables. It is usually solved on a finite frequency grid as follows:

1. Define a finite set of N frequency points $\{\omega_1, \ldots, \omega_N\}$,
2. Compute upper bounds β_i of $\mu_{\mathbf{\Delta}}(M(j\omega_i))$ for all $i = 1 \ldots N$,
3. Compute an estimate of the robustness margin $\overline{k}_{max} = 1/ \max\limits_{i\in[1,N]} (\beta_i)$.

However, a crucial problem appears in this procedure. Indeed, the frequency grid which is introduced in step 1 must contain the most critical frequency point for which the maximal value of $\mu_{\mathbf{\Delta}}$ is reached. If this is not the case, the robustness margin calculated in step 3 is under-evaluated and becomes useless. But if such a critical frequency point is known *a priori*, the problem is then half-solved and the frequency grid can be reduced to a single point!

To overcome the above difficulty, an alternative method is proposed in this chapter, which essentially consists in computing a reliable frequency segment around each ω_i for which the upper bound β_i remains valid (see Sect. 7.2.2). As it is clarified in Sect. 7.3, this removes any risk of under-evaluation of the robustness margin.

7.2.2 Validity of the Scaling Matrices

Thanks to the characterisation of Proposition 7.1, which is implemented in the standard Matlab routine `mussv.m` [10], a μ upper bound β_i and associated scaling matrices $D(\omega_i)$ and $G(\omega_i)$ can be easily computed for any given frequency point ω_i, even for high-order plants affected by mixed uncertainties (involving real and complex elements). Let us now slightly increase this upper bound, *i.e.* set $\beta_i \leftarrow (1+\epsilon)\beta_i$, so as to enforce the strict inequality

$$\overline{\sigma}\left(F(\omega_i)^{-\frac{1}{4}}\left(\frac{D(\omega_i)M(j\omega_i)D(\omega_i)^{-1}}{\beta_i} - jG(\omega_i)\right)F(\omega_i)^{-\frac{1}{4}}\right) < 1$$

The objective is then to compute the largest frequency interval $\mathcal{I}(\omega_i) \ni \omega_i$ for which the increased upper bound and the associated scaling matrices remain valid, *i.e.* such that $\forall\omega \in \mathcal{I}(\omega_i)$

$$\overline{\sigma}\left(F(\omega_i)^{-\frac{1}{4}}\left(\frac{D(\omega_i)M(j\omega)D(\omega_i)^{-1}}{\beta_i}-jG(\omega_i)\right)F(\omega_i)^{-\frac{1}{4}}\right)\leq 1 \tag{7.4}$$

As is shown in Proposition 7.2, the determination of $\mathcal{I}(\omega_i)$ boils down to a standard eigenvalues computation.

Proposition 7.2. *Let* (A_M, B_M, C_M, D_M) *denote a state-space representation of* $M(s)$*. Build the matrix*

$$\mathcal{H}=\begin{bmatrix} A_H & 0 \\ -C_H{}^*C_H & -A_H{}^* \end{bmatrix}+\begin{bmatrix} B_H \\ -C_H{}^*D_H \end{bmatrix}(I-D_H{}^*D_H)^{-1}\begin{bmatrix} D_H{}^*C_H & B_H{}^* \end{bmatrix}$$

where

$$\begin{bmatrix} A_H & B_H \\ C_H & D_H \end{bmatrix}=\begin{bmatrix} I & 0 \\ 0 & \frac{F^{-1/4}}{\sqrt{\beta_i}} \end{bmatrix}\begin{bmatrix} A_M-j\omega_i I & B_M D^{-1} \\ DC_M & DD_M D^{-1}-j\beta_i G \end{bmatrix}\begin{bmatrix} I & 0 \\ 0 & \frac{F^{-1/4}}{\sqrt{\beta_i}} \end{bmatrix} \tag{7.5}$$

Define $\underline{\omega}$ *and* $\overline{\omega}$ *as*

$$\begin{aligned}\underline{\omega} &= \max\{\lambda\in\mathbb{R}_-: \det(\lambda I + j\mathcal{H})=0\}\\ &= -\omega_i \text{ if } j\mathcal{H} \text{ has no positive real eigenvalue}\end{aligned}$$

$$\begin{aligned}\overline{\omega} &= \min\{\lambda\in\mathbb{R}_+: \det(\lambda I + j\mathcal{H})=0\}\\ &= \infty \text{ if } j\mathcal{H} \text{ has no negative real eigenvalue}\end{aligned}$$

Then condition (7.4) holds true $\forall\omega\in\mathcal{I}(\omega_i)$*, where*

$$\mathcal{I}(\omega_i)=[\omega_i+\underline{\omega},\ \omega_i+\overline{\omega}] \tag{7.6}$$

Proof: Define $H(j\omega)$ as

$$H(j\omega)=F(\omega_i)^{-\frac{1}{4}}\left(\frac{D(\omega_i)M(j(\omega_i+\omega))D(\omega_i)^{-1}}{\beta_i}-jG(\omega_i)\right)F(\omega_i)^{-\frac{1}{4}}.$$

The bounds defining $\mathcal{I}(\omega_i)$ are obtained by searching for both positive and negative ω of smallest magnitude such that $I - H(j\omega)^*H(j\omega)$ becomes singular, *i.e.* $\det(I-H(j\omega)^*H(j\omega))=0$. A state-space representation (A_H, B_H, C_H, D_H) of $H(s)$ is given by (7.5). A state-space representation (A_X, B_X, C_X, D_X) of $I - H^*(s)H(s)$ is then given by

$$A_X=\begin{bmatrix} A_H & 0 \\ -C_H{}^*C_H & -A_H{}^* \end{bmatrix} \qquad B_X=\begin{bmatrix} -B_H \\ C_H{}^*D_H \end{bmatrix}$$

$$C_X=\begin{bmatrix} D_H{}^*C_H & B_H{}^* \end{bmatrix} \qquad D_X=I-D_H^*D_H$$

Some standard manipulations finally conclude the proof

$$\begin{aligned}\det(I - H(j\omega)^* H(j\omega)) = 0 &\Leftrightarrow \det(I + C_X(j\omega I - A_X)^{-1} B_X D_X^{-1}) = 0 \\ &\Leftrightarrow \det(I + (j\omega I - A_X)^{-1} B_X D_X^{-1} C_X) = 0 \\ &\Leftrightarrow \det(j\omega I - (A_X - B_X D_X^{-1} C_X)) = 0 \\ &\Leftrightarrow \det(\omega I + j\mathcal{H}) = 0\end{aligned}$$

□

7.3 Computation of a Guaranteed Robustness Margin

7.3.1 Standard Version of the Algorithm

The following algorithm is proposed to compute a guaranteed robustness margin for a high-order uncertain LTI plant. It consists of an initialisation phase followed by an iterative procedure on a list of intervals.

Algorithm 7.1 (computation of a robustness margin).

1. *Initialisation phase:*
   a. *Define an initial value β_{max} for the $\mu_{\mathbf{\Delta}}$ upper bound. This initialisation can be achieved by a $\mu_{\mathbf{\Delta}}$ lower bound computation. Otherwise, set $\beta_{max} = 0$.*
   b. *Define the initial frequency range $[\omega_{min}\,,\ \omega_{max}]$ on which the maximal value of the $\mu_{\mathbf{\Delta}}$ upper bound is to be computed.*
   c. *Set a tolerance level ϵ on the $\mu_{\mathbf{\Delta}}$ upper bounds, which will be used to enforce strict inequalities as described in Sect. 7.2.2.*
   d. *Initialise the list of frequency intervals $\mathbb{I}$ to be investigated as follows:*

$$\mathbb{I} = \{\mathcal{I}_1\} = \{[\omega_{min}\,,\ \omega_{max}]\}$$

2. *While $\mathbb{I} \neq \emptyset$, repeat:*
   a. *Select an interval $\mathcal{I}_i$ from the list and choose a frequency $\omega_i \in \mathcal{I}_i$.*
   b. *Compute $\beta, D(\omega_i), G(\omega_i)$ so that condition (7.3) holds for $\omega = \omega_i$.*
   c. *Increase $\beta \leftarrow (1+\epsilon)\beta$ so that condition (7.3) becomes a strict inequality.*
   d. *Set $\beta_{max} \leftarrow \max(\beta, \beta_{max})$.*
   e. *Set $\beta_i = \beta_{max}$ and use Proposition 7.2 to compute the largest frequency interval $\mathcal{I}(\omega_i)$ for which condition (7.4) holds for all $\omega \in \mathcal{I}(\omega_i)$.*
   f. *Generate a new list $\mathbb{I}$ of frequency intervals to be investigated by eliminating the frequencies contained in $\mathcal{I}(\omega_i)$.*
3. *Compute a guaranteed robustness margin as follows:*

$$\overline{k}_{max} = 1/\beta_{max}$$

Remark 7.1. Several possibilities exist to select the frequency ω_i in step $2(a)$. The easiest one consists in systematically choosing the middle point of $\mathcal{I}_i$.

As already mentioned, the proposed algorithm is not based on a frequency grid to be defined *a priori*, with the risk of missing a critical frequency. On the contrary, it relies on a list of frequency intervals which is updated *automatically* during the iterations. By this approach, the robustness margin is guaranteed on the whole frequency range and no tricky initialisation is required.

The precision level of the $\mu_{\mathbf{\Delta}}$ upper bounds computed in step 2 is controlled by both the user-defined tolerance ϵ and the optional argument of the function `mussv.m`. Indeed, these parameters allow to handle the trade-off between high accuracy and reasonable computational cost. If ϵ is increased for example, the elimination step $2(f)$ is more efficient and fewer iterations are required, but accuracy slightly decreases. Moreover, a standard call to `mussv.m` without any option usually gives satisfactory results, but setting the optional argument to `'a'` can improve accuracy provided that the order of the considered system is not to high.

7.3.2 Extension to Modal Performance Analysis

In its standard version, Algorithm 7.1 is dedicated to the computation of a robust stability margin. In practice, however, the notion of robust performance is very important too. Indeed, it is often desirable to quantify the performance degradations, which are induced by model uncertainties and appear before instability. Fortunately, Algorithm 7.1 can be easily adapted to allow for modal performance analysis requirements: the structured singular value is not computed along the imaginary axis anymore, but on the borderline of a truncated sector defined by three parameters α, Φ and ω_c, as depicted in Fig. 7.2. In this context, the robustness margin has to be computed along two segments instead of one:

- segment 1: $s = -\alpha + j\omega$ for $\omega \leq \omega_c$,
- segment 2: $s = j\omega z$ for $\omega \geq \omega_c$,

where $z = [1 + j\tan(\Phi)]$ and $\omega_c = \alpha/\tan(\Phi)$. Let (A_M, B_M, C_M, D_M) denote a state-space representation of $M(s)$. It can then be easily observed that:

- $M(-\alpha + j\omega) = M_1(j\omega)$ on segment 1, where $(A_M + \alpha I, B_M, C_M, D_M)$ is a state-space representation of $M_1(s)$.
- $M(j\omega z) = M_2(j\omega)$ on segment 2, where $(A_M/z, B_M/\sqrt{z}, C_M/\sqrt{z}, D_M)$ is a state-space representation of $M_2(s)$.

The adaptation of Algorithm 7.1 to handle modal performance is thus rather straightforward.

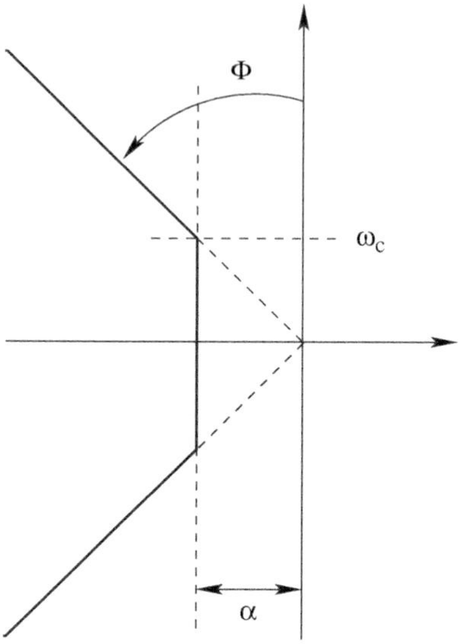

Fig. 7.2 Truncated sector

7.4 Computation of a Guaranteed Stability Domain

7.4.1 Standard Version of the Algorithm

Algorithm 7.1 allows to compute a guaranteed robustness margin for an uncertain LTI plant, provided that it can be modelled by a Linear Fractional Representation (LFR) as depicted in Fig. 7.1. A typical aeronautical application consists in considering an aircraft model for which the flight conditions are fixed, but several parameters such as the aerodynamic coefficients, the mass and the center of gravity are not known precisely.

Let us now consider a parameter dependent plant, which describes the behaviour of a given system on its whole operating domain Θ, and assume that it can be modelled by an LFR as depicted in Fig. 7.1. The operator Δ is composed of n real and possibly repeated parameters whose variations cover all possible operating conditions. In this context, Algorithm 7.1 can be used to compute a guaranteed stability domain, *i.e.* a subset of the whole parametric domain on which stability of the system is ensured.

More precisely, assume that the considered representation $(M(s), \Delta)$ is normalised, which means that the operating domain Θ of the plant is covered by all possible variations of Δ inside the structured unit ball $\mathcal{B}_{\mathbf{\Delta}}$. If $\overline{k}_{max} \geq 1$ when Algorithm 7.1 is applied, the problem is solved, since stability is guaranteed on the entire domain. Otherwise, the following recursive algorithm can be introduced, which allows to partition Θ and to investigate stability on smaller domains, so as to validate gradually the largest possible subset $\mathcal{S}$ of Θ.

Algorithm 7.2 (computation of a stability domain).

1. *Initialisation phase:*
   a. *Let* $\Theta = [a_1, b_1] \times \cdots \times [a_n, b_n] \subset \mathbb{R}^n$ *be the initial box on which stability is investigated.*
   b. *Let* $(M(s), \Delta)$ *be a normalised representation of the parameter dependent plant on* Θ *as defined above.*
   c. *Let* l *be the minimum length below which the edges of the investigated boxes cannot be bisected anymore.*
   d. *Let* $\mathcal{S} = \emptyset$.
2. *Recursive phase:*
   a. *If the plant* $M(s)$ *is stable, compute a guaranteed stability domain* Θ_{max} *using Algorithm 7.1 (see remark 7.3) and go to step 2(b). Otherwise, go to step 2(c).*
   b. *Set* $\mathcal{S} \leftarrow \mathcal{S} \cup \Theta_{max}$. *If* $\overline{k}_{max} \geq 1$, *interrupt the current execution of step 2. Otherwise, go to step 2(c).*
   c. *Let* $\boldsymbol{\Theta}$ *be the set of boxes obtained by bisecting* Θ *along each of its edges whose length is greater than* $2l$. *While* $\boldsymbol{\Theta} \neq \emptyset$:
      - *Select a box* $\tilde{\Theta} \in \boldsymbol{\Theta}$.
      - *Compute a normalised representation* $(\tilde{M}(s), \tilde{\Delta})$ *of the plant on* $\tilde{\Theta}$.
      - *Execute step 2 with* $M(s) \leftarrow \tilde{M}(s)$ *and* $\Theta \leftarrow \tilde{\Theta}$.
      - *Eliminate* $\tilde{\Theta}$ *from* $\boldsymbol{\Theta}$.

Remark 7.2. Algorithm 7.2 must be applied on a box. Thus, it can sometimes be necessary to consider a box approximation of the real operating domain.

Remark 7.3. The stability domain Θ_{max} computed in step 2(a) is similar to Θ. Both domains have the same center and the scaling factor between the two equals $\overline{k}_{max}$

$$\Theta_{max} = [\overline{a}_1, \overline{b}_1] \times \cdots \times [\overline{a}_n, \overline{b}_n] \tag{7.7}$$

where for all $j \in [1, n]$

$$[\overline{a}_j, \overline{b}_j] = \overline{k}_{max}[a_j, b_j] + \frac{a_j + b_j}{2}\left(1 - \overline{k}_{max}\right) \tag{7.8}$$

Remark 7.4. Algorithm 7.2 can be applied with slight modifications if the considered parameter dependent plant is uncertain:

- A skew-μ problem is solved in step 2(a). The idea is to maximise the size of the parametric domain, on which stability is guaranteed for all values of the uncertainties inside the structured unit ball. Algorithm 7.1 can be easily adapted accordingly.

- The bisection of the box Θ in step 2(c) is restricted to the edges associated to the system parameters, the ones associated to the uncertainties remaining unchanged.

7.4.2 Computation of the μ-Sensitivities

A single call to Algorithm 7.1 is usually not computationally expensive, even if a plant with numerous parameters or uncertainties is analysed (see Chapter 12). But 2^n calls to this algorithm are required each time the investigated domain needs to be bisected in Algorithm 7.2. The computational time thus grows exponentially with the number of parameters in the Δ block and can become totally unacceptable is n is quite high. This means that Algorithm 7.2 in its present form is mostly applicable to systems with only few but possibly highly repeated parameters.

If numerous parameters are considered, the computational burden can be alleviated by bisecting the investigated domain along a single edge corresponding to the parameter with the highest influence on stability. This can be achieved in step 2(a) of Algorithm 7.2 by computing the μ-sensitivities [11], which quantify the impact on system stability when the bounds on the parameters are increased one at a time, the others being fixed to their nominal values. To speed up this process, the μ-sensitivities can be evaluated only at the critical frequency for which the value of $\overline{k}_{max}$ computed by Algorithm 7.1 is obtained.

7.4.3 Other Algorithmic Variants

Algorithmic variants can be devised to reduce the computational costs, while still getting a large stability domain:

1) The investigated domain Θ is systematically bisected in step 2(c) of Algorithm 7.2 if $\overline{k}_{max} < 1$ in step 2(a). It is thus possible to interrupt Algorithm 7.1 as soon as $\beta_{max} > 1$ and go directly to step 2(c).
2) The higher the value of β_{max}, the larger the frequency interval $\mathcal{I}(\omega_i)$ computed in step 2(e) of Algorithm 7.1. Setting the initial value $\beta_{max} = 1$ in step 1(a) of Algorithm 7.1 thus allows to increase the efficiency of the elimination procedure, without for all that forcing the robustness margin $\overline{k}_{max}$ to become larger than 1.

If both variants are used, the domain Θ_{max} computed at each occurrence of step 2 of Algorithm 7.2 will be either empty if stability cannot be guaranteed on the whole investigated domain Θ, or equal to Θ even if stability could have been ensured on a larger domain.

Finally, note that it is possible to use the LMI formulation (7.9) below instead of (7.3) to compute β in Algorithm 7.1. Indeed, condition (7.3) is usually solved suboptimally by the function `mussv.m`, and (7.9) thus allows to improve accuracy at the price of an increase in the computational time.

Proposition 7.3. *Let β a positive scalar. If there exist scaling matrices $D(\omega) \in \mathcal{D}$ and $G(\omega) \in \mathcal{G}$ which satisfy*

$$M(j\omega)^* D(\omega) M(j\omega) + \\ j(G(\omega)M(j\omega) - M^*(j\omega)G(\omega)) \leq \beta^2 D(j\omega) \quad (7.9)$$

then $\mu_{\boldsymbol{\Delta}}(M(j\omega)) \leq \beta$.

7.5 Connection with Clearance of Flight Control Laws

The aforementioned analysis technique can be more widely exploited to evaluate some of the clearance criteria that need to be assessed during the certification process of an aircraft. More precisely, two criteria can be easily investigated using μ-analysis techniques: the eigenvalue criterion and the stability margin criterion. These criteria are defined in Chapter 2 from an industrial point of view, and the corresponding mathematical formulations are recalled in Sections 7.5.1 and 7.5.2. A special emphasis is notably put on how they can be efficiently evaluated using the algorithms introduced in Sections 7.3 and 7.4.

Notation: The interconnections depicted in Fig. 7.3 are called upper (a) and lower (b) linear fractional transformations (LFT). They are denoted by $\mathcal{F}_u(M, \Delta)$ and $\mathcal{F}_l(M, K)$ respectively. Moreover, it is assumed in the sequel that Δ is composed of real repeated parameters $\delta = (\delta_1, \ldots, \delta_n)$, *i.e.* $\Delta = \text{diag}(\delta_1 I_{i_1}, \ldots, \delta_n I_{i_n})$.

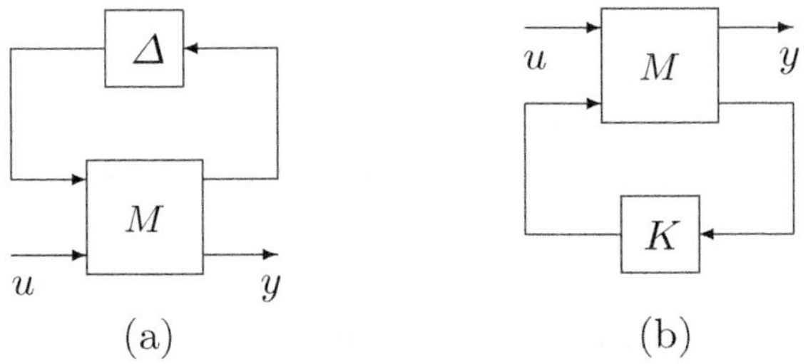

Fig. 7.3 Upper (a) and lower (b) LFTs

7.5.1 Eigenvalue Criterion

This criterion is intended to check over a given parametric domain that the eigenvalues of the considered closed-loop model do not become unstable. It can be computed at each point $\delta \in \mathbb{R}^n$ of the considered domain as the largest real part of the closed-loop system eigenvalues $\lambda_1(\delta), \ldots, \lambda_p(\delta)$

$$c_{eig}(\delta) = -\max_{j\in[1,p]} Re(\lambda_j(\delta)) \tag{7.10}$$

Assume now that δ is normalised, which means that the whole parametric domain is covered when each of the n parameters in δ varies between -1 and 1. It is thus desirable in a certification perspective that $c_{eig}(\delta) > 0$ for all $\delta \in \mathbb{R}^n$ such that $\delta_i \in [-1, 1]$. This is usually not the case for two reasons. First, the model can be unstable in specific regions of the parametric domain. Then, the method used to evaluate the largest parametric domain on which $c_{eig}(\delta) > 0$ can be conservative, and thus fail to prove stability for certain values of δ for which the model is actually stable.

In this context, let Θ_r denote the real stability domain, *i.e.* $\Theta_r = \{\delta \in \mathbb{R}^n,\ \delta_i \in [-1, 1] : c_{eig}(\delta) > 0\}$. The issue is to compute the largest subdomain $\Theta_s \subset \Theta_r$ on which stability of the closed-loop plant can be guaranteed. This can be achieved using the method introduced in Sect. 7.4, provided that an LFR describing the system behaviour on the whole parametric domain is available. Note that such an LFR can be generated using the method described in Chapters 4 and 5.

Several tests have revealed that the conservatism of the analysis method introduced in the previous sections remains quite moderate in practice. It means that most of the configurations that cannot be cleared are actually unstable, and thus that Θ_s is usually almost equal to Θ_r. A detailed study on conservatism is presented in Chapter 12.

The eigenvalue criterion (7.10) can be easily adapted to quantify not only the stability but also the performance properties of the closed-loop plant. The idea is to compute the minimum distance between the eigenvalues and a given truncated sector instead of the imaginary axis, as explained in Sect. 7.3.2. Such an extension may not be necessary for certification but it can help the control engineers to identify easily which are the most critical regions of Θ_r, *i.e.* the regions for which performance degradations or even loss of stability are most likely to occur if additional uncertainties or unmodelled dynamics are considered. It also allows to take into account the modelling error resulting from the use of simplified LFRs instead of full-order reference models to perform analysis (this question is discussed in detail in Chapter 12).

7.5.2 Stability Margin Criterion

This criterion is intended to check over a given parametric domain that the open-loop system obtained when breaking a feedback loop at the input of a given actuator has sufficient gain and phase margins. More precisely, the open-loop Nichols plot of the frequency response obtained when breaking the loop at the input of this actuator, while leaving the other loops closed, must avoid the trapezoidal exclusion region shown in Fig. 7.4 (left).

A preliminary task before such a criterion can be analised using μ-analysis tools is to generate an LFR $\mathcal{F}_u(M(s), \Delta)$ describing the aforementioned

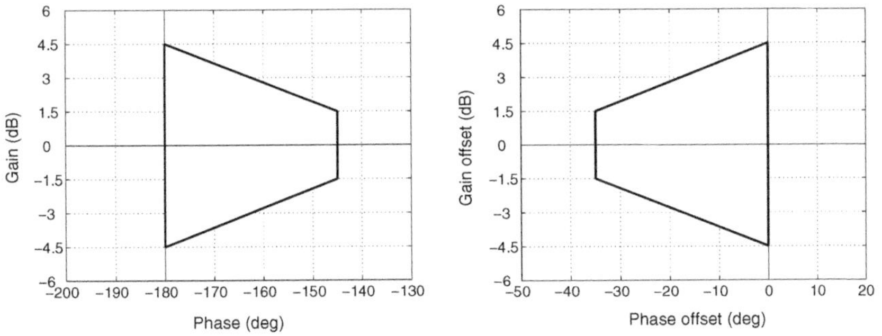

Fig. 7.4 Nichols plane exclusion region (left) - Gain and phase offsets (right)

open-loop system. This can be achieved using the method described in Chapters 4 and 5. A three-step procedure is then implemented:

1. the trapezoidal region shown in Fig. 7.4 (right) is represented by an LFR, which is denoted by $\mathcal{F}_u(T, \Delta_T)$ in the sequel,
2. an augmented open-loop LFR $\mathcal{F}_u(P(s), \Delta_P) = \mathcal{F}_u(M(s), \Delta)\mathcal{F}_u(T, \Delta_T)$ is created, where $\Delta_P = \text{diag}(\Delta, \Delta_T)$,
3. the stability of the closed-loop LFR $\mathcal{F}_l(\mathcal{F}_u(P(s), \Delta_P), 1)$ is evaluated using μ-analysis tools.

Such a strategy is based on the equivalence of the three following assertions:

1. the Nichols plot of the frequency response of the open-loop LFR $\mathcal{F}_u(M(s), \Delta)$ avoids the trapezoidal exclusion region shown in Fig. 7.4 (left) $\forall \Delta \in \mathcal{B}_{\mathbf{\Delta}}$,
2. the Nichols plot of the frequency response of the augmented open-loop LFR $\mathcal{F}_u(P(s), \Delta_P)$ avoids the critical point $(-180\,deg, 0\,dB)$ $\forall \Delta_P \in \mathcal{B}_{\mathbf{\Delta_P}}$,
3. the augmented closed-loop LFR $\mathcal{F}_l(\mathcal{F}_u(P(s), \Delta_P), 1)$ is stable $\forall \Delta_P \in \mathcal{B}_{\mathbf{\Delta_P}}$.

Two approaches exist to convert the trapezoidal region shown in Fig. 7.4 (right) into an LFR, which are based either on an elliptical or a first-order Padé approximation.

Elliptical Approximation

This approach was first introduced in [12] and further exploited in [13, 14]. The idea is to represent the trapezoidal region shown in Fig. 7.4 (right) by an ellipse $\mathcal{E}$

$$\mathcal{E} = \left\{ (\phi_{deg}, G_{dB}) \in \mathbb{R}^2 : \frac{\phi_{deg}^2}{\phi_m^2} + \frac{G_{dB}^2}{G_m^2} \leq 1 \right\} \tag{7.11}$$

The numerical values $\phi_m = 35\,deg$ and $G_m = 4.5\,dB$ offer a good compromise, as shown in Fig. 7.5. Such an approach is interesting, since Δ_T is only composed of one complex non-repeated uncertainty, but the resulting approximation is not very accurate.

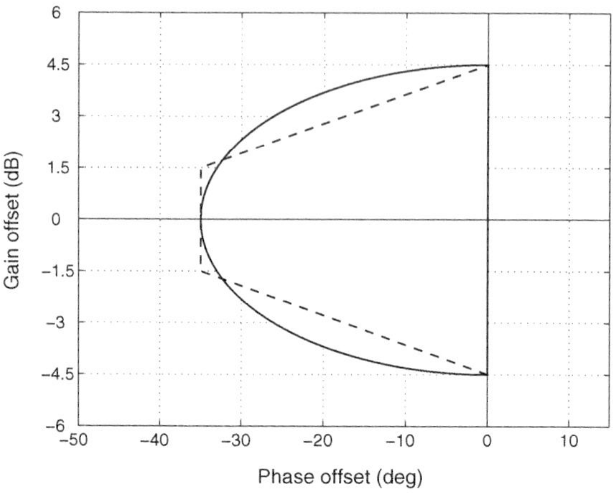

Fig. 7.5 Elliptical approximation

First-Order Padé Approximation

The second approach was introduced in [14, 15]. The idea is now to model the trapezoidal region shown in Fig. 7.4 (right) using a first-order Padé approximation. To this end, let us describe the phase and gain variations as

$$\phi_{deg} = \frac{\phi_{max} - \phi_{min}}{2}\,\delta_\phi + \frac{\phi_{max} + \phi_{min}}{2} \tag{7.12}$$

$$G_{dB} = \delta_G (t - m\delta_\phi) \tag{7.13}$$

where $\phi_{min} = 0\,deg$, $\phi_{max} = 35\,deg$, $t = 3$, $m = 1.5$, $\delta_G \in [-1,\, 1]$ and $\delta_\phi \in [-1,\, 1]$. The augmented open-loop frequency response is thus obtained by multiplying the nominal open-loop frequency response by $Ge^{-j\phi}$, where

$$\phi = \phi_{deg} \frac{\pi}{180} \tag{7.14}$$

$$G = e^{\frac{\ln(10)}{20} G_{dB}} \tag{7.15}$$

Let $c = \dfrac{\ln(10)}{20}$, $\gamma_1 = \dfrac{\phi_{max} - \phi_{min}}{2}\dfrac{\pi}{180}$ and $\gamma_2 = \dfrac{\phi_{max} + \phi_{min}}{2}\dfrac{\pi}{180}$. Then

$$Ge^{-j\phi} = e^{-j\gamma_2} e^{c\delta_G (t - m\delta_\phi) - j\gamma_1 \delta_\phi} \tag{7.16}$$

To obtain a rational expression in δ_G and δ_ϕ from (7.16), a first-order Padé approximation is used

$$e^x = 1 + \frac{x}{1 - x/2} \tag{7.17}$$

where $x = c\delta_G(t - m\delta_\phi) - j\gamma_1\delta_\phi$. The resulting LFR $\mathcal{F}_u(T, \Delta_T)$ is finally defined by

$$T \leftrightarrow \left[\begin{array}{cc|c} 0.5ct & -0.5c & c \\ m + 0.5j\gamma_1 t & -0.5j\gamma_1 & j\gamma_1 \\ \hline te^{-j\gamma_2} & -e^{-j\gamma_2} & e^{-j\gamma_2} \end{array}\right] \quad ; \quad \Delta_T = \begin{pmatrix} \delta_G & 0 \\ 0 & \delta_\phi \end{pmatrix} \tag{7.18}$$

Such an approach leads to a very accurate description of the trapezoidal region, as shown in Fig. 7.6. Furthermore, the size of Δ_T is only slightly larger than for the elliptical approximation. In addition, the result presented in [15] is improved here, since the repetition of δ_ϕ in Δ_T is avoided by the use of a suitable factorisation.

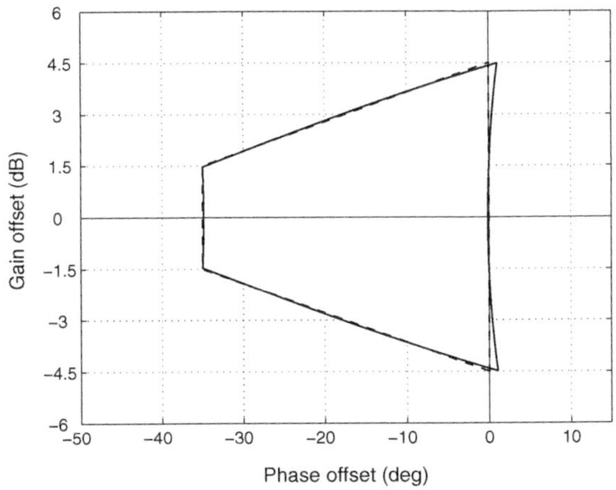

Fig. 7.6 First-order Padé approximation

The stability margin criterion can thus be evaluated using the same strategy as in Sect. 7.5.1, where $\lambda_1(\delta), \ldots, \lambda_p(\delta)$ now denote the eigenvalues of $\mathcal{F}_l(\mathcal{F}_u(P(s), \Delta_P), 1)$ at the considered point $\delta = (\delta_1, \ldots, \delta_n, \delta_G, \delta_\phi)$ of the parametric domain.

7.6 Conclusion

Two efficient algorithms are proposed in this chapter. They allow to compute either a guaranteed robustness margin for a high-order LTI plant with highly repeated parametric uncertainties, or a guaranteed stability domain for a (possibly uncertain) linear parameter dependent plant. The key point of the method consists in combining strategies for computing a guaranteed μ upper bound over a whole frequency interval, and for eliminating the frequency intervals inside which μ is less than a given threshold. As a result, no grid is required and the whole parametric domain is systematically investigated. Numerous tests presented in Chapter 12 show that the proposed methodology allows to handle high-order industrial models that cannot be analysed rigorously using classical methods. Indeed, it appears that conservatism can be easily mastered and that computational time remains quite reasonable, even if very demanding problems are considered. The algorithms introduced in the present chapter can also be used in an aeronautical context to evaluate some of the clearance criteria that need to be assessed during the certification process of an aircraft, such as the eigenvalue and the stability margin criteria introduced in Chapter 2. The modelling methodology proposed in Chapter 4 and the clearance technique presented here are thus the two main stages towards the development of a modelling and optimisation tool dedicated to the clearance of stability criteria, which is able to meet the industrial needs specified in Chapter 2. The way how they can be efficiently combined and implemented is discussed in Chapter 12.

References

1. Doyle, J.: Analysis of feedback systems with structured uncertainties. IEE Proceedings, Part D 129(6), 242–250 (1982)
2. Braatz, R.P., Young, P.M., Doyle, J.C., Morari, M.: Computational complexity of μ calculation. IEEE Transactions on Automatic Control 39(5), 1000–1002 (1994)
3. Fan, M.K.H., Tits, A.L., Doyle, J.C.: Robustness in the presence of mixed parametric uncertainty and unmodeled dynamics. IEEE Transactions on Automatic Control 36(1), 25–38 (1991)
4. Young, P.M., Newlin, M.P., Doyle, J.C.: Computing bounds for the mixed μ problem. International Journal of Robust and Nonlinear Control 5(6), 573–590 (1995)
5. Sideris, A.: Elimination of frequency search from robustness tests. IEEE Transactions on Automatic Control 37(10), 1635–1640 (1992)
6. Ly, J.H., Chiang, R.Y., Goh, K.C., Safonov, M.G.: LMI multiplier Km/μ analysis of the Cassini spacecraft. International Journal of Robust and Nonlinear Control 8(2), 155–168 (1998)
7. Magni, J.F., Doll, C., Chiappa, C., Frapard, B., Girouart, B.: Mixed μ–analysis for flexible systems - Part I: Theory. In: Proceedings of the 14th IFAC World Congress, Beijing, China, pp. 325–360 (July 1999)
8. Ferreres, G., Biannic, J.M.: Reliable computation of the robustness margin for a flexible transport aircraft. Control Engineering Practice 9, 1267–1278 (2001)
9. Ferreres, G., Magni, J.F., Biannic, J.M.: Robustness analysis of flexible structures: practical algorithms. International Journal of Robust and Nonlinear Control 13(8), 715–734 (2003)
10. Balas, G., Chiang, R., Packard, A., Safonov, M.: Robust Control Toolbox: User's Guide. The MathWorks, Inc. (2007)
11. Braatz, R.D., Morari, M.: μ-sensitivities as an aid for robust identification. In: Proceedings of the American Control Conference, Boston, Massachusetts, pp. 231–236 (June 1991)

12. Deodhare, G., Patel, V.V.: A 'modern' look at gain and phase margins: an H_∞/μ approach. In: Proceedings of the AIAA Guidance, Navigation and Control Conference, Boston, Massachusetts, pp. 323–335 (August 1998)
13. Kureemun, R., Bates, D.G., Postlethwaite, I.: Quantifying the robustness of flight control systems using nichols exclusion regions and the structured singular value. IMechE Journal of Systems and Control Engineering 215(16), 625–638 (2001)
14. Bates, D.G., Kureemun, R., Mannchen, T.: Improved clearance of a flight control law using μ-analysis techniques. AIAA Journal of Guidance, Control and Dynamics 26(6), 869–884 (2003)
15. Fielding, C., Varga, A., Bennani, S., Selier, M. (eds.): Advanced techniques for clearance of flight control laws. LNCIS, vol. 283. Springer, Heidelberg (2002)

Chapter 8
Worst-Case Parameter Search Based Clearance Using Parallel Nonlinear Programming Methods

Hans-Dieter Joos

Abstract. This chapter presents the theoretical background for several enhancements of the worst-case parameter search based clearance of flight control laws. The worst case search method aims to find combinations of parameters and flight conditions for which the clearance criteria are mostly violated or poorly satisfied. The two main aspects for the proposed enhancements are: (1) increasing the reliability of clearance by the application of global optimization methods in conjunction with established simulation and analysis tools; and (2) increasing the efficiency of clearance by applying parallel computation techniques. These enhancements are illustrated in Chapter 13 by the application of the selected optimization methods and parallelization techniques to several challenging clearance criteria.

8.1 Introduction

The clearance problem of flight control laws can be formulated as a robustness analysis problem, where a set of suitably defined clearance criteria must be checked to lie within certain limits for all admissible variations of aircraft parameters, pilot inputs and all flight conditions. The idea of optimisation based flight control law clearance is to use available and efficient optimisation methods in combination with established and reliable simulation and analysis tools, to find worst-case combinations of parameters and flight conditions for which the criteria are mostly violated or poorly satisfied.

For the first time, a comprehensive description of the optimization-based clearance was given in [17]. An analysis cycle was introduced which combines the traditionally used gridding-based approach with respect to the flight

Hans-Dieter Joos
Institute of Robotics and Mechatronics, DLR - Oberpfaffenhofen,
82234 Wessling, Germany
e-mail: dieter.joos@dlr.de

A. Varga et al. (Eds.): Opti. Based Clearance of Flight Control Laws, LNCIS 416, pp. 149–159.
springerlink.com

envelope with worst-case search using local parameter optimization at selected flight conditions. Parameter optimisation is used to determine worst-case parameter combinations around these flight conditions which lead to worst performance within the parameter range and flight conditions under investigation. Discretisation with respect to some parameters is used to overcome to some extent the problem of local minima and to find the overall worst case parameter combination.

In the aftermath, several investigations have been made regarding the suitability and effectiveness of different optimization algorithms [3], [12], [13] and [14]. It was shown that the reliability of the approach, i.e. the probability to find hidden weaknesses, is increased when global search algorithms are applied instead of local search methods. The loss of performance measured in computing time is expected to be compensated when parallel computation techniques are applied.

The strength of the optimization-based approach is its flexibility in that it can be used to check all linear or nonlinear clearance criteria as far as they can be expressed as mathematical objective functions. Within the COFCLUO project global optimisation methods are investigated for the clearance of manoeuvrability and protection violation criteria with pilot input. Among the various existing methods, an evolutionary algorithm called *evolution strategy* (ES) proved to be most effective. See [10] for a detailed comparison of the applied algorithms. An overview and preliminary results are also given in [9].

Worst case search implies finding the global optimum, but there is no theory guaranteeing that evolutionary algorithms can reach a global solution with a prescribed level of confidence within finite time. There exist deterministic optimisation algorithms, such as box decomposition methods, which guarantee to reach the global solution under some variational conditions but at the expense of exponentially long time as the number of variables increases [1].

In this chapter a clearance strategy is proposed based on the assumption that the worst case can be found with nonzero probability. This rate of success is in general unknown in advance and might even be zero for a given problem. However, the results achieved within the COFCLUO-project show that for this kind of optimisation problems reasonable values can be derived from the existing computational experience.

8.2 Theoretical Basis

8.2.1 Formulation as Global Optimisation Problem

For utilising optimisation methods the clearance problem has to be expressed as a scalar objective function $c(p, d)$ with optimisation parameters p that are uncertain or varying during operation (e.g. aerodynamic coefficients, wind or aircraft mass, inertia, speed, height, etc.) and discrete conditions d

(e.g. aircraft configuration, landing gear settings, etc.). The components of p are assumed to be bounded and continuously varying over known intervals, defining a hyper-box P. The clearance problem can now be formulated as a minimization problem. Let $c_0(d)$a lower acceptable value for c then

$$c_{\min}(d) = \min_{p \in P}(c(p,d)) \tag{8.1}$$

is a measure for the clearance performance. We assume that the clearance requirement is fulfilled for condition d if $c_{\min}(d) \geq c_0(d)$, otherwise the criterion is not cleared.

8.2.2 Level of Confidence

The difficulty with the optimisation based approach is not to find parameters/inputs/flight conditions such that a criterion is not satisfied, i.e. to demonstrate that the flight control system (FCS) is not cleared, but to confidently assert that all criteria are fulfilled. That means one has to solve a global optimisation problem where the global solution, i.e. the worst case can be found with some reasonable level of confidence. The level of confidence can be estimated from a series of optimisation runs with different initial conditions assuming a positive probability of success as outlined in the following.

In order to estimate the probability that in a series of n optimisation runs the worst case is detected for k times, a random variable X is defined which counts the number of successful optimisations. Such a random variable has a binomial cumulative distribution

$$\Pr(X \leq k) = B(k|n,q) := \sum_{i=0}^{k} \binom{n}{i} q^i (1-q)^{n-i}$$

depending on the number of runs n and the probability q of success. If we assume that the outcome of the different optimisation runs are independent trials, the probability to obtain at least k successful runs out of a set of n runs is given by

$$\Pr(X \geq k) = 1 - B(k-1|n,q). \tag{8.2}$$

As an example, for a given success rate of $q = 0.1$ about 40 runs are necessary to get the worst case at least once with a confidence level of $\Pr = 0.92$. For 80 runs the corresponding detection probability would be greater than 0.99.

8.2.3 Clearance Strategy

The consideration on confidence level leads to the following clearance strategy which has been applied to the COFCLUO clearance problems:

Assume the success rate for a specific criterion is q, specify the desired level of confidence Pr_0 which gives the requested number of optimisation runs in order to detect the worst case for example at least once. Start a sequence of independent optimisations runs with randomly chosen initial values and initialisations of random number generators. Stop this sequence when either the solution found is less than the lower limit c_0 or the required number of runs is reached.

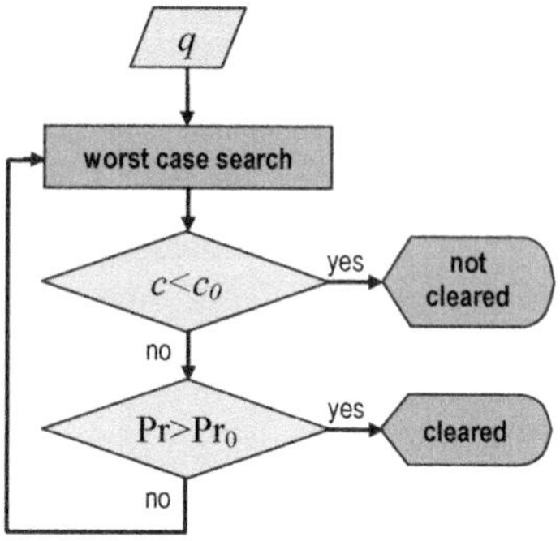

The crucial assumption for this strategy is the expected success rate, i.e. the probability to detect the global minimum. The success rate depends on the criterion properties, the selected optimisation algorithm and the number of considered parameters and of corresponding parameter ranges. It is a priori unknown. However, an estimate for success rate can be deduced from computational experiments. As it can be seen in Chapter 13, for the reported criteria more than 30 from 40 optimisation runs terminate with almost identical results (up to computing tolerance). Assuming this result to be the worst case the success rate can be estimated to be about 0.7. Analysis of other criteria confirms this estimate. For the COFCLUO clearance problems a value of $q = 0.1$ will be assumed as conservative and cautious working hypothesis. Of course there is no guarantee that some kind of 'singular' worst cases exist which can be found with vanishingly small probability only. In that case the proposed strategy will fail.

8.2.4 Transformation of Parameter Space

In equation (8.1) it is assumed that the optimisation variables p are elements of a hyper-box P and no other constraints are involved. However, when using physical variables directly as optimisation variables this is not generally valid. While parameters like mass m and height h can be simply bounded as $m \in [m_{min}, m_{max}]$ and $h \in [h_{min}, h_{max}]$, other parameters as the centre of gravity X_{cg} has lower and upper bounds depending on mass, thus $X_{cg} \in [X_{cg,min}(m), X_{cg,max}(m)]$, while the speed V_{cas} has bounds generally depending on h, m and X_{cg}, thus $V_{cas} \in [V_{min}(h, m, X_{cg}), V_{max}(h, m, X_{cg})]$. To work with such bounds, additional non-linear constraints would be necessary to restrict the parameters to the feasible range. Since the available implementations of the global search methods solve constrained problems by transforming them into unconstrained problems via penalty terms, non-feasible trial points would inevitably lead to questionable simulation or analysis results.

For the clearance problems in COFCLUO exact mathematical characterizations of the bounds are available, and these can be used to substitute physical variables by box bounded normalized variables. Assuming the normalized parameters for speed $\Delta V_{cas} \in [0,1]$ and centre of gravity $\Delta X_{cg} \in [0,1]$, the physical parameters X_{cg} and V_{cas} can be expressed in the form

$$\begin{aligned} X_{cg} &= X_{cg,min}(m) + \Delta X_{cg}\,(X_{cg,max}(m) - X_{cg,min}(m)) \\ V_{cas} &= V_{min}(h,m,X_{cg}) + \Delta V_{cas}(V_{max}(h,m,X_{cg}) - V_{min}(h,m,X_{cg})) \end{aligned} \tag{8.3}$$

Hence the normalized variables ΔV_{cas} and ΔX_{cg} can be used as optimisation parameters instead of V_{cas} and X_{cg} leading to a simple box bounded problem.

8.3 Applied Optimisation Methods

Several global search algorithms like *evolutionary strategy* (ES) [1], *genetic algorithm* (GA) [1], [4], *differential evolution* (DE) [15], *particle swarm optimisation* (PSO) [11] and deterministic box decomposition techniques [2] have been investigated for their usefulness in the analysis of several simulation based clearance criteria. All implementations have been taken from DLR's optimisation environment MOPS [7]. In terms of success rate, i.e. the probability to find the global minimum, the ES algorithm [1] outperformed all others and was therefore used for the analyses reported in Chapter 13. More details on the comparison of the applied methods and according results can be found in [10].

According to [1], the ES belongs to the family of evolutionary algorithms whose special feature is the self-adaptation of algorithm parameters. Self-adaptation is achieved by including additional parameters into the optimisation process and by co-evolving them with the solution. In the implementation used here, for each optimisation variable a self-adapting mutation step size is introduced. In detail the implementation comprises:

- representation of individuals as real valued vectors;
- uniform random selection of parents;
- discrete recombination of optimisation variables by choosing randomly elements from two different parents;
- intermediary recombination (arithmetic mean) of two different parents for the step size parameters;
- generation of a number of children which is much greater than the number of parents;
- mutation of step size parameters σ_i is done by applying a lognormal distribution

$$\sigma_i' = \sigma_i \cdot \exp(\tau_1 \mathcal{N}(0,1) + \tau_2 \mathcal{N}_i(0,1))\ , \quad \sigma_i' \geq \varepsilon_0 \tag{8.4}$$

  where τ_1 , τ_2 are internal constants depending on the number of variables;
- mutation of optimisation variables x_i is done by a Gaussian perturbation

$$x_i^{'} = x_i + \sigma_i^{'} \cdot \mathcal{N}(0,1) \cdot (x_{u,i} - x_{l,i}) \tag{8.5}$$

where $x_{u,i}$ and $x_{l,i}$ are the upper and lower bounds of the i-th variable x_i.
- forming the next generation optionally either from the best children, the best children including the overall best (survival of the fittest) or by elitists selection from offspring and parents.

The user has to specify only four algorithm parameters which of course can influence the success rate of detecting the worst case. These parameters are the population size, offspring size, type of survivor selection and accuracy tolerance. All clearance optimisations in Chapter 13 have been performed using parameter values recommended in [1] and adapted using own experience with various criteria. The chosen population size was 2 or 3 times larger than the number of optimisation variables, while the offspring size was 7 times larger than the population size. The elitists selection from parents and offspring has been used, together with a moderate accuracy tolerance of 10^{-3}, to comply with the noisy characteristics of the simulation based criteria. The user can additionally set three other termination conditions as the maximum number of generations (used value 200), maximum number of generations without progress in the criterion value (used value 10) and maximum number of steps with size less than the tolerance (used value 1). In all analyses, the optimisation runs have been terminated by one of the last two termination conditions indicating convergence of the algorithm.

After termination of the evolutionary global search the worst case found was used as initial value for a final local search applying a pattern search algorithm similar to [6] (see [7] for extensions). Performing a final local search provides the advantages that the optimisation result may be further improved with a comparatively small additional computational effort. In this way, the (local) minimum determined by the global algorithm can be approximated more accurately and hence identical solutions from different optimisation runs can be assessed easier. Moreover, the global search can be terminated earlier when no significant improvements are made or no further improvements can be expected due to small step sizes.

The additional computational effort of a final local search however is small compared to the global search, because the local method can be expected to converge fast in the neighborhood of a (local) minimum. The user has no additional parameters to specify.

The gradient free *pattern search* (PS) algorithm is a robust optimisation method. By experience it is less sensitive to noisy criteria than gradient based methods, but has still reasonable convergence properties. A global convergence theory for algorithms based on pattern search is established for instance in [16].

8.4 Parallel Computation

It is well known that solving clearance problems is cost intensive, requiring a huge amount of computational work [17]. The parallelisation of computations, by using distributed computing on a computer network or multiple processors, can help to drastically reduce the computing times. For this purpose, the problem must be broken into independent parts which can be concurrently executed. However, parallel computation introduces new software engineering aspects, like communication and synchronisation between different processors or computer nodes, which occasionally may significantly influence the resulting speedups.

There are several levels (or types) of parallel computing: bit-level, instruction level, data level or task level. While bit and instruction level parallelisations are inherently implemented in the processing elements (CPUs) and are exploited by modern numerical libraries or compilers, parallelisation on data or task level can be used on a higher programming layer, as solving optimisation problems. This is the level considered here.

Fortunately, worst-case search based on stochastic global search algorithms like ES can be parallelised efficiently in two straightforward ways using a simple master/slave concept (for more details see [8]). A first approach illustrated in Fig. 8.1 is the parallelization of criteria evaluations for each member of a population. This approach requires a 'parallel' implementation of the underlying algorithms.

The second approach shown in Fig. 8.2 performs individual optimisation tasks in parallel with different initial conditions to enforce the reliability of global search. In contrast to the first approach, this parallelization can be performed without any changes in the implementations of the algorithms.

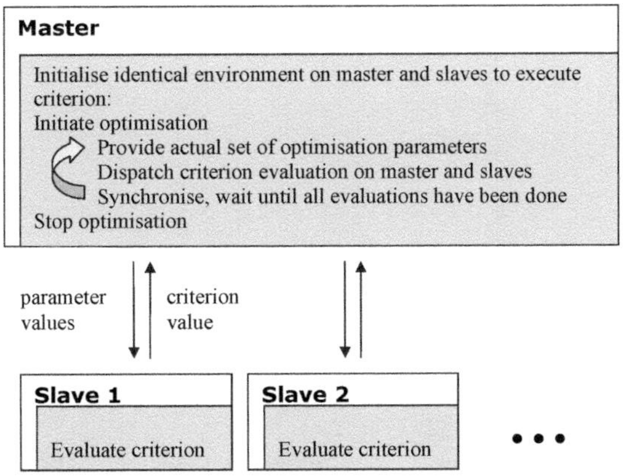

Fig. 8.1 Parallelisation of criteria evaluations by means of a master/slave concept.

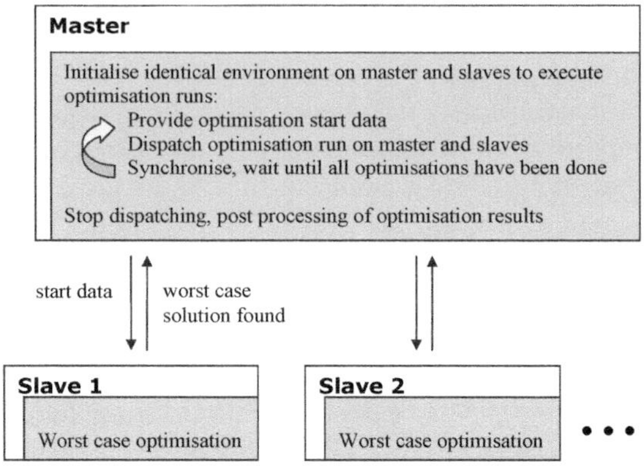

Fig. 8.2 Parallelisation of multiple optimisation runs by means of a master/slave concept.

The parallelisation can be made very efficient without producing too much overhead in communication or data transfer. The diagram in Fig. 8.3 shows the dependence of computing times on the number of used processors and the amount of time necessary for one criterion evaluation (simulation time). The results are obtained on a Linux-cluster (installed in 2002) consisting of 32 single-core, Intel Xeon processors (2.8GHz) with 2GB memory each. For each run 1200 evaluations have been performed.

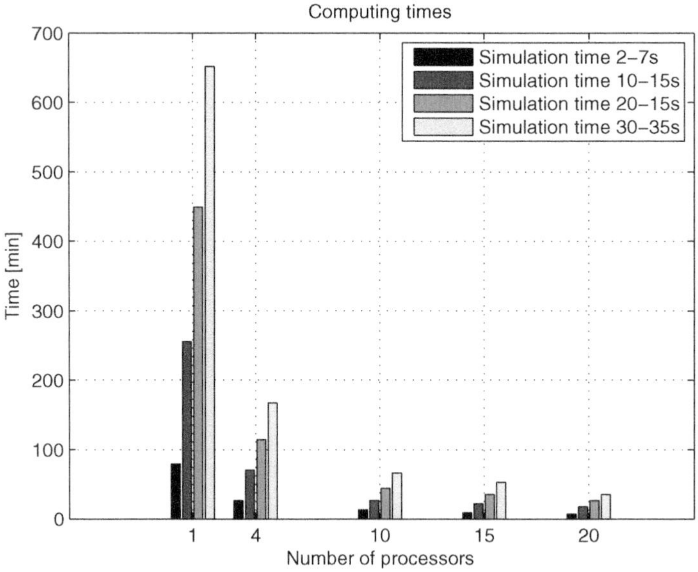

Fig. 8.3 Computing times for different number of processors and different amount of time for criterion computation (simulation time) based on 1200 evaluations.

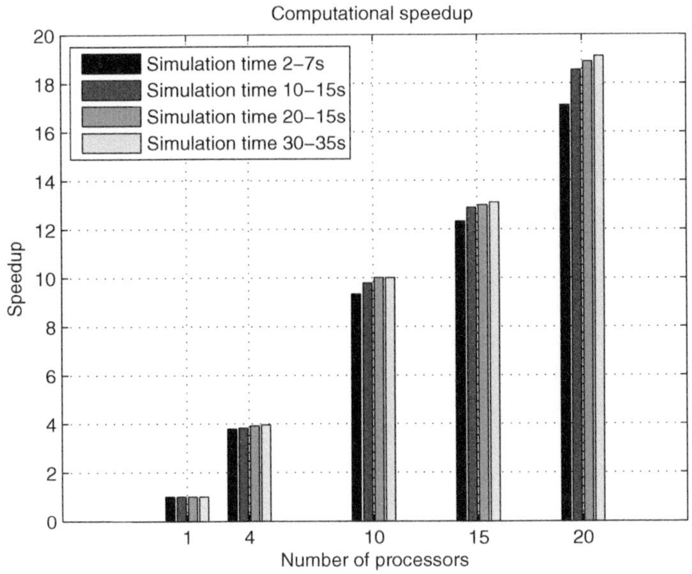

Fig. 8.4 Speedup achieved for different number of processors (4, 10, 15, 20) and different amount of time for criterion computation (simulation time) based on 1200 evaluations.

The gain in efficiency when using parallel computations can be measured by the speedup S defined as [5]

$$S = (s + N \cdot p)/(s + p) \tag{8.6}$$

where N is the number of processors, and s and p are, respectively, the times spent in the serial and parallel parts of the code. The theoretically achievable maximum speedup is N. According to measurements performed for s and p and shown in Fig. 8.4, this maximum value can be almost reached in all cases under investigation.

8.5 Conclusions

As already mentioned the main advantage of the optimization-based clearance method is its flexibility regarding its applicability to virtually all clearance problems and all classes of underlying models. The expected benefits of this approach are:

- increased reliability in finding weaknesses compared to traditional gridding based methods or Monte Carlo analysis,
- reduction of overall analysis time compared to traditional methods
- possibility to deal simultaneously with a greater number of parameters,

- applicability to any kind of clearance problem without adding any conservatism to the method
- finding worst-case combinations of parameter and flight conditions which can be useful to improve the tuning of control laws.

A known weakness of the approach is that there is no absolute guarantee to find the global worst case with a prescribed level of confidence because the proposed clearance strategy relies on an estimate for the success rate which is in general unknown. However, the type of problems considered in the COFCLUO-project allow to conclude that a non-zero success rate exists because the sensitivity of the dynamics of a controlled aircraft with respect to clearance parameters is bounded for physical reasons. However, a reliable estimate of the success rate can only be hypothesised from a series of optimisation runs.

In the absence of theoretical guarantees, the proposed worst case search methodology needs possibly benchmarking with classical methods like Monte Carlo analysis. The results documented in Chapter 13 show that bad cases (not necessarily the worst case) can be found more reliable by means of the proposed search algorithms and clearance strategy (see also [10] for more details).

References

1. Eiben, A.E., Smith, J.E.: Introduction to Evolutionary Computing. Springer, Heidelberg (2007)
2. Finkel, D.E.: DIRECT Optimization Algorithm User Guide, North Caroline State University, Department of Mathematics, Center for Research in Scientific Computation, Raleigh, NC 27695-8205 (March 2003)
3. Forssell, L.S.: Flight clearance analysis using global nonlinear optimisation-based search algorithms. In: AIAA Guidance, Navigation, and Control Conference and Exhibit, Austin,Texas, pp. 1023–1030 (2003)
4. Goldberg, D.E.: Genetic Algorithms in Search, Optimization and Machine Learning. Addison-Wesley, Reading (1989)
5. Gustafson, J.L.: Reevaluating Amdahl's law. Communication of the ACM 31(5), 532–533 (1988)
6. Hooke, R., Jeeves, T.A.: "Direct Search" Solution of Numerical and Statistical Problems. Journal of the ACM 8, 212–229 (1961)
7. Joos, H.-D., Bals, J., Looye, G., Schnepper, K., Varga, A.: A multi-objective optimisation based software environment for control systems design. In: Proc. of 2002 IEEE International Conference on Control Applications and International Symposium on Computer Aided Control Systems Design, CCA/CACSD, Glasgow, Scotland, U.K (2002)
8. Joos, H.-D.: Specification of an optimisation based clearance tool based on parallel computation. Technical Report, D2.2.1, COFCLUO, EU (January 2008)
9. Joos, H.-D.: Flight control law clearance using optimization-based worst-case search. Preprints of the 6th IFAC Symposium on Robust Control Design, Haifa, Israel (2009)

10. Joos, H.-D.: Final report with analysis results. Technical Report, D2.2.3, COFCLUO, EU (January 2010)
11. Kennedy, J., Eberhart, R.: Particle swarm optimization. In: Proc. of the 1995 IEEE International Conference on Neural Networks, pp. 1942–1948 (1995)
12. Menon, P.P., Kim, J., Bates, D.G., Postlethwaite, I.: Clearance of nonlinear flight control laws using hybrid evolutionary optimization. IEEE Transactions on Evolutionary Computation 10(6), 689–699 (2006)
13. Menon, P.P., Bates, D., Postlethwaite, I.: Optimisation-based flight control law clearance. In: Bates, D., Hagström, M. (eds.) Nonlinear Analysis and Synthesis Techniques for Aircraft Control. LNCIS, vol. 365, Springer, Berlin (2007)
14. Menon, P.P., Bates, D., Postlethwaite, I.: Nonlinear Robustness Analysis of Flight Control Laws for Highly Augmented Aircraft. Control Engineering Practice 15(6), 655–662 (2007)
15. Storn, R., Price, K.: Differential Evolution - a Simple and Efficient Adaptive Scheme for Global Optimization over Continuous Spaces. Technical Report TR-95-012, ICSI (March 1995)
16. Torczon, V.: On the convergence of pattern search algorithms. SIAM Journal of Optimization 7(1), 1–25 (1997)
17. Varga, A.: Optimisation-based clearance. In: Fielding, C., Varga, A., Benani, S., Sellier, M. (eds.) Advanced Techniques for Clearance of Flight Control Laws. LNCIS, vol. 283, Springer, Berlin (2002)

Chapter 9
Lyapunov-Based Robustness Analysis Techniques for Clearance

Andrea Garulli, Alfio Masi, Simone Paoletti, and Ercüment Türkoğlu

Abstract. This chapter considers different techniques for robustness analysis of uncertain systems modeled by linear fractional representations, based on the use of parameter-dependent Lyapunov functions and multipliers. Several sufficient conditions for robust stability, relying on the choice of simplified structures for the Lyapunov function and the multiplier matrices, are proposed. These conditions provide a useful tool for trading off conservatism and computational burden, which is a key issue when addressing robustness analysis of high dimensional uncertain systems arising from clearance problems.

9.1 Introduction

Clearance of flight control laws is an integral part of the certification of an aircraft, and is characterized by the evaluation of its robustness against predefined stability or performance criteria within a given region of interest, the flight/uncertainty domain. A commonly practiced approach in industrial aircraft clearance is the grid-based approach: a finite set of grid points are selected within the flight/uncertainty domain and the clearance criterion of interest is tested for each point. The main drawbacks of this approach are the high computational times required (which depend on the density of the grid), and the fact that only the grid points are actually cleared and not the entire domain. Indeed, this has motivated intensive research on alternative approaches, combining intensive computer simulations, optimization and robustness analysis techniques (see e.g. [1, 2, 3, 4] and references therein).

Andrea Garulli · Alfio Masi · Simone Paoletti · Ercüment Türkoğlu
Dipartimento di Ingegneria dell'Informazione, Università di Siena,
via Roma 56, 53100 Siena, Italy
e-mail: garulli@dii.unisi.it, masi@dii.unisi.it,
paoletti@dii.unisi.it, turkoglu@dii.unisi.it

A. Varga et al. (Eds.): Opti. Based Clearance of Flight Control Laws, LNCIS 416, pp. 161–178.
springerlink.com

Robustness analysis of linear systems depending on uncertain parameters has been a subject of notable interest in the systems and control community for several decades, and Lyapunov theory has played a pivotal role in this context. The development of computationally efficient techniques for solving convex optimization problems involving Linear Matrix Inequalities (LMIs) has motivated research efforts towards more and more sophisticated sufficient conditions for assessing robust stability of uncertain systems [5].

Within the context of Lyapunov theory, robustness of linear systems affected by parametric uncertainty can be analyzed through the search for Lyapunov functions which are either parameter independent, or parameter dependent. The notion of *quadratic stability*, namely the existence of a common quadratic Lyapunov function for all the admissible systems, allows one to address also time-varying uncertainties, but it usually leads to conservative results. This shortcoming has instigated research efforts on Lyapunov functions that are parameter-dependent. Sufficient conditions for the existence of parameter-dependent Lyapunov functions depending affinely or polynomially on the uncertain parameters have been proposed in the literature (see e.g. [6,7,8,9]). The main drawback of most of these approaches is that they either assume systems to depend affinely on the uncertain parameters, or they consider polytopic uncertainty models given as convex combinations of known nominal models. However, in real-world systems, like flight control schemes, the dependence on the uncertain parameters is much more involved (usually rational) and is modeled by linear fractional representations (LFRs) [10].

In the literature, sufficient conditions for robust stability of uncertain LFR systems, based on parameter-dependent Lyapunov functions, have been proposed by several authors. In this chapter, we will focus on three specific techniques. The first one, proposed in [11], combines the constant scaling technique for LFR systems with the use of (rationally) parameter-dependent Lyapunov functions. This technique can also deal with time-varying parameters with bounded variation rate. The other two techniques considered rely on ideas from the classical multiplier approach adopted in the absolute stability theory, which also finds useful applications in robustness analysis of uncertain systems [12]. While several authors propose the use of constant multipliers, less conservative conditions are obtained if parameter-dependent multipliers are adopted. Two different approaches along this line, which employ multiaffine parameter-dependent Lyapunov functions, have been presented in [13] and [14]. Relationships between these approaches have been investigated in [15].

The three techniques recalled above have been employed in two benchmark clearance problems defined within the COFCLUO project: the aeroelastic stability criterion, for integral models including rigid and flexible modes, and the un-piloted stability criterion, for nonlinear aircraft models (see Chapter 2). Robustness analysis is performed on the LFR models derived in Chapters 3, 4 and 5. Direct application of the considered techniques for such models turns out to be computationally unfeasible. Therefore, several relaxations relying on the choice of specific structures for the Lyapunov function and for the multipliers, are proposed. Under the assumption that the uncertain parameters and trim flight values are time-invariant, the uncertainty domain is partitioned in order to obtain accurate approximations of the robust stability region. In this respect, two alternative approaches, denoted as progressive and adaptive partitioning, are introduced. The proposed relaxations and partitioning strategies allow one to properly trade off conservatism and computational workload of the considered techniques.

The chapter is organized as follows. In Sect. 9.2, the sufficient conditions for robust stability analysis of systems in LFR form are described. Application of the above conditions to the specific clearance problems at hand is addressed in Sect. 9.3. In order to compare the potential and the computational workload of each method, the considered techniques are tested on a simple numerical example in Sect. 9.4. Finally, some conclusions are presented in Sect. 9.5.

9.2 Robustness Analysis Using Lyapunov Functions

In this section, several methods are presented for testing the robust stability of systems whose dependence on the uncertain parameters is rational, i.e. systems in LFR form. Sufficient conditions for the existence of a common or parameter-dependent quadratic Lyapunov function are provided in terms of a finite number of LMIs. Simplified structures of Lyapunov matrices and multipliers are introduced, in order to yield more conservative but computationally less demanding sufficient conditions (hereafter referred to as *relaxations*).

9.2.1 Problem Statement in LFR Framework

Consider the autonomous system

$$\dot{x}(t) = \mathbf{A}(\delta)x(t), \tag{9.1}$$

where $x \in \mathbb{R}^n$ is the state vector and $\mathbf{A}(\delta)$ is a function of the parameter $\delta \in \mathbb{R}^{n_\delta}$ according to the relation

$$\mathbf{A}(\delta) = A + B\Delta(\delta)(I - D\Delta(\delta))^{-1}C, \tag{9.2}$$

with

$$\Delta(\delta) = \operatorname{diag}(\delta_1 I_{s_1}, \ldots, \delta_{n_\delta} I_{s_{n_\delta}}). \tag{9.3}$$

and δ_i denotes the i-th component of vector δ. An equivalent linear fractional representation of the system (9.1)-(9.3) is given by:

$$\begin{cases} \dot{x}(t) = Ax(t) + Bq(t) \\ p(t) = Cx(t) + Dq(t) \\ q(t) = \Delta(\delta)p(t), \end{cases} \tag{9.4}$$

where $q \in \mathbb{R}^d$, $p \in \mathbb{R}^d$, with $d = \sum_{i=1}^{n_\delta} s_i$, and A, B, C, D are real matrices of appropriate dimensions. In view of the stability analysis of the LFR system (9.4), matrix A is assumed to be Hurwitz.

The uncertain parameter vector δ is supposed to belong to a hyper-rectangle Θ. Hereafter, we shall refer to the 2^{n_δ} vertices of Θ as Ver[Θ]. The uncertain parameters are assumed to be constant, i.e. $\dot{\delta} = 0$. System (9.4) is said to be:

- *quadratically stable*, if there exists a common quadratic Lyapunov function $V(x) = x^T P x$, for all matrices $\mathbf{A}(\delta)$, $\delta \in \Theta$;
- *robustly stable*, if $\mathbf{A}(\delta)$ is Hurwitz, for all $\delta \in \Theta$.

Clearly, quadratic stability implies robust stability, while the converse is not true. Moreover, quadratic stability is also a sufficient condition for global exponential stability of the equilibrium $x = 0$ when the uncertain parameters δ_i are time-varying.

In the following, we review three different approaches to robust stability of system (9.4), which are based on the use of parameter-dependent Lyapunov functions.

9.2.2 Wang-Balakrishnan Relaxations

In [11], several sufficient conditions for robust stability analysis of LFR systems are proposed. Here, two of them are recalled. A sufficient condition for quadratic stability of system (9.4) is stated as follows.

Proposition 9.1 (WBQ). *System* (9.4) *is quadratically stable if there exist* $P \in \mathbb{R}^{n\times n}$, $P = P^T > 0$ *and* $M \in \mathbb{R}^{d\times d}$, $M = M^T > 0$, *such that*

$$\begin{bmatrix} A^T P + PA + C^T MC & PB(\delta) + C^T MD(\delta) \\ B(\delta)^T P + D(\delta)^T MC & -M + D(\delta)^T MD(\delta) \end{bmatrix} < 0 \tag{9.5}$$

for all $\delta \in$ *Ver*[Θ], *with* $B(\delta) = B\Delta(\delta)$ *and* $D(\delta) = D\Delta(\delta)$.

To establish whether condition (9.5) is verified, one has to check feasibility of a set of 2^{n_δ} LMI constraints of dimension $n + d$, one LMI of dimension d and one LMI of dimension n. This results in a family of LMIs with a total number of free variables $n_{var} = \frac{d(d+1)}{2} + \frac{n(n+1)}{2}$.

A less conservative condition for robust stability of system (9.4), exploits parameter-dependent Lyapunov functions of the form $V(x) = x^T Q(\delta)^{-1} x$ with $Q(\delta) = Q_0 + \sum_{j=1}^{n_\delta} \delta_j Q_j$.

Proposition 9.2 (WB). *System* (9.4) *is robustly stable if there exist* $n_\delta + 1$ *symmetric matrices* $Q_0, \ldots, Q_{n_\delta} \in \mathbb{R}^{n\times n}$, *and* $N \in \mathbb{R}^{d\times d}$, $N = N^T > 0$, *such that*

$$
\begin{cases} Q(\delta) > 0 \\ \begin{bmatrix} AQ(\delta)+Q(\delta)A^T+B(\delta)NB(\delta)^T & Q(\delta)C^T+B(\delta)ND(\delta)^T \\ CQ(\delta)+D(\delta)NB(\delta)^T & D(\delta)ND(\delta)^T-N \end{bmatrix} < 0 \end{cases} \tag{9.6}
$$

for all $\delta \in Ver[\Theta]$.

In order to verify the condition in Proposition 9.2, it is necessary to solve a set of 2^{n_δ} LMIs of dimension n, 2^{n_δ} LMIs of dimension $n+d$, and one LMI of dimension d. The number of free variables is $n_{var} = (n_\delta + 1)\frac{n(n+1)}{2} + \frac{d(d+1)}{2}$.

It is important to note that when the uncertainty set Θ is symmetric with respect to the origin, the feasibility of the condition in Proposition 9.2 implies that the system is also quadratically stable, i.e. the LMIs (9.6) admit also a solution with $Q_i = 0$, $i = 1, \ldots, n_\delta$.

In order to obtain a condition which is computationally less demanding, one can reduce the number of free variables by imposing a structure on the scaling matrices M and N. For example, the choice of a diagonal M in the WBQ condition, or a diagonal N in the WB condition, can significantly speed up the solution of the LMI feasibility problem when d is much larger than n, however, at the price of a higher conservatism. We refer to these conditions as *WBQ-dM relaxation* and *WB-dN relaxation*, respectively.

9.2.3 Fu-Dasgupta Relaxations

Another approach, based on parameter-dependent Lyapunov functions, provides sufficient conditions for robust stability of LFR systems. It is usually referred to as the *parametric multiplier* approach. It can be seen as a generalization of the traditional multiplier approach, used in absolute stability theory [16], employing parameter-dependent multipliers. The main idea of the approach presented in [13], is to seek a transfer matrix with affine structure, called affine multiplier, which, when cascaded with another matrix related to the uncertain system, will result in a strictly positive real transfer matrix. This affine multiplier can be found by solving a set of LMIs, whose feasibility guarantees the existence of a parameter-dependent Lyapunov function for the uncertain LFR system.

In order to cast the problem into the framework adopted in [13] let us introduce the matrices

$$
\begin{aligned} C(\delta) &:= \Delta(\delta)C = \textstyle\sum_{i=1}^{n_\delta} \delta_i C_i \\ D(\delta) &:= -I + \Delta(\delta)D = D_0 + \textstyle\sum_{i=1}^{n_\delta} \delta_i D_i, \end{aligned}
$$

where $C_i = T_i C$, $D_0 = -I$, $D_i = T_i D$, for $i = 1, \ldots, n_\delta$, and

$$
T_i = \mathrm{diag}\,(0_{s_1}, \ldots, 0_{s_{i-1}}, I_{s_i}, 0_{s_{i+1}}, \ldots, 0_{s_{n_\delta}}).
$$

A sufficient condition for robust stability of system (9.4) is introduced in the next result, by using a multi-affine parameter-dependent Lyapunov function (mapdlf), i.e. $V(x) = x^T P(\delta) x$, where

$$P(\delta) = P_0 + \sum_{j=1}^{n_\delta} \delta_j P_j + \sum_{i=1}^{n_\delta} \sum_{j=i+1}^{n_\delta} \delta_i \delta_j P_{ij} + \dots \tag{9.7}$$

is a generic symmetric multi-affine matrix function of δ.

Proposition 9.3 (FD). *System* (9.4) *is robustly stable if there exist* $2n_\delta + 2$ *matrices* $C_{\mu,i} \in \mathbb{R}^{d\times n}$, $D_{\mu,i} \in \mathbb{R}^{d\times d}$ *for* $i = 0, \dots, n_\delta$ *such that*

$$\begin{bmatrix} C_i^T \\ D_i^T \end{bmatrix} \begin{bmatrix} C_{\mu,i} & D_{\mu,i} \end{bmatrix} + \begin{bmatrix} C_{\mu,i}^T \\ D_{\mu,i}^T \end{bmatrix} \begin{bmatrix} C_i & D_i \end{bmatrix} \le 0 \tag{9.8}$$

for $i = 1, \dots, n_\delta$, *and*

$$\begin{cases} P(\delta) > 0 \\ \begin{bmatrix} \mathbf{A}^T(\delta)P(\delta) + P(\delta)\mathbf{A}(\delta) & \Pi_{1,2}(\delta) \\ \Pi_{1,2}^T(\delta) & \Pi_{2,2}(\delta) \end{bmatrix} < 0, \end{cases} \tag{9.9}$$

for all $\delta \in Ver[\Theta]$, *where* $P(\delta)$ *is given by* (9.7) *and*

$$\begin{aligned} \Pi_{2,2}(\delta) &= -\left(D_\mu(\delta)D^{-1}(\delta) + D^{-T}(\delta)D_\mu^T(\delta)\right) \\ \Pi_{1,2}(\delta) &= P(\delta)BD^{-1}(\delta) - C_\mu^T(\delta) + C^T(\delta)D^{-T}(\delta)D_\mu^T(\delta) \end{aligned}$$

with

$$\begin{aligned} C_\mu(\delta) &= C_{\mu,0} + \textstyle\sum_{i=1}^{n_\delta} \delta_i C_{\mu,i} \\ D_\mu(\delta) &= D_{\mu,0} + \textstyle\sum_{i=1}^{n_\delta} \delta_i D_{\mu,i} \ . \end{aligned}$$

The family of LMIs (9.8)-(9.9) is composed by $n_\delta + 2^{n_\delta}$ constraints of dimension $n + d$, and 2^{n_δ} constraints of dimension n. The total number of free variables is $n_{var} = (n_\delta + 1)(nd + d^2) + 2^{n_\delta}\left(\frac{n(n+1)}{2}\right)$, the rightmost term being due to the multi-affine parametrization of $P(\delta)$ in (9.7).

The number of free variables in the sufficient condition of Proposition 9.3 can be reduced by simplifying the structure of the affine parametric multiplier and/or the structure of the Lyapunov function, at the price of a higher degree of conservatism, but with significant benefits in terms of reduced computational complexity. In this paper, we will consider two possible relaxations of the FD condition:

- *FD-cμ relaxation*: full constant multipliers $C_\mu(\delta) = C_{\mu,0}$, $D_\mu(\delta) = D_{\mu,0}$ (i.e., $C_{\mu,i} = 0$, $D_{\mu,i} = 0$, for $i = 1, \dots, n_\delta$);
- *FD-cdμ relaxation*: constant diagonal multipliers, which is the same as FD-cμ, but with diagonal matrices $C_{\mu,0}$ and $D_{\mu,0}$.

Moreover, we will consider the following simplified structures for the Lyapunov function:

- common Lyapunov function (clf), $P(\delta) = P_0$;
- affine parameter-dependent Lyapunov function (apdlf),

$$P(\delta) = P_0 + \sum_{j=1}^{n_\delta} \delta_j P_j.$$

It is worth remarking that the combined use of FD-cμ or FD-cdμ relaxations with the clf structure, guarantees quadratic stability of the LFR system.

9.2.4 Dettori-Scherer Relaxations

Another way to assess robust stability of system (9.4) by jointly using multi-affine parameter-dependent Lyapunov functions and parameter-dependent multipliers has been proposed in [14]. The resulting sufficient condition can be stated as follows.

Proposition 9.4 (DS). *System* (9.4) *is robustly stable if there exist two matrices* $S_0, S_1 \in \mathbb{R}^{d\times d}$ *such that*

$$\begin{cases} P(\delta) > 0 \\ \left[\begin{array}{cc} I & 0 \\ A & B \\ \hline 0 & I \\ C & D \end{array}\right]^T \left[\begin{array}{cc|c} 0 & P(\delta) & 0 \\ P(\delta) & 0 & 0 \\ \hline 0 & 0 & W(\delta) \end{array}\right] \left[\begin{array}{cc} I & 0 \\ A & B \\ \hline 0 & I \\ C & D \end{array}\right] < 0 \end{cases} \tag{9.10}$$

for all $\delta \in \mathrm{Ver}[\Theta]$, *where* $P(\delta)$ *is given by* (9.7) *and*

$$W(\delta) = \begin{bmatrix} S_1 + S_1^T & -S_0 - S_1 \Delta(\delta) \\ -S_0^T - \Delta(\delta) S_1^T & S_0^T \Delta(\delta) + \Delta(\delta) S_0 \end{bmatrix}, \tag{9.11}$$

with $\Delta(\delta)$ *given by* (9.3).

The set of constraints (9.10) consists of 2^{n_δ} LMIs of dimension $(n+d)$ and 2^{n_δ} LMIs of dimension n, with a total number of free variables $n_{var} = 2d^2 + 2^{n_\delta}\left(\frac{n(n+1)}{2}\right)$. Once again, one can introduce relaxed versions of the DS condition by choosing simplified structures for the matrices S_0 and S_1 parameterizing the multiplier $W(\delta)$, and for the Lyapunov matrix $P(\delta)$. We will refer to the robustness stability test (9.10) with diagonal matrices S_0 and S_1 in (9.11), as *DS-dS relaxation.* Moreover, we will consider both common (clf) and affine parameter-dependent (apdlf) Lyapunov functions, as for the FD relaxations.

Remark 9.1. In [15], it has been proved that the FD-cμ condition with full constant multipliers $C_{\mu,0}$, $D_{\mu,0}$ is in general less conservative than the DS condition with full free matrices S_0, S_1. The two conditions turn out to be equivalent when the C matrix of the LFR system is full column rank. The existence of relationships between these techniques, when a more restrictive structure of the multipliers is imposed (e.g. between FD-cdμ and DS-dS) is still an open issue. Numerical comparisons between the above methods and relaxations on a simple numerical example are reported in Sect. 9.4.

9.2.5 Special Case: Affine Parameter-Dependent Systems

In Chapter 6, an H_2 approximation technique has been employed to derive reduced-order LFR models of the aircraft dynamics. In particular, some of these models turned out to depend linearly on the uncertain parameters, i.e. to be of the form (9.1) with

$$\mathbf{A}(\delta) = A_0 + A_1\delta_1 + \ldots + A_{n_\delta}\delta_{n_\delta}. \tag{9.12}$$

Clearly, (9.12) is a special case of the general LFR (9.2)-(9.3), where $s_1 = s_2 = \cdots = s_n = n$ and

$$A = A_0\ , \quad B = [A_1\ A_2\ \ldots\ A_{n_\delta}]\ , \quad C = \begin{bmatrix} I_n \\ \vdots \\ I_n \end{bmatrix}\ , \quad D = 0.$$

A number of sufficient conditions for dealing with robust stability of systems with affine parameter-dependence has been derived in the literature (the interested reader is referred to [9] and references therein). Here, we briefly recall two of them.

Proposition 9.5. *System* (9.1), *with* $\mathbf{A}(\delta)$ *given by* (9.12), *is quadratically stable if and only if there exists* $P \in \mathbb{R}^{n\times n}$, $P = P^T > 0$ *such that*

$$A^T(\delta)P + PA(\delta) < 0\ , \qquad \forall \delta \in \mathrm{Ver}[\Theta]. \tag{9.13}$$

The next proposition is based on the multi-convexity condition given in [6].

Proposition 9.6. *System* (9.1), *with* $\mathbf{A}(\delta)$ *given by* (9.12), *is robustly stable if there exist* $n_\delta + 1$ *symmetric matrices* $P_0, P_1, \ldots, P_{n_\delta}$ *such that*

$$A_i^T P_i + P_i A_i \geq 0 \qquad \forall i = 1, \ldots, n_\delta \tag{9.14}$$

and

$$\begin{cases} P(\delta) = P_0 + \delta_1 P_1 + \ldots + \delta_{n_\delta} P_{n_\delta} > 0 \\ A^T(\delta)P(\delta) + P(\delta)A(\delta) - P_0 < 0, \end{cases} \tag{9.15}$$

for all $\delta \in \mathrm{Ver}[\Theta]$.

The set of constraints (9.13) consists of $2^{n_\delta} + 1$ LMIs of dimension n with $n_{var} = \frac{n(n+1)}{2}$, while (9.14)-(9.15) amounts to $2^{n_\delta+1} + n_\delta$ LMIs of dimension n with a number of free variables equal to $n_{var} = (n_\delta + 1)\frac{n(n+1)}{2}$.

9.3 Clearance of Flight Control Schemes

In this section, the robustness analysis techniques presented in Sect. 9.2 are applied to two specific flight clearance problems introduced in Chapter 2: the aeroelastic stability and the un-piloted stability criteria.

9.3.1 Aeroelastic Stability Criterion

Within the COFCLUO project, several clearance criteria have been proposed for the certification process of the integral model of an aircraft, accounting for both rigid and flexible modes. The *aeroelastic stability* criterion (also called *eigenvalue criterion*) requires that the largest real part of the closed-loop eigenvalues $(\lambda_1, \lambda_2, \cdots, \lambda_n)$ of the linearized aircraft model is negative, i.e.

$$\max_{j\in\{1,\ldots,n\}} \Re(\lambda_j) < 0, \tag{9.16}$$

for all admissible values taken by the uncertain parameters (describing the aircraft mass configuration) and the trimmed flight variables (namely, Mach number and calibrated air speed).

In order to address this criterion, LFR models describing the closed-loop aeroelastic behavior of the aircraft have been derived in Chapters 4 and 5. With the mentioned uncertain parameters and flight variables as the entries of the δ vector in (9.3), the techniques presented in Sect. 9.2 can be directly applied to verify which of these LFR models satisfy the eigenvalue criterion in (9.16) for all uncertain parameters and flight variables $\delta \in \Theta$, i.e. to prove that

$$\max_{\delta\in\Theta} \max_{j\in\{1,\ldots,n\}} \Re(\lambda_j(\delta)) < 0,$$

where $\lambda_j(\delta)$, $j = 1, \ldots, n$, denote the eigenvalues of $\mathbf{A}(\delta)$ in (9.2).

The aeroelastic stability criterion can be tackled also by considering the reduced-order LFR models derived in Chapter 6. When the dependence on the uncertain parameters is affine, the conditions presented in Propositions 9.5 and 9.6 can be employed to verify robust stability of such LFR models.

9.3.2 Un-piloted Stability Criterion

Several clearance criteria have also been presented for the certification process of the nonlinear aircraft model considered in the COFCLUO benchmark. The *un-piloted stability* criterion requires that the augmented system remains stable, for all admissible values of the position of the center of gravity assumed by the control law, and for all possible trimmed point values. Because of the presence of pilot-in-the-loop control, the stability requirement is relaxed to include slowly divergent modes, provided that the time of doubling of the divergent variables is more than 6 sec.

LFR models representative of the nonlinear aircraft and controller dynamics within given regions in the flight envelope, have been derived in Chapters 3 and 5. Due to the presence of saturations in the aircraft actuators, the Δ block of the resulting LFR models contains not only uncertain parameters (mass, position of center of gravity assumed by the controller) and trimmed flight variables (Mach, calibrated air speed), but also memoryless nonlinearities (dead-zones). Therefore, two different un-piloted stability clearance problems are addressed:

i) Robustness analysis of LFR models containing only flight envelope variables and/or uncertain parameters in the Δ block, with dead-zones set to their nominal values (which corresponds to assuming that actuators do not saturate). In this case, all the robustness analysis techniques described in Sect. 9.2 can be applied directly to check the relaxed robust stability requirement, i.e. that the largest real part of the eigenvalues of the LFR matrix $\mathbf{A}(\delta)$ does not exceed $\log(2)/6$. Notice that, due to the structure of $\mathbf{A}(\delta)$ in (9.2), this can be imposed by simply replacing the open-loop matrix A by the shifted matrix $A - \frac{\log(2)}{6} I$. Hereafter, this relaxed stability condition will be referred to as *shifted* stability.
ii) Robustness analysis of LFR models containing dead-zones in the Δ block: in this case, dead-zones are treated as sector-bounded time-varying uncertainties, within the sector $[0, 1]$. The techniques of Sect. 9.2 can still be applied to check shifted stability, provided that a common Lyapunov function (clf) and parameter-independent multipliers are employed.

Lyapunov-based conditions for dealing with LFR models with dead-zones have been recently proposed in [17, 18, 19, 20]. Such conditions are usually less conservative than those treating dead-zones as sector-bounded uncertainties. However, they do not consider the presence of time-invariant parametric uncertainties in the LFR structure, and therefore can be applied to the considered LFR models only after setting the uncertain parameters and trimmed flight values to a constant nominal value (see Chapter 14).

9.3.3 Partitioning of the Uncertainty Domain

Since all the conditions presented in Sect. 9.2 are only sufficient, they may be too conservative to clear the entire uncertainty domain, however they may succeed in clearing portions of this domain. This is still acceptable for the considered criteria, as the uncertain parameters and the trim flight values are assumed to be time-invariant. This has motivated the formulation of two approaches, namely, the *progressive tiling* and the *adaptive tiling*, which rely on partitioning of the uncertainty domain. These are described next.

In the progressive tiling, the idea is to progressively partition the uncertainty domain Θ into hyper-rectangular regions (hereafter, *tiles*), and then, apply the robustness analysis conditions presented in Sect. 9.2 to each tile. The clearance procedure is carried out by successively reducing the sizes of the uncleared tiles; this step involves bisection of each side of an uncleared tile. The size of each uncleared tile is reduced until the whole domain is cleared, or the predefined maximum number of bisection steps is reached. To initialize the procedure, the initial tile is set equal to the whole uncertainty domain (or a hyper-rectangle containing the actual domain, when this is a generic polytope). For each tile, the LFR system (9.4) is re-parameterized, so that the resulting uncertainty region is centered in the origin of the normalized uncertainty space (recall that this implies that the conditions WB and WBQ are equivalent). If the LFR model contains dead-zones, such as in the un-piloted stability criterion, these are treated as sector-bounded time-varying uncertainties. Hence, partitioning is not performed with respect to the corresponding parameters, but only with respect to the time-invariant flight parameters.

With the objective of improving the efficiency of the clearance process, one may choose to combine progressive tiling with an *adaptive* choice of the robustness condition. The idea is to proceed as in the progressive tiling approach, but to change the condition whenever a predefined partitioning level has been reached. For example, one may employ conservative (but fast) methods in the first steps of the progressive tiling, with the aim to clear large regions of the uncertainty domain which are "easy" (in the sense that they can be cleared even by the most conservative techniques). Then, as soon as a predefined number of partitions is reached, one can employ more powerful (and more computationally demanding) techniques only for the "difficult" tiles, that have not been cleared in the first steps. In order to test this idea, adaptation with respect to the choice of the Lyapunov function has been performed (clearly, adaptation with respect to the multipliers structure can also be considered).

9.3.4 Gridding

The flight/uncertainty domain of interest may contain models which are unstable. Since a model corresponds to an aircraft configuration with fixed values for the uncertain parameters and the trimmed flight variables, such unstable models will render the entire flight/uncertainty domain unclearable. Hence, in order to save computational time, before attempting to clear each tile by applying one of the methods presented in Sect. 9.2, a coarse gridding of the tile is performed. This involves the selection of a set of uniformly spread models within the tile. If any one of these models is found to be unstable, clearance of the tile is not attempted and the tile containing the model is temporarily marked as unstable. Conversely, if there are no unstable models within the tile grid, then the selected clearance method is applied. Notice that as the partitioning of the domain proceeds, tiles that have been marked as unstable are successively partitioned and portions of them can be later cleared. When the maximum number of partitions is reached, the tiles containing unstable models found by gridding are finally marked as unstable. Clearly, by suitably defining the maximum number of partitions (and hence the minimum tile size), one can obtain an approximation to the desired precision of the domain which can be cleared by the considered robustness analysis technique.

9.3.5 A Graphical User Interface for Robust Stability Clearance Problems

A Graphical User Interface (GUI) has been developed to facilitate the selection of the various options in the clearance problem, and to present the results of the robustness analysis process. The GUI provides a user interactive set-up for the two considered clearance criteria, and for a collection of closed-loop LFR models. The software package comprises of a collection of routines coded in Matlab and exploits three publicly available software packages: YALMIP [21], SDPT3 [22] and LFR Toolbox [23]. Figure 9.1 shows a screenshot of the GUI panel.

Figure 9.2 is an example of graphical presentation of a clearance analysis result for an LFR model with $n_\delta = 2$. The light gray areas represent regions where the corresponding model has been cleared; dark gray areas denote regions where the LFR model could not be cleared (due to the presence of unstable models found by gridding); white areas represent regions where no unstable models were found by gridding, but stability of the model could not be certified with the chosen robustness analysis technique.

Besides graphical results, the GUI returns also detailed information about all the outcomes of the clearance process, and in particular:

- the number of optimization problems that have been solved (corresponding to the number of tiles attempted to be cleared);
- the time elapsed in the course of the clearance procedure;
- the percentage of the cleared domain with respect to the whole uncertainty domain;

- the percentage of the uncertainty domain which, at the end of the tiling procedure, was found to contain closed-loop unstable models by gridding;
- the percentage of the uncertainty domain, which could neither be defined as unstable after gridding, nor be certified as cleared by the tested method;
- the *clearance rate*, defined as ratio between the cleared uncertainty region and the *clearable* domain (i.e the domain that did not contain closed-loop unstable models found by gridding).

The latter indicator can be considered as a measure of performance of the chosen clearance technique (with the precision allowed by the selected minimum tile size).

An extensive collection of results obtained by applying the proposed robustness analysis tools to the considered clearance problems are reported in Chapter 14.

9.4 Numerical Example

In this section, the methods previously discussed are applied to a simple numerical example, in order to highlight the potential of the considered techniques and to offer insight into the computational workload one may expect when techniques are applied to high dimensional systems.

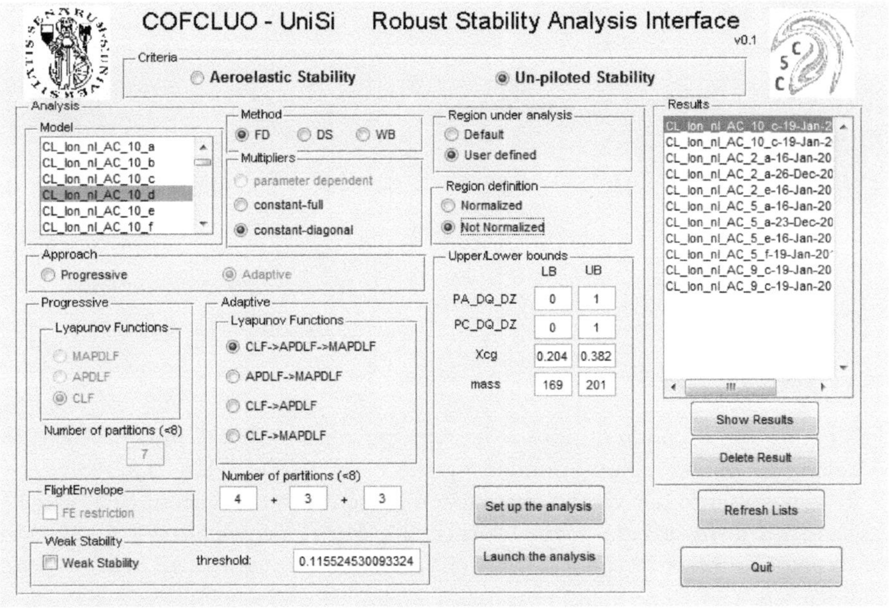

Fig. 9.1 GUI panel for robust stability analysis.

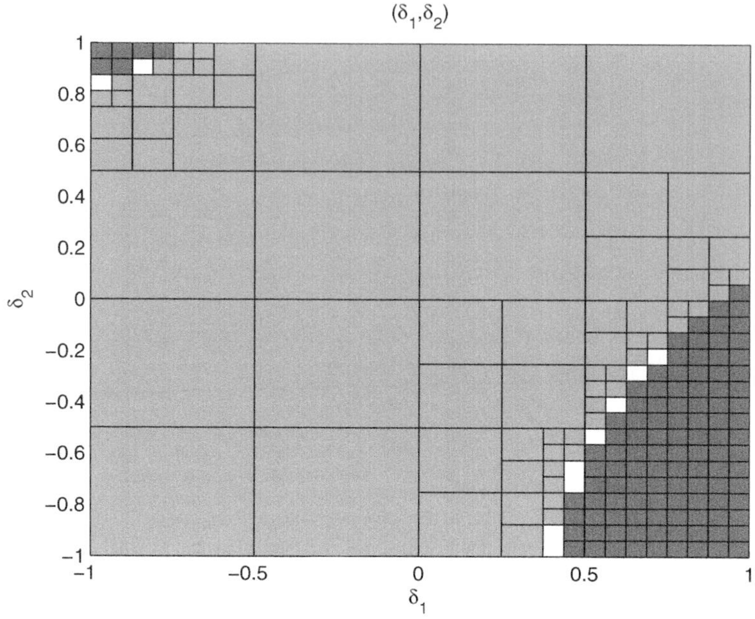

Fig. 9.2 Example of clearance result: cleared (light gray), unstable (dark gray), unknown (white).

Consider the system

$$\dot{x}(t) = f(\delta)x(t), \tag{9.17}$$

with $x \in \mathbb{R}$ and

$$\begin{aligned} f(\delta) = {} & 672.2\delta_1^6 + 401.7\delta_1^5\delta_2 - 75\delta_1^5 - 727.9\delta_1^4\delta_2^2 + 233.6\delta_1^4\delta_2 - 30.5\delta_1^4 \\ & -254.5\delta_1^3\delta_2^3 - 71\delta_1^3\delta_2^2 - 19.6\delta_1^3\delta_2 + 26.6\delta_1^3 - 233.5\delta_1^2\delta_2^3 + 40.2\delta_1^2\delta_2^2 \\ & +64.3\delta_1^2\delta_2^4 + 2\delta_1^2\delta_2 + 6.6\delta_1^2 - 282.1\delta_1\delta_2^5 + 152\delta_1\delta_2^4 - 9.3\delta_1\delta_2^3 \\ & -19.7\delta_1\delta_2^2 + 3.7\delta_1\delta_2 + 1.9\delta_1 + 251.8\delta_2^6 + 234.6\delta_2^5 - 64.8\delta_2^4 \\ & -19.9\delta_2^3 + 9\delta_2^2 + 1.7\delta_2 - 3.1 \end{aligned} \tag{9.18}$$

is a bivariate polynomial in $\delta = (\delta_1, \delta_2)$, such that $f(0) < 0$ and $f(\delta) < 0$ for δ belonging to the region bounded by the black curve in Fig. 9.3.

It is possible to derive an equivalent representation of system (9.17)-(9.18) in the LFR form (9.4) by using tools in [23]. For the considered $f(\delta)$, the size of the $\Delta(\delta)$ matrix in the LFR system turns out to be $d = 27$, with $s_1 = 21$ and $s_2 = 6$.

Table 9.1 reports the results obtained by applying the robustness analysis techniques presented in Sect. 9.2 to the considered system. The GUI software described in Sect. 9.3.5 has been run on a PC equipped with an Intel XEON

Table 9.1 Comparison among the techniques on the analytical example.

Method	Clearance rate	NOPs	Time (h:m:s)
WBQ	1	75	0 : 04 : 45
WBQ-dM	0.8655	147	0 : 05 : 59
WB	1	75	0 : 05 : 57
WB-dM	0.8655	147	0 : 08 : 07
DS (clf)	1	71	0 : 15 : 10
DS (apdlf)	1	71	0 : 16 : 12
DS (mapdlf)	1	71	0 : 16 : 54
DS-dS (clf)	0.9793	79	0 : 03 : 28
DS-dS (apdlf)	0.9828	79	0 : 04 : 31
DS-dS (mapdlf)	0.9828	75	0 : 04 : 55
FD-cμ (clf)	1	71	0 : 03 : 41
FD-cμ (apdlf)	1	71	0 : 03 : 54
FD-cμ (mapdlf)	1	71	0 : 04 : 04
FD-cdμ (clf)	0.9138	111	0 : 01 : 46
FD-cdμ (apdlf)	0.9138	111	0 : 01 : 49
FD-cdμ (mapdlf)	0.9138	111	0 : 01 : 51

Table 9.2 Adaptive tiling DS-dS.

Method	clf	apdlf	Clearance rate	NOPs	Time (h:m:s)
DS-dS	0	5	0.9827	79	0 : 04 : 31
	1	4	0.9827	79	0 : 04 : 46
	2	3	0.9827	79	0 : 04 : 51
	3	2	0.9827	79	0 : 04 : 42
	4	1	0.9827	79	0 : 04 : 22

5150 processor and 4 Gbyte of DDRII RAM. In Table 9.1, the clearance rate is defined according to Sect. 9.3.5, `NOPs` is the number of LMI optimization problems solved and `Time` is the total elapsed time required by the clearance process. All relaxations have been applied with progressive partitioning and maximum number of partitions equal to 5. Figure 9.3 illustrates the cleared areas and the resulting tiling patterns for the relaxations FD-cμ, FD-cdμ, DS-dS and WBQ-dM, all applied with a common quadratic Lyapunov function. Tile colors are defined according to the example in Figure 9.2.

The different tiling patterns in Fig. 9.3 testify the different potential of the relaxations within the progressive partitioning approach. In particular, it can be observed that the techniques employing diagonal scaling matrices or multipliers fail to clear some tiles that are close to the boundary of the stability region. On the other hand, they often manage to significantly reduce the

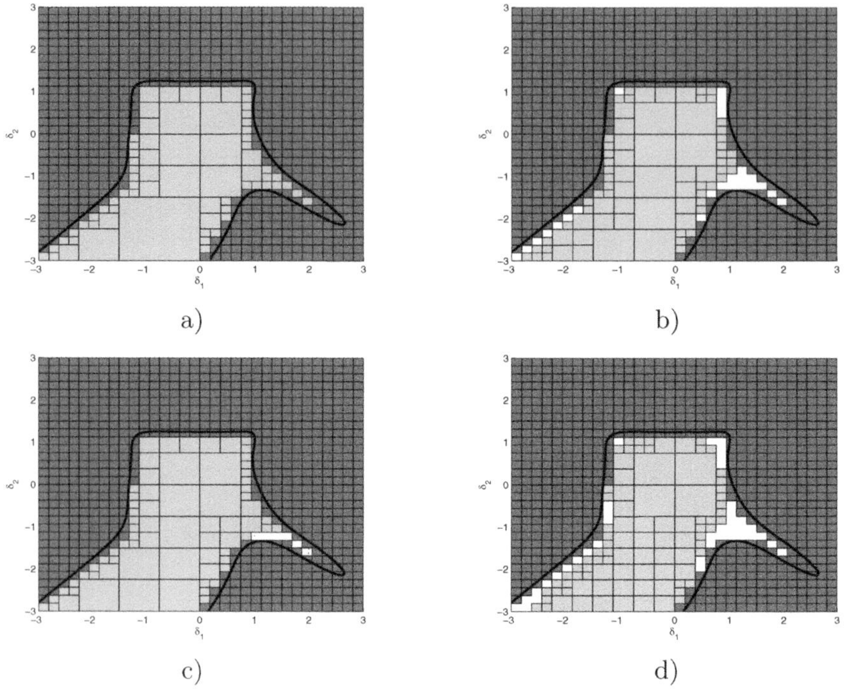

Fig. 9.3 Stability regions: a) FD-cμ (clf). b) FD-cdμ (clf). c) DS-dS (clf). d) WBQ-dM.

computational time; notice however that this is not the case for WBQ-dM and WB-dN due to the much larger number of optimization problems to be solved. It is also interesting to note that, being the C matrix full column rank, the conditions DS and FD-cμ are equivalent (see Remark 9.1). However, the times required by DS are always larger than those employed by FD-cμ, which seems to suggest that the parametrization of the FD-cμ condition is more efficient. On the whole, one may conclude that in this example, the FD-cdμ and DS-dS relaxations offer a reasonable compromise between conservativeness and computational workload. This will be confirmed in most clearance problems tackled in Chapter 14, although it seems difficult to establish a general hierarchy between the methods with respect to this fundamental trade-off.

Table 9.2 shows the results obtained by applying the adaptive tiling procedure with the DS-dS relaxation. The second and third column report the number of partitioning steps in which the clf and apdlf structures of the Lyapunov function have been employed. As expected, the clearance rate is always the same because apdlf is always applied in the last partitioning step. The minimum computational time is obtained with clf in the first 4 steps and apdlf only in the last step. Though the benefit here is minor, the potential of the adaptive tiling strategy is more evident when addressing robust stability of large LFR models, for which the computational time may vary significantly with different choices of the Lyapunov matrix structure (see applications to clearance problems in Chapter 14).

9.5 Conclusions

This chapter has presented several relaxations of sufficient conditions for robust stability of uncertain systems in LFR form. The relaxations rely on the combined choice of simplified structures for the Lyapunov function and the multiplier or scaling matrices employed in the sufficient conditions. The motivation for the introduction of such relaxations is the possibility of trading off conservativeness and computational load in the robustness analysis of high dimensional systems with rational dependence on the uncertain parameters. The proposed techniques have been applied to a set of LFR models representing the longitudinal dynamics of an augmented civil aircraft, used as a benchmark within the COFCLUO project. Two clearance criteria, aeroelastic stability and un-piloted stability, have been cast as robustness analysis problems. The results of the clearance processes are reported in Chapter 14.

References

1. Fielding, C., Varga, A., Bennani, S., Selier, M. (eds.): Advanced Techniques for Clearance of Flight Control Laws. LNCIS, vol. 283. Springer, Heidelberg (2002)
2. Bates, D.G., Kureemun, R., Mannchen, T.: Improved clearance of a flight control law using μ-analysis techniques. Journal of Guidance, Control and Dynamics 26(6), 869–884 (2003)
3. Menon, P.P., Bates, D., Postlethwaite, I.: Nonlinear robustness analysis of flight control laws for highly augmented aircraft. Control Engineering Practice 15, 655–662 (2007)
4. Juliana, S., Chu, Q.P., Mulder, J.A.: Reentry flight clearance using interval analysis. Journal of Guidance, Control and Dynamics 31(5), 1295–1306 (2008)
5. Boyd, S., El Ghaoui, L., Feron, E., Balakrishnan, V.: Linear Matrix Inequalities in System and Control Theory. Studies in Applied Mathematics. SIAM, Philadelphia (1994)
6. Gahinet, P., Apkarian, P., Chilali, M.: Affine parameter-dependent Lyapunov functions and real parametric uncertainty. IEEE Trans. on Automatic Control 41(3), 436–442 (1996)
7. Peaucelle, D., Arzelier, D., Bachelier, O., Bernussou, J.: A new robust $\mathcal{D}$-stability condition for real convex polytopic uncertainty. Systems and Control Letters 40, 21–30 (2000)
8. Leite, V.J.S., Peres, P.L.D.: An improved LMI condition for robust $\mathcal{D}$-stability of uncertain polytopic systems. IEEE Trans. on Automatic Control 48(3), 500–504 (2003)

9. Chesi, G., Garulli, A., Tesi, A., Vicino, A.: Polynomially parameter-dependent Lyapunov functions for robust stability of polytopic systems: an LMI approach. IEEE Trans. on Automatic Control 50(3), 365–370 (2005)
10. Zhou, K., Doyle, J.C., Glover, K.: Robust and Optimal Control. Prentice-Hall Inc., Englewood Cliffs (1996)
11. Wang, F., Balakrishnan, V.: Improved stability analysis and gain-scheduled controller synthesis for parameter-dependent systems. IEEE Trans. on Automatic Control 47(5), 720–734 (2002)
12. Dasgupta, S., Chockalingam, G., Anderson, B.D.O., Fu, M.: Lyapunov functions for uncertain systems with applications to the stability of time varying systems. IEEE Transactions on Circuits and Systems - I 41(2), 93–106 (1994)
13. Fu, M., Dasgupta, S.: Parametric Lyapunov function for uncertain systems: The multiplier approach. In: El Ghaoui, L., Niculescu, S.-I. (eds.) Advances in Linear Matrix Inequality Methods in Control. SIAM, Philadelphia (2000)
14. Dettori, M., Scherer, C.: New robust stability and performance conditions based on parameter dependent multipliers. In: Proc. of 39th IEEE Conf. on Decision and Control, Sydney, Australia, pp. 4187–4192 (2000)
15. Ebihara, Y., Hagiwara, T.: A dilated LMI approach to robust performance analysis of linear time-invariant uncertain systems. Automatica 41(11), 1933–1941 (2005)
16. Popov, V.M.: Absolute stability of nonlinear systems of automatic control. Automation and Remote Control 22, 857–875 (1962)
17. Gomes da Sliva Jr., J.M., Tarbouriech, S.: Anti-windup design with guaranteed regions of stability: an LMI-based approach. IEEE Transactions on Automatic Control 50(1), 106–111 (2005)
18. Hu, T., Teel, A.R., Zaccarian, L.: Stability and performance for saturated systems via quadratic and nonquadratic Lyapunov functions. IEEE Transactions on Automatic Control 51(11), 1770–1786 (2006)
19. Dai, D., Hu, T., Teel, A.R., Zaccarian, L.: Piecewise-quadratic Lyapunov functions for systems with deadzones or saturations. Systems & Control Letters 58(5), 365–371 (2009)
20. Biannic, J.-M., Tarbouriech, S.: Optimization and implementation of dynamic anti-windup compensators with multiple saturations in flight control systems. Control Engineering Practice 17(6), 707–713 (2009)
21. Löfberg, J.: YALMIP: A toolbox for modeling and optimization in MATLAB. In: Proceedings of the CACSD Conference, Taipei, Taiwan (2004), `http://users.isy.liu.se/johanl/yalmip/`
22. Toh, K.C., Todd, M.J., Tütüncü, R.H.: SDPT3 - a Matlab software package for semidefinite programming. Optimization Methods and Software 11, 545–581 (1999)
23. Magni, J.-F.: Linear Fractional Representation Toolbox (version 2.0) for use with Matlab. Technical report (2006), `http://www.onera.fr/staff-en/jean-marc-biannic/docs/lfrtv20s.zip`

Chapter 10
IQC-Based Analysis Techniques for Clearance

Christos Papageorgiou, Rikard Falkeborn, and Anders Hansson

Abstract. This chapter presents the use of Integral Quadratic Constraints (IQCs) for solving flight control clearance problems. The theory of IQCs provides a powerful framework for the robustness analysis of control systems with respect to a very broad range of uncertainties and nonlinearities. The clearance criterion of robust stability with respect to parameter variations is addressed by employing the standard robust stability theorem of IQCs and by a suitable choice of an IQC for real parametric uncertainty. In addition, we use IQCs to solve two specific flight control clearance problems which are formulated as robust performance problems with respect to real parameter variations. These problems are the stability margins criterion and the comfort criterion with respect to turbulence which are formulated as robust $\mathcal{H}_\infty$ and $\mathcal{H}_2$ problems respectively. The formulation of a flight control clearance problem using IQCs results in a convex optimization problem involving Linear Matrix Inequalities (LMIs) for which there exist efficient, numerical solvers. Even so, there exist limitations related to increased computational complexity in case of optimization problems resulting from the analysis of large systems.

10.1 Introduction

The clearance of a flight control law is defined as the procedure which proves that the control law will satisfy the clearance criteria for a given flight envelope and in the presence of uncertainty [1]. The clearance criteria are measures to quantify stability and performance under all possible flight conditions. The flight envelope is defined as the set of possible flight conditions and is

Christos Papageorgiou · Rikard Falkeborn · Anders Hansson
Division of Automatic Control, Linköping University, SE-581 83 Linköping, Sweden
e-mail: papageorgiou@isy.liu.se, falkeborn@isy.liu.se,
hansson@isy.liu.se

A. Varga et al. (Eds.): Opti. Based Clearance of Flight Control Laws, LNCIS 416, pp. 179–201.
springerlink.com

characterized by flight parameters such as speed, altitude and angle of attack. The uncertainty is defined as the set of uncertain parameters whose values are not known but they have known ranges of variation and bounded rates of variation.

In this chapter we propose the formulation of a control law clearance task as a convex optimization problem by using ideas from robust control theory [4]. We formulate the clearance task either as a robust stability or a robust performance problem with respect to variations in the uncertain parameters and the flight envelope parameters. The powerful paradigm of Linear Fractional Representations (LFR) is used to avoid the gridding of the flight envelope and the uncertain parameter space. Using this approach, the uncertain closed-loop system is embedded with some conservatism in an LFR representation.

A solution to a lot of robustness analysis problems such as robust stability with respect to uncertain parameters or robust $\mathcal{H}_2$ and $\mathcal{H}_\infty$ minimization can be obtained using convex optimization and in particular Semidefinite Programming (SDP). An SDP problem is an optimization problem with the cost function being a linear function of the decision variables and the constraints being Linear Matrix Inequalities (LMIs) [5]. The formulation of a robustness analysis problem as an SDP results by searching for Lyapunov functions (either constant or parameter-dependent) to prove stability and performance or by assessing robust stability using the theory of IQCs [6] in conjunction with the Kalman-Yakubovich-Popov (KYP) Lemma.

In this chapter the IQC theory is used to evaluate the robust stability of the flight control system with respect to variations in real uncertain parameters. In conjunction with IQC theory, a tool from robust control theory is used in order to solve the stability margins problem. This tool is the generalized stability margin b_{PC} which is associated with the $\mathcal{H}_\infty$ loop-shaping methodology [7] for designing robust controllers. Finally, we use an IQC formulation of a robust $\mathcal{H}_2$ performance problem [22] in order to evaluate the comfort criterion with respect to excitation from turbulence.

10.2 Analysis of Robust Stability and Robust Performance Using IQCs

This section will give a brief introduction to IQCs and describe some of the theory needed in order to formulate the clearance criteria. The robust stability criterion with respect to parametric uncertainty is a direct outcome of the IQC robust stability condition of Theorem 10.1 and is not treated in more detail. The clearance criteria which are formulated as robust performance problems are presented in the following sections.

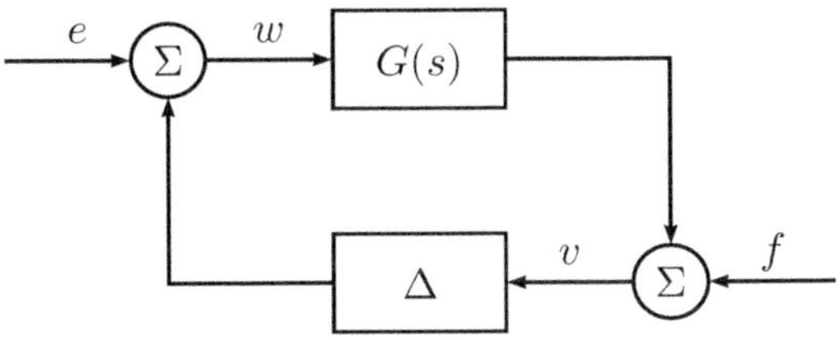

Fig. 10.1 Basic feedback interconnection considered.

The uncertain system under consideration is defined by

$$\begin{aligned} v &= Gw + f, \\ w &= \Delta(v) + e, \end{aligned} \tag{10.1}$$

and is illustrated in Fig. 10.1. We assume that G is a known stable linear time-invariant (LTI) system and that the operator Δ is causal and has bounded gain. The signals f and e correspond to finite energy external disturbances. The operator Δ could represent a parametric uncertainty, a time-delay or a general nonlinear function such as a saturation or a deadzone. This is precisely the system studied in [6] where much of the theory that we present here can be found.

Before we continue with the definition of IQCs, we remark that when we say that $v(t) \in \mathcal{L}_2^l$, we mean that $v(t)$ is an l-dimensional signal with bounded energy.

The definition of an IQC follows the definition in [6] which states:

Definition 10.1 (IQC). Two signals $v(t) \in \mathcal{L}_2^l$ and $w(t) \in \mathcal{L}_2^m$ are said to satisfy the IQC defined by $\Pi(j\omega)$ if

$$\int_{-\infty}^{\infty} \begin{bmatrix} \hat{v}(j\omega) \\ \hat{w}(j\omega) \end{bmatrix}^* \Pi(j\omega) \begin{bmatrix} \hat{v}(j\omega) \\ \hat{w}(j\omega) \end{bmatrix} d\omega \geq 0. \tag{10.2}$$

Here, $\hat{v}(j\omega)$ and $\hat{w}(j\omega)$ denote the Fourier transforms of $v(t)$ and $w(t)$ respectively. In principle, $\Pi : j\mathbb{R} \to \mathbb{C}^{(l+m)\times(l+m)}$ can be any measurable Hermitian-valued function. For practical purposes, it is usually sufficient to consider rational functions that are bounded on the imaginary axis.

The following theorem from [6] characterizes the stability of the interconnection in Fig. 10.1 when the uncertain operator Δ is described using an IQC.

Theorem 10.1 ([6]). *Let G be a stable LTI system, and let Δ be a bounded causal operator. Assume that:*

i) for every $\tau \in [0,1]$, the interconnection of G and $\tau\Delta$ is well-posed [4]*;*
ii) for every $\tau \in [0,1]$, the IQC defined by $\Pi(j\omega)$ is satisfied by $\tau\Delta$;
iii) there exists $\epsilon > 0$ such that

$$\begin{bmatrix} G(j\omega) \\ I \end{bmatrix}^* \Pi(j\omega) \begin{bmatrix} G(j\omega) \\ I \end{bmatrix} \leq -\epsilon I, \quad \forall \omega \in \mathbb{R}. \tag{10.3}$$

Then, the feedback interconnection of G and Δ is stable.

Two famous theorems for proving stability of nonlinear systems are the *small gain theorem* and the *passivity theorem* [10]. By using suitable choices for $\Pi(j\omega)$ in the above theorem, we can recover the small gain and passivity theorems. Namely, if we use

$$\Pi_1(j\omega) = \begin{bmatrix} I & 0 \\ 0 & -I \end{bmatrix}, \qquad \Pi_2(j\omega) = \begin{bmatrix} 0 & I \\ I & 0, \end{bmatrix}, \tag{10.4}$$

we get, with $\Pi(j\omega) = \Pi_1(j\omega)$, the small gain theorem and, with $\Pi(j\omega) = \Pi_2(j\omega)$, the passivity theorem.

We can see in (10.4), that both Π_1 and Π_2 have the upper left corner positive semi-definite and the lower right corner negative semi-definite. This is often the case in applications [6] and simplifies assumption *ii)* since $\tau\Delta$ then satisfies the IQC defined by $\Pi(j\omega)$ for $\tau \in [0, 1]$ if and only if Δ does.

10.2.1 Parametrized and Multiple IQCs

The IQCs described by the multipliers in (10.4) were both fixed, i.e. there were no degrees of freedom in the choice of the multiplier $\Pi(j\omega)$. In practice, a parametrization is chosen for the multiplier with free variables which need to be calculated through an optimization problem in order to satisfy the frequency-domain inequality and thus prove robust stability. This is illustrated with an example.

Let us assume we have two signals $v(t)$ and $w(t)$ in $\mathcal{L}_2$ and their relation is such that

$$\hat{w}(j\omega) = \Delta(j\omega)\hat{v}(j\omega), \tag{10.5}$$

where Δ is an unknown LTI system with bounded $\mathcal{H}_\infty$-norm satisfying

$$||\Delta||_\infty < g \tag{10.6}$$

for some positive constant g. This uncertainty characterization can be used if we have some unmodeled or uncertain dynamics in our system. Then, a multiplier that satisfies condition (10.2) for this pair of signals is given by

$$\Pi(j\omega) = \begin{bmatrix} g^2 x(\omega) I & 0 \\ 0 & -x(\omega) I \end{bmatrix}, \tag{10.7}$$

for any scalar $x(\omega) \geq 0$ such that $\Pi(j\omega)$ is a bounded, self-adjoint operator. To verify this we insert the multiplier in the definition of the IQC and we get,

$$\int_{-\infty}^{\infty} \begin{bmatrix} \hat{v}(j\omega) \\ \hat{w}(j\omega) \end{bmatrix}^* \Pi(j\omega) \begin{bmatrix} \hat{v}(j\omega) \\ \hat{w}(j\omega) \end{bmatrix} d\omega =$$
$$\int_{-\infty}^{\infty} \hat{v}(j\omega)^* \hat{v}(j\omega) g^2 x(\omega) - \hat{w}(j\omega)^* \hat{w}(j\omega) x(\omega)\, d\omega >$$
$$\int_{-\infty}^{\infty} \hat{v}(j\omega)^* \hat{v}(j\omega) g^2 x(\omega) - \hat{v}(j\omega)^* \hat{v}(j\omega) g^2 x(\omega)\, d\omega = 0. \quad (10.8)$$

The inequality is obtained by first inserting the relation between $\hat{w}(j\omega)$ and $\hat{v}(j\omega)$ and then using the fact that $|\Delta(j\omega)|$ is bounded by g.

In theory, we can choose $x(\omega)$ as any positive bounded hermitian function, but in order to facilitate the verification of the stability condition, we will usually take $x(\omega)$ to be a rational transfer function where we leave the coefficients of the nominator as free variables; the reason for this will be clear in Section 10.2.2. As an example, for the IQC specified in (10.7), the IQCβ-toolbox [11] uses $x(\omega) = X(j\omega) + X(j\omega)^*$, where X^* is the conjugate transpose of X and

$$X(j\omega) = x_0 + \frac{x_1}{j\omega + a_1} + \cdots + \frac{x_N}{j\omega + a_N}, \quad (10.9)$$

where the a_i's are positive constants fixed by the user and the x_i's are free variables.

In a similar way, one can derive IQCs for other uncertainties. We mention another common uncertainty which will be used in the later chapters of this book, namely the relation

$$w(t) = \delta v(t), \quad (10.10)$$

where δ is an unknown constant with values between -1 and 1. An IQC for this relation is given by

$$\int_{-\infty}^{\infty} \begin{bmatrix} \hat{v}(j\omega) \\ \hat{w}(j\omega) \end{bmatrix}^* \begin{bmatrix} P(j\omega) & Q(j\omega) \\ Q^*(j\omega) & -P(j\omega) \end{bmatrix} \begin{bmatrix} \hat{v}(j\omega) \\ \hat{w}(j\omega) \end{bmatrix} d\omega \geq 0, \quad P > 0, \quad (10.11)$$

where $P(j\omega) = R(j\omega) + R^*(j\omega)$ and

$$R(j\omega) = X_0 + \frac{1}{j\omega + a_1} X_1 + \cdots + \frac{1}{j\omega + a_N} X_N, \quad (10.12)$$

$$Q(j\omega) = Z_0 + \frac{j\omega}{-\omega^2 - a_1^2} Y_1 + \frac{a_1}{-\omega^2 - a_1^2} Z_1 + \ldots + \frac{j\omega}{-\omega^2 - a_N^2} Y_N + \frac{a_N}{-\omega^2 - a_N^2} Z_N, \quad (10.13)$$

where X_i are arbitrary square matrices, Y_i are arbitrary symmetric matrices and Z_i are arbitrary skew-symmetric matrices. The verification of this IQC can be derived in a similar way as for the IQC for unmodeled dynamics.

A list of several useful IQC's for different kinds of uncertainties, delays and nonlinearities can be found in Sect. VI in [6]. Other examples of IQC's can be found in the manual of the toolbox IQCβ [11] and in papers such as [12] (delays with bounded rate of variation).

A nice property of IQCs is that if an uncertain operator Δ satisfies several IQC's defined by $\Pi_1, \ldots, \Pi_n$, then a sufficient condition for stability is the existence of $\gamma_1, \ldots, \gamma_n \geq 0$ such that (10.3) holds for $\Pi = \gamma_1 \Pi_1 + \cdots + \gamma_n \Pi_n$. Hence, one can reduce conservatism by using more IQC's for Δ at the expense of increasing the computational complexity since a larger number of free variables are included in the verification of the stability condition.

10.2.2 Verifying the Robust Stability Condition

Verifying condition (10.3) requires the solution of an infinite dimensional feasibility problem since we need to verify it for all ω. Furthermore, it is not known how we should choose $\Pi(j\omega)$ if we have parametrized or multiple IQCs as in Section 10.2.1.

We choose to parametrize our IQCs as in (10.9), i.e. the IQC depends on the free variables x_i in an affine way. If that is the case, we can use suitable state-space realizations for $G(s)$ and $\Pi(s)$ which yields

$$\begin{bmatrix} G(j\omega) \\ I \end{bmatrix}^* \Pi(j\omega) \begin{bmatrix} G(j\omega) \\ I \end{bmatrix} = \begin{bmatrix} (j\omega - A)^{-1}B \\ I \end{bmatrix}^* M(x) \begin{bmatrix} (j\omega - A)^{-1}B \\ I \end{bmatrix}, \quad (10.14)$$

where $M(x)$ depends affinely on the decision variables, i.e. $M(x) = M_0 + \sum_{i=1}^{N} M_i x_i$. By using the KYP Lemma [13], we get that the condition in (10.3) is equivalent to the existence of a positive definite symmetric matrix P such that

$$\begin{bmatrix} A^T P + PA & PB \\ B^T P & 0 \end{bmatrix} + M_0 + \sum_{i=1}^{N} x_i M_i < 0, \quad (10.15)$$

where N is the number of variables required to specify the IQCs. Here we stress that the dimension of A is *not* the number of states in the original system, since A also contains the dynamics of the multipliers used in the IQCs for characterizing the uncertainties. Frequency domain conditions on IQCs or variables that define IQCs can be handled in a similar way. For example, the positivity condition on $x(\omega)$ in (10.7) can be transformed into a constraint similar to (10.15).

Verifying the feasibility of the LMI (10.15) is a finite-dimensional, convex optimization problem for which there exist tailor-made algorithms and solvers, see for example [14, 15, 16, 17, 18, 19, 20].

In [6], it is established that a system that is quadratically stable can always be proven stable by using IQC's. For more complex Lyapunov functions the connections have not been established yet.

In this section we have only addressed the robust stability of our system. Robust performance can be addressed in a similar way if the performance criteria can be formulated in the form of (10.2). One such example is the induced $\mathcal{L}_2$-norm

$$\int_0^{\infty} (|z|^2 - \gamma^2 |e|^2)\, dt, \tag{10.16}$$

which, by the use of Parseval's Theorem, can be put in the form of (10.2). The clearance criteria which are formulated as robust performance problems are presented in the sequel.

10.3 Stability Margins Clearance with IQCs

10.3.1 Specifying Nichols Exclusion Regions

In this section we present how the generalized stability margin b_{PC} can be used to analyze the stability of closed-loop systems and how the results can be related to exclusion regions in the Nichols diagrams.

Given the closed-loop system $[P, C]$ in Fig. 10.2 consisting of a nominal plant P and a controller C (designed with any control design method), we calculate the closed-loop transfer matrix $T_{cl}(s)$ as

$$T_{cl}(s) := T_{\begin{bmatrix} v_y \\ v_u \end{bmatrix} \to \begin{bmatrix} y \\ u \end{bmatrix}} = \begin{bmatrix} P \\ I \end{bmatrix} (I - CP)^{-1} \begin{bmatrix} -C & I \end{bmatrix}. \tag{10.17}$$

This closed-loop transfer matrix corresponds to the transfer matrix from external disturbances (v_u, v_y) injected at the inputs and outputs of the plant to the inputs and outputs (u, y) of the plant. Its stability is equivalent to the internal stability of the closed-loop system and it contains all the important closed-loop transfer functions in terms of characterizing nominal performance, like the sensitivity function, the complementary sensitivity function, the control sensitivity function and the plant sensitivity function. Additionally, the size of the closed-loop transfer matrix $T_{cl}(s)$ characterizes the robust stability of the closed-loop system with respect to coprime factor uncertainty in the plant [21].

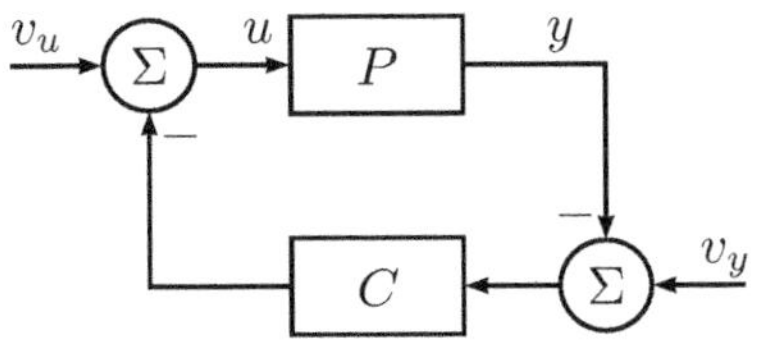

Fig. 10.2 Closed-loop interconnection for defining the generalized stability margin.

For the analysis of the closed-loop system in Fig. 10.2, it is desirable to introduce stable, minimum-phase weights in the loop so that the loop transfer matrix remains unchanged. Therefore, the weights are applied at the inputs and outputs of the plant with their inverses applied at the inputs and outputs of the controller as shown in Fig. 10.3, with the artificial disturbance signals defined as $\bar{v}_u = W_i^{-1} v_u$ and $\bar{v}_y = W_o v_y$.

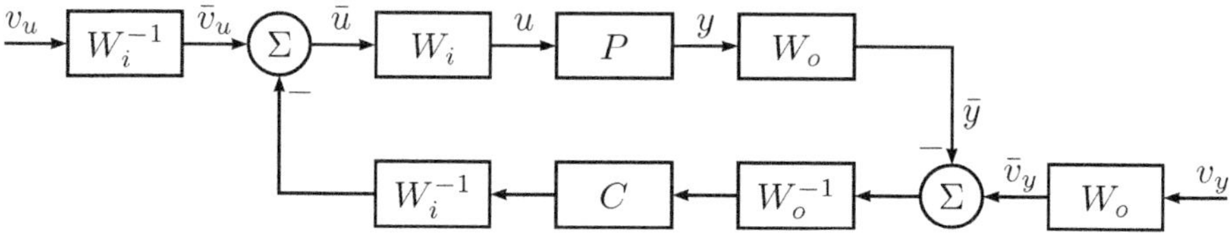

Fig. 10.3 The placement of weights for the calculation of the weighted generalized stability margin.

The placement of weights does not affect the actual properties of the closed-loop system, it is merely an analysis technique for obtaining more meaningful information regarding the robust stability and nominal performance of the closed-loop system.

The weighted generalized stability margin is defined as

$$b_{W_o P W_i, W_i^{-1} C W_o^{-1}} := \left\| \begin{bmatrix} W_o & 0 \\ 0 & W_i^{-1} \end{bmatrix} T_{\begin{bmatrix} v_y \\ v_u \end{bmatrix} \to \begin{bmatrix} y \\ u \end{bmatrix}} \begin{bmatrix} W_o^{-1} & 0 \\ 0 & W_i \end{bmatrix} \right\|_{\infty}^{-1} \tag{10.18}$$

which is the inverse of the size of the closed-loop transfer matrix from the weighted disturbances to the weighted internal signals of the system.

For the analysis of the closed-loop system, the objective is to choose weights W_o and W_i to maximize the stability margin, or minimize the weighted $\mathcal{H}_\infty$ norm. We therefore define the optimization problem:

$$b_{PC} := \max_{W_o, W_i} b_{W_o P W_i, W_i^{-1} C W_o^{-1}}. \tag{10.19}$$

Furthermore, it was shown in [8] that if we maximize the weighted generalized stability margin over **diagonal** weights, we can relate the optimal value to the robust stability of the closed-loop system with respect to input and output independent and simultaneous multiplicative perturbations applied at the plant using the following theorem.

Theorem 10.2 ([8]). *Let Δ_1 and Δ_2 be complex* **diagonal** *matrices which perturb a nominal plant P to $P_\Delta = (I + \Delta_1) P (I - \Delta_2)^{-1}$. If the weighted stability margin satisfies*

$$b_{W_o P W_i, W_i^{-1} C W_o^{-1}} \geq \beta \tag{10.20}$$

for any **diagonal** *input and output analysis weights,* W_i*,* W_o*, then* $[P_\Delta, C]$ *is stable for any perturbations satisfying* $||\Delta_2||_\infty < \beta$ *and* $||\Delta_1||_\infty < \beta$*.*

The above theorem presents the motivation for solving the optimization problem in (10.19) over a more restricted class of stable and minimum-phase weights, those which are diagonal. In this way we can guarantee the robust stability of the closed-loop system with a perturbed plant P_Δ, where the uncertainty enters in a special way, as diagonal input and output multiplicative perturbations. A further step can be taken to establish a connection between the optimal value β to the Nichols stability margins by the characterization of the perturbed plant P_Δ, $(||\Delta_1||_\infty, ||\Delta_2||_\infty < \beta)$ using multiplicative, input-output, gain-phase perturbations. Expressing the perturbed plant as

$$P_\Delta = \frac{1}{\sqrt{1-\beta^2}}(I + \Delta_1)P(I - \Delta_2)^{-1}\sqrt{1-\beta^2} \tag{10.21}$$

and taking into account that $\beta < 1$ (which is the case since the generalized stability margin is bounded between 0 and 1), we observe that the sets $\{\frac{1+\delta_1}{\sqrt{1-\beta^2}} : |\delta_1| < \beta\}$ and $\{\frac{\sqrt{1-\beta^2}}{1-\delta_2} : |\delta_2| < \beta\}$ are identical. Therefore the closed-loop system is robust with respect to simultaneous and independent multiplicative perturbations at the inputs and outputs of the plant (MIMO case) as shown in Fig. 10.4. The set $\{\frac{1+\delta}{\sqrt{1-\beta^2}} : |\delta| < \beta\}$ describes an ellipse if we consider the logarithm of the magnitude against the phase of the complex number. An example for $\beta = 0.3$ is shown in Fig. 10.5 where the smaller ellipse corresponds to the MIMO case uncertainty. If we wish to consider an

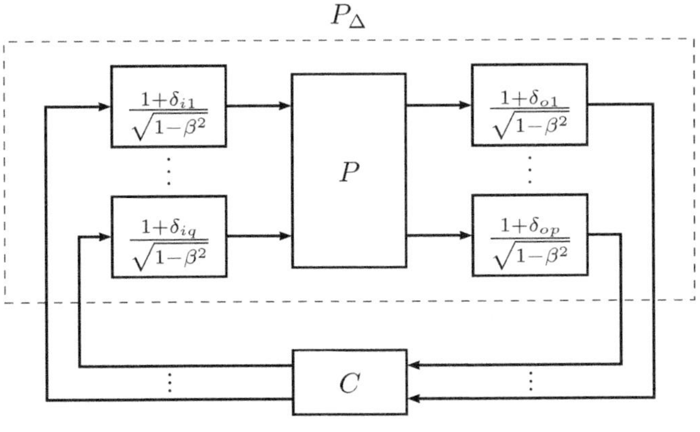

Fig. 10.4 The interconnection of the multiplicatively perturbed plant and the controller. If $b_{W_o P W_i, W_i^{-1} C W_o^{-1}} \geq \beta$, the closed-loop system can tolerate the multiplicative perturbations shown in the figure for any $|\delta| < \beta$.

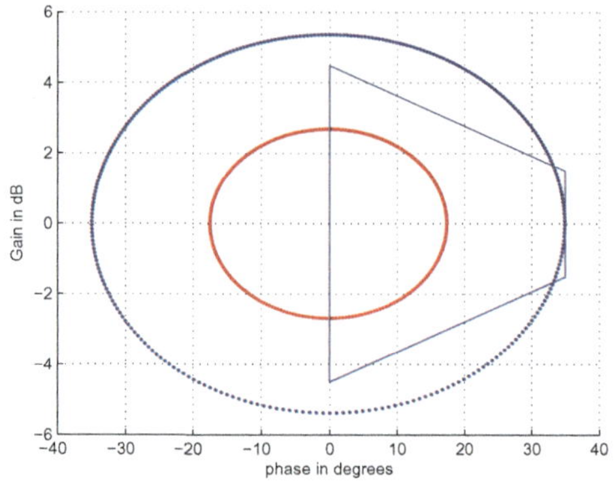

Fig. 10.5 The region described by the multiplicative perturbation set for $\beta = 0.3$ both in the SISO (larger ellipse) and MIMO (smaller ellipse) cases.

uncertainty acting either at the input or the output of a single channel (SISO case), the allowable uncertainty set is given by $\{\frac{(1+\delta)^2}{1-\beta^2} : |\delta| < \beta\}$ [8]. This gives an ellipse twice as big as the one obtained for the MIMO case. It is important to remember that although in the MIMO case the perturbation is smaller, it can be tolerated simultaneously at both the input and output of a certain channel. The trapezoidal region specifies the standard SISO gain and phase margin requirements from an industrial point of view [2]. It is easy to see that a value of $\beta > 0.3$ will guarantee the clearance of the SISO gain and phase margin criterion.

10.3.2 Lower Bound Calculation

In this section we propose an optimization method to calculate a lower bound on the value of b_{PC}. The lower bound is obtained using IQC analysis of a certain closed-loop interconnection with respect to a diagonal fictitious uncertainty.

Following the definition of b_{PC} in (10.19) we wish to solve the equivalent optimization problem

$$b_{PC}^{-1} = \min_{\text{diag } W(s)} ||W(s)T_{cl}(s)W^{-1}(s)||_{\infty}, \tag{10.22}$$

where $W(s) = \text{diag}\,(W_o(s), W_i^{-1}(s))$. For the lower bound calculation, consider the problem of analyzing the stability of the closed-loop system in Fig. 10.6 using IQCs. The first step is to use IQCs to characterize the fictitious, diagonal uncertainty. Assume that the structured uncertainty $\Delta = \text{diag}\,(\Delta_o(s), \Delta_i(s))$ satisfying $||\Delta_i(s)||_{\infty} \leq \gamma$, $||\Delta_o(s)||_{\infty} \leq \gamma$ satisfies the following IQC

$$\int_{-\infty}^{\infty} \begin{bmatrix} \hat{y}(j\omega) \\ \hat{u}(j\omega) \\ \hat{v}_y(j\omega) \\ \hat{v}_u(j\omega) \end{bmatrix}^* \begin{bmatrix} \Pi(j\omega) & 0 \\ 0 & -\frac{1}{\gamma^2}\Pi(j\omega) \end{bmatrix} \begin{bmatrix} \hat{y}(j\omega) \\ \hat{u}(j\omega) \\ \hat{v}_y(j\omega) \\ \hat{v}_u(j\omega) \end{bmatrix} d\omega \geq 0, \tag{10.23}$$

where

$$\Pi(j\omega) = \mathrm{diag}\left(\Pi_o(j\omega), \Pi_i(j\omega)\right), \; \Pi(j\omega) = \Pi(j\omega)^* \geq 0. \tag{10.24}$$

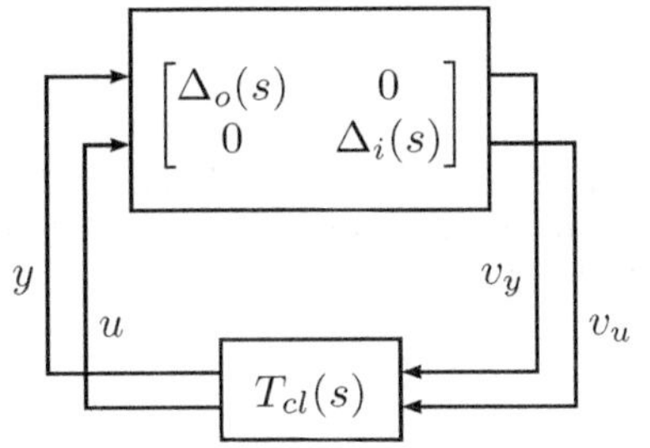

Fig. 10.6 Robustness analysis with respect to a structured, diagonal, fictitious uncertainty, for the purpose of using IQC analysis for obtaining a lower bound on b_{PC}.

The sufficient IQC condition for the closed-loop stability of the interconnection in Fig. 10.6 is the existence of a multiplier $\Pi(j\omega)$ such that,

$$\begin{bmatrix} T_{cl}(j\omega) \\ I \end{bmatrix}^* \begin{bmatrix} \Pi(j\omega) & 0 \\ 0 & -\frac{1}{\gamma^2}\Pi(j\omega) \end{bmatrix} \begin{bmatrix} T_{cl}(j\omega) \\ I \end{bmatrix} < 0 \; \forall \; \omega \Leftrightarrow$$
$$\Leftrightarrow T_{cl}^*(j\omega)\Pi(j\omega)T_{cl}(j\omega) - \frac{1}{\gamma^2}\Pi(j\omega) < 0 \; \forall \; \omega. \tag{10.25}$$

Given that we can factorize $\Pi(j\omega) = W^*(j\omega)W(j\omega)$, we substitute in the previous inequality and also right-multiply by $W^{-1}(j\omega)$ and left-multiply by $W^{-*}(j\omega)$ to obtain the equivalent inequality

$$(X(j\omega))^*(X(j\omega)) - \frac{1}{\gamma^2}I < 0 \; \forall \; \omega, \tag{10.26}$$

with

$$X(j\omega) = W(j\omega)T_{cl}(j\omega)W^{-1}(j\omega). \tag{10.27}$$

The above inequality is equivalent to

$$\sup_{\omega} \bar{\sigma}(X(j\omega)) < \frac{1}{\gamma} \Leftrightarrow ||W(s)T_{cl}(s)W^{-1}(s)||_\infty < \frac{1}{\gamma}. \tag{10.28}$$

We know that $b_{PC}^{-1} = \min_{\mathrm{diag}\; W(s)} ||W(s)T_{cl}(s)W^{-1}(s)||_\infty$ and we have found a candidate diagonal weight that satisfies an upper bound on the $\mathcal{H}_\infty$

norm of the weighted closed-loop transfer matrix. This implies that $b_{PC} > \gamma$. Therefore, checking the robust stability of the closed-loop system in Fig. 10.6 for increasing values of the uncertainty size γ will provide successively improved lower bounds on b_{PC}. Thus, the optimization problem for obtaining the lower bound is,

$$\min_{\Pi(j\omega)} \frac{1}{\gamma^2} \qquad \text{subject to (10.25)} \tag{10.29}$$

Using a bisection algorithm on $\frac{1}{\gamma^2}$ and using appropriate state-space representations for $\Pi(s)$ and $T_{cl}(s)$, we can apply the KYP-Lemma on condition (10.25) to transform it into an LMI problem. It is also possible to use the IQC-toolbox which automates the step of transforming the frequency domain inequality into an LMI problem.

10.3.3 Formulation of the Perturbed Stability Margin Problem

In certain clearance problems it is required to estimate not only the nominal stability margins but also the perturbed stability margins in the presence of real, parametric uncertainty. This parametric uncertainty is either due to uncertain physical parameters or due to flight envelope parameters. Assume that the set of the parametric uncertainty is given by

$$\boldsymbol{\Delta_u} = \{\Delta_u : \Delta_u = \operatorname{diag}(\delta_1 I_{k_1}, \ldots, \delta_m I_{k_m}), \delta_i \in \mathbb{R}, |\delta_i| \le \epsilon\}. \tag{10.30}$$

The problem of calculating the worst-case stability margin $b_{PC,WC}$ for a given uncertainty size $0 < \epsilon \le 1$ can be formulated as follows with reference to Fig. 10.7

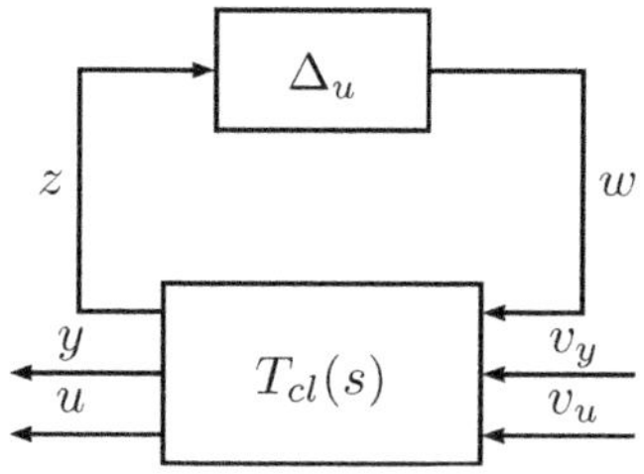

Fig. 10.7 The closed-loop system for the formulation of the perturbed stability margin problem. Note that the transfer matrix $T_{cl}(s)$ is not the same as the one used for the calculation of the nominal stability margin.

$$b_{PC,WC}^{-1}(\epsilon) := \min_{\text{diag } W(s)} \sup_{\Delta_u \in \boldsymbol{\Delta_u}} ||W(s)\mathcal{F}_u(T_{cl}(s), \Delta_u)W^{-1}(s)||_\infty, \quad (10.31)$$

where $\mathcal{F}_u(T_{cl}(s), \Delta_u)$ is the perturbed closed-loop system expressed as an upper Linear Fractional Transformation (LFT) of the nominal closed-loop transfer matrix $T_{cl}(s)$ with respect to the structured, parametric uncertainty Δ_u. Note that the transfer matrix $T_{cl}(s)$ is not the same as the one used for the calculation of the nominal stability margin. It also contains terms that describe how the real, parametric uncertainty is interconnected to the known system dynamics. Clearly, as $\epsilon \to 0$ we have $\Delta_u \to 0$ and we approach the nominal stability margin. The idea here is to gradually increase ϵ to 1 (the normalization of Δ_u captures the physical parameter variations) and investigate the possible degradation in the stability margin value as a function of the uncertainty size.

A lower bound on the perturbed stability margin can be obtained by utilizing the robust performance analysis result of the IQC theory and applying it for the case where the desired performance is expressed as a weighted $\mathcal{H}_\infty$ norm. Assume that the parametric uncertainty Δ_u satisfies the IQC given by

$$\int_{\infty}^{\infty} \begin{bmatrix} \hat{z}(j\omega) \\ \hat{w}(j\omega) \end{bmatrix}^* \begin{bmatrix} \Pi_{11}(j\omega) & \Pi_{12}(j\omega) \\ \Pi_{12}^*(j\omega) & \Pi_{22}(j\omega) \end{bmatrix} \begin{bmatrix} \hat{z}(j\omega) \\ \hat{w}(j\omega) \end{bmatrix} d\omega \geq 0. \quad (10.32)$$

Theorem 10.3 (Weighted robust $\mathcal{L}_2$-gain performance). *Assume that any uncertainty $\Delta_u \in \boldsymbol{\Delta_u}$ satisfies the IQC given in (10.32). If there exists a multiplier $\Pi(j\omega)$ and a γ such that the following frequency-domain inequality holds*

$$\begin{bmatrix} T_{cl}(j\omega) \\ I \end{bmatrix}^* \begin{bmatrix} \Pi_{11}(j\omega) & 0 & \Pi_{12}(j\omega) & 0 \\ 0 & \Pi(j\omega) & 0 & 0 \\ \Pi_{12}^*(j\omega) & 0 & \Pi_{22}(j\omega) & 0 \\ 0 & 0 & 0 & -\frac{1}{\gamma^2}\Pi(j\omega) \end{bmatrix} \begin{bmatrix} T_{cl}(j\omega) \\ I \end{bmatrix} < 0 \ \forall\omega \quad (10.33)$$

then

$$||W(s)\mathcal{F}_u(T_{cl}(s), \Delta_u)W^{-1}(s)||_\infty \leq \frac{1}{\gamma} \ \forall \Delta_u \in \boldsymbol{\Delta_u}$$
$$\text{with } \Pi(j\omega) = W^*(j\omega)W(j\omega). \quad (10.34)$$

The satisfaction of the inequality in (10.34) implies the following:

$$b_{PC,WC}(\epsilon) \geq \gamma. \quad (10.35)$$

A proof of the above theorem can be found in [9]. As with the nominal case, the optimization problem will involve a bisection algorithm on $\frac{1}{\gamma^2}$ and a search for multipliers $\Pi(j\omega)$, $\Pi_{11}(j\omega)$, $\Pi_{12}(j\omega)$ and $\Pi_{22}(j\omega)$ to satisfy the frequency-domain inequality in (10.33).

10.4 Formulation of the Comfort Clearance Criterion Using IQCs

10.4.1 State-Space Formulation of the Comfort Criterion

The standard method for the calculation of the comfort criterion is to perform a frequency-wise, numerical integration of the power spectral density of the filtered acceleration signals at a specific point of interest on the aircraft. More specifically, the comfort criterion at a specific point on the aircraft is defined in the frequency domain as

$$J_c = \sqrt{J_{feet}^2 + J_{seat}^2 + J_{back}^2}, \tag{10.36}$$

$$J_{feet} = \sqrt{\frac{1}{\pi}\int_{\omega_1}^{\omega_2} \Phi_{a_{fw}a_{fw}}(\omega)d\omega}, \tag{10.37}$$

$$J_{seat} = \sqrt{\frac{1}{\pi}\int_{\omega_1}^{\omega_2} \Phi_{a_{sw}a_{sw}}(\omega)d\omega}, \tag{10.38}$$

$$J_{back} = \sqrt{\frac{1}{\pi}\int_{\omega_1}^{\omega_2} \Phi_{a_{bw}a_{bw}}(\omega)d\omega}. \tag{10.39}$$

The comfort criterion is composed of three components, for example, the quantity $\Phi_{a_{fw}a_{fw}}(\omega)$ denotes the power spectral density of the filtered acceleration at the feet position of the specific point of interest.

The calculation of the comfort criterion can also be formulated in the state-space domain as an $\mathcal{H}_2$ analysis problem of the linear system shown in Fig. 10.8.

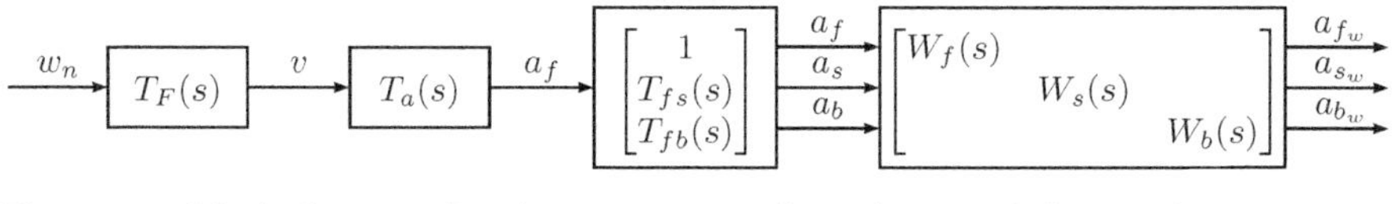

Fig. 10.8 Block diagram for the state-space formulation of the comfort criterion.

The signal w_n denotes the white noise signal and the transfer function $T_F(s)$ denotes the turbulence filter whose output's power spectral density approximates the Von Kármán turbulence spectrum. Both for the symmetric and anti-symmetric turbulence case the filter is given by

$$T_F(s) = \frac{\sigma\sqrt{\frac{L}{V\pi}}\left(1 + 2.7478\frac{L}{V}s + 0.3398\left(\frac{L}{V}\right)^2 s^2\right)}{1 + 2.9958\frac{L}{V}s + 1.9754\left(\frac{L}{V}\right)^2 s^2 + 0.1539\left(\frac{L}{V}\right)^3 s^3} \tag{10.40}$$

with $\sigma = 1$ and $L = 762$ m. The parameter σ denotes the turbulence intensity and the parameter L denotes the turbulence scale length and both depend on altitude [3]. The parameter V denotes the true airspeed of the aircraft.

The signal v denotes the stochastic wind velocity (either vertical or lateral) and $T_a(s)$ denotes the open-loop or closed-loop transfer function from the wind velocity to the acceleration at the feet position of a specific measurement point on the fuselage. In the general case $T_a(s)$ will be expressed as an LFT with respect to flight and uncertain parameters. The signal a_f denotes the acceleration at the feet position. The transfer functions $T_{fs}(s)$ and $T_{fb}(s)$ relate the feet acceleration to the seat and back accelerations respectively. They are independent of the specific measurement point but they change depending on whether the turbulence field is symmetric or anti-symmetric. The transfer functions $W_f(s)$, $W_s(s)$ and $W_b(s)$ correspond to the comfort filters for the feet, seat and back positions respectively and depend on whether the turbulence field is symmetric or anti-symmetric. Let the total transfer function relating the white noise signal w_n to the weighted accelerations be given by

$$\begin{bmatrix} a_{f_w} \\ a_{s_w} \\ a_{b_w} \end{bmatrix} = \begin{bmatrix} G_1(s) \\ G_2(s) \\ G_3(s) \end{bmatrix} w_n. \tag{10.41}$$

It can be shown that the $\mathcal{H}_2$ norm of this transfer function is equal to

$$||G||_2 = \sqrt{\frac{1}{\pi} \int_0^\infty \left(\Phi_{a_{f_w} a_{f_w}} + \Phi_{a_{s_w} a_{s_w}} + \Phi_{a_{b_w} a_{b_w}} \right) d\omega}. \tag{10.42}$$

The frequency-domain comfort measure J_c should tend to $||G||_2$ when $\omega_1 = 0$ and $\omega_2 \to \infty$ and the number of frequency points tends to ∞, thus $||G||_2 \geq J_c$.

Possible sources of discrepancy between the frequency-domain and the state-space domain comfort criteria, apart from the finite frequency range used in the frequency-domain formulation, are the approximation of several frequency-domain responses with rational transfer functions. This is the case for the Von Kármán spectrum which is approximated by the transfer function in (10.40) and for the frequency responses that relate the feet acceleration to the seat and back accelerations.

10.4.2 Robust $\mathcal{H}_2$ Performance Analysis Using IQCs

This section presents the implementation of an $\mathcal{H}_2$ robust performance test developed in [22] based on IQC theory for the solution of the comfort criterion.

Consider the interconnection shown in Fig. 10.9, where $G(s)$ represents the transfer matrix of the generalized plant and Δ represents the uncertainty. The

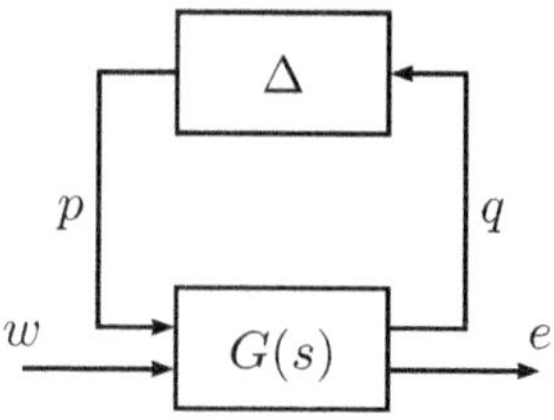

Fig. 10.9 Robustness analysis with respect to a structured uncertainty Δ.

uncertainty Δ is characterized using an IQC with an associated multiplier Π such that

$$\int_{-\infty}^{\infty} \begin{bmatrix} \hat{q}(j\omega) \\ \hat{p}(j\omega) \end{bmatrix}^* \Pi(j\omega) \begin{bmatrix} \hat{q}(j\omega) \\ \hat{p}(j\omega) \end{bmatrix} d\omega \geq 0, \quad \forall q \in \mathcal{L}_2^{n_q}. \tag{10.43}$$

It is assumed that Π is a dynamic multiplier and can be factorized as $\Pi = \Psi(s)^* M \Psi(s)$ where $\Psi(s) = \begin{bmatrix} \Psi_A(s) \; \Psi_B(s) \end{bmatrix}$ is partitioned accordingly with the dimensions $[n_q, n_p]$ and admits the minimal realization

$$\begin{bmatrix} \Psi_A(s) \; \Psi_B(s) \end{bmatrix} = \left[\begin{array}{cc|cc} A_{11} & A_{12} & B_{11} & B_{12} \\ 0 & A_{22} & 0 & B_{22} \\ \hline C_1 & C_2 & D_1 & D_2 \end{array}\right]. \tag{10.44}$$

Assume also that the generalized plant has a realization given by

$$G(s) = \begin{pmatrix} G_{qp}(s) \; G_{qw}(s) \\ G_{ep}(s) \; G_{ew}(s) \end{pmatrix} = \left[\begin{array}{c|cc} A & B_p & B_w \\ \hline C_q & D_{qp} & D_{qw} \\ C_e & D_{ep} & D_{ew} \end{array}\right]. \tag{10.45}$$

The generalized plant $G(s)$ is defined according to the analysis in the previous section, i.e the performance signals w and e correspond to the white noise and weighted acceleration signals respectively and the signals p and q characterize the uncertainty interaction. Define the realization of

$$\begin{pmatrix} \Psi_A(s)G_{qp}(s) + \Psi_B(s) & \Psi_A(s)G_{qw}(s) \\ G_{ep}(s) & G_{ew}(s) \end{pmatrix} = \left[\begin{array}{c|cc} \mathcal{A} & \mathcal{B}_p & \mathcal{B}_w \\ \hline \mathcal{C}_\Psi & \mathcal{D}_{\Psi p} & \mathcal{D}_{\Psi w} \\ \mathcal{C}_e & \mathcal{D}_{ep} & \mathcal{D}_{ew} \end{array}\right]. \tag{10.46}$$

Theorem 10.4. *The interconnection in Fig. 10.9 is stable and the squared* $\mathcal{H}_2$ *norm from* $w \to e$ *is less than* γ *for all uncertainties characterized by the IQC with multiplier* Π*, if* $\mathcal{D}_{\Psi w} = 0$, $\mathcal{D}_{ew} = 0$ *and if there exist matrices* $M = M^T$, $\mathcal{X} = \mathcal{X}^T$ *and* $Q = Q^T$ *such that,*

$$\begin{bmatrix} I & 0 \\ \mathcal{A} & \mathcal{B}_p \\ \mathcal{C}_\Psi & \mathcal{D}_{\Psi p} \\ \mathcal{C}_e & \mathcal{D}_{ep} \end{bmatrix}^T \begin{bmatrix} 0 & \mathcal{X} & 0 & 0 \\ \mathcal{X} & 0 & 0 & 0 \\ 0 & 0 & M & 0 \\ 0 & 0 & 0 & I \end{bmatrix} \begin{bmatrix} I & 0 \\ \mathcal{A} & \mathcal{B}_p \\ \mathcal{C}_\Psi & \mathcal{D}_{\Psi p} \\ \mathcal{C}_e & \mathcal{D}_{ep} \end{bmatrix} < 0, \tag{10.47}$$

$$\mathcal{B}_w^T \mathcal{X} \mathcal{B}_w < Q, \tag{10.48}$$

$$\text{trace}\,(Q) < \gamma. \tag{10.49}$$

Minimizing the trace of the matrix $\mathcal{B}_w^T \mathcal{X} \mathcal{B}_w$ directly and expanding (10.47) results in the following optimization problem which involves a KYP-LMI with possibly additional constraints on M depending on the type and structure of the uncertainty block Δ.

$$\min_{\mathcal{X},M} \text{trace}\,(\mathcal{B}_w^T \mathcal{X} \mathcal{B}_w) = \text{trace}\,(\mathcal{B}_w \mathcal{B}_w^T \mathcal{X})$$

subject to

$$\begin{bmatrix} \mathcal{X}\mathcal{A} + \mathcal{A}^T \mathcal{X} & \mathcal{X}\mathcal{B}_p \\ \mathcal{B}_p^T \mathcal{X} & 0 \end{bmatrix} + \begin{bmatrix} \mathcal{C}_e^T \\ \mathcal{D}_{ep}^T \end{bmatrix} \begin{bmatrix} \mathcal{C}_e & \mathcal{D}_{ep} \end{bmatrix} + \begin{bmatrix} \mathcal{C}_\Psi^T \\ \mathcal{D}_{\Psi p}^T \end{bmatrix} M \begin{bmatrix} \mathcal{C}_\Psi & \mathcal{D}_{\Psi p} \end{bmatrix} < 0. \tag{10.50}$$

The implementation of the above KYP-LMI problem requires the completion of the following tasks which are presented in detail in the sequel. Initially, we specify a general form of the uncertainty matrix Δ such that it is more suitable to flight control clearance problems. For the specific choice of Δ, we specify the structure of the dynamic multiplier used in the IQC and construct a state-space representation of the dynamic part of the multiplier as presented in (10.44). Finally, we specify the state-space realization of (10.46) in terms of the state-space realizations of (10.45) and of (10.44).

10.4.2.1 Characterization of Real Parametric Uncertainty with an IQC

In the context of flight control clearance we are mainly interested in real, parametric uncertainties and we wish to characterize the structured, parametric uncertainty given by

$$\Delta = \begin{bmatrix} \delta_1 I_{m_1} & \dots & 0 & 0 \\ 0 & \delta_2 I_{m_2} & \dots & 0 \\ \vdots & \vdots & \ddots & \vdots \\ 0 & 0 & \dots & \delta_r I_{m_r} \end{bmatrix}, \qquad I_{m_i} \in \mathbb{R}^{m_i \times m_i}, \qquad \delta_i \in \mathbb{R}, \qquad |\delta_i| \leq 1 \tag{10.51}$$

using an IQC with a suitable multiplier $\Pi(j\omega)$. For the sub-block $\delta_i I_{m_i}$, it is suggested in [22] to use the dynamic multiplier $\Pi_i(j\omega) = \Psi_i(j\omega)^* M_i \Psi_i(j\omega)$, where

$$\Psi_i(j\omega) = \begin{bmatrix} \psi_i(j\omega) & 0 \\ 0 & \psi_i(j\omega) \end{bmatrix}, \tag{10.52}$$

with $\psi_i(j\omega) \in \mathbf{C}^{[(n_\psi+1)\times m_i]\times m_i}$ of the form

$$\psi_i(j\omega) = \begin{bmatrix} I_{m_i} & \frac{j\omega-p}{j\omega+p} I_{m_i} & \dots & \left(\frac{j\omega-p}{j\omega+p}\right)^{n_\psi} I_{m_i} \end{bmatrix}^T \tag{10.53}$$

and

$$M_i = \begin{bmatrix} P_i & Q_i \\ Q_i^T & -P_i \end{bmatrix}, \tag{10.54}$$

with $P_i = P_i^T > 0$, $P_i \in \mathbb{R}^{[(n_\psi+1)m_i]\times[(n_\psi+1)m_i]}$ and $Q_i + Q_i^T = 0$, $Q_i \in \mathbb{R}^{[(n_\psi+1)m_i]\times[(n_\psi+1)m_i]}$.

Consider that the uncertainty Δ in Fig. 10.9 consists of only one sub-block as $p = \delta_i I_{m_i} q$. Substituting in the LHS of the IQC of (10.43) results in

$$\begin{aligned} &\int_{-\infty}^{\infty} \begin{bmatrix} \hat{q}(j\omega) \\ \hat{p}(j\omega) \end{bmatrix}^* \begin{bmatrix} \psi(j\omega)^* P_i \psi(j\omega) & \psi(j\omega)^* Q_i \psi(j\omega) \\ \psi(j\omega)^* Q_i^T \psi(j\omega) & -\psi(j\omega)^* P_i \psi(j\omega) \end{bmatrix} \begin{bmatrix} \hat{q}(j\omega) \\ \hat{p}(j\omega) \end{bmatrix} d\omega \\ &= \int_{-\infty}^{\infty} \hat{q}(j\omega)^* \psi(j\omega)^* \left((1-\delta_i^2)P_i + \delta_i(Q_i + Q_i^T)\right) \psi(j\omega)\hat{q}(j\omega) d\omega. \end{aligned} \tag{10.55}$$

Considering the positivity condition on P_i and the skew-symmetry condition on Q_i and the fact that $|\delta_i| \le 1$, it is obvious that for the specific choice of the multiplier the IQC is satisfied. We remark the condition $P_i > 0$ can be relaxed to $\psi_i^*(j\omega)P_i\psi_i(j\omega) > 0$ according to [22], but this would introduce another KYP-LMI for each i, with an associated extra matrix valued variable. In the application considered, we have not seen the need for this, i.e. the accuracy was good enough. See Chapter 13 for more details.

For the case where we have $i = 1, \dots, r$ different uncertain parameters each with multiplicity m_i, the multiplier $\Pi(j\omega)$ is given by

$$\Pi(j\omega) = \begin{bmatrix} \operatorname{diag}(\psi_i) & 0 \\ 0 & \operatorname{diag}(\psi_i) \end{bmatrix}^* \begin{bmatrix} \operatorname{diag}(P_i) & \operatorname{diag}(Q_i) \\ \operatorname{diag}(Q_i^T) & \operatorname{diag}(-P_i) \end{bmatrix} \underbrace{\begin{bmatrix} \operatorname{diag}(\psi_i) & 0 \\ 0 & \operatorname{diag}(\psi_i) \end{bmatrix}}_{\Psi(j\omega)}.$$

The notation $\operatorname{diag}(\psi_i)$ denotes a diagonal matrix whose entries are the ψ_i's for the corresponding range of the index i.

10.4.2.2 State-Space Representation for $\Psi(s)$

We start with a state-space representation of

$$\psi_i(s) = \begin{bmatrix} I_{m_i} & \frac{s-p}{s+p} I_{m_i} & \dots & \left(\frac{s-p}{s+p}\right)^{n_{\psi_i}} I_{m_i} \end{bmatrix}^T. \tag{10.56}$$

Given the order n_{ψ_i} we have

$$\begin{aligned}
\dot{x}_{\psi_i} &= A_{\psi_i} x_{\psi_i} + B_{\psi_i} q_i, \\
y_{\psi_i} &= C_{\psi_i} x_{\psi_i} + D_{\psi_i} q_i, \\
A_{\psi_i} &\in \mathbb{R}^{(n_{\psi_i} \times m_i) \times (n_{\psi_i} \times m_i)}, \\
B_{\psi_i} &\in \mathbb{R}^{(n_{\psi_i} \times m_i) \times (m_i)}, \\
C_{\psi_i} &\in \mathbb{R}^{((n_{\psi_i}+1) \times m_i) \times (n_{\psi_i} \times m_i)}, \\
D_{\psi_i} &\in \mathbb{R}^{((n_{\psi_i}+1) \times m_i) \times (m_i)},
\end{aligned} \tag{10.57}$$

with

$$A_{\psi_i} = \begin{bmatrix} -pI_{m_i} & 0 & \cdots & \cdots & 0 \\ -2pI_{m_i} & -pI_{m_i} & 0 & \cdots & 0 \\ -2pI_{m_i} & -2pI_{m_i} & -pI_{m_i} & \cdots & 0 \\ \vdots & \vdots & \vdots & \ddots & \vdots \\ -2pI_{m_i} & \cdots & \cdots & -2pI_{m_i} & -pI_{m_i} \end{bmatrix}, \quad B_{\psi_i} = \begin{bmatrix} -2pI_{m_i} \\ -2pI_{m_i} \\ -2pI_{m_i} \\ \vdots \\ -2pI_{m_i} \end{bmatrix}, \tag{10.58}$$

$$C_{\psi_i} = \begin{bmatrix} 0 & 0 & \cdots & \cdots & 0 \\ I_{m_i} & 0 & 0 & \cdots & 0 \\ I_{m_i} & I_{m_i} & 0 & \cdots & 0 \\ \vdots & \vdots & \ddots & \ddots & \vdots \\ I_{m_i} & I_{m_i} & \cdots & I_{m_i} & 0 \\ I_{m_i} & I_{m_i} & \cdots & I_{m_i} & I_{m_i} \end{bmatrix}, \qquad D_{\psi_i} = \begin{bmatrix} I_{m_i} \\ I_{m_i} \\ I_{m_i} \\ \vdots \\ I_{m_i} \\ I_{m_i} \end{bmatrix}. \tag{10.59}$$

The state-space realization of Ψ_A results by diagonally augmenting the realizations for the ψ_i for $i = 1, \ldots, r$ and by introducing an equal amount of outputs corresponding to the zero transfer matrix as shown below

$$\dot{x}_{\Psi_A} = [\operatorname{diag}(A_{\psi_i})]\, x_{\Psi_A} + [\operatorname{diag}(B_{\psi_i})] \begin{bmatrix} q_1 \\ \vdots \\ q_r \end{bmatrix}, \tag{10.60}$$

$$y_{\Psi_A} = \begin{bmatrix} \operatorname{diag}(C_{\psi_i}) \\ 0 \end{bmatrix} x_{\Psi_A} + \begin{bmatrix} \operatorname{diag}(D_{\psi_i}) \\ 0 \end{bmatrix} \begin{bmatrix} q_1 \\ \vdots \\ q_r \end{bmatrix}. \tag{10.61}$$

Considering a similar state-space realization for Ψ_B and combining the two so that $\Psi = \begin{bmatrix} \Psi_A & \Psi_B \end{bmatrix}$ we have the following state-space realization for Ψ where $i = 1, \ldots, r$.

$$\begin{aligned} \begin{bmatrix} \dot{x}_{\Psi_A} \\ \dot{x}_{\Psi_B} \end{bmatrix} &= \begin{bmatrix} \text{diag}\,(A_{\psi_i}) & 0 \\ 0 & \text{diag}\,(A_{\psi_i}) \end{bmatrix} \begin{bmatrix} x_{\Psi_A} \\ x_{\Psi_B} \end{bmatrix} + \begin{bmatrix} \text{diag}\,(B_{\psi_i}) & 0 \\ 0 & \text{diag}\,(B_{\psi_i}) \end{bmatrix} \begin{bmatrix} q_i \\ p_i \end{bmatrix} \\ y_{\Psi_A} + y_{\Psi_B} &= \begin{bmatrix} \text{diag}\,(C_{\psi_i}) & 0 \\ 0 & \text{diag}\,(C_{\psi_i}) \end{bmatrix} \begin{bmatrix} x_{\Psi_A} \\ x_{\Psi_B} \end{bmatrix} + \begin{bmatrix} \text{diag}\,(D_{\psi_i}) & 0 \\ 0 & \text{diag}\,(D_{\psi_i}) \end{bmatrix} \begin{bmatrix} q_i \\ p_i \end{bmatrix}. \end{aligned} \tag{10.62}$$

It is then straightforward to establish the correspondence to the minimal state-space realization of $\Psi(s)$ in (10.44).

It remains to describe the matrices involved in the KYP-LMI problem of Theorem 10.4 in terms of the matrices of the state-space realizations of $\Psi(s)$ and of the generalized plant $G(s)$. Defining $C_1 = \begin{bmatrix} C_{11} \\ 0 \end{bmatrix}$, $C_2 = \begin{bmatrix} C_{12} \\ C_{22} \end{bmatrix}$, $D_1 = \begin{bmatrix} D_{11} \\ 0 \end{bmatrix}$ and $D_2 = \begin{bmatrix} D_{12} \\ D_{22} \end{bmatrix}$ we have,

$$\left[\begin{array}{c|cc} \mathcal{A} & \mathcal{B}_p & \mathcal{B}_w \\ \hline \mathcal{C}_\Psi & \mathcal{D}_{\Psi p} & \mathcal{D}_{\Psi w} \\ \mathcal{C}_e & \mathcal{D}_{ep} & \mathcal{D}_{ew} \end{array}\right] = \left[\begin{array}{ccc|cc} A_{11} & A_{12} & B_{11}C_q & B_{11}D_{qp} + B_{12} & B_{11}D_{qw} \\ 0 & A_{22} & 0 & B_{22} & 0 \\ 0 & 0 & A & B_p & B_w \\ \hline C_{11} & C_{12} & D_{11}C_q & D_{11}D_{qp} + D_{12} & D_{11}D_{qw} \\ 0 & C_{22} & 0 & D_{22} & 0 \\ 0 & 0 & C_e & D_{ep} & D_{ew} \end{array}\right]. \tag{10.63}$$

The conditions of Theorem 10.4 require that $D_{ew} = 0$ and $D_{11}D_{qw} = 0$ so that there is no direct feedthrough term from the disturbance signal to the performance signal and thus the $\mathcal{H}_2$ norm is finite.

10.5 Preliminary Steps of the Analysis Algorithms

In this section we present the preliminary analysis steps required to obtain the appropriate generalized plant for the analysis of the aforementioned criteria. This preliminary analysis refers to the flight control system depicted in Fig. 10.10. The flight control system contains all the systems and signals required to define the three clearance criteria. The parameters ρ and Δ denote the flight parameters and uncertain parameters respectively. The controller is also presented as an LFT with respect to certain parameters to include the case of a scheduled controller whose scheduling parameters are measured with some error when compared to the actual ones.

- **Robust stability:** The disturbance signal w and performance signal e are disregarded along with the turbulence and performance filters. The loop is not cut at the plant input. The aircraft model LFT is combined with the controller LFT along with the sensor and actuator models to create the overall LFT of the uncertain flight control system.

- **Stability margins (nominal):** The disturbance signal w and performance signal e are disregarded along with the turbulence and performance filters. The uncertainty blocks of the aircraft and controller LFTs are disregarded and the flight parameters are fixed at certain values. Decide whether to investigate the margins in the MIMO case or in the SISO case.
  - **First method:** Introduce fictitious disturbance signals v_u, v_y at the appropriate input and output channels of the plant as shown in Fig. 10.2. Calculate $T_{cl}(s)$ as shown in (10.17) and use it in the analysis depicted by Fig. 10.6.
  - **Second method:** Interrupt the loop at the appropriate input or output channels of the plant. An example is shown in Fig. 10.10 where the SISO case stability margin is considered at the third input of the plant. Calculate the loop transfer matrix as $L(s) = T_{u \to u_c}$. Let $P(s) = L(s)$ and $C(s) = I$ and form the closed-loop transfer matrix $Tcl(s)$ as shown in (10.17).

  For both methods it must be ensured that $T_{cl}(s)$ satisfies the requirement of internal stability and that the appropriate state-space manipulations are used in order to avoid the presence of redundant states when forming it.
- **Stability margins (perturbed):** The flight parameters are not fixed but are allowed to vary in their pre-defined ranges. The uncertain parameters are not disregarded. The same steps can be taken as for the case of the nominal stability margins in order to form $T_{cl}(s)$. The transfer matrix $T_{cl}(s)$ can be used in the analysis diagram of Fig. 10.7.
- **Comfort criterion:** The loop is not interrupted at any of the plant inputs or outputs. The external disturbance w corresponds to the white noise signal w_n of Fig. 10.8 and the performance signal e corresponds to the vector of weighted accelerations. The generalized plant $G(s)$ in Fig. 10.9 is formed by expressing the transfer function from w to e as an LFT with respect to the flight and uncertain parameters.

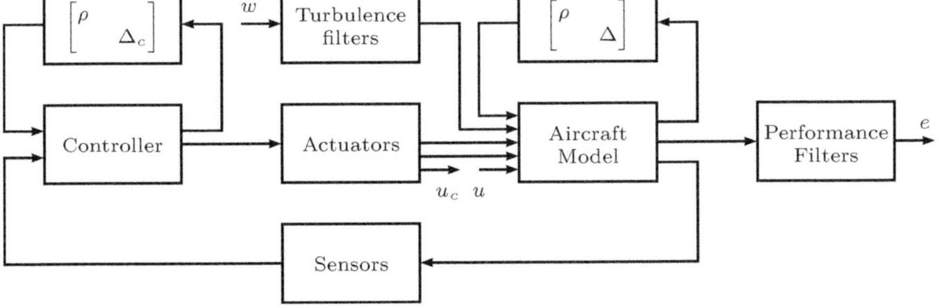

Fig. 10.10 Flight control system used for calculating the generalized plants used in the flight control clearance of the aforementioned criteria.

10.6 Conclusion

This chapter presents the use of IQCs and convex optimization in order to examine three flight clearance criteria. These are the robust stability criterion with respect to variations in real, uncertain parameters, the stability margins criterion and the comfort criterion with respect to excitation from turbulence. The robust stability criterion with respect to variations in uncertain parameters is a standard stability criterion and is naturally formulated using IQCs.

The classical stability margins criterion is analyzed using a technique based on the use of the generalized stability margin b_{PC} which has already been shown to have a connection to an exclusion region in the Nichols plane that the loop transfer function does not enter. A non-conservative lower bound on b_{PC} allows the construction of a larger ellipse that the loop transfer function does not enter and therefore implies the satisfaction of better phase and gain margins. We show in this chapter how to calculate the lower bound on b_{PC} by solving a semidefinite program based on IQC theory. We are also able to formulate the problem of calculating a lower bound on the perturbed stability margin with respect to variations in parametric uncertainty. In this way we can investigate the classical stability margins in the presence of parametric uncertainty.

The comfort criterion can be formulated as a robust $\mathcal{H}_2$ performance analysis problem. The objective is to obtain an upper bound on the energy of the acceleration signals at certain points on the fuselage under the excitation of turbulence. Again, the calculation of the upper bound of the energy of the acceleration signals over variations in parameters is performed using IQCs.

References

1. Fielding, C., Varga, A., Bennani, S., Selier, M. (eds.): Advanced Techniques for Clearance of Flight Control Laws. LNCIS, vol. 283. Springer, Heidelberg (2002)
2. Anon. General Specification for Flight Control Systems: Design, Installation and Test of Piloted Aircraft. USAF, Mil. Spec. MIL-F-9490D (1975)
3. Anon. Military Specification: Flying Qualities of Piloted Airplanes. USAF, Mil. Spec. MIL-F-8785C (1980)
4. Zhou, K., Doyle, J.C., Glover, K.: Robust and Optimal Control. Prentice-Hall, Englewood Cliffs (1996)
5. Boyd, S., El Ghaoui, L., Feron, E., Balakrishnan, V.: Linear Matrix Inequalities in System and Control Theory. Studies in Applied Mathematics. SIAM, Philadelphia (1994)
6. Megretski, A., Rantzer, A.: System analysis via integral quadratic constraints. IEEE Transactions on Automatic Control 42(6), 819–830 (1997)
7. McFarlane, D.C., Glover, K.: A loop shaping design procedure using $\mathcal{H}_\infty$ synthesis. IEEE Transactions on Automatic Control 37(6), 759–769 (1992)

8. Glover, K., Vinnicombe, G., Papageorgiou, G.: Guaranteed multi-loop stability margins and the gap metric. In: Proceedings of the 39th IEEE Conference on Decision and Control, pp. 4084–4085. IEEE, Sydney (2000)
9. Papageorgiou, C., Falkeborn, R., Hansson, A.: Formulation of the stability margins clearance criterion as a convex optimization problem. In: Proceedings of the 6th IFAC Symposium on Robust Control Design, Haifa, Israel, June 16-18, IFAC (2009)
10. Khalil, H.K.: Nonlinear Systems. Prentice Hall, Englewood Cliffs (2002)
11. Jönsson, U., Kao, C.-Y., Megretski, A., Rantzer, A.: A Guide To IQCβ: A MATLAB Toolbox for Robust Stability and Performance Analysis
12. Kao, C.-Y., Rantzer, A.: Stability analysis of systems with uncertain time-varying delays. Automatica 43(6), 959–970 (2007)
13. Rantzer, A.: On the Kalman-Yakubovich-Popov lemma. Systems and Control Letters 28(1), 7–10 (1996)
14. Wallin, R., Hansson, A.: KYPD: A solver for semidefinite programs derived from the Kalman-Yakubovich-Popov lemma. In: Proceedings of the IEEE International Symposium on Computer-Aided Control System Design, Taipei, Taiwan, pp. 1–6 (2004)
15. Kao, C.-Y., Megretski, A., Jönsson, U.: Specialized fast algorithms for IQC feasibility and optimization problems. Automatica 40(2), 239–252 (2004)
16. Wallin, R., Hansson, A., Johansson, J.H.: A structure exploiting preprocessor for semidefinite programs derived from the Kalman-Yakubovich-Popov lemma. IEEE Transactions on Automatic Control 54(4), 697–704 (2009)
17. Wallin, R., Kao, C.-Y., Hansson, A.: A cutting plane method for solving KYP-SDPs. Automatica 44(2), 418–429 (2008)
18. Hansson, A., Vandenberghe, L.: Efficient solution of linear matrix inequalities for integral quadratic constraints. In: IEEE Conference, Citeseer, vol. 5, pp. 5033–5034 (2000)
19. Hansson, A., Vandenberghe, L.: A primal-dual potential reduction method for integral quadratic constraints. In: 2001 American Control Conference, Arlington, Virginia, pp. 3013–3018 (June 2001)
20. Vandenberghe, L., Balakrishnan, V.R., Wallin, R., Hansson, A., Roh, T.: Interior-point algorithms for semidefinite programming problems derived from the KYP lemma. In: Positive Polynomials in Control. LNCIS. Springer, Heidelberg (2004)
21. Vinnicombe, G.: Uncertainty and Feedback: $\mathcal{H}_\infty$ loop-shaping and the ν-gap metric. Imperial College Press, London (2000)
22. Scherer, C.W., Köse, I.E.: Robust $\mathcal{H}_2$ estimation with dynamic IQCs: A convex solution. In: Proceedings of the 45th IEEE Conference on Decision and Control, December 13–15, pp. 4746–4751. IEEE, San Diego (2006)

Chapter 11
Nonlinear Programming Methods for Worst-Case Pilot Input Determination

Daniel Skoogh and Fredrik Berefelt

Abstract. This chapter describes optimisation algorithms suitable for searching of worst-case pilot manoeuvres. The worst-case behaviour is determined by optimization of suitable criteria capturing the maximum departure of important flight parameters (e.g., angle of attack). The search must be performed over a parameter space consisting of pilot inputs, wind perturbations, aerodynamic coefficient's uncertainties, and initial flight conditions, by applying global and local optimisation techniques. We present an overview of several local methods (pattern search, cyclic coordinate descent, quasi-Newton) and global methods (genetic algorithms, differential evolution, evolution strategies, dividing rectangles). All these methods have been implemented to cope with the special features of the underlying optimization problem (e.g., noisy functions, large parameter space). The application of these methods is described in Chapter 16.

11.1 Introduction

Modern aircraft use fly-by-wire Flight Control Systems (FCS) to enhance handling qualities, safety, and controllability. Nevertheless, the introduction of fly-by-wire systems also carries the risk of unwanted dynamical properties like loss of stability or presence of uncontrollable modes through an inadequate design of Flight Control Laws (FCL). The process of clearing a FCS becomes an increasingly expensive and time consuming task as the FCS becomes more complex.

Daniel Skoogh · Fredrik Berefelt
Swedish Research Defence Agency (FOI), System Technology,
SE-16490 Stockholm, Sweden
e-mail: daniel.skoogh@foi.se, fredrik.berefelt@foi.se

A. Varga et al. (Eds.): Opti. Based Clearance of Flight Control Laws, LNCIS 416, pp. 203–217.
springerlink.com

For the clearance of flight control laws it must be proven to the aircraft certification authorities, that for each point of the flight envelope, for all possible aircraft configurations and for all combinations of parameter variations and uncertainties, there should be no possibility to drive the aircraft into an uncontrollable state. This is however very difficult since there are infinitely many combinations of flight modes, aircraft configurations and pilot inputs to analyse. Current flight clearance schemes employed by the aerospace industry try to face this challenge by performing the search for hidden weaknesses by using gridding-based search techniques of the parameter space in combination with statistical analysis methods (Monte Carlo simulations).

There is a need to improve flight clearance methods by reducing analysis times and guaranteeing a full coverage of the parameter space. A promising approach is to use optimisation based methods to find the parameters that cause the worst cases of the used clearance criteria. This chapter describes an effort to enhance the current clearance methods for determining worst-case pilot manoeuvres by using nonlinear programming techniques and a suitable parameterisation of the pilot input signals, flight envelope and uncertainty set. For this, we give an overview of several local and global optimisation methods which have been specially adapted for the needs of the COFCLUO project [1] to solve the worst-case pilot input problem. The application of these methods is described in Chapter 16.

Early works on optimisation based clearance were done within the GARTEUR Flight Mechanics Action Group FM(AG11) [2], where methods based on formulating linear and nonlinear clearance criteria as nonlinear programming problems have been applied. Besides that, methods based on the μ-analysis technique, parameter gridding and the ν-gap metrics have been investigated.

Several works of optimisation based clearance used simulation of the nonlinear aircraft model together with a fixed pilot signal over a parametric uncertainty set. The evaluated criterion was the normal load factor, n_z or the angle of attack, α, exceedance criterion. In [3,4,5,6] the pilot signal was a fixed ramp or step function and in [7] the pilot signal was the "Klonk" signal, a fixed sequence designed to cause the worst possible behaviour. None of the above methods cover the case of a general pilot input signal.

Few works have been done with a parameterised pilot input signal of pulse type [9,10,8]. It should be noted that in these studies the signal values were confined to a few discrete levels.

In the present work, the pilot signals are spanned by a finite number of parameters. However, differently from [9,10,8], the parameters are *real* valued and can take any values in a finite interval. Some preliminary results are presented in [11]. This Chapter is built on results obtained in the COFCLUO-project, see also [1,12,13].

11.2 Optimisation-Based Search of Worst-Case Pilot Inputs

11.2.1 Definition of Clearance Criteria

To capture the worst possible aircraft behaviour produced by a pilot input, we need to define a way to measure such a behaviour. For this purpose, simple measures can be used, as for example, taking the maximum of a relevant flight variable, e.g., the angle of attack, over a given time period $[0, T]$:

$$c(\boldsymbol{p}, \boldsymbol{e}, \boldsymbol{u}) := \max_{t \in [0,T]} \alpha(t; \boldsymbol{p}, \boldsymbol{e}, \boldsymbol{u}) \tag{11.1}$$

The angle of attack depends on the pilot signals via a set of parameters $\boldsymbol{p}$ used for the signal parametrization, the vector of Flight Condition (FC) parameters $\boldsymbol{e}$ specified at $t = 0$, and the uncertain parameters $\boldsymbol{u}$. By maximizing $c(\boldsymbol{p}, \boldsymbol{e}, \boldsymbol{u})$ over the combination of these parameters, the worst-case behaviour is determined, which also defines the corresponding worst-case pilot input.

11.2.2 Parametrization of Pilot Signals

The parametrization of the pilot signal must allow a sufficiently rich signal dynamics to break, if possible, the flight domain protections laws (e.g., by violating the angle of attack protection) in the case of inadequate designs (see Chapter 2 for the definition of several protection criteria). We chose a simple parametrization via low degree piecewise polynomials. In fact, as it will be shown in Chapter 16, a parametrization relying on piecewise constant and linear signals as shown in Figure 11.1 is sufficient for the determination of worst-case pilot signals (for the longitudinal and lateral stick inputs, or for the pedal signal) which violate the protection laws.

As decision variables $\boldsymbol{p}$ for the optimisation problem relying on criterion (11.1) we have taken the pilot signal values on a fixed time grid. For simulation purposes, these values are interpolated using constant or linear interpolation. The special structure induced by this parametrization on the optimisation problem can be exploited by the optimization algorithms. Such enhancements have been implemented in the quasi-Newton and cyclic coordinate descent method, see Sect. 11.3.3 and [1, 12] for more details.

11.2.3 Definition of Optimisation Problem

The optimisation problem can be set up as follows. We simply put the parameter vector $\boldsymbol{p}$, Flight Condition (FC) vector $\boldsymbol{e}$ and the uncertainty vector $\boldsymbol{u}$ into a single vector $\boldsymbol{x}$

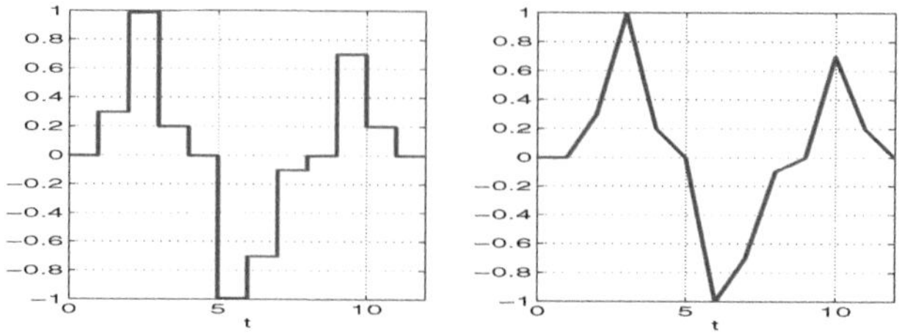

Fig. 11.1 Piecewise constant and linear pilot signals

$$\boldsymbol{x} := \begin{bmatrix} \boldsymbol{p} \\ \boldsymbol{e} \\ \boldsymbol{u} \end{bmatrix} \tag{11.2}$$

and, using the criterion definition (11.1), we define the objective function f as

$$f(\boldsymbol{x}) := c(\boldsymbol{p}, \boldsymbol{e}, \boldsymbol{u}) \tag{11.3}$$

The resulting optimisation problem will be

$$\max_{\boldsymbol{x} \in \mathcal{X}} f(\boldsymbol{x}) \tag{11.4}$$

where $\mathcal{X}$ is the set of feasible $\boldsymbol{x}$. Here it is assumed that the set $\mathcal{X}$ can be described by simple bound constraints, i.e.

$$\boldsymbol{x}_{\min}(i) \leq \boldsymbol{x}(i) \leq \boldsymbol{x}_{\max}(i) \tag{11.5}$$

where $\boldsymbol{x}(i)$ denotes the ith component of the vector $\boldsymbol{x}$. Thus the optimisation problem is a real valued optimisation problem with a nonlinear objective function and simple bound constraints.

The flight clearance problems covered in this Chapter are stated in [14]. A background on the formulation of the worst-case pilot input problem in terms of global optimisation is given in [15], and [16] presents a summary of the benefits and drawbacks of optimisation based clearance. In the analysis performed in Chapter 16, we included in $\boldsymbol{x}$ only the pilot signal parameters in the set of decision variables, because protection violations were possible even without considering the additional effects of variations in flight conditions and uncertain parameters.

The optimisation methods discussed in what follows have been applied to determine worst-case pilot inputs for the nonlinear Airbus model described in [17] This is a nonlinear simulation model, which is implemented in a MATLAB/Simulink environment. The main challenges of solving the worst-case pilot input problem are the presence of local minima and the noisy objective function evaluations. These challenges are addressed in Chapter 16, where comparative results are presented for all methods described in the rest of this chapter.

11.3 Local Optimisation Methods

Three methods for local optimisation in a bounded real-valued domain have been implemented and investigated. The methods called *Pattern Search* (PS) and *Cyclic Coordinate Descent method* (CCD) do not need any computation of derivative information. The *Quasi-Newton method* (QN) uses the gradient of the objective function (analytic or an approximation via finite differences) and an approximation of the Hessian. Although the quasi-Newton method has a theoretically higher convergence rate on smooth objective functions than the other two methods, this method is however considered less robust in determining minima of non-convex problems [18].

The existing convergence theory of these methods assumes smooth objective functions. Noisy objective function evaluations (e.g., due to the presence of truncation errors, finite difference approximations, or numerical integrations) put more strain on the optimisation methods, and the convergence conditions of the methods will not be fulfilled. Nevertheless, with a proper implementation, it can be achieved that these methods perform satisfactory even on problems with noisy functions.

Pattern search has been used in optimisation based flight clearance, see [19]. Further, the quasi-Newton method has been used in the work [19], and also in the works [3,4,27] through the MATLAB function `fmincon`[1].

11.3.1 Pattern Search

Pattern search (PS) is a deterministic local optimisation method, based only on sampling of the objective function in a neighborhood of a current point. The term *pattern search* is used in [20] for a collection of direct search (derivative free) methods of similar structure and origin. The particular method employed in the present implementation is the classical algorithm of Hooke and Jeeves [21].

Somewhat simplified, what happens is the following. Take steps forward and backward in each coordinate direction around the current iterate and see if the objective function value improves. Every time it does, accept the new point as current point and continue to investigate the remaining directions. If all directions are tried without success, shrink the step size by a factor θ, which is an option parameter (a good default value is 0.5), and make a new round of this so-called *coordinate search*. However, if a better point *is* found, declare it as new current iterate and take a step *ahead* of this better point (in the hope that the direction from the old current iterate to the better one is profitable) and around the point that is then reached, a new coordinate search is performed.

---

[1] The MATLAB function `fmincon` uses a BFGS quasi-Newton update scheme if medium scale algorithm is selected.

The situation concerning convergence results is somewhat better as compared to the global optimizers. In [20], proofs of convergence to a stationary point, $\nabla f(x) = 0$, are given for continuously differentiable functions. In [22] a very interesting result is given about the possibility of using the step size as a convergence measure. So, for continuously differentiable objective functions, the decrease in step size provides an actual indicator of convergence.

11.3.2 Cyclic Coordinate Descent Method

We implemented a cyclic coordinate descent method inspired by [23], that minimises the objective function $f(\boldsymbol{x})$ with respect to one component of $\boldsymbol{x}$ at a time in a cyclic way. This method requires no derivative information and is more robust than the quasi-Newton method, but has considerably slower convergence rate. This is why, we use this method mainly in conjunction with the quasi-Newton method to refine the outcome of a coordinate descent search.

The search takes place on a discrete set of grid points determined by the coordinate direction $\boldsymbol{e}_i$, defined by the i-th unit vector, and the corresponding step length h_i. The algorithm tries to find a point $\boldsymbol{x} \in \mathcal{X}$ such that the objective function value $f(\boldsymbol{x})$ reaches the lowest value compared to the surrounding points on the grid, i.e., such that the following holds

$$f(\boldsymbol{x}) < f(\boldsymbol{x} + h_i\boldsymbol{e}_i) \text{ and } f(\boldsymbol{x}) < f(\boldsymbol{x} - h_i\boldsymbol{e}_i) \tag{11.6}$$

for all i such that $\boldsymbol{x}(i)$ is not on the boundary. For a coordinate $\boldsymbol{x}(i)$ that touches the boundary described by the simple bound constraints (11.5), the following holds

$$\begin{aligned} f(\boldsymbol{x}) &< f(\boldsymbol{x} + h_i\boldsymbol{e}_i) \text{ and } \boldsymbol{x}(i) = \boldsymbol{x}_{\min}(i) \\ f(\boldsymbol{x}) &< f(\boldsymbol{x} - h_i\boldsymbol{e}_i) \text{ and } \boldsymbol{x}(i) = \boldsymbol{x}_{\max}(i) \end{aligned} \tag{11.7}$$

If a point $\boldsymbol{x} \in \mathcal{X}$ is found that (11.6) and (11.7) hold, then the algorithm terminates.

Let $\boldsymbol{h}$ be the step length vector, whose i-th component is h_i. The conditions (11.6) and (11.7) guarantee that a local minimiser $\boldsymbol{x}$ is found on the grid determined by $\boldsymbol{h}$, but it does not guarantee that $\boldsymbol{x} \in \mathcal{X}$ is a local minimiser. If the step length $\boldsymbol{h}$ is sufficiently small and the objective function is smooth the above criterion assures that $\boldsymbol{x} \in \mathcal{X}$ is a sufficient good approximation to a local minimiser. It is also important to select $\boldsymbol{h}$ large enough to avoid converging prematurely to a "false" local minima.

For optimisation problems where the objective function evaluations are based on simulations, it can be very hard to check if a point $\boldsymbol{x} \in \mathcal{X}$ is a local minimiser based on gradient and Hessian information. The accuracy in the gradient and Hessian evaluations might not be sufficiently high. Further, the approximation of the Hessian can be very expensive through finite differences. In such cases the above described strategy offers a good alternative.

An algorithm description of the cyclic coordinate descent method is given in [1, 12].

11.3.3 Quasi-Newton Method

Quasi-Newton methods constitute a family of methods which use at each iteration $\boldsymbol{x}_k$ an approximation $\boldsymbol{B}_k$ of the Hessian $\boldsymbol{H}_k$ of the objective function $f(\boldsymbol{x}_k)$ in a quadratic approximation which serves for step length determination. The approximate Hessian $\boldsymbol{B}_k$ is updated in every iteration by using gradient information. For a more in-depth discussion about quasi-Newton methods, see [24, 15] and the literature cited therein.

Local gradient based methods with Hessian approximation using quasi-Newton updating schemes are interesting candidates for addressing flight clearance problems. For a typical clearance criterion involving n optimization parameters, the main advantage is that only gradient information is needed, which can be computed using finite difference approximations with $\mathcal{O}(n)$ function evaluations. In this way, the direct computation of the more expensive Hessian can be avoided, which would involve $\mathcal{O}(n^2)$ function evaluations using finite difference approximations. The disadvantage is that their performance degrades when the function evaluations can not be done accurately enough (e.g., in the case of noisy function evaluations).

The quasi-Newton method has superlinear convergence under certain smoothness conditions [24]. The challenge we encounter in applying optimization-based search techniques to noisy clearance criteria is to preserve the same high convergence rate on these much harder problems.

We will just briefly describe the ideas behind our implementation of the quasi-Newton method for the determination of worst-case pilot inputs. For further details see [1, 12]. One of the major points is that the objective function evaluations are based on simulation runs, which totally dominates the computational costs. This has several implications.

First, the cost of numerical linear algebra computations, like eigendecomposition of the approximate Hessian, is negligible in comparison with the objective function evaluations. Second, the number of objective function evaluations can be reduced by considering causality aspects and boundary conditions. Due to causality, small changes of the pilot signal that occur at a time $t > t_{max}$, where at t_{max} the flight variable in question (e.g., angle of attack) achieves it maximum value, will not affect the computed objective function. This implies that all components of the current gradient $\boldsymbol{g}_k$ corresponding to time values $t > t_{max}$ will be zero[2]. Third, lazy evaluation is used, i.e. computation involving objective function evaluations like the gradient computation

[2] Here we have assumed that the maximum occurs at only one location, and that the objective function is Lipschitz continuous.

through finite difference is not done before it is needed. Another of the major points is that our method is adapted to handling noisy objective functions. Here, the line search plays an important role.

11.4 Global Optimisation Methods

Four methods for global optimisation in a bounded real-valued domain have been implemented and investigated. *Differential Evolution* (DE), *Genetic Algorithm* (GA), and *Evolution Strategies* (ES) are population based stochastic search algorithms, collectively referred to as *evolutionary algorithms.* A deterministic sampling method, *Dividing Rectangles* (DIRECT), has been included for comparison. All four methods are *derivative free*, i.e. they do not use gradient information, but only evaluations (sampling) of the objective function.

Evolutionary algorithms share a common framework and terminology borrowed from biology. A population of feasible solutions evolves into a series of generations by the action of recombination, mutation and selection operators. These operators have parameters on which the search performance depends. These parameters may be monitored during an optimization run in order to incorporate adaptation into the algorithm. Such self-adaptation has become a main feature of ES.

Global optimisation methods have been proven to be robust and efficient on several problems in science and engineering including several works in optimisation based flight clearance. For further information about DE see [25], and specifically in optimisation based flight clearance [11, 27]. Genetic Algorithms are discussed, for example, in [28] and applied to optimisation based flight clearance in [3, 4, 8, 9, 19, 27]. Further, Evolution Strategies are discussed in [29], while the deterministic sampling method DIRECT is discussed in [30, 31] and in connection to optimisation based flight clearance in [27].

11.4.1 Genetic Algorithms

Genetic algorithms are perhaps the most well known evolutionary algorithms. The term *genetic algorithms* (GA) denotes not just one optimisation procedure, but a whole family of almost endless variability. The present presentation captures just some possibilities, and is the basis of our implementation of the GA. For a more detailed account, see, for example, [28].

A *population* is a sequence of N vectors $\mathcal{P} = (\boldsymbol{x}_i)_{i=1,\ldots,N}$ representing candidate solutions from a given bounded region $\mathcal{X}$, called the *search space.* The D coordinates of the vectors are the *decision variables.* The population iteratively evolves into a series of new *generations*, by applying stochastic *variation* and *selection* operators. The terms of $\mathcal{P}$ are often referred to as "members", or "individuals".

The sequence of the best individual from each generation will hopefully converge to a global optimizer ("best" in terms of objective function value, also called *fitness*). The initial population is seeded randomly within the bound constraints, but may also be a "recycled" population from a previous run.

Variation

There are two main kinds of variation operators, called *mutation* and *crossover*. A mutation operator takes *one* individual and perturbs some of its coordinates randomly. In crossover (also called *recombination*), *two* individuals, usually referred to as *parents*, are combined by some random method of mixing/exchanging the values of the parents' coordinates. The resulting vector is in both cases called a *child*, or *offspring*. Here, we will refer to the single operand of mutation also as a "parent".

Three crossover operators have been implemented, resulting in children that lie somewhere within a hypercube with the two parents positioned in opposite corners. This introduces a certain degree of correlation between the child and its parents. The two implemented mutation operators, on the other hand, create children that are (more or less) uncorrelated with the parent in the mutated coordinates. The role of mutation is to create diversity in the population and to explore the whole search space, whereas crossover may find an even better solution in the vicinity of two already "good" parents (i.e., it is more local in its character).

In our implementation, each produced child is the result of applying only *one* variation operator (*either* crossover *or* mutation). The number of children coming from each operator is determined by a set of option parameters. The choice of a specific blend of variation operators is crucial for a successful result.

Selection

The explorative activity of the variation operators is counteracted by a selection mechanism whereby the individuals "compete for survival". This creates a force towards convergence, which has to be carefully balanced against the striving for diversification from variation. This takes place at two stages.

First, *parent selection*: the parents used for creating the offspring are randomly selected from $\mathcal{P}$ on the basis of their fitness; individuals with better fitness have higher probability of being selected for mutation or crossover.

Second, *survivor selection*: from the combined sequence of current population and offspring, individuals for the next generation are selected also based on their fitness but this time deterministically.

Termination Criteria

Unfortunately, there exist no useful proofs of convergence for the kind of population based stochastic search algorithms to which GA, DE, and ES belong. Only recently, there have emerged some partial results [32]. The proofs given are for convergence in a probabilistic sense. It is impossible to give a measure of the "degree of convergence" achieved so far, one can only rely on a sensible judgment of the state of affairs. The termination criterion used in our implementation just limits the number of generations.

The size of the population is an important parameter. It influences not only the total computation time, but also the quality of convergence. A large population maintains a greater diversity, explores larger areas of search space and may even lead to faster convergence. A too small population may give premature convergence. In [26], a population size of five to ten times the number of search space dimensions is advocated as reasonable for DE, but this is presumably a good recommendation also for GA. More difficult optimisation problems will certainly need even larger populations. It is mandatory not to be too greedy about this. This entails in turn the necessity of parallel computing for large search spaces.

11.4.2 Differential Evolution

Differential evolution (DE) is a relatively recent optimisation method that handles the real valued representation and its variational operators very efficiently with few control parameters. A more detailed account is given in [25].

As in GA, the population $\mathcal{P}$ is a sequence of N vectors in a given bounded search space $\mathcal{X}$. After initialisation, $\mathcal{P}$ is iteratively evolved by applying quite simple variation and selection operators.

Variation and Selection

The population is updated by running through three steps, *mutation*, *crossover*, and *survivor selection*, for each member $\boldsymbol{x}_i$ of $\mathcal{P}$. Thus, for each index $i = 1, \ldots, N$, pick the member $\boldsymbol{x}_i$ and do the following:

Mutation: Randomly choose three different individuals $(\boldsymbol{x}_{i_1}\ \boldsymbol{x}_{i_2}\ \boldsymbol{x}_{i_3})$ from $\mathcal{P}$ other than the current member $\boldsymbol{x}_i$. Define a *mutant vector* $\boldsymbol{v}_i$ as

$$\boldsymbol{v}_i = \boldsymbol{x}_{i_1} + F(\boldsymbol{x}_{i_2} - \boldsymbol{x}_{i_3}),$$

where $F \in (0, 2)$ is a fixed scale factor (an option parameter).

Crossover: From the current vector $\boldsymbol{x}_i$ and the mutant $\boldsymbol{v}_i$, a *trail vector* $\boldsymbol{u}_i$ is formed component-wise by selecting with probability p_{cross} the coordinate from $\boldsymbol{v}_i$, otherwise from $\boldsymbol{x}_i$. Here, $p_{cross} \in (0, 1]$ is another option parameter.

Survivor selection: The trail vector $\boldsymbol{u}_i$ is then compared with the current vector $\boldsymbol{x}_i$, and if $\boldsymbol{u}_i$ has a better fitness value, it takes the place of $\boldsymbol{x}_i$ in the next generation, otherwise $\boldsymbol{x}_i$ is retained.

Compared to GA, there is no *parent selection* in that the "parents" in the mutation/crossover steps are not chosen on the basis of their fitness values. Further, the survivor selection takes place by a simple direct tournament between the current target vector and its offspring, implying that there is no need for collecting and sorting a whole sequence of children.

A mutation operator in GA usually depends on some fixed probability distribution. In DE, on the other hand, mutation is influenced by the population itself through the base and difference vectors, which confers a kind of "self-adapting" character. Another attractive feature of DE is its dependence on only two parameters. Good default values are $F = 0.5$ and $p_{cross} = 0.5 - 0.8$.

11.4.3 Evolution Strategies

Evolution strategies (ES) are algorithms for global optimisation in real-valued domains. They have been around for as long as GA, but were developed independently. ES share many features with GA. What distinguishes them is the inclusion of self-adaptive control parameters. In GA, the mutation operators typically depend on fixed (exogenous) parameters. By letting those parameters themselves evolve during the search, subjecting them to mutation and crossover, the search process may become more efficient. A thorough presentations of ES can be found, for example, in [29], to which we refer the reader for further details. Some useful information is also found in [28].

Here, the population $\mathcal{P}$ is a sequence of μ individuals $\boldsymbol{a} = (\boldsymbol{x}, \boldsymbol{\sigma})$ in an extended search space $\mathcal{X} \times \mathbb{R}_+^{n_\sigma}$. An individual's first component $\boldsymbol{x}$ is a vector of D decision variables. The second component $\boldsymbol{\sigma}$ is a vector of n_σ standard deviations, n_σ is either 1 or D. In each generation, λ new individuals (children) are produced by recombination and mutation, from which the new generation is selected. A parameter ρ determines the number of parents involved in recombination. Our implementation deals with multi-member ES where $\mu \geq 2$.

Self-adaptation

The task of mutation is to create diversity in the population. In ES, the decision variables are mutated by adding random perturbations z_k, scaled by the lengths of the search space edges according to

$$\boldsymbol{x}'(k) = \boldsymbol{x}(k) + (\boldsymbol{x}_{\max}(k) - \boldsymbol{x}_{\min}(k)) \cdot z_k, \quad k = 1, \ldots, D.$$

The D random variables (z_k) have a joint normal (Gaussian) distribution with zero mean. In the simplest case of *isotropic* mutation, the covariance matrix of the joint distribution is proportional to the unit matrix, i.e. the components are independent and share the same standard deviation σ. In *non-isotropic* mutation, the covariance matrix is diagonal, the components are independent but have individual standard deviations σ_k, $k = 1, \ldots, D$. This gives a better adaptability to the local fitness landscape, but also demands more parameters to control. A third variant, giving even better adaptability, is *correlated* mutation, where the covariance matrix also includes non-zero off-diagonal elements. We have so far implemented only the first two cases.

Instead of being fixed from the beginning, the standard deviations are included in the extended search space of ES and undergo recombination and mutation in every iteration cycle. The idea is to have the search process for the decision variables adjusting itself to the topology of the objective function in the course of evolution. The objective function depends only on the decision variables, but the standard deviations get evaluated indirectly: good σ produce good $\boldsymbol{x}$ that survives selection.

11.4.4 Dividing Rectangles Method

The *Dividing Rectangles* (DIRECT) algorithm is a deterministic and derivative free method for global optimisation. It systematically samples the search space in a clever way, that includes both global and local explorations simultaneously. It was introduced in [30] as a generalisation of so-called *Lipschitzian optimization*. It contains only one parameter, the *Jones factor*, of which the algorithm is fairly insensitive. Unfortunately, the computational complexity scales unfavorable with the number of variables, which is limited to problems with no more than about 20 decision variables.

We refrain from giving a detailed description and instead redirect the user to [30]. Our implementation is based on a free software developed by Daniel Finkel [31]. We have adapted the code to the framework of the optimisation package for worst-case pilot input search.

DIRECT always starts by sampling the center of the search space, which is a hyper-rectangle (a "box") due to the bound constraints. This box is then iteratively partitioned into a series of sub-boxes, all of which are sampled at their center points where the objective function is evaluated. In each iteration, some of the current boxes are selected for further division along their longest edges.

It is possible to show that, in the limit of infinite number of such divisions, the set of all samples forms a dense subset of the search space [30]. If the objective function is continuous in a neighborhood of the global optimum, then convergence of the algorithm is guaranteed.

However, for a real world application, it is not possible to quantify how far a given result is located from the true global optimum. The "state of convergence" after a finite number of iterations must be judiciously estimated or assessed.

11.5 Conclusions

In this chapter, by employing simple parametrizations of the pilot inputs, we formulated the worst-case pilot input determination problem as a standard nonlinear programming problem with simple bounds. An overview of optimisation methods suited for searching of a worst case manoeuvre has been presented, by considering both local and global search techniques. The selection of the best methods highly depends on the concrete problem to be solved.

Of the presented local methods, it is difficult to select the best one. For all of them, the existing convergence proofs depend on the smoothness of the objective functions. Nevertheless it is possible to make them run quite well on noisy problems, at least if the noise is not too large. The pattern search method does not rely on any derivative information in its execution, which makes it less sensitive to noise than the quasi-Newton method. The latter method has higher theoretical convergence rate than the former. It relies on an estimate of the gradient through finite difference. How well it works on a noisy problem depends very much upon the implementation, as seen in Chapter 16. On a noisy problem, it is recommended that a run with the quasi-Newton method is supplemented with a run with a method that converges on a grid of points like our implementation of the cyclic coordinate descent method.

Of the presented global methods, differential evolution seams to be most suitable for the Airbus clearance problem as seen in Chapter 16. However, one cannot rule out the other stochastic optimisation methods. Since there are many internal option parameters of these methods, and many different "flavors", it is difficult to select a winner. The computational complexity of the DIRECT method grows very fast with the dimension of the optimisation problem which makes it less suitable for the Airbus clearance problem, at least for higher dimensions. All global methods presented use only objective function evaluations, i.e. no derivative information, which makes them less sensitive to noisy objective functions.

The piecewise constant and piecewise linear parametrisation of pilot signals are sufficient for clearance purposes, in the sense that, they allow to determine worst-case maneuvers which break the protection laws, as shown in Chapter 16. A natural extension would be to use higher order piecewise polynomials for interpolation. If this is done, a careful inspection of the properties of the resulting parameterised pilot signal needs to be done. For example, the use of cubic splines would ensure a certain smoothness of first and second order derivatives, but will destroy the causality property which can be exploited by gradient based local methods. This happens because any change in the interpolation parameters will affect the whole curvature of the cubic spline. Another possible extension is to restrict the movements of the pilot signals between interpolation points, by bounding derivatives of the pilot signals, or add an appropriate penalty function to the objective function.

References

1. Skoogh, D., Berefelt, F.: Final report WP2.5. COFCLUO Technical Report D2.5.5, FOI (January 2010)
2. Fielding, C., Varga, A., Bennani, S., Selier, M. (eds.): Advanced Techniques for Clearance of Flight Control Laws. LNCIS, vol. 283. Springer, Berlin (2002)
3. Forssell, L., Sandblom, A.: Optimisation-based clearance: The nonlinear analysis. In: Fielding, C., Varga, A., Bennani, S., Selier, M. (eds.) Advanced Techniques for Clearance of Flight Control Laws. LNCIS, vol. 283, pp. 415–430. Springer, Berlin (2002)
4. Forssell, L.: Flight clearance analysis using global nonlinear optimisation-based search algorithms. In: Proceedigns of AIAA Guidance, Navigation, and Control Conference and Exhibit, Austin, Texas, AIAA paper AIAA-2003-5481 (2003)
5. Menon, P.P., Kim, J., Bates, D.G., Postlethwaite, I.: Clearance of nonlinear flight control laws using hybrid evolutionary optimisation. IEEE Transactions on Evolutionary Computation 10(6), 689–699 (2006)
6. Menon, P.P., Bates, D.G., Postletwaite, I., Marcos, A., Fernandez, V., Bennani, S.: Worst-case analysis of flight control laws for re-entry vehicles. In: Proceedings of the IFAC Symposium on Automatic Control in Aerospace, Toulouse, France (2007)
7. Forssell, L., Hyden, A.: Optimisation based worst case search of ADMIRE. GARTEUR Technical report, TP-119-26 (2002)
8. Menon, P.P., Bates, D.G., Postlethwaite, I.: Computation of worst-case pilot inputs for nonlinear flight control systems analysis. Journal of Guidance, Control and Dynamics 29(1), 195–199 (2006)
9. Ryan III., G.W.: A genetic search technique for identification of aircraft departures. In: Proceedings of the AIAA Atmosferic Flight Mechanics Conference, Baltimore, USA, AIAA paper AIAA-95-3453 (1995)
10. Menon, P.P., Pashilkar, A.A., Sudhakar, K.: Identification of departure susceptibility for design of carefree maneuverable control scheme. In: Proceedings of the MSO-DMES Conference, Goa, India, vol. 1 (2003)
11. Skoogh, D., Eliasson, P., Berefelt, F., Amiree, R., Tourde, D.: Clearance of flight control laws for time varying pilot input signals. In: 6th IFAC Symposium on Robust Control Design, Haifa, Israel (2009)
12. Skoogh, D.: Final software - Part 2: Local methods for search of worst case manoeuvre, Version 2.0. COFCLUO Technical Report D2.5.4, FOI (January 2010)
13. Berefelt, F.: Final software - Part 1: COFOPT, a worst case maneuver search package version 2.1. COFCLUO Technical Report D2.5.4, FOI (January 2010)
14. Puyou, G.: Selected clearance problems. Part 1 - nonlinear model. COFCLUO Technical Report D1.1.1, Airbus France (2007)
15. Skoogh, D., Berefelt, F., Hagström, M., Eliasson, P., Robinson, J.: Report on comparative study of optimisation and Lyapunov methods. COFCLUO Technical Report D2.5.2, FOI (October 2007)
16. Papageorgiou, C., Joos, H.-D., Garulli, A., Skoogh, D., Hagström, M.: Survey of existing methods for flight control clearance, COFCLUO Report D2.1.1 (2007)
17. Puyou, G.: Models delivery: Part 1 - Nonlinear model. COFCLUO Technical Report D1.1.2, Airbus France (2007)
18. Heath, M.T.: Scientific Computing, An Introductory Survey, 2nd edn. Mc Graw Hill, New York (2002)

19. Varga, A.: Optimisation-based clearance: The linear analysis. In: Fielding, C., Varga, A., Bennani, S., Selier, M. (eds.) Advanced Techniques for Clearance of Flight Control Laws. LNCIS, vol. 283, pp. 385–413. Springer, Berlin (2002)
20. Torczon, V.: On the convergence of pattern search algorithms. SIAM J. Optim. 7(1), 1–25 (1997)
21. Hooke, R., Jeeves, T.A.: Direct search solution of numerical and statistical problems. J. Assoc. Comput. Mach. 8, 212–229 (1961)
22. Dolan, E.D., Lewis, R.M., Torczon, V.: On the local convergence of pattern search. SIAM J. Optim. 14(2), 567–583 (2003)
23. Luenberger, D.G., Ye, Y.: Linear and Nonlinear Programming, 3rd edn. Springer, Heidelberg (2008)
24. Nocedal, J., Wright, S.J.: Numerical Optimization. Springer, Heidelberg (1999)
25. Price, K.V., Storn, R.M., Lampinen, J.A.: Differential Evolution A Practical Approach to Global Optimization. Springer, Berlin (2005)
26. Storn, R., Price, K.: Differential evolution: a simple and efficient heuristic for global optimization over continuous space. Journal of Global Optimization 11, 341–359 (1997)
27. Menon, P.P., Bates, D.G., Postlethwaite, I.: Optimisation-based flight control law clearance. In: Bates, D., Hagström, M. (eds.) Nonlinear Analysis and Synthesis Techniques for Aircraft Control. LNCIS, vol. 365, pp. 259–300. Springer, Berlin (2007)
28. Eiben, A.E., Smith, J.E.: Introduction to Evolutionary Computing. Springer, Berlin (2003)
29. Beyer, H.-G., Schwefel, H.P.: Evolution strategies, Natural Computing 1. Kluwer Academic Publishers, Dordrecht (2002)
30. Jones, D.R., Perttunen, C.D., Stuckman, B.E.: Lipschitzian optimization without the Lipschitz constant. Journal of Optimization Theory and Application 79(1) (October 1993)
31. Finkel, D.E.: DIRECT Optimization Algorithm User Guide, March 2 (2003), `http://www.ncsu.edu/crsc/reports/ftp/pdf/crsc-tr03-11.pdf`
32. François, O.: Convergence analysis of evolution strategies with random numbers of offspring. In: Rudolph, G., Jansen, T., Lucas, S., Poloni, C., Beume, N. (eds.) PPSN 2008. LNCS, vol. 5199, pp. 21–30. Springer, Heidelberg (2008)

Part IV
Analysis Results

Chapter 12
Application of μ-Analysis Techniques to Clearance

Clément Roos and Jean-Marc Biannic

Abstract. An efficient algorithm based on an enhanced μ-analysis technique has been proposed in Chapter 7 to compute a guaranteed robust stability domain for a linear parameter dependent plant. This algorithm is applied in this chapter to the clearance of flight control laws, more precisely, for the robustness analysis of the eigenvalue and the stability margin criteria. Numerous performed tests revealed that the proposed methodology allows to handle very complex flexible plant models, that cannot be fully handled using traditional methods as grid-based worst-case search or Monte-Carlo simulation. The low conservatism of the method and the reasonably low computational effort allow its application to highly demanding problems. Thus, the enhanced μ-analysis technique represents an attractive alternative to traditional approaches to analyse stability, loads and comfort related clearance criteria.

12.1 Introduction

An efficient method based on μ-analysis is described in Chapter 7 to analyse some of the clearance criteria that need to be assessed during the certification process of an aircraft, such as the eigenvalue and the stability margin criteria introduced in Chapter 2. A MATLAB package has been implemented, which allows even non-expert users to fully benefit from the last theoretical advances in using μ-analysis techniques. These tools, presented in detail in [1], are validated by the analysis results described in the present chapter.

Starting from a set of high-order reference models describing both the rigid and the flexible dynamics of a civil passenger aircraft, several linear parameter

Jean-Marc Biannic · Clément Roos
ONERA, Systems Control and Flight Dynamics Department,
2 avenue Edouard Belin, 31055 Toulouse Cedex 4, France
e-mail: jean-marc.biannic@onera.fr, clement.roos@onera.fr

A. Varga et al. (Eds.): Opti. Based Clearance of Flight Control Laws, LNCIS 416, pp. 221–232.
springerlink.com

dependent closed-loop models have been developed in Chapters 4 and 5. These models have then been converted into linear fractional representations (LFR), which serve here as a basis for the aforementioned validation. Their features are summarised in Table 12.1, where n_f and n_s denote the number of flexible modes and the total number of states, respectively. The n parameters in Δ are normalised and characterise either the flight point (Mach number $\overline{M}$, calibrated airspeed $\overline{V}_{cas}$) or the aircraft mass configuration (center and outer tanks filling levels $\overline{\delta}_{CT}$ and $\overline{\delta}_{OT}$, embarked payload $\overline{\delta}_{PL}$). The robustness analysis is thus always investigated for parameter variations inside the unit hypercube $[-1, 1]^n$.

When trying to validate the theoretical tools proposed in Chapter 7, it is reasonable to assume that these LFRs are fully representative of the true aircraft behaviour. But in the perspective of incorporating μ-analysis techniques into an industrial clearance strategy, such an assumption becomes unacceptable, since it can lead to over-optimistic analysis results. Indeed, several reductions and other simplifying operations are usually required to generate LFRs, whose sizes allow the use of the proposed analysis tools. The price to pay is that they can be quite rough approximations of the original reference models. Fortunately, the modelling errors can be taken into account during the analysis step, as explained in Sect. 12.2.4.

The chapter is organised as follows. Analysis results are presented in Sections 12.2 and 12.3 for the eigenvalue and the stability margin criteria, respectively. Accuracy and computational times are evaluated for different values of the tuning parameters. The way μ-analysis techniques can be integrated in an industrial clearance process is then briefly discussed in Sect. 12.4.

Table 12.1 Description of the closed-loop LFRs used to validate the clearance tool

LFR number	Parameters in Δ	n_f	n_s	Size of Δ
1	$\overline{M}, \overline{V}_{cas}$	10	32	87×87
2	$\overline{M}, \overline{V}_{cas}$	4	20	51×51
3	$\overline{M}, \overline{V}_{cas}, \overline{\delta}_{OT}$	4	20	104×104
4	$\overline{M}, \overline{V}_{cas}, \overline{\delta}_{OT}, \overline{\delta}_{CT}$	4	20	203×203
5	$\overline{M}, \overline{V}_{cas}, \overline{\delta}_{OT}, \overline{\delta}_{CT}, \overline{\delta}_{PL}$	4	20	290×290

12.2 Analysis of the Eigenvalue Criterion

12.2.1 Direct Application of the Analysis Method and Variations

The standard version of Algorithm 7.2 of Chapter 7 is first applied to LFR number 1, for which the Mach number and the calibrated airspeed are

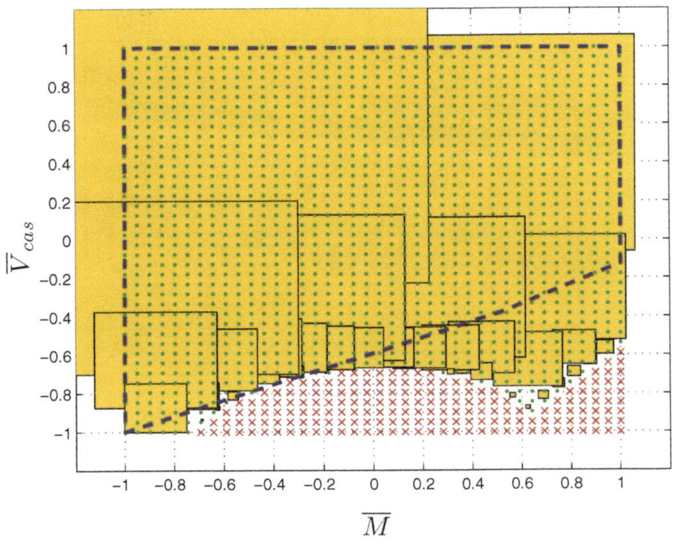

Fig. 12.1 Guaranteed stability domain - Standard computation

repeated 50 and 37 times in Δ. The domain $\mathcal{S}$ on which stability can be guaranteed is depicted in Fig. 12.1. Each of its constituting subdomains is represented by a rectangle and the shape of the flight envelope is drawn with dashed lines. Note that β_{max} is allowed to be smaller than 1 in Algorithm 7.1 of Chapter 7. Some configurations are thus cleared, which are outside the investigated box $[-1, 1] \times [-1, 1]$. The stability domain obtained using a standard grid-based approach is also plotted for the sake of comparison (stable and unstable configurations are represented by dots and x-marks respectively). Results are very conclusive. Indeed, 98.6% of the aforementioned stable configurations are included in the domain $\mathcal{S}$, which shows that the proposed μ-analysis based method is not very conservative. It can also be seen that the flight envelope is almost entirely cleared. The reference computational time is 2830 seconds on a standard 3 GHz PC with 4 GB memory.

The first two algorithmic variants introduced in Sect. 7.4.3 are then employed. The domain $\mathcal{S}$, on which stability can be guaranteed, is depicted in Fig. 12.2. The stability is proved for 96.0% of the stable flight conditions computed using a standard grid-based approach. This is almost as good as the results obtained before. The main gain is however a drastic reduction of the computational time to 342 seconds.

The μ-sensitivities are finally computed, as explained in Sect. 7.4.2. The resulting domain $\mathcal{S}$, on which stability can be guaranteed, is depicted in Fig. 12.3. The stability is proved for 96.0% of the stable flight conditions computed using a standard grid-based approach. As it can be expected, the number of stability subdomains is lower than before and the computational

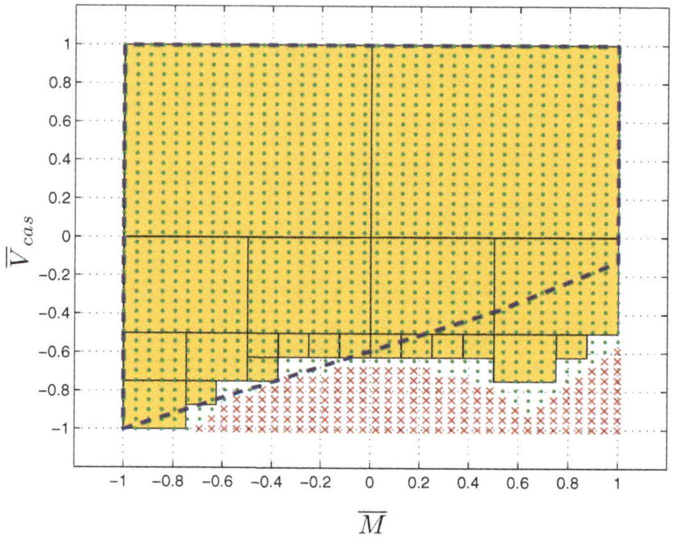

Fig. 12.2 Guaranteed stability domain - Variations on the algorithms

time is 463 seconds. It thus appears that in this specific case, the evaluation of the μ-sensitivities is quite demanding and does not bring a significant improvement. This is mainly due to the fact that a 2-D domain is investigated here. Nevertheless, benefits can be expected when the number n of parameters increases, since only 2 subdomains instead of 2^n must be investigated each time a box is bisected in Algorithm 7.2 of Chapter 7.

12.2.2 Influence of the Shape of the Initial Domain

The flight envelope represented in Figs. 12.1–12.3 is not square. Fortunately, it is almost trapezoidal, as can be seen in Fig. 12.4 (left), where the real envelope and its trapezoidal approximation are represented by dashed and solid lines respectively. Its shape can thus be easily modified to obtain a quasi-square domain. It suffices to replace $\overline{V}_{cas}$ by a new parameter V_n defined as

$$V_n = \frac{2\overline{V}_{cas} - 1 - f(\overline{M})}{1 - f(\overline{M})} \tag{12.1}$$

where $f(\overline{M}) = (7\overline{M} - 9)/16$. The modified envelope is represented in Fig. 12.4 (right) and is well suited to the analysis method introduced in Chapter 7, which can only investigate stability on a box. But the price to pay is an increase in the size of the Δ block of the resulting LFR, which is now 124×124.

Using this transformation to LFR number 1, we obtain the result depicted in Fig. 12.5, which shows that stability can be guaranteed for all flight conditions within the investigated domain $[-1, 1] \times [-1, 1]$. This means that the

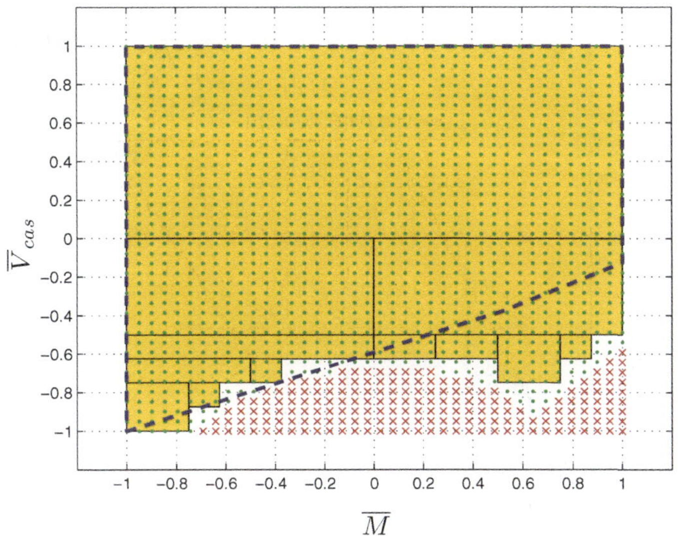

Fig. 12.3 Guaranteed stability domain - Computation of the μ-sensitivities

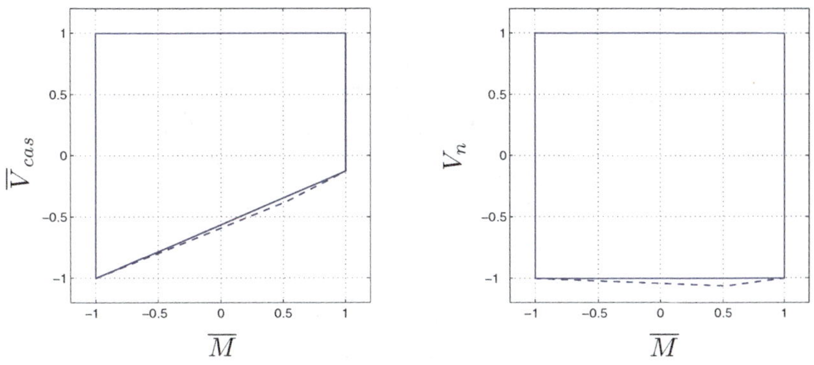

Fig. 12.4 Transformation of the initial flight envelope into a square domain

flight envelope is entirely cleared. The computational time is 349 seconds, which is quite low, although the Mach number is repeated 87 times in the Δ block of the LFR. This substantial reduction of computational time is primary due to the fact that, as it can be seen in Fig. 12.5, the clearance tool has only divided the whole envelope into seven subdomains. Thus, the preliminary transformation of the initial domain into a square domain is a worthwhile option which can substantially ease the analysis.

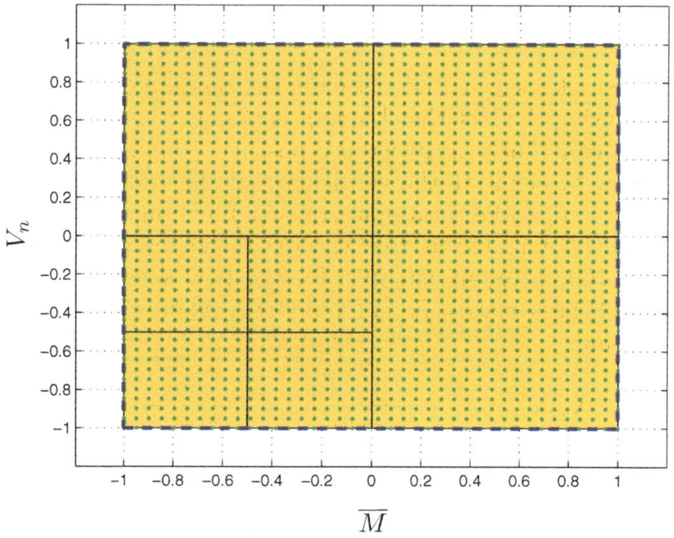

Fig. 12.5 Guaranteed stability domain - Influence of the shape of the initial domain

12.2.3 Evaluation of Conservatism

The proposed analysis method is conservative, since only μ upper bounds are computed, and therefore the corresponding robustness margins are underestimated values. It follows that, proving stability for certain configurations can fail, although the system may actually be stable. Fortunately, it is possible to efficiently handle the trade-off between conservatism and computational time. This can be achieved by appropriately tuning the parameter l introduced in Algorithm 7.2 of Chapter 7, which defines the minimum length below which the edges of the investigated boxes cannot be bisected anymore. Analysis results are shown in Fig. 12.6 for LFR number 1 and for values of l ranging from 0.01 to 0.5.

The conservatism of a method can be measured as the ratio R between the size of the domain on which stability can be guaranteed using that method and the size of the actual stability domain. As it can be seen in Fig. 12.7, the conservatism of the proposed method tends to become negligible as soon as l is chosen sufficiently small. As it can be expected, the price to pay is the exponential growth of computational time (see Fig. 12.8). Nevertheless, the computational time remains acceptable when l is chosen higher than 0.1. In this case, all computations can indeed be achieved in less than 350 seconds, although the size of Δ is 87×87. Moreover, up to 96% of the actual stability domain can be validated, as can be seen in Fig. 12.7.

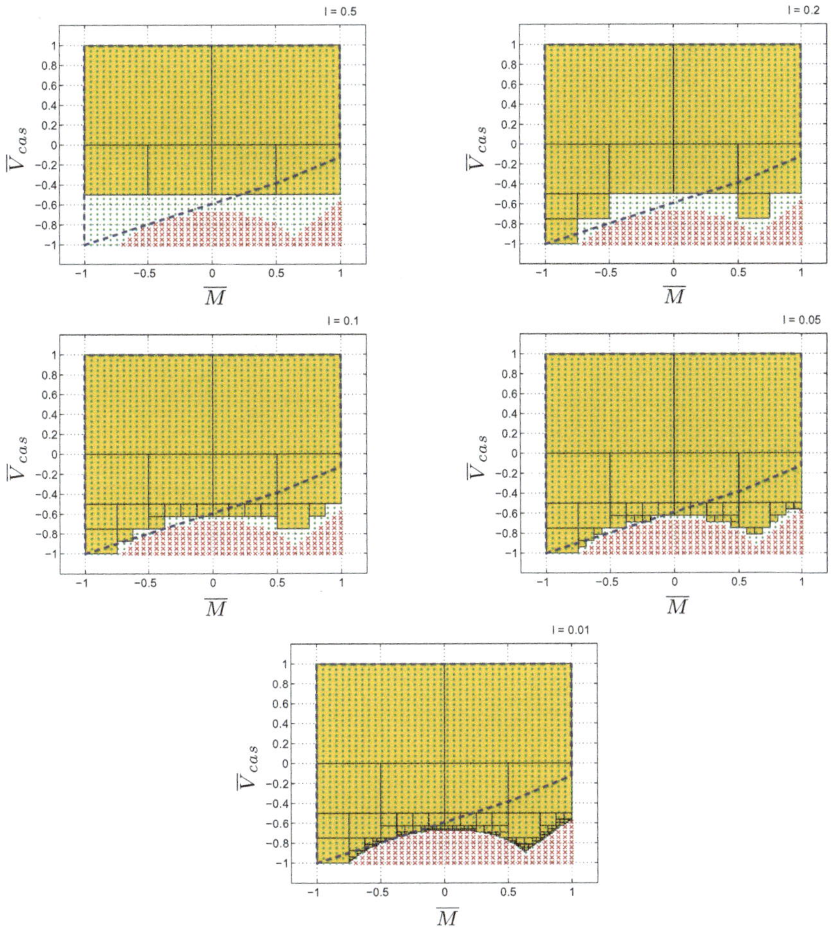

Fig. 12.6 Guaranteed stability domain for several values of l

The conservatism of the analysis method thus remains very reasonable in practice, which means that most of the configurations that cannot be validated are actually unstable.

12.2.4 Stability with Respect to the Boundary of a Truncated Sector

Algorithm 7.1 of Chapter 7 can be used to evaluate not only the stability but also the modal performance properties of a closed-loop system. The idea is to investigate stability with respect to the boundary of a truncated sector instead of the imaginary axis, as explained in Sect. 7.3.2. Such an extension can help the control engineers to identify easily the most critical configurations,

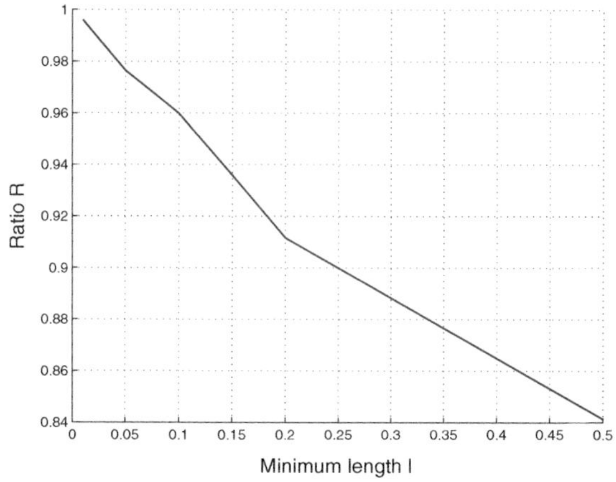

Fig. 12.7 Ratio R between the size of the domain on which stability can be guaranteed and the size of the actual stability domain

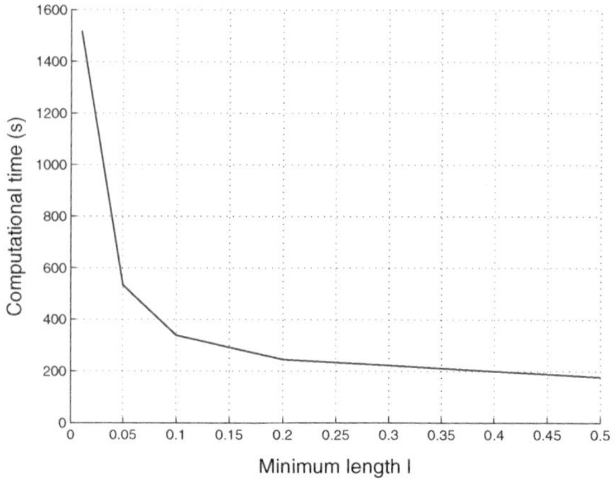

Fig. 12.8 Computational time

for which performance degradations or even loss of stability are most likely to occur if additional uncertainties or unmodelled dynamics are considered. It also allows to take into account the modelling error resulting from the use of simplified LFRs that are not always fully representative of the real aircraft behaviour. Results are shown in Fig. 12.9 for LFR number 1 and for a set of sectors characterised by a relative stability degree $\alpha = -0.168k$ and a damping factor $\xi = \sin(\phi) = 0.0168k$, where $k \in [\,0,\ 1\,]$.

Note that setting $k = 0$ amounts to investigating stability with respect to the imaginary axis. For example, it can be observed that the configuration $(\overline{M} = -1, \overline{V}_{cas} = -1)$ cannot be cleared for $k \geq 0.6$. It is thus more critical

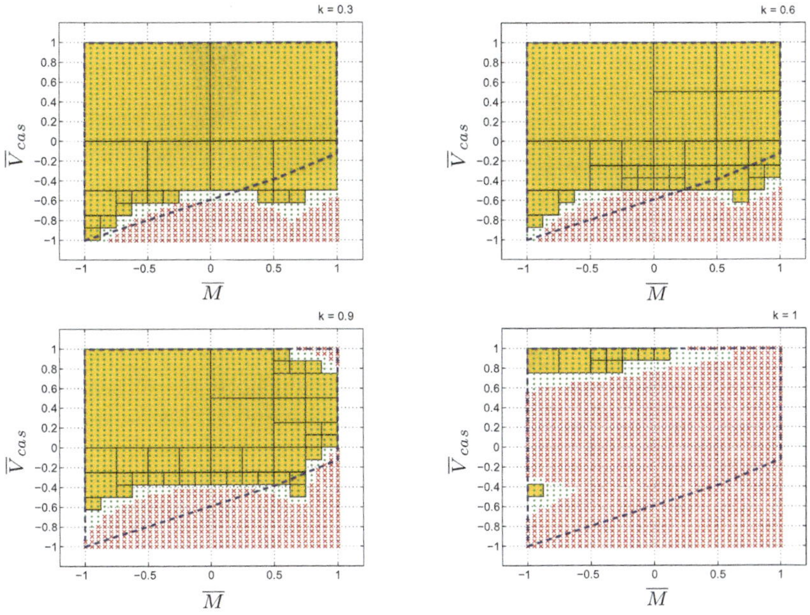

Fig. 12.9 Guaranteed stability domain with respect to the boundary of a truncated sector

than ($\overline{M} = 1, \overline{V}_{cas} = 1$), which is in turn more critical than ($\overline{M} = -1, \overline{V}_{cas} = 1$). The size of the stability subdomains also provides a confidence level: the larger a stability subdomain, the higher the guarantee that the real system is actually stable.

12.2.5 Application to Systems with Highly Repeated Parameters

To evaluate the ability of the analysis tool to handle systems with highly repeated parameters, the Algorithm 7.2 of Chapter 7 is applied successively to LFRs number 2–5. The obtained results are summarised in Table 12.2.

The ratio R between the sizes of the guaranteed and the actual stability domains remains quite high, even if 4 or 5 parameters are considered. Thus, it appears that the proposed method allows to tackle challenging real-world problems. Moreover, the computational time remains reasonable compared to the complexity of the data. Indeed, the parametric domain is very large, since the whole flight domain as well as many mass configurations are investigated. Finally, note that the algorithm can be interrupted at any time and always provides at least a partial stability domain.

Table 12.2 Application to systems with highly repeated parameters

LFR number	Size of the Δ block	Value of l	Computational time	Ratio R
2	51×51	0.1	79 seconds	0.961
3	104×104	0.1	90 minutes	0.953
4	203×203	0.2	30 hours	0.784
5	290×290	0.5	170 hours	0.652

12.3 Analysis of the Stability Margin Criterion

We analysed the stability margin criterion using both the elliptical and the first-order Padé approximation of the Nichols plane exclusion region, as described in Sect. 7.5.2. For this purpose, an LFR has been used, which resulted from LFR number 1 by cutting the feedback loop at the input of the elevator. With this LFR, two augmented closed-loop LFRs have been generated for the two above mentioned approximations of the exclusion region. The domains, on which it can be guaranteed that the criterion is satisfied, are depicted in Figs. 12.10 and 12.11.

As expected, the first-order Padé approximation is much less conservative than the elliptical one. It leads to quite satisfactory results that are summarised in Table 12.3, where R denotes the ratio between the areas of the cleared domain and the domain on which the stability margin criterion is actually satisfied.

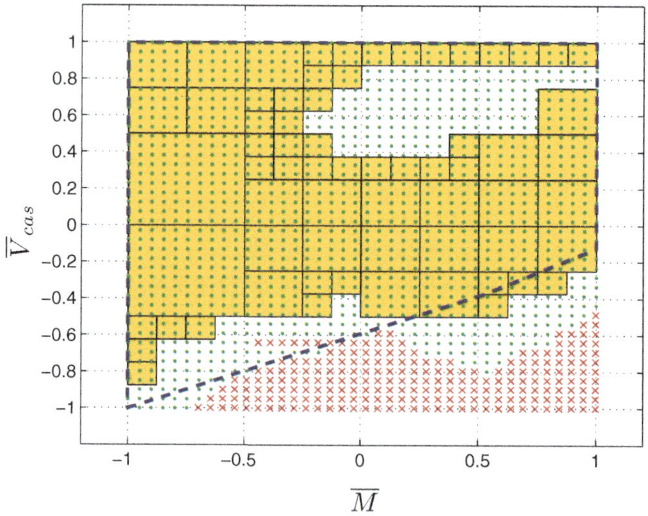

Fig. 12.10 Domain on which the stability margin criterion is satisfied - Elliptical approximation

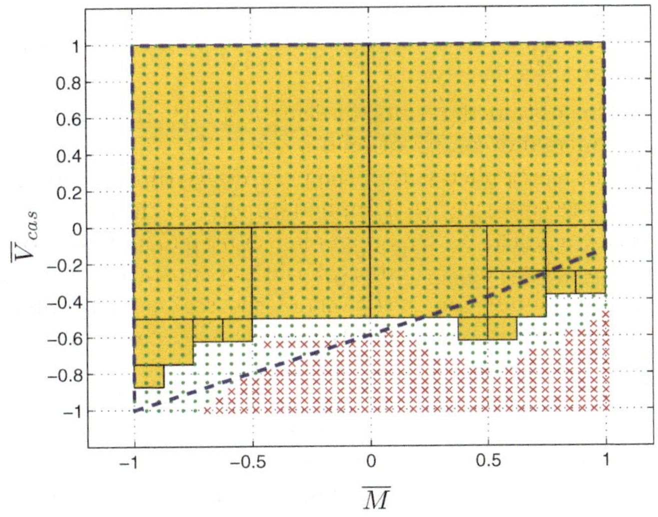

Fig. 12.11 Domain on which the stability margin criterion is satisfied - First-order Padé approximation

Table 12.3 Evaluation of the stability margin criterion

Type of approximation	Computational time (s)	Ratio R
Elliptical	732	0.758
First-order Padé	448	0.903

12.4 Conclusions and Future Perspectives

The numerous tests performed in this chapter demonstrate that several clearance criteria can be analysed more rigorously using the proposed methodology than using classical methods. Moreover, it is possible to efficiently handle the trade-off between conservatism and computational time, even if very demanding problems are considered. Enhanced μ-analysis techniques thus appear as an attractive alternative to Monte-Carlo simulations or grid-based approaches in the perspective to assess stability, loads and comfort criteria. Finally, it is worth being emphasised that these techniques could be efficiently incorporated in an industrial process. At first, the proposed clearance tools could be seen as indicators, which allow to determine quickly the most critical parametric configurations in terms of system stability or performance. In the light of the obtained results, it would then be possible to perform Monte Carlo simulations only on reduced parametric domains, thus decreasing the computational cost. Subsequently, it could be considered to use the proposed analysis tools as real validation means, thus making it possible to get rid of

Monte Carlo simulations to compute guaranteed stability domains. Nevertheless, a thorough validation of the closed-loop LFR would become necessary to quantify the modelling error as rigorously as possible. This is one of the next challenges for the future years.

Reference

1. Masi, A., Türkoğlu, E., Benedettelli, D., Paoletti, S., Garulli, A., Roos, C., Biannic, J.-M.: Final software developed. Technical Report COFCLUO D2.3.4, University of Siena, Italy (January 2010)

Chapter 13
Application of Parallel Nonlinear Programming Methods for Worst-Case Parameter Search

Hans-Dieter Joos

Abstract. This chapter presents results obtained by applying optimisation-based worst case search for flight control law clearance problems. The focus is on clearance of aircraft manoeuvrability and compliance to protection limits. Such kind of assessment problems requires sophisticated non-linear simulation models with sufficient accuracy even at extreme flight conditions. The results obtained for for several clearance criteria demonstrate the potential of combined global and local optimisation techniques in conjunction with parallel computation.

13.1 Introduction

The worst case search techniques presented in Chapter 8 are applied to clearance problems regarding simulation based manoeuvrability and flight domain protection criteria. These clearance problems are described in [1]. During the COFCLUO-project 8 different criteria based on non-linear simulation have been analysed and documented in the final project report, regarding low and high speed manoeuvrability, protections of load factor, pitch attitude and angle of attack as well as un-piloted stability. In this chapter three representative criteria are discussed: the low speed manoeuvrability criterion briefly denoted here as C_{lsm}, and two flight domain protection criteria, namely, the load factor protection criteria denoted as $C_{n_z,long}$ and $C_{n_z,lat}$ for longitudinal and lateral manoeuvres, respectively. For all criteria the clearance strategy introduced in Chapter 8 is applied to find the overall worst case up to a prescribed level of confidence based on an assumed success rate to find the global minimum.

Hans-Dieter Joos
Institute of Robotics and Mechatronics, DLR - Oberpfaffenhofen,
82234 Wessling, Germany
e-mail: dieter.joos@dlr.de

A. Varga et al. (Eds.): Opti. Based Clearance of Flight Control Laws, LNCIS 416, pp. 233–252.
springerlink.com

All results are obtained by applying the *evolution strategy* (ES) optimisation method implemented in DLR's optimisation environment MOPS [4]. It is a combined use of a stochastic global search algorithm based on evolutionary strategy concepts [2] and a local *pattern search* (PS) method [3]. More details are outlined in Chapter 8. Method ES was selected from four stochastic global search algorithms available in MOPS because it outperformed all others regarding the achieved quality of the solution and the reliability to repeatedly find the minimum.

The chapter is organised as follows: Sect. 13.2 presents the clearance results for the selected criteria. In Sect. 13.3 different optimsation methods are compared and the choice of method ES is motivated. In Sect. 13.4 computation times are reported and the effectiveness of the parallel computation is summarised. Final remarks and future perspectives are provided in Sect. 13.5.

13.2 Simulation Based Clearance Criteria

13.2.1 Low Speed Manoeuvrability Criterion

13.2.1.1 Criterion Description

This criterion is intended to check that the flight domain protections implemented in the control laws do not limit the avoidance manoeuvrability of the aircraft for low speed. Starting from any equilibrium point in the peripheral low speed flight domain, the pilot must be able to modify significantly the short term flight trajectory by pulling the longitudinal stick.

The aircraft is trimmed for a normal level flight for the purpose of simulation of the following pilot manoeuvre: the longitudinal side-stick is set to its minimum value $\delta q_m = -16deg$ for a time interval $\Delta t = 5s$. There is no lateral stick or pedal input and the thrust is set constant to the trim level with auto-thrust not engaged [1]. The low speed manoeuvrability criterion C_{lsm} can be mathematically defined as

$$C_{lsm} = \frac{1}{\Delta t \cdot C_{z,\alpha_{max}}} \int_0^{\Delta t} C_z(t)dt$$

where $C_z(t)$ is the time response of the lift coefficient and $C_{z,\alpha_{max}}$ is the maximum lift coefficient value of the current aircraft configuration at the current flight point. The criterion is satisfied when $C_{lsm} > C_{0,lsm} = 1/\sqrt{2}$.

The worst case is searched with respect to the parameters described in Table 13.1.

The parameter ΔX_{cg} defines the normalised location of the centre of gravity between its minimum and maximum value depending on mass m. The corresponding centre of gravity location $X_{cg}(m)$ is calculated using linear interpolation between the minimum and maximum values provided by

Table 13.1 Optimisation parameters of low speed manoeuvrability criterion

Name	Unit	Min	Max	Comment
h	ft	0	41000	altitude
m	t	120	233	mass
ΔX_{cg}	-	0	1	$X_{cg}(m) = X_{cg,\min}(m) + \Delta X_{cg} \cdot (X_{cg,\max}(m) - X_{cg,\min}(m))$
ΔV_{cas}	-	0	1	$V_{cas} = V_{s1g} + \Delta V_{cas} \cdot (V_{ls} - V_{s1g})$
δC_x	-	-0.1	0.1	relative uncertainty in C_x
δC_z	-	-0.1	0.1	relative uncertainty in C_z
δC_m	-	-0.1	0.1	relative uncertainty in C_m

the weight and balance functions included in the non-linear COFCLUO-simulation model [1].

Similar holds for the parameter ΔV_{cas} which defines the aircraft speed in the peripheral flight envelope between *stall speed* V_{s1g} and *minimum selectable speed* V_{ls}. These quantities are depending on height, centre of gravity and mass. Functions to compute V_{ls} and V_{s1g} are also available with the COFCLUO-simulation model [1].

The parameters δC_x, δC_z and δC_m describe relative uncertainties of the longitudinal aerodynamic coefficients C_x, C_z and C_m, respectively.

The auxiliary parameters ΔX_{cg} and ΔV_{cas} have been introduced explicitly in order to transform nonlinear parameter ranges to independent parameter intervals leading to a simpler unconstrained parameter optimisation problem, see Chapter 8. Independent from that, the applied optimisation algorithm automatically scales all optimisation parameters according to their range to improve efficiency. This allows to define parameters with physical meaning even if their ranges of values differ significantly. Compare e.g. the range of altitude h and uncertainty δC_x in Table 13.1.

13.2.1.2 Worst Case Search Results

According to the clearance strategy proposed in Chapter 8, a series of 40 independent optimisation runs has been started with different initial parameter values and different sequences of random numbers. Assuming a success rate to detect the worst case of 0.1, the worst case should be achieved at least once with a confidence level of about 0.9. Contrary to the proposed strategy, the series of optimisations is not terminated after a solution not satisfying the clearance level has been detected. For the first clearance criterion considered here, the complete series of 40 optimisations is computed in order to ensure and certify the assumption on the success rate.

The optimisation runs are performed applying the following options for the method ES listed in Table 13.2.

Table 13.2 Method options applied for clearance of criterion C_{lsm}

`popsize: 20`	`size of population`
`ngen: 100`	`number of generations`
`offspringsize: 140`	`offspring size`
`elite: 2`	`elitists selection from parents and children`
`randseed: -1`	`no reset of random number generator`
`nfunval: 10`	`termination after nfunval generations without improvement of the criterion`
`nsteps: 1`	`termination after nsteps generations with mutation step size smaller than ftol`
`ftol: 1.0e-03`	`accuracy tolerance`
`hybrid: 0`	`start local optimisation for refinement after termination of gobal search`

The initial populations are generated randomly with uniform distribution of the parameters, guaranteeing the statistical independency of the individual runs.

The worst case criterion value found is $C_{lsm} = 0.66235 < 0.7071$ and hence the criterion is not cleared. Up to numerical tolerances this value was found in all 40 optimisation runs. The corresponding worst case parameter values are listed in Table 13.3.

Table 13.3 Worst case parameter values found for criterion C_{lsm}

Name	Unit	Worst Case Value
h	ft	1.4489
m	t	120
ΔX_{cg}	-	1
ΔV_{cas}	-	1
δC_x	-	0.1
δC_z	-	0.1
δC_m	-	0.1

The results obtained are illustrated in Fig. 13.1. The diagrams show that all achieved criterion and depicted parameter values are graphically identical. All 40 optimisations end up with almost the same solution which can be assumed as worst case. A more detailed analysis of all results yields that only the parameters h and δC_x are not exactly the same for all solutions. The worst case values of altitude h lye between 1.4ft and 29ft, which represent

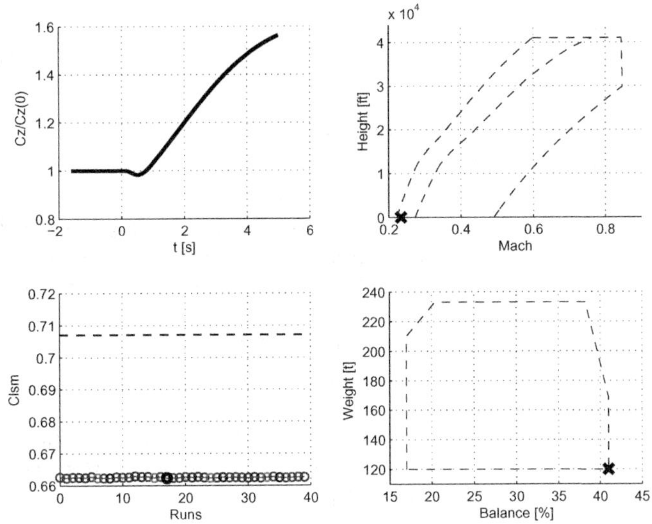

Fig. 13.1 Results of the low speed manoeuvrability criterion C_{lsm} found by 40 runs of ES with different initial settings (initial values, random numbers). All optimisation runs end up with graphically identical parameter values and identical time responses of normalised C_z.

practically the same values regarding the total parameter range of 41000ft and the used optimisation tolerances. The worst case values of the uncertainty δC_x vary from -0.04 to 0.1, indicating less sensitivity of the criterion with respect to this parameter. The resulting normalised time responses $C_z/C_z(0)$ are also graphically identical and can not be distinguished for the different optimisation runs.

The contour of the flight envelope in Fig. 13.1 is sketched for mean values of weight and position of the centre of gravity, shortly denoted as balance. The corresponding mean low speed bounds of the normal and peripheral envelope are also depicted.

The outcome of a single optimisation run is illustrated in Fig. 13.2. The lower left diagram shows the optimisation progress, where the values of the criterion are correlated with shading. The diagrams on the right hand side show the corresponding trial points in the flight envelope and the mass and balance diagram, respectively. The shading of the points corresponds to the shading of the criteria values.

The data produced during optimisation can be further analysed. Even though the optimisation is goal oriented towards worst case, most of the trial points have a criterion value $C_{lsm} > C_{0,lsm}$ and hence are satisfactory. Only 14% of the trial points have criterion values below the clearance level. Unsatisfactory solutions only occur for low altitude ($h < 70ft$), low weight ($m < 128t$), high values of the centre of gravity location ($\Delta X_{cg} > 0.52$) and

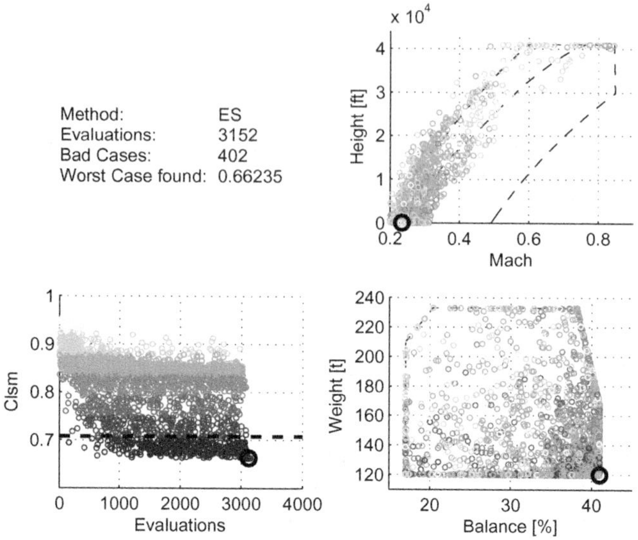

Fig. 13.2 Outcome of a single C_{lsm} optimisation run utilising evolutionary strategy. The diagrams on the right show parameter trials of Mach, height, weight and balance examined during optimisation. The shading is correlated with the criterion values. The worst case found is encircled. The left diagram illustrates the progress of the criterion value over evaluations.

higher speed values ($\Delta V_{cas} > 0.79$). Since this analysis is only based on the trial points of a single optimisation, an exact separation of the flight points into cleared and not cleared is not possible. Nevertheless, the results clearly indicate regions in the parameter space with unsatisfactory behavior.

Regarding the optimisation progress, Fig. 13.2 shows that unsatisfactory solutions are found very early. The first unsatisfactory result is obtained after about 200 trials. However, the evolutionary strategy algorithm does not stop before 3000 evaluations. Otherwise a sufficient examination of the parameter space would not be possible. In all runs the stochastic search terminates because of the exceedance of option parameter `nfunval`, see Table 13.3, i.e. the criterion value did not improve for more than 10 succeeding generations. The value of `nfunval` is therefore crucial for optimisation success on one hand and computational effort on the other hand. The value of `nfunval` = 10 is chosen to give a high success rate.

The final optimisation using a pattern search algorithm provides only a small improvement of the overall result. The criterion value was improved from about 0.6632 to the final value of 0.6624 at the expense of about 150 evaluations.

13.2.1.3 Confidence of Worst Case Results

Additional 40 independent runs have been performed to get higher confidence in the worst case found up to now, although the minimum value is below the clearance bound already. The achieved solutions of the new runs are the same as before, see Fig. 13.3. This confirms the minimum solution to be the worst case. Assuming a success rate of $q = 0.1$ gives now a confidence of $Pr = 0.99$.

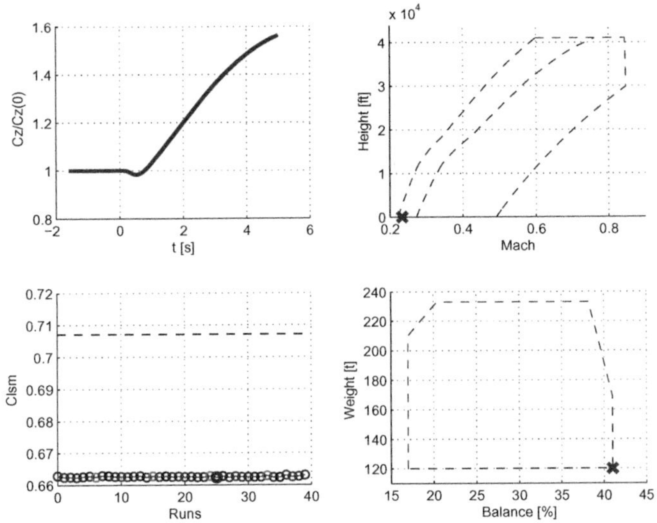

Fig. 13.3 Results of the 2nd series of 40 C_{lsm} optimisation runs applying ES; the minimum solutions are the same as in Fig. 13.1.

From the results obtained one can conclude that the assumed value of the success rate, $q = 0.1$, underestimates the true value. Experiences show that other COFCLUO-clearance criteria have similar characteristics regarding the worst case search as the low speed manoeuvrability criterion has. Hence, for further investigations a higher success rate can be assumed. The worst case search for the remaining criteria is done now by assuming a value of $q = 0.5$, which leads to at most 10 runs for detecting the worst case at least once with a confidence level of 0.99.

13.2.1.4 Comparison with Monte Carlo Analysis

A Monte Carlo analysis of criterion C_{lsm} was performed for comparison purpose. All worst case search parameters are assumed to be uniformly distributed over the defined intervals. A sample size of 200000 was chosen. This is more than the total number of evaluations needed for 40 optimisation runs. For this about 160000 evaluations have been necessary. Fig. 13.4 shows the frequencies of criterion values and the corresponding cumulative distribution

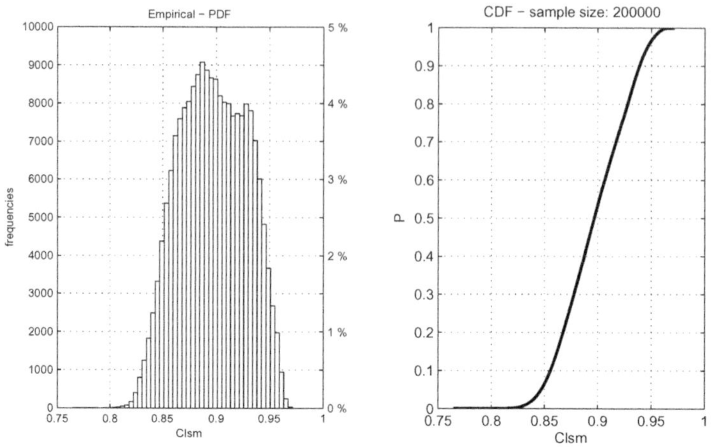

Fig. 13.4 Empirical probability and cumulative distribution of criterion C_{lsm} for a Monte-Carlo analysis run with uniformly distributed parameters and a sample size of 200000.

function. No trial point with criterion value less than the clearance bound $C_{0_{lsm}} = 0.7071$ was detected.

From the Monte Carlo analysis we can conclude that unsatisfactory events are very unlikely having a probability less than $5 \cdot 10^{-6}$. To detect such an event at least once with confidence level of, say, 0.9 more than 460000 sample points would be necessary. Thus unsatisfactory values can be detected more efficiently and reliably by worst case search than by Monte Carlo simulation. However, Monte Carlo simulation is still relevant to asses the probability of unsatisfactory cases.

13.2.2 Load Factor Protection: Longitudinal Manoeuvre

13.2.2.1 Criterion Description

For a specific longitudinal manoeuvre it must be shown that for all parameter combinations in a special flight condition the normal acceleration n_z does not exceed the prescribed bounds, i.e., -1g $\leq n_z \leq$ 2.5g in clean configuration (i.e., wing flaps, slats, spoilers and landing gear are retracted) and $0 \leq n_z \leq$ 2g in high lift configurations, see [1].

The indicator function is the time response of the normal acceleration n_z over a finite time interval $[0, T]$. The criterion considered here evaluates the maximum load factor in clean configuration. The criterion manoeuvre is performed for the maximum allowed weight at an altitude of 20000ft as suggested in [1]. The manoeuvre starts from IDLE (i.e., minimum thrust level) and a flight path corresponding to the correct speed. Then a 2.5g pull-up is performed while applying full thrust. The pitch angle is stabilized at 20° until the activation of the alpha protection occurs. The criterion $C_{n_z,long}$ is simply computed as:

$$C_{n_z,long} = -\max_{t \le T} n_z(t)$$

where $T = 30s$. The value of the criterion should not be below -2.5, i.e. normal acceleration should not exceed 2.5g.

The COFCLUO nonlinear aircraft model provides several pilot models to facilitate simulation of pilot manoeuvres. The so called "flight path angle mode" is used to perform the specific longitudinal manoeuvre. The required pitch angle target $\theta_t = 20°$ is achieved by transforming it to a flight path angle target γ_t according to:

$$\gamma_t = \theta_t + (\gamma - \theta).$$

The simulation model has been augmented by this flight path angle target law. Full thrust has been achieved by setting the speed target to the maximum value, while the auto thrust is engaged during the whole manoeuvre.

The parameters for worst case search are described in Table 13.4. The parameter ΔV_{cas} is introduced to cover the speed range of the flight envelope between minimum selectable speed V_{ls} and *maximum operating velocity* V_{MO}. The parameter ΔX_{cg} defines the location of the centre of gravity between its possible minimum and maximum value depending on mass.

Table 13.4 Optimisation parameters for longitudinal load factor protection criterion $C_{n_z,long}$.

Name	Unit	Min	Max	Comment
ΔV_{cas}	-	0	1	$V_{cas} = V_{ls} + \Delta V_{cas} \cdot (V_{maxop} - V_{ls})$
ΔX_{cg}	-	0	1	$X_{cg}(m) = X_{cg,\min}(m) +$ $\Delta X_{cg} \cdot (X_{cg,\max}(m) - X_{cg,min}(m))$
δC_x	-	-0.1	0.1	relative uncertainty in C_x
δC_z	-	-0.1	0.1	relative uncertainty in C_z
δC_m	-	-0.1	0.1	relative uncertainty in C_m
w_{x,t_i}	-	0	1	wind initiation time, x direction, fraction of simulation time
$w_{x,a}$	kts	-20	20	wind amplitude, x direction
$w_{x,g}$	kts/s	1	5	wind gradient, x direction
w_{z,t_i}	-	0	1	wind initiation time, z direction, fraction of simulation time
$w_{z,a}$	kts	-20	20	wind amplitude, z direction
$w_{z,g}$	kts/s	1	5	wind gradient, z direction

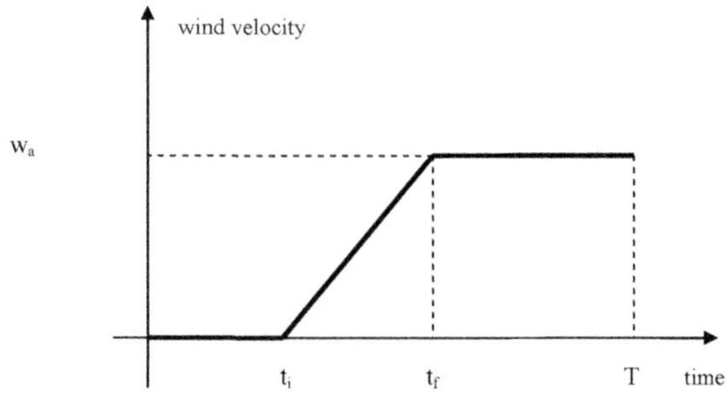

Fig. 13.5 Shape of the wind profile and the corresponding parameters.

Wind velocity components in x, and z-directions are specified by 3 parameters for each direction. The profile of each component is sketched in Fig. 13.5 according to [1]. The time instances t_i and t_f are computed for the x-axis from the optimisation parameters wind initiation time w_{x,t_i}, wind amplitude $w_{x,a}$ and wind gradient $w_{x,g}$ as follows:

$$t_i = w_{x,t_i} T, \qquad t_f = \frac{w_{x,a}}{w_{x,g}} + t_i$$

and similarly for the z-axis.

13.2.2.2 Worst Case Search Results

A series of independent optimisations has been started applying search method ES and stopped after 10 runs. With our assumptions on the success rate, made in subsection 13.2.1.3, we can conclude that the worst case has been found at least once with probability $Pr = 0.99$.

The optimisation runs are performed applying the options for method ES listed in Table 13.5.

Compared to criterion C_{lsm}, the population size has been increased because of the higher number of parameters. The number of generated offspring is adapted accordingly.

The outcomes of the 10 optimisations are depicted in Figs. 13.6 and 13.7. All optimisation runs end with maximum values for speed and balance. The small variation of the criterion values result from slightly different wind parameters which also causes the differences in the time responses of n_z and θ (Fig. 13.7). The most effective parameters are speed and balance, wind is less effective.

The right diagram in Fig. 13.7 shows that the response of the pitch angle is perfectly kept at 20° by the implemented control law as demanded for the manoeuvre.

Table 13.5 Method options applied for clearance of criterion $C_{n_z,long}$

popsize:	30	size of population
ngen:	100	number of generations
offspringsize:	180	offspring size
elite:	2	elitists selection
randseed:	-1	no reset of random number generator
nfunval:	10	termination after nfunval generations without improvement of the criterion
nsteps:	1	termination after nsteps generations with mutation step size smaller than ftol
ftol:	1.0e-03	accuracy tolerance
hybrid:	0	start local optimisation after termination of global search

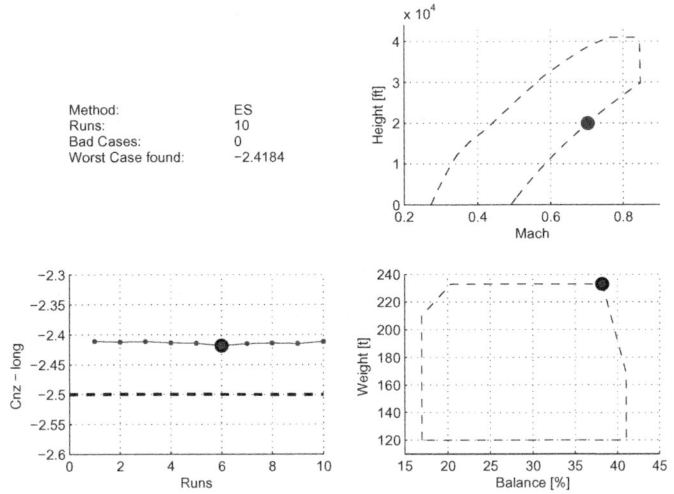

Fig. 13.6 Results of the longitudinal load factor protection criterion $C_{n_z,long}$ found by 10 runs of ES with different initial settings (initial values, random numbers).

The criterion values are greater than -2.5 in all cases and hence the lower bound is not violated. Thus the criterion can be considered as cleared with a high level of confidence. The parameter values of the worst case can be found in Table 13.6.

Analysis of a single optimisation run shows, that low criterion values are found very early in the optimisation run and low values also occur very frequently, but the clearance bound was never reached, see Fig. 13.8.

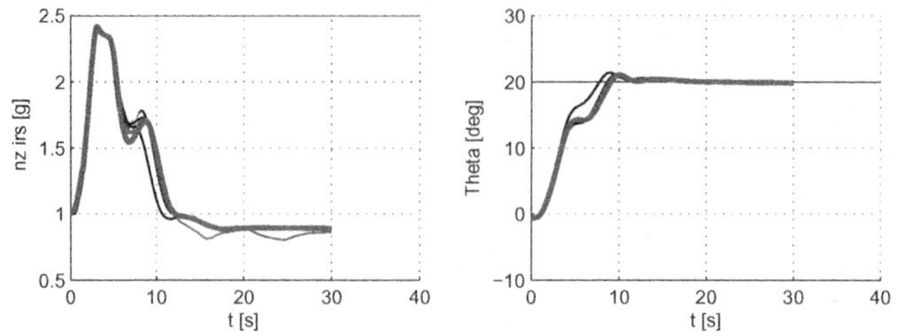

Fig. 13.7 Time responses of n_z and θ for the 10 results found for criterion $C_{n_z,long}$.

Table 13.6 Worst case parameters found for criterion $C_{n_z,long}$

Name	Unit	Worst Case Value
ΔV_{cas}	-	1
ΔX_{cg}	-	1
δC_x	-	0.065
δC_z	-	0.1
δC_m	-	0.1
w_{x,t_i}	-	0.06
$w_{x,a}$	kts	1.18
$w_{x,g}$	kts/s	3.95
w_{z,t_i}	-	0.07
$w_{z,a}$	kts	13.93
$w_{z,g}$	kts/s	5

13.2.3 Load Factor Protection: Lateral Manoeuvre

13.2.3.1 Criterion Description

For a specific lateral manoeuvre it must be shown that for all parameter combinations in a special flight condition the normal acceleration does not exceed the prescribed bound of 2.5g in clean configuration (i.e., wing flaps, slats, spoilers and landing gear are retracted), see [1].

The indicator function used is the time response of the normal acceleration n_z over a finite time interval $[0, T]$. The manoeuvre is performed for the maximum allowed weight at an altitude of 20000ft as suggested in [1]. The aircraft is trimmed for a straight level flight. A turn is performed at constant speed with bank angle slowly increasing up to 66°. The bank angle should be hold at 66° for 10 seconds and then is decreased to zero. The criterion $C_{n_z,lat}$ is computed as

$$C_{n_z,lat} = - \max_{t \in [0,\, T]} n_z(t)$$

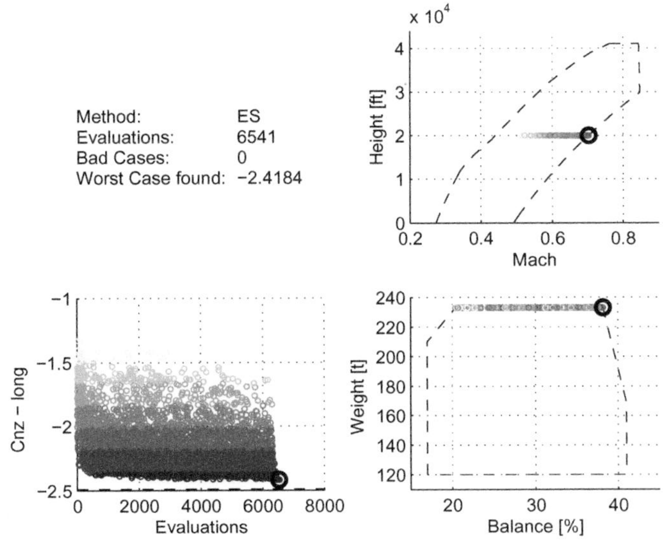

Fig. 13.8 Outcome of a single $C_{n_z,long}$ optimisation run utilising evolutionary strategy. The diagrams on the right show parameter trials of Mach and balance examined during optimisation. The shading is correlated with the criterion values. The worst case found is encircled. The left diagram illustrates the progress of the criterion value over evaluations.

where $T = 40s$. The value of the criterion should not be below -2.5 for clean configuration which is considered here.

The pilot model "yaw angle target mode", provided with the COFCLUO-simulation model [1], is used to perform the specific lateral manoeuvre where the yaw angle target is generated by an additional feedback law. The increasing bank angle ϕ up to 66° is achieved by commanding an increasing yaw angle χ as output of an integrator where the integrator input $\dot{\chi}$ is derived from the kinematic relation of load factor and yaw rate:

$$n_z = \sqrt{\cos^2\theta + (\dot{\chi}V/g)^2}\,.$$

For small pitch angle θ it can be assumed that $\cos\theta \approx 1$ and hence

$$\dot{\chi} \approx \frac{g}{V}\sqrt{n_z^2 - 1}.$$

Here, V denotes the ground speed. When the maximum achievable bank angle was detected the value of $\dot{\chi}$ is held for 10 seconds. Afterwards the integrator is reset to a constant value providing a new constant yaw angle target such that ϕ goes back to zero. In addition the flight path angle hold mode and the auto thrust were engaged during the manoeuvre in order to hold altitude and speed.

Table 13.7 Optimisation parameters for lateral load factor protection criterion $C_{n_z,lat}$

Name	Unit	Min	Max	Comment
ΔV_{cas}	-	0	1	$V_{cas} = V_{ls} + \Delta V_{cas} \cdot (V_{maxop} - V_{ls})$
ΔX_{cg}	-	0	1	$X_{cg}(m) = X_{cg,\min}(m) +$ $\Delta X_{cg} \cdot (X_{cg,\max}(m) - X_{cg,\min}(m))$
δC_x	-	-0.1	0.1	relative uncertainty in C_x
δC_y	-	-0.1	0.1	relative uncertainty in C_y
δC_z	-	-0.1	0.1	relative uncertainty in C_z
δC_l	-	-0.1	0.1	relative uncertainty in C_l
δC_m	-	-0.1	0.1	relative uncertainty in C_m
δC_n	-	-0.1	0.1	relative uncertainty in C_n
w_{x,t_i}	-	0	1	wind initiation time, x direction, fraction of simulation time
$w_{x,a}$	kts	-20	20	wind amplitude, x direction
$w_{x,g}$	kts/s	1	5	wind gradient, x direction
w_{y,t_i}	-	0	1	wind initiation time, y direction, fraction of simulation time
$w_{y,a}$	kts	-20	20	wind amplitude, y direction
$w_{y,g}$	kts/s	1	5	wind gradient, y direction
w_{z,t_i}	-	0	1	wind initiation time, z direction, fraction of simulation time
$w_{z,a}$	kts	-20	20	wind amplitude, z direction
$w_{z,g}$	kts/s	1	5	wind gradient, z direction

The optimisation parameters for worst case search are described in Table 13.7. For this protection criterion all longitudinal and lateral uncertainties and wind components are applied. This results in a total of 17 optimisation parameters.

13.2.3.2 Worst Case Search Results

Again a series of 10 independent optimisation runs has been performed applying search method ES . The method options listed in Table 13.8 have been used in each run.

Despite the larger number of parameters, the same value of 30 has been assigned to the population size as for the longitudinal criterion. Only the offspring number has been increased to 7 times the population size in order to produce more variation in the successive generations.

The results obtained are depicted in Figs. 13.9 and 13.10. As in the longitudinal case, all optimisation runs result in maximum values for speed and

Table 13.8 Method options applied for clearance of criterion $C_{n_z,lat}$.

popsize: 30	size of population
ngen: 100	number of generations
offspringsize: 210	offspring size
elite: 2	elitists selection
randseed: -1	no reset of random number generator
nfunval: 10	termination after nfunval generations without improvement of the criterion
nsteps: 1	termination after nsteps generations with mutation step size smaller than ftol
ftol: 1.0e-03	accuracy tolerance
hybrid: 0	start local optimisation after termination of global search

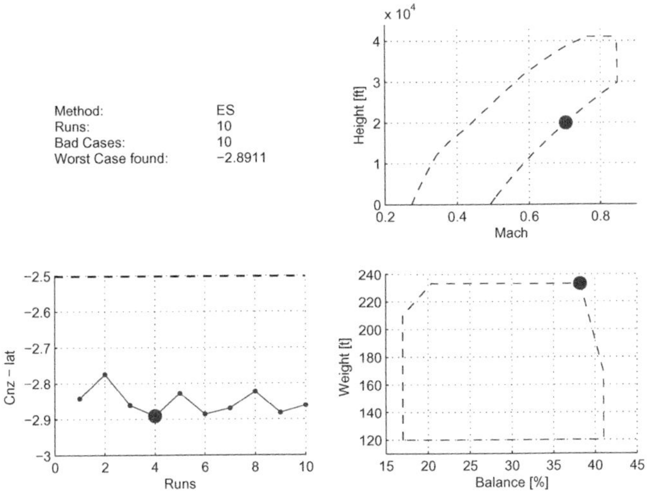

Fig. 13.9 Results of the lateral load factor protection criterion $C_{n_z,lat}$ found by 10 runs of ES with different initial settings (initial values, random numbers).

balance. The variation of the criterion values are mainly caused by different wind parameters. The most effective parameters are speed and balance.

The manoeuvre causes in all runs an exceedance of the roll angle envelope boundary of 66° and a criterion value which is less than -2.5. Hence the criterion is not cleared. The parameter values of the overall worst case found are shown in Table 13.9.

The analysis of a single optimisation run shows, that low criterion values are found very early in the optimisation run and low values also occur very

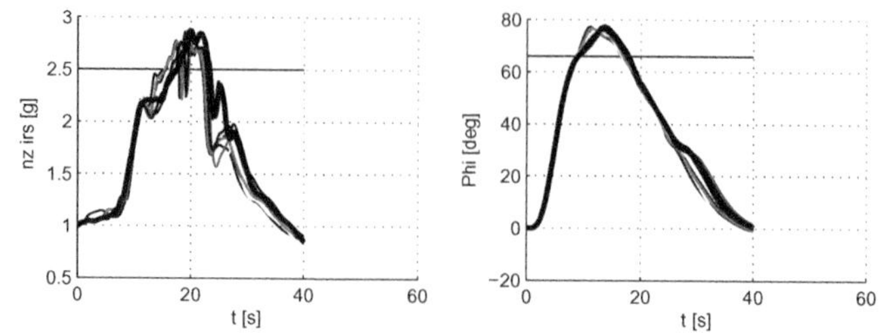

Fig. 13.10 Time responses of n_z and Φ for the 10 results found for criterion $C_{n_z,lat}$.

Table 13.9 Worst case parameters for criterion $C_{n_z,lat}$

Name	Unit	Worst Case Value
ΔV_{cas}	-	1
ΔX_{cg}	-	1
δC_x	-	0.1
δC_y	-	0.1
δC_z	-	0.1
δC_l	-	0.1
δC_m	-	0.1
δC_n	-	0.1
w_{x,t_i}	-	0.246
$w_{x,a}$	kts	20
$w_{x,g}$	kts/s	3.43
w_{y,t_i}	-	0.409
$w_{y,a}$	kts	20
$w_{y,g}$	kts/s	4.03
w_{z,t_i}	-	0.152
$w_{z,a}$	kts	20
$w_{z,g}$	kts/s	4.268

frequently, see Fig. 13.11. However, trial points with speed and balance not at the maximum value are mostly satisfactory.

The global search was terminated after about 8000 criterion evaluations and the local search algorithm applied subsequently was quite successful in further improving the criterion value (see also the "tail" of the optimisation progress diagram in Fig. 13.11). The final local optimisation using a PS algorithm provides an improvement of about -0.1 in the criterion value. The high number of evaluations compared to the former optimisations can be explained with the higher number optimisation parameters. However, a simple screening of the parameter space by only inspecting the min-max parameter values, would result in $2^{17} = 131072$ criterion evaluations which is even more than the evaluations needed for all 10 optimisation runs.

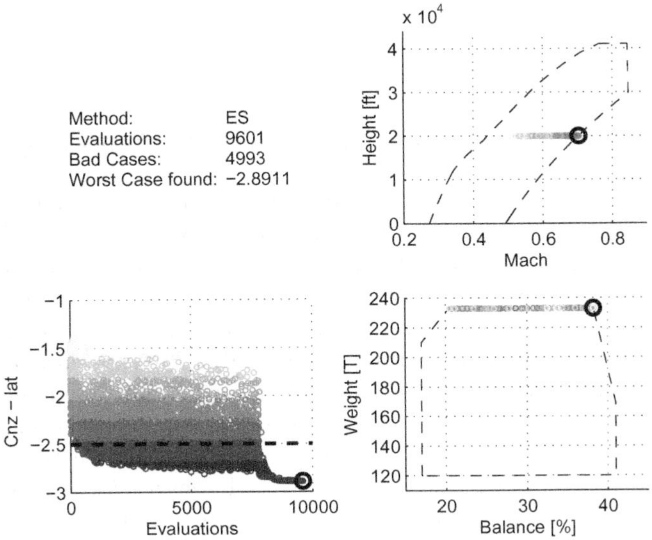

Fig. 13.11 Outcome of a single $C_{n_z,lat}$ optimisation run utilising evolutionary strategy. The diagrams on the right show parameter trials of Mach and balance examined during optimisation. The shading is correlated with the criterion values. The worst case found is encircled. The left diagram illustrates the progress of the criterion value over evaluations.

13.3 Performance Comparison of Optimisation Methods

The optimisation environment MOPS [4] provides the implementation of four global search algorithms: *evolutionary strategy* (ES), *genetic algorithm* (GA), *differential evolution* (DE) and *particle swarm optimisation* (PSO)). For more details and the corresponding references see Chapter 8. Several preliminary tests have been made in order to select the method most suitable for clearance problems of the type investigated here. The outcome of such a test is shown in Fig. 13.12, where only the results for the low speed manoeuvrability criterion (C_{lsm}) are presented achieved in 20 optimisation runs for each optimisation method.

For each method the default option parameter values have been taken as a reasonable choice. The population size for the methods DE, PSO and GA was set to 40 each. For ES a population size of 15 and a offspring size of 100 was assumed as recommended in [2] for a default choice. The optimisation runs differ in randomly chosen initial populations.

The ES is the only algorithm which was able to achieve the worst case solution in all runs and for which all results are equal within computing tolerances. Besides ES, the DE algorithm is also able to find solutions lower than the clearance level in all runs. However the results differ in the achieved worst

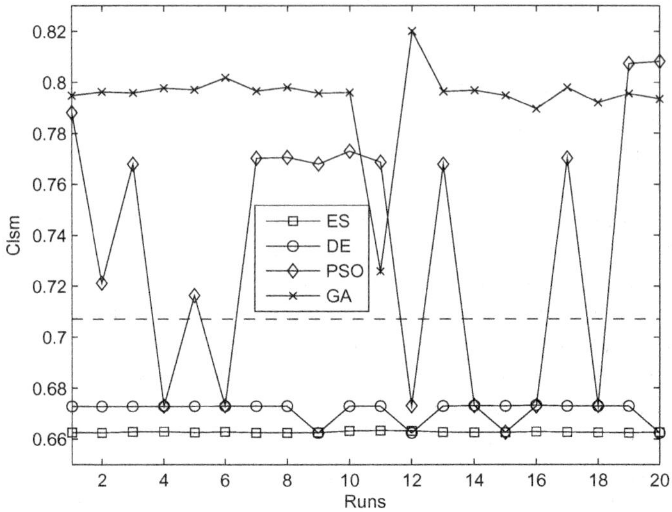

Fig. 13.12 Optimisation results achieved for the low speed manoeuvrability criterion (C_{lsm}) by 4 different optimisation methods (ES, GA, DE, PSO). 20 optimisation runs have been performed for each method with different initial populations.

case solution. Two different solutions are obtained depending on the initial population. The genetic algorithm provides results all above the clearance level. The results achieved by the particle swarm optimisation are varying very much while the smallest solution, obtained by ES, was reached only once (run 15 in Fig. 13.12).

The average number of necessary criterion evaluations per optimisation run is 2516 for ES, 2267 for DE, 1693 for PSO and 2249 for GA. The ES method needs only about 11% more criterion evaluations than DE but gives the highest success rate and accuracy. Since emphasis is put on the reliability in finding worst cases, the method ES was chosen for the investigations done here.

13.4 Computation Times

Along the general computation time analysis of Chapter 8 we investigated the time effort required by the ES method for the considered three criteria. The applied parallelisation technique is described in Chapter 8, where the evaluation of sample points or search patterns was done in parallel, while using serial optimisation runs. Computations are made on a high performance Linux-cluster with superior performance compared to the tests reported in Chapter 8. The cluster consists of 16 dual-core, dual-processor blades with Intel® Xeon® processors (CPU 5160, 3GHz) and 16GByte memory per blade. 20 CPUs have been utilised for all computations.

Table 13.10 summarizes the elapsed computing times for the different criteria. The second column of the table contains the averaged computing time for one criterion evaluation. In the third column the average numbers of evaluations for one optimisation run are listed. For serial and parallel optimisations the average times of a single optimisation are given in columns 4 and 5. The ratio of column 4 and 5 gives the speedup as introduced in Chapter 8. Dividing the speedup by the number of CPUs gives the effectiveness shown in column 6.

Table 13.10 Average computing times for serial and parallel optimisations and resulting speedup

Criterion	Single evaluation time [s]	Average number of evaluations	Average time of one clearance optimisation [h:m:s] serial [h:m:s]	20 CPUs [h:m:s]	Speedup/ effectiveness
C_{lsm}	4.5	3972	4:56:26	0:16:04	18.46 / 0.923
$C_{n_z,long}$	7.2	4342	8:38:21	0:26:23	19.65 / 0.982
$C_{n_z,lat}$	8.5	7676	18:04:17	1:01:16	17.70 / 0.885

Speedup or effectiveness of parallelisation increases when the single evaluation time is larger, because serial part of the computation and parallelisation overhead are more or less constant. This effect can be observed for criteria C_{lsm} and $C_{n_z,long}$. However, in spite of a larger evaluation time, criterion $C_{n_z,lat}$ has a smaller speedup. This is due to the comparatively large local optimisation after the evolutionary strategy has terminated, see e.g. Fig. 13.11. The available implementation of the local pattern search allows a theoretical maximum speedup of factor 3 only. Hence, local optimisation reduces the speedup achievable by algorithm parallelisation of population based optimisation methods.

13.5 Summary

The major conclusion from the clearance results achieved is that optimisation based worst case search proved to be a general, direct and reliable approach to solve clearance problems. Applying global stochastic search methods makes it applicable to any kind of clearance problem (not only simulation based) without adding conservatism involved in the method. With an assumption on the success rate reasonable confidence values can be determined. An estimate of the success rate can be found empirically.

The direct formulation of clearance criteria from simulation results can lead to noisy or even discontinuous objective functions with multiple minima which are difficult to detect. However, the proposed search techniques, global and local, are insensitive to such kind of criteria and hence are able to solve the worst case search effectively and reliably.

Especially the applied evolutionary strategy proved to be most successful in finding solutions which are sufficiently close to the global minimum. A reasonable value for the success rate can be assumed for that method even for criteria difficult to be optimised. Experiences with the presented clearance criteria showed, that a value of 0.5 for the success rate is a precautious assumption. Such a value yields a manageable amount of necessary optimisation runs for a high level of confidence.

Since the proposed clearance strategy is an iterative procedure the amount of computational work can be kept small at least in those cases where unsatisfactory criteria values could be found. Computation time can be reduced effectively when parallel computation is available. Since the amount of computational effort is still large, at least for manoeuvres with long simulation times, parallel computation is a prerequisite for applying optimisation based clearance efficiently in production work.

Optimisation based worst case search seems to be very "aggressive" in the sense that it can take advantage out of any possibly incomplete or incorrect modelling of the clearance criterion. Hence, unsatisfactory criterion values are not necessarily due to control law weaknesses. Before accepting a criterion as not cleared, a careful examination of potential error sources must be performed, as for example: validity of parameter ranges; completeness and correctness of the models involved (aircraft, controller, criterion); reasonable and realistic manoeuvres and the corresponding implementations as simulation tasks. Since worst case search often operates at extreme flight conditions in the flight envelope, the validity of the aerodynamic models in those regions is of crucial importance. Therefore worst case search can not only be applied for clearance of flight control laws, but also for verification of design and assessment models.

References

1. Puyou, W.: Selected clearance problems: Part 1 - Nonlinear model COFCLUO Report D1.1.1, AIRBUS (2007)
2. Eiben, A.E., Smith, J.E.: Introduction to Evolutionary Computing. Springer, Heidelberg (2007)
3. Hooke, R., Jeeves, T.A.: "Direct Search" Solution of Numerical and Statistical Problems. Journal of the ACM 8, 212–229 (1961)
4. Joos, H.-D., Bals, J., Looye, G., Schnepper, K., Varga, A.: A multi-objective optimisation based software environment for control systems design. In: Proc. of 2002 IEEE International Conference on Control Applications and International Symposium on Computer Aided Control Systems Design, Glasgow, Scotland, UK (2002)
5. Joos, H.-D.: Flight control law clearance using optimization-based worst-case search. Preprints of the 6th IFAC Symposium on Robust Control Design, Haifa, Israel (2009)

Chapter 14
Applications of Lyapunov-Based Analysis Techniques for Clearance

Andrea Garulli, Alfio Masi, Simone Paoletti, and Ercüment Türkoğlu

Abstract. This chapter presents results obtained by applying the Lyapunov-based robustness analysis techniques developed in Chapter 9 to two clearance problems: aeroelastic stability and un-piloted stability. The considered techniques have been employed to certify robust stability of a number of LFR models, derived from the physical models of a benchmark civil aircraft. The results illustrate the potential and limitations of Lyapunov-based analysis in the clearance context.

14.1 Introduction

In this chapter, the Lyapunov-based robustness analysis techniques presented in Chapter 9 are applied to two benchmark clearance problems concerning the closed-loop longitudinal dynamics of a civil aircraft: the aeroelastic stability criterion for integral models including rigid and flexible modes, and the un-piloted stability criterion for nonlinear aircraft models. The analysis is performed on the uncertainty models in linear fractional representation (LFR) form, developed by applying the methodologies presented in Chapters 3, 4 and 5. The considered techniques, briefly denoted as FD [1], DS [2] and WB [3], provide sufficient conditions for robust stability of systems in LFR form, in terms of LMI optimization problems. Several relaxations of these techniques are employed, based on different choices of the structure of the Lyapunov functions, multipliers and scaling matrices. The abbreviations used to denote the relaxations are recalled in Table 14.1. Two types of quadratic Lyapunov functions are used in combination with the mentioned relaxations:

Andrea Garulli · Alfio Masi · Simone Paoletti · Ercüment Türkoğlu
Dipartimento di Ingegneria dell'Informazione, Università di Siena,
via Roma 56, 53100 Siena, Italy
e-mail: garulli@dii.unisi.it, masi@dii.unisi.it,
paoletti@dii.unisi.it, turkoglu@dii.unisi.it

A. Varga et al. (Eds.): Opti. Based Clearance of Flight Control Laws, LNCIS 416, pp. 253–275.
springerlink.com

i) common Lyapunov function (clf); ii) affine parameter-dependent Lyapunov function (apdlf). The reader is referred to Chapter 9 for the technical details.

Additional results on the aeroelastic stability criterion have been obtained by employing reduced-order models with affine parameter dependence, derived by applying the H_2-norm approximation technique proposed in Chapter 6.

The chapter is organized as follows. Sect. 14.2 presents the results on the aeroelastic stability criterion, while Sect. 14.3 reports results concerning the un-piloted stability criterion. Results on affine parameter-dependent models are given in Sect. 14.4. Final remarks and future perspectives are provided in Sect. 14.5.

Table 14.1 Methods for robust stability.

Relaxation	Characteristics
FD-cμ	FD method with constant full multipliers
FD-cdμ	FD method with constant diagonal multipliers
DS	DS method with constant full multipliers
DS-dS	DS method with diagonal multipliers
WBQ-dM	WB method for quadratic stability with diagonal scaling

14.2 Results on Aeroelastic Stability Criterion

In this section, results on the robustness analysis of models developed for the aeroelastic stability criterion (also called eigenvalue criterion) are presented. The considered robust stability methods are applied to a set of LFR models, by employing two different strategies for partitioning the uncertainty domain.

14.2.1 LFR Models for Aeroelastic Stability

Several LFR models of the open-loop longitudinal dynamics of a civil aircraft have been derived in Chapter 4, from a set of linear aeroelastic models dependent on:

- the mass configuration (expressed in terms of fullness of two fuel tanks and a payload);
- the trim flight point (characterized by Mach number and calibrated air speed).

In order to generate closed-loop models, these LFRs have been combined with the LFRs describing the longitudinal axis actuators, sensors and the controller dynamics (see Chapter 5). The position of the center of gravity along the longitudinal axis, assumed by the controller, is considered as an additional uncertain parameter.

Depending on the parameters δ appearing in the uncertainty block Δ, several different closed-loop LFR models have been generated to be representative of the aircraft rigid and flexible dynamics in the frequency range $[0, 15]$ rad/sec. The models are named according to the uncertain parameters or trimmed flight variables appearing in the Δ block, which are denoted by the symbols reported in Table 14.2. The number of states n, the size d of the Δ block, and the dimension s_i of the uncertainty sub-blocks associated with each uncertain parameter or trim flight value, are summarized in Table 14.3. Each one of the fuel load parameters and/or flight point values not appearing among the uncertain parameters δ in Table 14.3, are fixed to the nominal values reported in Table 14.2.

Table 14.2 Symbols used in the model description.

Symbol	Description	Nominal value
δ_{CT}	central tank	0.5
δ_{OT}	outer tank	0.5
δ_{PL}	payload	0
V_{cas}	calibrated air speed	310 kt
M	Mach number	0.86
X_{cg}	center of gravity	0.5

Table 14.3 LFR models for aeroelastic stability criterion: uncertainty structure (s_i denotes the dimension of the Δ block corresponding to parameter δ_i).

Model	n	d	δ_1, s_1	δ_2, s_2	δ_3, s_3
C	20	16	δ_{CT}, 16	–	–
CX	20	18	δ_{CT}, 14	X_{cg}, 4	–
OC	20	50	δ_{CT}, 26	δ_{OT}, 24	–
OCX	20	50	δ_{CT}, 24	δ_{OT}, 22	X_{cg}, 4
POC	20	79	δ_{CT}, 42	δ_{OT}, 24	δ_{PL}, 13
MV	20	54	M, 26	V_{cas}, 28	–

For the MV model, bounds on the flight parameters δ are available in terms of a polytope Θ, representing the considered flight envelope (the polytope bounded by the thick line in Figures 14.1-14.6). The robustness analysis has been carried out on the smallest rectangle including the polytope Θ. For all the other models, the fuel loads (δ_{CT} and δ_{OT}), payload (δ_{PL}) and position of center of gravity (X_{cg}) take normalized values between 0 and 1. Hence, the corresponding uncertainty domains are hyper-boxes in the appropriate dimensions.

14.2.2 Robustness Analysis by Progressive Tiling

A collection of results on robustness analysis for the LFR models reported in Table 14.3 is presented in the following. All clearance problems have been solved by employing the GUI software presented in Chapter 9, within 64-bit Matlab 2007*b*, running under Linux Ubuntu, on a PC equipped with an Intel XEON 5150 processor and 4 Gbyte of DDRII RAM.

We first present results obtained by applying the techniques described in Chapter 9 with progressive tiling and maximum number of partitions set to 7, corresponding to a minimum tile side equal to $\frac{1}{128}$ of the related initial side of the flight/uncertainty domain.

C, CX, OC and OCX Models

Table 14.4 summarizes clearance results for model C obtained by applying the progressive tiling approach with different methods, and different structure of the Lyapunov function (clf and apdlf). The first column (*Cleared*) represents the percentage of the uncertainty domain which has been cleared; the second column (*NOPs*) denotes the number of convex optimization problems that have been solved (corresponding to the number of tiles attempted to be cleared), while the third column provides the corresponding computational times (in seconds). Similar results are reported in Table 14.5 for the CX model.

Table 14.4 Progressive tiling: C model.

Method	Cleared (%)	NOPs	Time (sec)
DS (clf)	100	1	9.78
DS (apdlf)	100	1	10.97
DS-dS (clf)	100	3	10.15
DS-dS (apdlf)	100	1	4.15
FD-cμ (clf)	100	1	5.81
FD-cμ (apdlf)	100	1	8.77
FD-cdμ (clf)	100	3	8.43
FD-cdμ (apdlf)	100	1	4.93
WBQ	100	3	31.45
WBQ-dM	100	3	4.38

Results from the robustness stability analysis carried out on the models OC and OCX are shown in Tables 14.6 and 14.7, respectively. Here, times are shown in hours:minutes:seconds format (h:m:s).

It can be observed that the employed methods and relaxations have managed to certify robust stability of all models considered so far, within the whole uncertainty domain. As expected, the conditions involving affine parameter-dependent Lyapunov functions (apdlf) have to solve in general a smaller number of optimization problems, with respect to those formulated

Table 14.5 Progressive tiling: CX model.

Method	Cleared (%)	NOPs	Time (sec)
DS (clf)	100	1	30.88
DS (apdlf)	100	1	44.47
DS-dS (clf)	100	5	48.59
DS-dS (apdlf)	100	1	16.33
FD-cμ (clf)	100	1	18.60
FD-cμ (apdlf)	100	1	32.30
FD-cdμ (clf)	100	5	32.37
FD-cdμ (apdlf)	100	1	18.91
WBQ	100	5	107.32
WBQ-dM	100	5	9.16

Table 14.6 Progressive tiling: OC model.

Method	Cleared (%)	NOPs	Time (h:m:s)
DS-dS (clf)	100	73	0 : 26 : 32
DS-dS (apdlf)	100	41	0 : 31 : 51
FD-cμ (clf)	100	33	2 : 55 : 30
FD-cμ (apdlf)	100	1	0 : 06 : 11
FD-cdμ (clf)	100	85	0 : 22 : 00
FD-cdμ (apdlf)	100	49	0 : 31 : 50
WBQ	100	37	7 : 55 : 14
WBQ-dM	100	169	0 : 14 : 48

Table 14.7 Progressive tiling: OCX model.

Method	Cleared (%)	NOPs	Time (h:m:s)
DS (clf)	100	185	327 : 00 : 28
DS (apdlf)	100	1	3 : 07 : 22
DS-dS (clf)	100	745	11 : 10 : 09
DS-dS (apdlf)	100	265	10 : 53 : 36
FD-cμ (clf)	100	185	41 : 31 : 34
FD-cμ (apdlf)	100	1	0 : 17 : 47
FD-cdμ (clf)	100	841	8 : 52 : 25
FD-cdμ (apdlf)	100	385	13 : 34 : 34
WBQ	100	201	61 : 20 : 32
WBQ-dM	100	2129	4 : 34 : 12

with common Lyapunov functions (clf). Nevertheless, this leads to a reduction of the computational time only if the number of solved optimization problems turns out to be significantly smaller (in this respect, see e.g. the different behavior of FD-cμ and FD-cdμ for models OC and OCX).

Relaxations employing diagonal multipliers or scaling matrices have managed to significantly reduce the computational times in several cases. Compare for example FD-cμ and FD-cdμ with clf, or WBQ and WBQ-dM, for models OC and OCX. However, the choice of structurally simple multipliers can increase the time required for robustness certification if the number of optimization problems to be solved grows too much (for example, this occurs for FD-cμ and FD-cdμ with apdlf, for models OC and OCX).

It is also interesting to note that in Tables 14.4, 14.5 and 14.7, the times required by DS is always larger than that employed by FD-cμ, although the latter method is in general less conservative than the former one (see Remark 1 in Chapter 9). This seems to suggest that the parametrization of the FD-cμ method is more efficient.

POC Model

Table 14.8 presents performance indicators of the stability analysis performed on the POC model by adopting the progressive tiling approach. It can be observed that, as the number of uncertain parameters δ and the size of the uncertainty block $\Delta(\delta)$ grow, the computational workload increases significantly. Nevertheless, the considered techniques have managed to clear the entire uncertainty domain. Other techniques have been tested, e.g. using apdlf instead of clf, but they turned out to be significantly more computationally demanding for this model.

Table 14.8 Progressive tiling: POC model.

Method	Cleared (%)	NOPs	Time (h:m:s)
DS-dS (clf)	100	993	33 : 42 : 48
FD-cμ (clf)	100	105	142 : 38 : 10

MV Model

Table 14.9 presents results on robust stability analysis for the MV model. Progressive tiling has been employed in the assessment of the flight envelope. In this example the *clearance rate* is not equal to 1 for all the considered techniques. Such a rate, reported in the second column of Table 14.9, is computed as the ratio of the cleared domain and the portion of the domain which does not contain unstable models found by gridding (see Chapter 9 for details). It can be interpreted as a measure of effectiveness of the considered technique: when the rate is less than 1, there are regions that could not be

Table 14.9 Progressive tiling: MV model.

Method	Clearance rate	NOPs	Time (h:m:s)
DS-dS (clf)	0.9875	1282	12 : 32 : 13
DS-dS (apdlf)	0.9931	1030	24 : 35 : 25
FD-cμ (clf)	0.9993	218	34 : 00 : 29
FD-cμ (apdlf)	1	174	30 : 09 : 45
FD-cdμ (clf)	0.9895	1346	8 : 28 : 02
FD-cdμ (apdlf)	0.9921	1202	20 : 59 : 54
WBQ-dM	0.9759	2798	27 : 30 : 40

cleared even if the gridding procedure did not return unstable models within them.

By inspecting the results in Table 14.9 one can observe that, while an increase in the number of free variables has led to a reduction in the number of optimization problems solved (e.g. from FD-cdμ to FD-cμ), this has not translated into computationally less demanding optimization problems. On the other hand, the least conservative and most computationally demanding method (FD-cμ with apdlf) has been the only one able to achieve a clearance rate equal to 1, for the considered minimum tile size.

Figures 14.1-14.6 offer a detailed picture of the certified regions provided by the considered techniques. The tiles in light gray represent the areas which have been cleared, while the dark gray tiles contain unstable models found by gridding. The white tiles indicate the regions which have neither been cleared nor found to contain unstable models by gridding. The different tiling patterns testify the different conservatism level of the relaxations. It can be noticed that the condition FD-cμ in Figures 14.3-14.4 provides the largest cleared region, and was the only one to completely clear the actual flight envelope (denoted by the polytope).

14.2.3 Robustness Analysis by Adaptive Tiling

The adaptive tiling procedure described in Chapter 9 has been tested on the model MV, and the adaptation has been performed on the structure of the Lyapunov function. The conditions DS-dS, FD-cμ and FD-cdμ have been applied with progressive tiling with a common Lyapunov function in the first partitioning steps; then, the same relaxations with an affine parameter-dependent Lyapunov function have been applied in the subsequent partitioning steps. Table 14.10 reports the clearance results obtained for the model MV by employing the adaptive tiling. Columns 2 and 3 in Table 14.10 indicate the number of bisections for which the clf and apdlf conditions have been employed, respectively. For example, clf=2 and apdlf=5 means that in the first two partitions we applied the corresponding condition with clf, while apdlf has

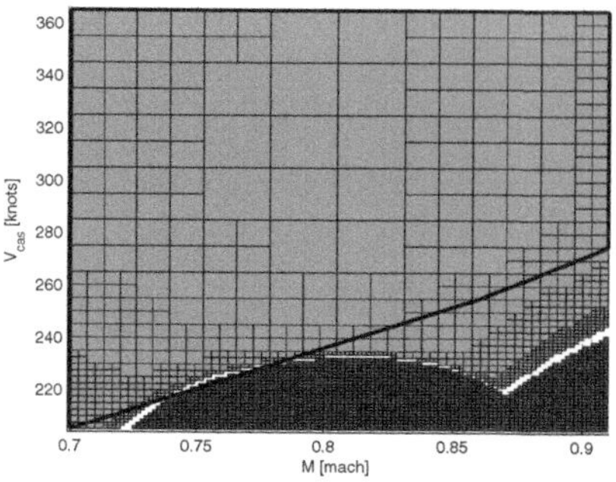

Fig. 14.1 Clearance by partitioning using DS-dS with clf.

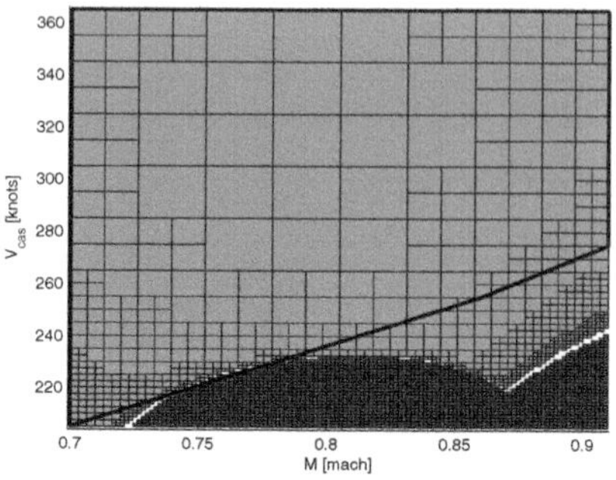

Fig. 14.2 Clearance by partitioning using DS-dS with apdlf.

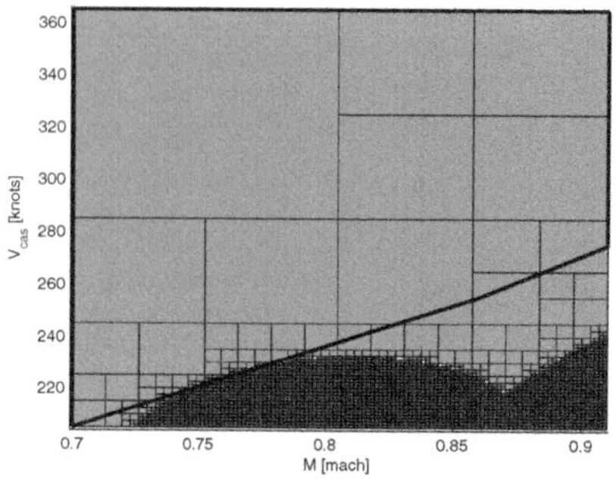

Fig. 14.3 Clearance by partitioning using FD-cμ with clf.

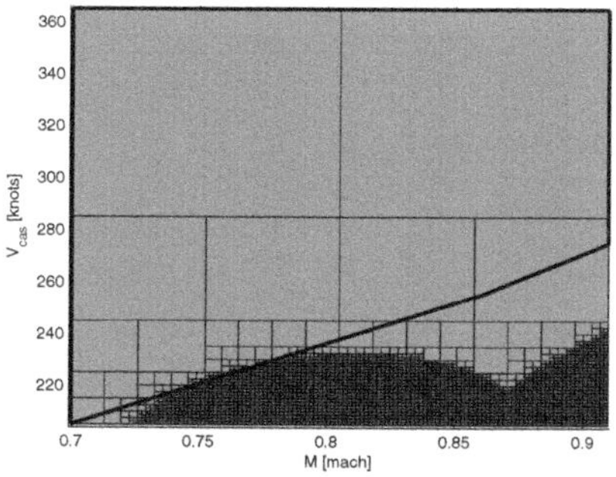

Fig. 14.4 Clearance by partitioning using FD-cμ with apdlf.

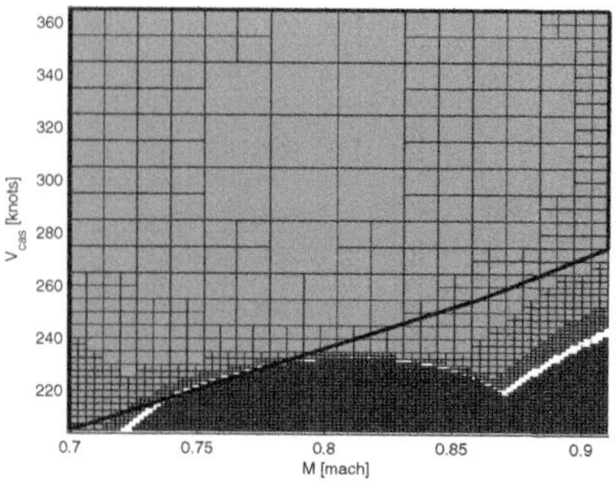

Fig. 14.5 Clearance by partitioning using FD-cdμ with clf.

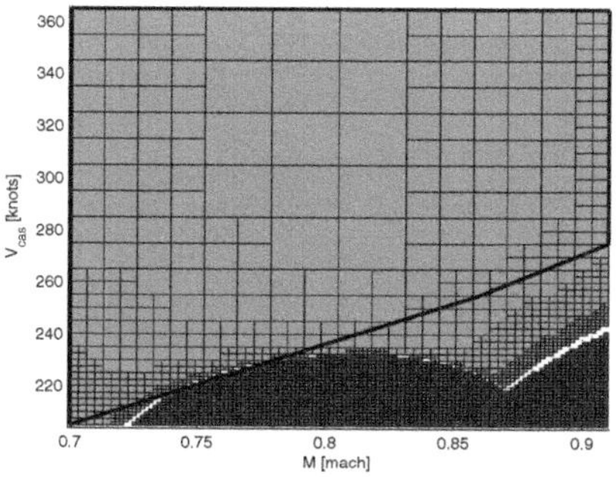

Fig. 14.6 Clearance by partitioning using FD-cdμ with apdlf.

been employed in the subsequent 5 partitions (the choice clf=0 and apdlf=7 is equivalent to the progressive approach with apdlf reported in Table 14.9, and is repeated here for comparison).

As expected, it is confirmed that the FD-cμ relaxation is the only one that has been able to clear all the clearable tiles. The adaptation has proved to be effective in reducing the computational time for both DS-dS and FD-cdμ. On the contrary, it has led to an increase of computational times for FD-cμ.

Table 14.10 Adaptive tiling: `MV` model.

Method	clf	apdlf	Clearance rate	NOPs	Time (h:m:s)
DS-dS	0	7	0.9931	1030	24 : 35 : 25
	2	5	0.9931	1042	27 : 27 : 54
	4	3	0.9931	1110	25 : 20 : 00
	6	1	0.9931	1236	20 : 10 : 37
FD-cμ	0	7	1	174	30 : 09 : 45
	2	5	1	179	31 : 19 : 19
	4	3	1	188	32 : 21 : 30
	6	1	1	206	36 : 05 : 11
FD-cdμ	0	7	0.9921	1202	20 : 59 : 54
	2	5	0.9921	1214	23 : 13 : 35
	4	3	0.9921	1266	21 : 18 : 30
	6	1	0.9921	1362	15 : 39 : 18

14.2.4 Discussion

On the whole, the results show that it is not possible to establish a priori which combination of robustness condition, relaxation and structure of the Lyapunov function will give the best compromise in terms of clearance rate and overall computational time, because this depends on the trade-off between conservatism and complexity of each condition, whose impact may in turn depend on the specific problem at hand. The FD-cμ relaxation is the most powerful one among those implemented, but it may sometime require an excessive computational effort. Experience accumulated in employing the sufficient conditions in the clearance process indicates that the FD-cdμ, DS-dS and WBQ-dM relaxations often provide a good trade-off between computational workload and clearance rate (see e.g., both progressive and adaptive results on the `MV` model). Concerning the selection of the structure of the Lyapunov function, starting with clf and than switching to apdlf to clear only the "most difficult" tiles, as in the adaptive approach, has proven to be useful in several cases.

14.3 Results on Un-piloted Stability Criterion

In this section, results on the robustness analysis of LFR models developed for the un-piloted stability criterion are presented.

14.3.1 LFR Models for Un-piloted Stability

The longitudinal nonlinear aircraft dynamics is described by LFR models developed according to the methodology proposed in Chapter 3. The resulting closed-loop LFR models, including the longitudinal actuators, sensors and controller dynamics, have been presented in Chapter 5. There are 16 different models corresponding to different regions of the flight envelope. The flight parameters are treated as linear time-invariant uncertain parameters. The actuator saturations are transformed into dead-zones in the LFR models. The full model considered here includes four flight parameters (mach number, calibrated air speed, aircraft mass, position of center of gravity) and four dead-zones (related to position and rate limiters in the elevator, both in the aircraft and in the controller). Table 14.11 summarizes the dimensions s_i of the Δ block corresponding to the flight parameters in the complete closed-loop model, for a collection of available LFRs. All the dead-zones enter in the Δ block with dimension 1. The meaning of parameter symbols is explained in Table 14.12. The uncertain parameters X_{cg} and m belong to given rectangular regions, while the values of the trim flight variables M and V_{cas} are bounded in a convex polytope (different for each model). For the latter variables, the robustness analysis is carried out in the smallest rectangle containing such a polytope.

Since the full models have Δ blocks of dimension up to 145, including the dead-zones, a collection of simpler models has been considered in order to obtain more tractable robustness analysis problems. Six different simplified LFR model classes have been defined, by considering only a subset of uncertain flight parameters and/or elevator dead-zones in the full LFRs, and

Table 14.11 LFR models for nonlinear longitudinal dynamics: uncertainty block dimensions.

Model number	n	M, s_1	V_{cas}, s_2	X_{cg}, s_3	m, s_4
1	14	46	37	37	21
2	14	38	45	23	35
5	14	38	45	23	35
6	14	38	45	37	21
9	14	38	40	22	35
10	14	38	41	36	21
13	14	38	40	36	21
14	14	30	48	22	35

Table 14.12 Symbols used in the model description.

Symbol	Description
m	mass
V_{cas}	calibrated air speed
X_{cg}	center of gravity
M	Mach number
P_a	dead-zone related to elevator position limiter in the aircraft
P_c	dead-zone related to elevator position limiter in the controller
R_a	dead-zone related to elevator rate limiter in the aircraft
R_c	dead-zone related to elevator rate limiter in the controller

setting the remaining ones to the corresponding nominal values. The set of considered simplified models are name-coded in the following format:

`class_number`

where `class` characterizes the set of uncertain flight parameters or dead-zones δ_i appearing in the Δ block, according to Table 14.13, while `number` refers to the model number reported in Table 14.11.

Table 14.13 Simplified LFR model classes for un-piloted stability.

Name	δ_1	δ_2	δ_3	δ_4
XW	X_{cg}	m		
MV	M	V_{cas}		
MVW	M	V_{cas}	m	
$\mathtt{P}_a\mathtt{P}_c\mathtt{R}_a\mathtt{R}_c$	P_a	P_c	R_a	R_c
$\mathtt{XWP}_a\mathtt{P}_c$	X_{cg}	m	P_a	P_c
$\mathtt{MVP}_a\mathtt{P}_c$	M	V_{cas}	P_a	P_c

14.3.2 Robustness Analysis Results

In the following, the results obtained by applying the proposed techniques on the LFR models described in Sect. 14.3.1 are reported. All tests have been performed with progressive tiling and maximum number of partitions set to 6 (except where otherwise specified).

`XW` Models

The uncertain flight parameters in this class of models are X_{cg} and m. The other flight parameters (M, V_{cas}) are set to their corresponding nominal values, and the dead-zones are set to zero. The analysis utilized the FD-$c\mu$ relaxation, with affine parameter-dependent Lyapunov functions (apdlf).

Tables 14.14 and 14.15 summarize the analysis results for the considered LFRs. Table 14.14 refers to the standard robust stability analysis, while Table 14.15 concerns *shifted* stability analysis, allowing for the presence of slowly divergent modes (corresponding to right shifting the imaginary axis by log(2)/6). The first column (*Model*) represents the model analyzed. The second column (*NOPs*) denotes the number of optimization problems that have been solved (i.e., the number of tiles attempted to be cleared). The ratio (in %) of the cleared domain to the whole uncertainty domain is given in column *Cleared*. The percentage of the whole uncertain parameter domain, which after gridding was found to host closed-loop unstable models, is given in column *Unstable*. The column *Unknown* shows the percentage of the whole uncertainty domain, which could not be defined as unstable after gridding, yet, with the tested method it could not be certified as cleared either. The column (*Time*) provides the time elapsed in the course of the clearance procedure.

It can be observed that when using the standard notion of stability, unstable models have been found almost everywhere within the flight uncertainty

Table 14.14 Clearance analysis for XW models: standard stability (FD-cμ with apdlf).

Model	NOPs	Cleared (%)	Unstable (%)	Unknown (%)	Time (h:m:s)
XW_1	8	0.195	99.805	0.0	0:49:08
XW_2	69	3.442	96.558	0.0	4:26:28
XW_5	91	9.839	90.161	0.0	6:35:25
XW_6	57	2.20	97.80	0.0	3:40:39
XW_9	1	0.024	99.976	0.0	0:04:24
XW_10	5461	0.0	100.0	0.0	0:06:55
XW_13	14	0.488	99.512	0.0	1:01:16
XW_14	5461	0.0	100.0	0.0	0:06:20

Table 14.15 Clearance analysis for XW models: *shifted* stability (FD-cμ with apdlf).

Model	NOPs	Cleared (%)	Unstable (%)	Unknown (%)	Time (h:m:s)
XW_1	1	100.0	0.0	0.0	0:04:13
XW_2	1	100.0	0.0	0.0	0:04:17
XW_5	1	100.0	0.0	0.0	0:03:27
XW_6	1	100.0	0.0	0.0	0:03:17
XW_9	1	100.0	0.0	0.0	0:03:13
XW_10	1	100.0	0.0	0.0	0:07:04
XW_13	1	100.0	0.0	0.0	0:03:13
XW_14	1	100.0	0.0	0.0	0:04:22

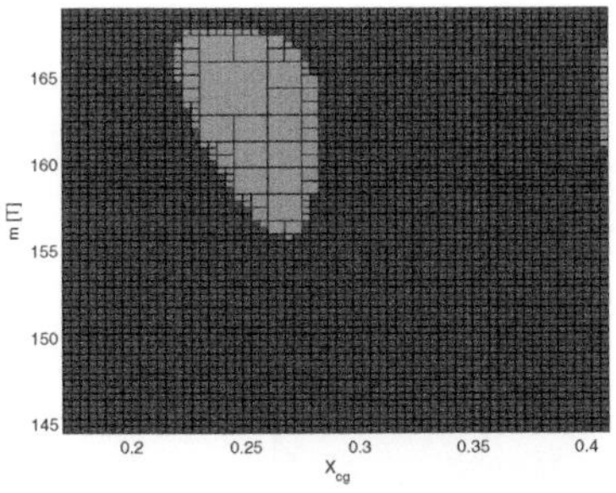

Fig. 14.7 Standard robust stability analysis for model XW_5.

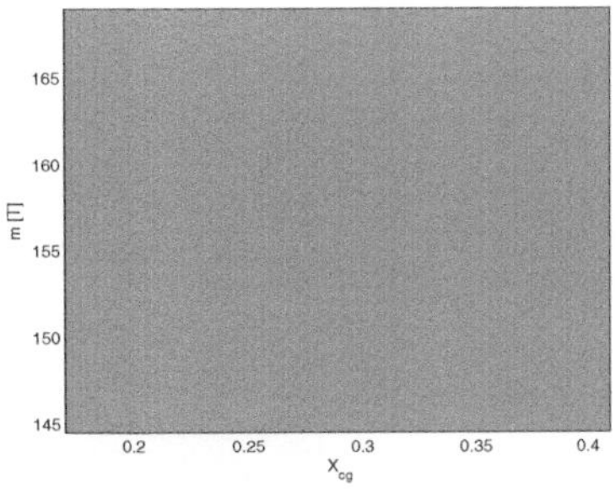

Fig. 14.8 Robust *shifted* stability analysis for model XW_5.

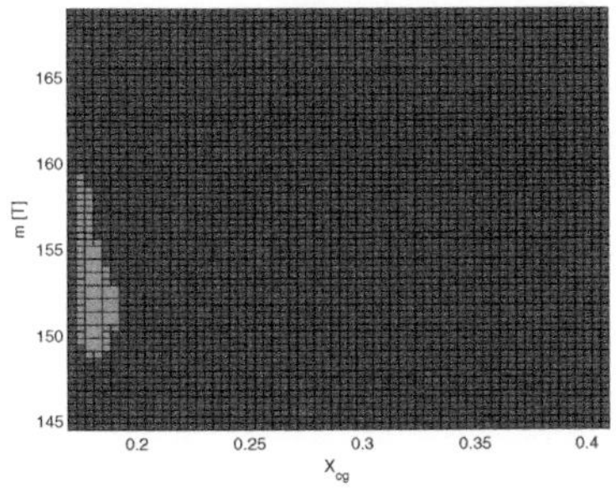

Fig. 14.9 Standard robust stability analysis for model XW_6.

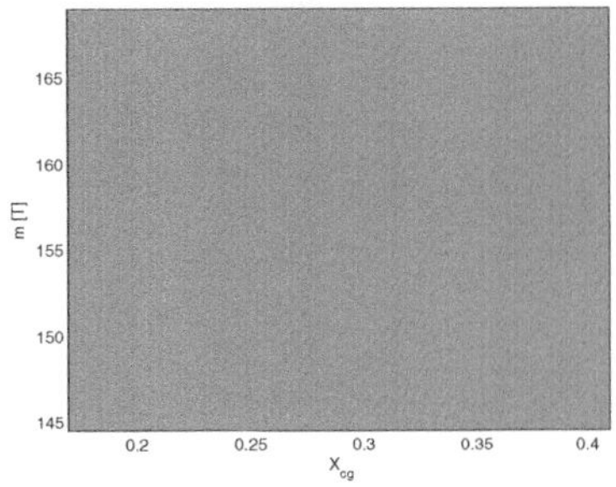

Fig. 14.10 Robust *shifted* stability analysis for model XW_6.

domain. For some models, the clearance process lasted hours as a fine tiling has been required (see Fig. 14.7 and Fig. 14.9). Conversely, when shifted stability is considered (i.e., slowly divergent modes are allowed), all models are fully cleared by solving one single optimization problem, without tiling the uncertainty domain (see Fig. 14.8 and Fig. 14.10). The computational times are in the order of few minutes. This testifies that slowly divergent modes are indeed present in the closed-loop system and motivated us to address only shifted stability in the subsequent analysis.

MV Models

This class of models treats M and V_{cas} as uncertain flight parameters, the other uncertain parameters (X_{cg}, m) are set to their corresponding nominal values, and the dead-zones are set to zero. The shifted stability analysis employed the FD-cμ relaxation with apdlf. Table 14.16 summarises the analysis results for the models considered. Figures 14.11-14.12 report the results of the analysis and the actual flight envelope boundaries (thick lines). Notice that the *Cleared* and *Unstable* rates in Table 14.16 refer to the entire rectangle in Fig. 14.11 and Fig. 14.12 and not only to the flight envelope.

Table 14.16 Clearance analysis for MV models (FD-cμ with apdlf).

Model	NOPs	Cleared (%)	Unstable (%)	Unknown (%)	Time (h:m:s)
MV_2	110	96.14	3.86	0.0	114:31:19
MV_5	33	97.93	2.07	0.0	47:23:31

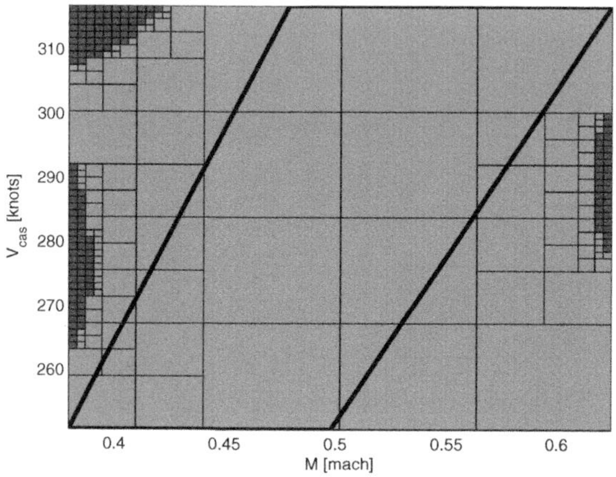

Fig. 14.11 Robust shifted stability analysis for model MV_2.

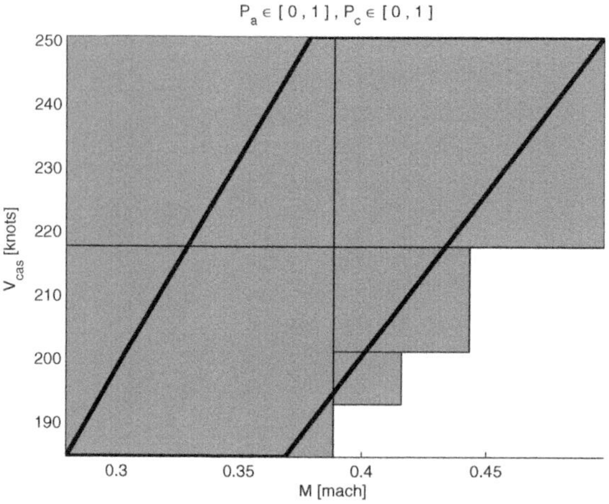

Fig. 14.12 Robust shifted stability analysis for model MV_5.

It can be observed that the computational times have significantly increased with respect to XW models, due to the larger dimension of the LFR Δ block and to the much higher number of optimization problems resulting from the progressive tiling. However, it can be noticed that most of the time is spent in the attempt to clear regions close to the stability boundary, requiring a much finer partitioning, which lie outside the actual flight envelope of interest. This information has been kept in this example, in order to show that the partitioning technique allows one to obtain a detailed approximation of the true robust stability domain. In the following analysis tests, the tiles which are completely outside the polytopic flight envelope will be skipped, in order to reduce the computational times.

MVW Models

This model class treats M, V_{cas}, and m, as uncertain flight parameters, the other uncertain parameter (X_{cg}) is set to its corresponding nominal value, and the dead-zones are set to zero. The shifted stability analysis utilized progressive tiling approach employing DS-dS method with clf, with maximum number of partitions set to 4 (other relaxations have proven to be computationally unfeasible for this example). Table 14.17 summarizes the analysis results for the considered models. Figures 14.13-14.15 report the projection of the cleared uncertainty domain on the M-V_{cas} plane, for different intervals of m. It can be observed that different cleared regions are obtained for different "slices" with respect to m (dark gray tiles, containing unstable models represented by dots, show up in Fig. 14.14). In particular, unstable models have been found within the considered flight envelope, in the region: $260 \leq V_{cas} \leq 267\, knots$, $0.395 \leq M \leq 0.41\, mach$, $142.95 \leq m \leq 144.5\, T$.

Table 14.17 Clearance analysis for MVW models (DS-dS with clf).

Model	NOPs	Cleared (%)	Unstable (%)	Unknown (%)	Time (h:m:s)
MVW_2	562	99.0420	0.7001	0.2579	34:48:49
MVW_5	173	100.0	0.0	0.0	12:29:10

$P_aP_cR_aR_c$ Models

In this class of models, all flight parameters are set to their corresponding nominal values, except the four dead-zones P_a, P_c, R_a and R_c. The dead-zones are treated as *sector-bounded nonlinearities*: this amounts to consider in the Δ-block all the nonlinearities belonging to the sector $[0, 1]$, i.e. the gray sector in Fig. 14.16. It is apparent that the results of the analysis are in general conservative, because of the large set of nonlinearities with respect to which stability is assessed.

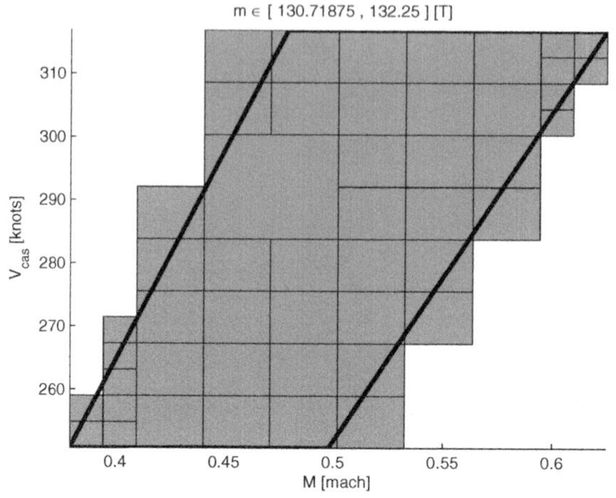

Fig. 14.13 Robust shifted stability analysis for model MVW_2: M–V_{cas} plane for $m \in [130.71875, 132.25]$ T.

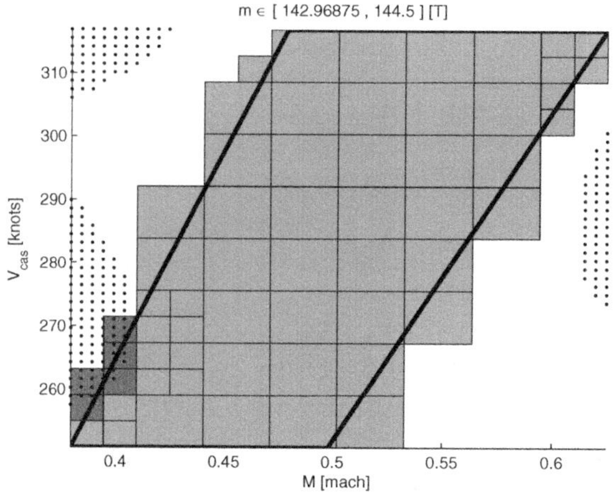

Fig. 14.14 Robust shifted stability analysis for model MVW_2: M–V_{cas} plane for $m \in [142.96875, 144.5]$ T. Dots represent unstable models.

In the case of the $\mathtt{P}_a\mathtt{P}_c\mathtt{R}_a\mathtt{R}_c$ models, after performing analysis with FD-$c\mu$ method, common Lyapunov function (clf) and dead-zones considered as sector-bounded uncertainties in the sector $[0, 1]$, none of the considered models have been cleared. Indeed, it can be checked that there exist constant values of the dead-zones parameters, such that the resulting models have eigenvalues with real part greater than $\log(2)/6$. This corresponds to the existence of linear gains in the $[0, 1]$ sector in Fig. 14.16, which make the closed-loop system *shifted* unstable. This means that any method

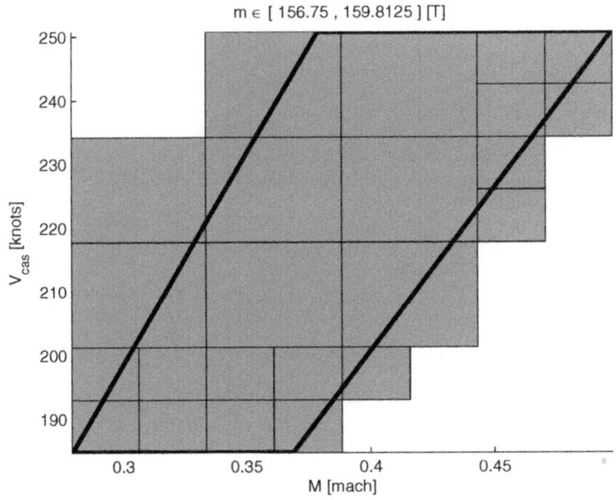

Fig. 14.15 Robust shifted stability analysis for model MVW_5.

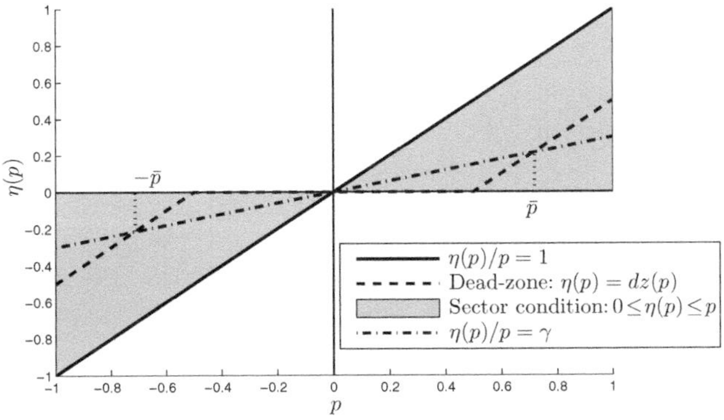

Fig. 14.16 Dead-zones as sector-bounded uncertainties.

considering sector-bounded, or even slope-bounded nonlinearities, cannot clear the considered $\mathtt{P}_a\mathtt{P}_c\mathtt{R}_a\mathtt{R}_c$ models (obviously, this does not mean that such models cannot be cleared by other methods).

A further analysis has been performed, by considering smaller sector bounds $[0,\gamma]$, with $\gamma < 1$. This corresponds to assuming that the signals entering the dead-zones are limited. In Fig. 14.16, the dash-dotted line is an example of a reduced sector $[0,\gamma]$. It can be observed that if the signals p entering the dead-zones are such that $|p| \leq \bar{p}$, then the dead-zones are correctly covered by the reduced sector. Table 14.18 reports the maximum values of γ for which robust shifted stability has been certified by applying the FD-cμ method. These values of γ allow one to compute the maximum range $\bar{p}$ of the input signals of the dead-zones for which the system remains stable.

Table 14.18 Maximum values of γ for which $\mathtt{P}_a\mathtt{P}_c\mathtt{R}_a\mathtt{R}_c$ models have been cleared.

Model	γ
$\mathtt{P}_a\mathtt{P}_c\mathtt{R}_a\mathtt{R}_c$_1	0.62
$\mathtt{P}_a\mathtt{P}_c\mathtt{R}_a\mathtt{R}_c$_2	0.64
$\mathtt{P}_a\mathtt{P}_c\mathtt{R}_a\mathtt{R}_c$_5	0.62
$\mathtt{P}_a\mathtt{P}_c\mathtt{R}_a\mathtt{R}_c$_9	0.60
$\mathtt{P}_a\mathtt{P}_c\mathtt{R}_a\mathtt{R}_c$_13	0.60

The same clearance procedure has been applied to LFR models with dead-zones P_a and P_c set as uncertain parameters, sector-bounded in sector $[0, 1]$, and all other parameters (including rate limiter dead-zones: R_a, R_c) set to their nominal values. All such models have been cleared. On the other hand, LFR models accommodating only R_a, R_c dead-zones, and with all the remaining parameters (including the actuator dead-zones P_a and P_c) set to their nominal values, have not been cleared. This suggests that the rate limiter dead-zones play a critical role in the clearance analysis.

In [4, 5] several approaches have been proposed to address robust stability of systems with dead-zones. In particular, [5] provides LMI conditions for global exponential stability, which exploit information on the time derivative of the saturated signals and a generalized Lur'e-Postnikov Lyapunov function. Unfortunately, also this technique was not able to clear any model in the $\mathtt{P}_a\mathtt{P}_c\mathtt{R}_a\mathtt{R}_c$ class. Then, motivated by the local stability results of Table 14.18, the regional analysis techniques proposed in [4] have been applied, in order to estimate the region of attraction of the origin in the state space. Two approaches have been considered, which are based on the embedding of the LFR with dead-zones either in a polytopic differential inclusion (PDI), or a norm-bounded differential inclusion (NDI). By using these approaches, it has been possible to certify the regional stability of the systems reported in Table 14.18 and to obtain a non trivial (spherical) estimate of the region of attraction of the trim point. The radius α of the spherical stability region is reported in Table 14.19, for the two considered approaches. As expected, the approach based on PDI embedding turned out to be slightly less conservative.

Table 14.19 Estimation of the stability radius α of $\mathtt{P}_a\mathtt{P}_c\mathtt{R}_a\mathtt{R}_c$ models.

Model	α (PDI)	α (NDI)
$\mathtt{P}_a\mathtt{P}_c\mathtt{R}_a\mathtt{R}_c$_1	1.7998	1.7927
$\mathtt{P}_a\mathtt{P}_c\mathtt{R}_a\mathtt{R}_c$_2	2.9506	2.9372
$\mathtt{P}_a\mathtt{P}_c\mathtt{R}_a\mathtt{R}_c$_5	1.7867	1.7795
$\mathtt{P}_a\mathtt{P}_c\mathtt{R}_a\mathtt{R}_c$_9	1.9234	1.9143
$\mathtt{P}_a\mathtt{P}_c\mathtt{R}_a\mathtt{R}_c$_13	1.9200	1.9107

$\mathtt{XWP}_a\mathtt{P}_c$ Models

This class of models treats X_{cg} and m as uncertain parameters, and the dead-zones related to the elevator position limiters P_a and P_c as sector-bounded uncertainties. The results obtained by applying the FD-cμ method with clf, for robust shifted stability with threshold log(2)/6, are given in Table 14.20.

Table 14.20 Clearance analysis for $\mathtt{XWP}_a\mathtt{P}_c$ models (FD-cμ with clf).

Model	NOPs	Cleared (%)	Unstable (%)	Unknown (%)	Time (h:m:s)
$\mathtt{XWP}_a\mathtt{P}_c$_1	1	100.0	0.0	0.0	0:13:45
$\mathtt{XWP}_a\mathtt{P}_c$_5	1	100.0	0.0	0.0	0:11:07
$\mathtt{XWP}_a\mathtt{P}_c$_9	1	100.0	0.0	0.0	0:09:56

It is worth remarking that these models have been fully cleared by solving one single optimization problem, i.e., no partitioning of the uncertainty domain, including X_{cg} and m, has been necessary (although up to 6 partitions have been allowed for such parameters).

$\mathtt{MVP}_a\mathtt{P}_c$ Models

These LFR models include M and V_{cas} as uncertain flight parameters, and the dead-zones related to the elevator position limiters P_a and P_c as sector-bounded uncertainties. Results of robust shifted stability analysis, using FD-cμ method with clf, are reported in Table 14.21.

Table 14.21 Clearance analysis for $\mathtt{MVP}_a\mathtt{P}_c$ models (FD-cμ with clf).

Model	NOPs	Cleared (%)	Unstable (%)	Unknown (%)	Time (h:m:s)
$\mathtt{MVP}_a\mathtt{P}_c$_2	21	100.0	0.0	0.0	31:45:29
$\mathtt{MVP}_a\mathtt{P}_c$_5	5	100.0	0.0	0.0	6:34:19

Both the considered models have been fully cleared, by exploiting partitioning with respect to the time-invariant flight parameters M and V_{cas}. The resulting tiling patterns are drawn in Fig. 14.17 and Fig. 14.18. By comparing the results obtained for LFR models in the classes $\mathtt{XWP}_a\mathtt{P}_c$ and $\mathtt{MVP}_a\mathtt{P}_c$ with those concerning XW and MV models, it is confirmed that the position limiter dead-zones P_a and P_c do not affect the stability of the system, within the considered flight envelopes.

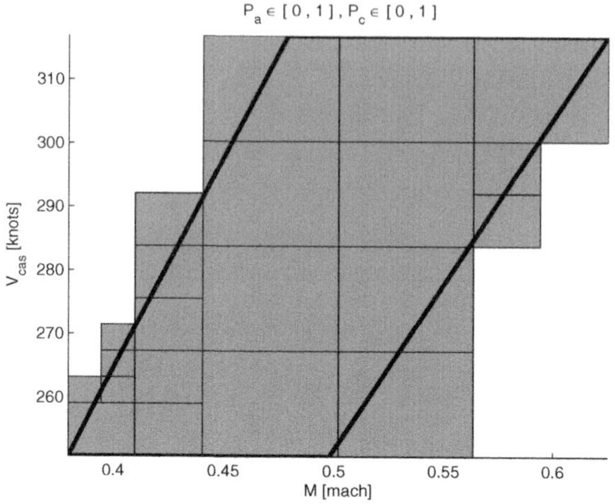

Fig. 14.17 Robust shifted stability analysis of $\mathtt{MVP}_a\mathtt{P}_c\mathtt{\_2}$.

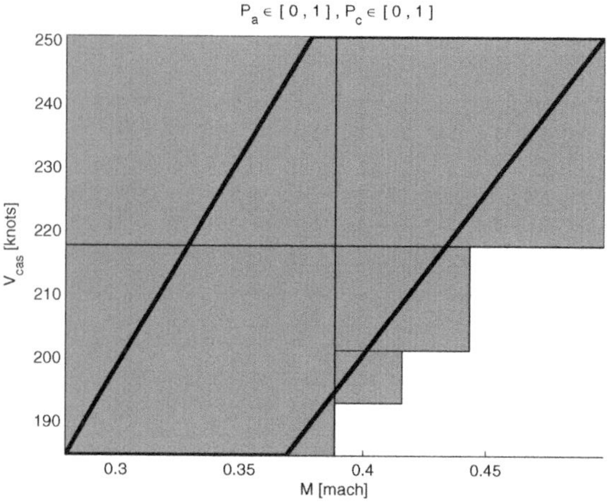

Fig. 14.18 Robust shifted stability analysis of $\mathtt{MVP}_a\mathtt{P}_c\mathtt{\_5}$.

14.4 Robustness Analysis of Affine Parameter-Dependent Models

By applying the H_2-norm approximation technique presented in Chapter 6, reduced-order LFR models have been derived which are representative of the rigid and flexible dynamics of the considered benchmark aircraft. In particular, models with affine dependence on the uncertain parameters related to the fuel loads (δ_{CT} and δ_{OT}) and the payload (δ_{PL}) have been provided, to address the aeroelastic stability criterion. These models have 22 states and the state matrix takes the form $\mathbf{A}(\delta) = A_0 + A_1\delta_1 + A_2\delta_2 + A_3\delta_3$. The accuracy of the H_2-norm approximation has been tuned by selecting the rank of the matrices A_i, $i = 1, 2, 3$. Robust stability analysis of such models has been performed by applying the standard quadratic stability condition and the multi-convexity robust stability condition proposed in [6] (see Chapter 9 for details). The techniques have been applied with progressive tiling and maximum number of partitions set to 5. Results are reported in Table 14.22, for models with different ranks of the A_i matrices.

As expected, the multi-convexity condition is much less conservative and hence more effective in completing the clearance process. For the model with full rank A_i matrices, quadratic stability required an excessively high computational effort.

Table 14.22 Results for affine parameter-dependent POC model.

Stability condition	Rank of A_i	NOPs	Cleared (%)	Unstable (%)	Unknown (%)	Time (h:m:s)
Quadratic	6	1545	99.927	0	0.0732	4 : 12 : 54
Multi-convexity	6	1	100.0	0	0	0 : 01 : 39
Quadratic	7	305	100.0	0	0	0 : 42 : 30
Multi-convexity	7	1	100.0	0	0	0 : 01 : 33
Multi-convexity	22	385	100.0	0	0	9 : 18 : 27

For comparison, also some of the robustness analysis methods for general LFRs have been applied to the affine parameter-dependent POC model. The results are reported in Table 14.23, for the cases in which the computational time has not exceeded 3 days. It can be observed that the techniques tailored to the affine parameter-dependent models have proven to be more efficient.

Table 14.23 FD relaxations for affine parameter-dependent POC model.

Stability condition	Rank of A_i	NOPs	Cleared (%)	Unstable (%)	Unknown (%)	Time (h:m:s)
FD-cμ (clf)	7	305	100.0	0	0	5 : 56 : 00
FD-cμ (apdlf)	7	1	100.0	0	0	0 : 03 : 20
FD-cμ (apdlf)	22	73	100.0	0	0	51 : 05 : 32

14.5 Conclusions

Lyapunov-based robustness analysis techniques have been applied to two clearance problems concerning a benchmark civil aircraft. The clearance problems have been cast as robust stability problems and several different techniques have been considered, which are able to cope with uncertainty models in LFR form. The obtained results indicate that there is a key trade-off between performance and computational burden. This is apparent when robustness conditions are applied within the progressive tiling strategy. Relaxed conditions, which are conservative if applied directly to the entire uncertainty region, may perform much better on smaller subregions. This, however, requires the solution of a large number of LMI optimization problems. The size of the tiles, the number of partitions of the region under analysis, the structure of the multipliers and/or the Lyapunov matrices turn out to be key tuning knobs in this respect.

Although it is difficult to devise an a priori strategy for choosing the "best" robustness analysis technique, based on experience accumulated in the testing of the clearance software and in view of the analysis results, a combined use of structurally simple multipliers in either the FD or DS method, with an affine parameter-dependent Lyapunov function, seems to offer a reasonable compromise between conservatism and computational feasibility. Adaptation of the Lyapunov function structure within the partitioning scheme has also proven to be beneficial in some cases.

The results obtained are by no means exhaustive and there are several open issues to be addressed in future research activities. A major limitation of Lyapunov-based techniques is the high computational burden, which scales exponentially with the number of uncertain parameters and trimmed flight variables. When these are assumed to be time-invariant, robustness analysis in the frequency domain has proven to be significantly more efficient (see Chapter 12). On the other hand, the main potential of Lyapunov-based techniques is that they can easily handle both time-invariant and time-varying uncertainties (including also bounds on the variation rate), as well as memoryless nonlinearities accounting for actuators saturations. When such complex clearance problems are faced, reduction of the conservatism is a key issue: this is testified, for example, by the models in the class $\mathtt{P}_a\mathtt{P}_c\mathtt{R}_a\mathtt{R}_c$, for which global exponential stability is still an open problem. In this respect, the development of less conservative sufficient conditions, which can tackle robustness analysis of LFR models containing both uncertain parameters and dead-zones in the Δ block, is a subject of ongoing research. Combined use of Lyapunov-based techniques and worst-case search based on optimization (see Chapter 8) for assessing the conservatism of clearance results, is another topic that deserves more investigation.

The robustness techniques employed in this chapter may provide a valuable tool for validating the LFR modelling process. Indeed, one of the key points to be addressed is the reliability of these models, i.e. their potential in representing all the aircraft dynamics of interest, within the considered flight envelope and for all admissible values of the uncertain parameters. By comparing the results of Lyapunov-based robust stability analysis with those obtained by applying the baseline industrial solution (usually based on gridding of the fligh/uncertainty domain), it will be possible to single out the most significant discrepancies between the physical aircraft models used so far in industrial clearance, and the models developed within the COFCLUO project. This will provide guidelines for trading off model accuracy and complexity, which is a key step towards a full integration of robustness analysis techniques within the industrial clearance process.

References

1. Fu, M., Dasgupta, S.: Parametric lyapunov function for uncertain systems: The multiplier approach. In: El Ghaoui, L., Niculescu, S.-I. (eds.) Advances in Linear Matrix Inequality Methods in Control. SIAM, Philadelphia (2000)
2. Dettori, M., Scherer, C.: New robust stability and performance conditions based on parameter dependent multipliers. In: Proc. of 39th IEEE Conf. on Decision and Control, Sydney, Australia, pp. 4187–4192 (2000)
3. Wang, F., Balakrishnan, V.: Improved stability analysis and gain-scheduled controller synthesis for parameter-dependent systems. IEEE Trans. on Automatic Control 47(5), 720–734 (2002)
4. Hu, T., Teel, A.R., Zaccarian, L.: Stability and performance for saturated systems via quadratic and nonquadratic Lyapunov functions. IEEE Transactions on Automatic Control 51(11), 1770–1786 (2006)
5. Dai, D., Hu, T., Teel, A.R., Zaccarian, L.: Piecewise-quadratic Lyapunov functions for systems with deadzones or saturations. Systems & Control Letters 58(5), 365–371 (2009)
6. Gahinet, P., Apkarian, P., Chilali, M.: Affine parameter-dependent Lyapunov functions and real parametric uncertainty. IEEE Trans. on Automatic Control 41(3), 436–442 (1996)

Chapter 15
Applications of IQC-Based Analysis Techniques for Clearance

Ragnar Wallin, Sina Khoshfetrat Pakazad, Anders Hansson,
Andrea Garulli, and Alfio Masi

Abstract. Results for stability analysis of the nonlinear rigid aircraft model and comfort and loads analysis of the integral aircraft model are presented in this chapter. The analysis is based on the theory for integral quadratic constraints and relies on linear fractional representations (LFRs) of the underlying closed-loop aircraft models. To alleviate the high computational demands associated with the usage of IQC based analysis to large order LFRs, two approaches have been employed aiming a trade-off between computational complexity and conservatism. First, the partitioning of the flight envelope in several smaller regions allows to use lower order LFRs in the analysis, and second, IQCs with lower computational demands have been used whenever possible. The obtained results illustrate the applicability of the IQCs based analysis techniques to solve highly complex analysis problems with an acceptable level of conservativeness.

15.1 Introduction

The setup for stability analysis using integral quadratic constraints (IQCs) is very similar to the setup used in μ-analysis. In both cases linear fractional representation (LFR) models of the system under analysis are needed. Such a model is depicted in Fig. 15.1. AIRBUS and ONERA have developed approximate models of this type for the nonlinear rigid aircraft model and the integral aircraft model including sensors, actuators, and controller, see

Ragnar Wallin · Sina Khoshfetrat Pakazad · Anders Hansson
Division of Automatic Control, Linköping University, SE-581 83 Linköping, Sweden
e-mail: {ragnarw,sina.kh.pa,hansson}@isy.liu.se

Andrea Garulli · Alfio Masi
Dipartimento di Ingegneria dell'Informazione, Università di Siena,
via Roma 56, 53100 Siena, Italy
e-mail: {garulli,masi}@dii.unisi.it

A. Varga et al. (Eds.): Opti. Based Clearance of Flight Control Laws, LNCIS 416, pp. 277–297.
springerlink.com

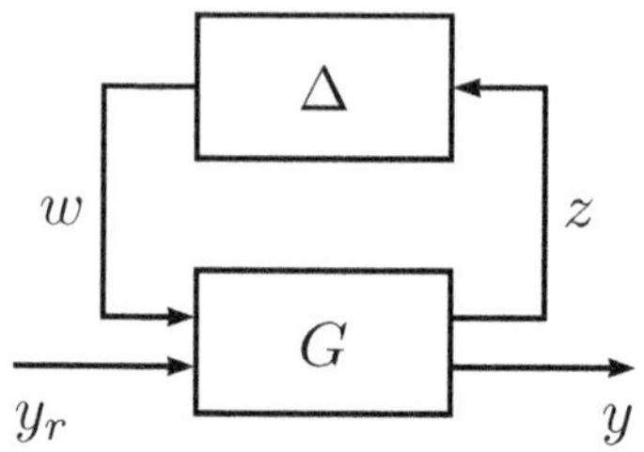

Fig. 15.1 A linear fractional representation

Chapters 3 and 5 for details. For the LFR approximation of the integral aircraft model the Δ block contains repeated constant real uncertainties, and for the LFR approximation of the rigid aircraft model the Δ block contains repeated constant real uncertainties as well as nonlinearities in the form of deadzones. The Δ blocks for these models have a diagonal structure.

The theory behind IQCs is briefly reviewed in Chapter 10. For a thorough description read the original paper by Megretski and Rantzer [1] or the more detailed lecture notes by Jönsson [2]. The theory for μ-analysis, see [3] and [4], was developed for linear time-invariant uncertainties but has been extended to handle also linear time-varying uncertainties, see [5]. Performing IQC analysis we can use an even wider variety of uncertainties in the Δ block. There can be uncertain linear time-invariant dynamics, constant parametric uncertainty, time-varying parameters, various nonlinearities, or combinations of the above. Furthermore, using IQCs it is possible to solve a wider class of analysis problems. One example is robust $\mathcal{H}_2$ performance analysis. However, if only stability analysis of the system is considered and the Δ block contains repeated constant real uncertainties μ-analysis is in most cases superior to IQC analysis in the sense that the latter method will deliver the same results but at a much lower computational cost. If the multipliers used in the IQCs are parameterized using transfer functions as basis functions, which is common practice, μ-analysis may actually yield slightly less conservative results.

The main drawback with IQC analysis is that, even though it results in convex optimisation problems which are considered tractable, the models used in advanced applications may be too complex to analyse. The resulting optimisation problems become very large. Much work and time has been invested in making it possible to solve such large optimisation problems, see for example [6], [7], [8], [9], [10], [11], [12], [13] and [14]. However, none of these methods are applicable in our case. For the methods described in [12], [13] and [14] the IQC analysis problem has been turned into a semidefinite program (SDP) via the Kalman-Yakubovich-Popov lemma. Sometimes SDPs with this particular structure are called KYP-SDPs. To actually speed up computations these methods require that the number of inputs to the linear time-invariant system M in Fig. 15.1 are few compared to the number of states. This is not the case for the models developed in the COFCLUO project. As we have one input for each repetition of a real uncertainty and one input for each nonlinearity in the Δ block the number of inputs is usually much greater than the number of states. The methods described in [6], [7], [8], [9], [10] and [11], based on outer approximation methods, on the other hand require that the multipliers in the IQCs can be described using rather few variables. This is not true either for the models at hand. Hence, we have used other methods than structure exploitation to cope with computational complexity. These methods are described in Section 15.2.

In this study the following two criteria are considered:

Stability analysis of the nonlinear rigid aircraft model. We consider longitudinal motion only. This criterion is described in Chapter 2 and the analysis results are presented in Sect. 15.3.

Comfort and loads analysis of the integral aircraft model. We consider longitudinal motion only. This criterion is also described in Chapter 2 and the analysis results are presented in Sect. 15.4. In Chapter 10 it is shown how this criterion can be formulated as an IQC-based analysis problem.

For the first criterion only the closed-loop system is considered. The closed loop system in principle has to be stable, but since the pilot can counteract slowly divergent modes such modes are allowed as long as the time of doubling of the signals is less than six seconds. Allowing such modes is taken care of by shifting the stability region, if necessary, from being the left half plane to being the half plane to the left of $\frac{\ln 2}{6}$.

15.2 Coping with Computational Complexity

As was mentioned in the introduction many successful approaches to reduce computational cost for IQC analysis are based on exploitation of the model structure. It is far from obvious if there is a way to do so in our case. This is a possible topic of future research. Hence, we have to rely on other methods to reduce the computational complexity. Three possibilities are described below. The techniques can be used separately or in combination.

15.2.1 Choice of IQCs

The combination of a complex model with complex IQCs may result in optimisation problems that are too large to solve. Fortunately, we sometimes have a choice between several possible IQCs to use. Both for the constant parameters and the deadzones we can use more than one IQC to prove stability. The more information we incorporate in the IQC the more complex the resulting optimisation problem gets but the less conservative the results are. On the other hand, if we succeed to prove stability using a low information IQC this is often good enough since we are sometimes only considering a sufficient condition for clearance.

The most informative IQC we can use for constant real parameters is the one called `iqc_ltigain` in the IQC toolbox [15]. This IQC is specifically designed for constant real parameters and is the basis for standard upper bounds on structured singular values [16], [17]. The multiplier has the structure

$$\Pi(j\omega) = \begin{bmatrix} X(j\omega) & Y(j\omega) \\ Y(j\omega)^* & -X(j\omega) \end{bmatrix}, \tag{15.1}$$

where $X(j\omega) = X(j\omega)^* \geq 0$ and $Y(j\omega) = -Y(j\omega)^*$ are bounded and measurable matrix functions. However, we can also use the IQC that is called `iqc_tvscalar` in the IQC toolbox [15]. This IQC is designed for parameters that are allowed to vary arbitrarily fast. It is clear that if we can prove stability for this case we have also proved stability for the case with constant parameters. The multiplier for this case looks like

$$\Pi(j\omega) = \begin{bmatrix} X & Y \\ Y^* & -X \end{bmatrix}, \tag{15.2}$$

where $X = X^T \geq 0$ and $Y = -Y^T$ are real matrices. A third option, with an even simpler multiplier, is to use `iqc_diag`. Also here the parameters are allowed to vary arbitrarily fast but here we ignore that the parameters are possibly repeated. This will of course introduce extra conservatism as it allows for more difficult parameter combinations. The corresponding multiplier has the structure

$$\Pi(j\omega) = \begin{bmatrix} X & 0 \\ 0 & -X \end{bmatrix}, \tag{15.3}$$

with $X > 0$ being a diagonal real matrix. It is obvious that each of these multipliers can be considered a special case of the previous one. Which one you choose is a trade-off between conservatism and computational complexity.

Also for the deadzones we have multiple choices of IQCs we can use. The first one is the IQC for sector bound functions. We say that a nonlinear function $\phi : \mathcal{R}^n \longmapsto \mathcal{R}^n$ belongs to a sector $[\alpha, \beta]$ $0 \leq \alpha \leq \beta < \infty$, if it satisfies the following condition,

$$\alpha x^2 \leq \phi(x)x \leq \beta x^2. \tag{15.4}$$

For the deadzone we have $\alpha = 0$ and $\beta = 1$. This condition uses very little information about the nonlinearity and hence often yields very conservative results. Functions that vary much more than a deadzone can also fit into such a sector. The IQC used for sector bound nonlinearities is `iqc_sector` which is characterised by the multiplier

$$\Pi(j\omega) = \begin{bmatrix} -2\alpha\beta & \alpha+\beta \\ \alpha+\beta & -2 \end{bmatrix}. \tag{15.5}$$

An IQC that incorporates a little bit more information about the nonlinearity is the IQC for odd slope-bound functions `iqc_slope_odd`. We say that the monotonic odd function $\phi : \mathcal{R}^n \longmapsto \mathcal{R}^n$ has a slope restricted to the interval $[0, k]$ if it satisfies the following condition

$$0 \leq \frac{\phi(x_1) - \phi(x_2)}{x_1 - x_2} \leq k \tag{15.6}$$

for all x_1 and x_2. The corresponding IQC is based on the work in [18] and is characterised by the multiplier

$$\Pi(j\omega) = \begin{bmatrix} 0 & 1 + H(j\omega) \\ 1 + H(-j\omega) & -2(1 + \mathrm{Re}H(j\omega))/k \end{bmatrix}. \tag{15.7}$$

The rational proper transfer function $H(j\omega)$ has real coefficients and satisfies

$$\int_{-\infty}^{\infty} |h(t)| dt \leq 1,$$

where $h(t)$ is the impulse response corresponding to $H(j\omega)$.

It is worth noting that if the multipliers are not mere special cases of each other it is often a very good idea to use a combination of several IQCs in the analysis. This will yield more precise results but of course the computational complexity will be higher. Hence, we will not do this.

15.2.2 Partitioning of the Parameter Space

For static nonlinearities in combination with constant real parameters or real parameters that can vary arbitrarily fast it is possible to clear the flight envelope part by part. This enables us to get a good picture of the shape of the stable and unstable regions.

It is possible to decide beforehand how large the smallest partition should be. However, it is not a good idea to divide the flight envelope into partitions of this size already from the beginning as it would result in a huge number of optimisation problems. It is much better to start with large partitions and then, if required, divide them into smaller ones. If a region of the flight envelope is proved to be stable, further partitioning of that region will always result in stable regions. Hence, no further partitioning is necessary. This may save us considerable amounts of time and computational effort.

15.2.3 Model Reduction

To further reduce the computational effort we may also perform model reduction in each partition. It is very likely that smaller regions of uncertainty can be accurately described by simpler models. As the computational effort needed grows very fast with model complexity there is much to be gained by this. Of course, the approximation error should be included in the analysis.

When it is possible to find low order models that approximate very large regions accurately the method described in Chapter 6 may be considered. However if we have to use many partitions a less computationally heavy method is required. It is a trade-off between how much time it takes to compute a reduced order model and how much time that is gained in the analysis.

15.3 Stability Analysis of the Nonlinear Rigid Aircraft Model

In this section we present a method for analysing the stability of the unpiloted nonlinear rigid aircraft model. We describe the choices made in the proposed method, and discuss the achieved results. The data supplied to the controller is uncertain and possibly erroneous. Hence, the robustness with respect to the uncertain variables should be checked, see Chapter 2 for details. The uncertain variables in the Δ block for the nonlinear rigid aircraft model are

$$\Delta = \text{diag}\left[MI_{k_M} \; V_{cas}I_{k_{V_{cas}}} \; X_{cg}I_{k_{cg}} \; mI_{k_m} \; \Delta_{NL}I_{k_{NL}}\right]. \tag{15.8}$$

where M is the Mach number, V_{cas} is the calibrated air speed, X_{cg} represents the uncertainty in the position of the centre of gravity, m is the mass of the aircraft and Δ_{NL} represents the nonlinearities due to deadzones in the actuators and control laws.

Generally, aircraft models transformed into LFRs, come with very high dimensional Δ blocks, which makes the stability analysis very challenging and time consuming. In order to make the process less computationally demanding it is often wise to divide the analysis into several smaller subproblems. One possible way is to perform the stability analysis in three separate stages which we here call analysis of *nominal stability*, analysis of *robust stability* and analysis of *nonlinear stability*. What we mean by this is explained next.

1. *Robust linear stability with nominal X_{cg}*: In this stage of the stability analysis, we only consider uncertainty over mass, calibrated airspeed and Mach number and neglect the nonlinearities and the uncertainty over the position of centre of gravity. Then we look for regions where stability of the system with respect to the considered uncertainties can be shown. In the analysis we use the IQC for diagonal real uncertainties that are allowed to vary arbitrarily fast. As we partition the flight envelope and clear it part by part this results in a good trade-off between how many partitions we need and the computational time for each partition.
2. *Robust linear stability*: In this stage, we investigate if the stability of the system is robust with respect to the uncertainty over the position of centre of gravity. We only consider regions with cleared robust linear stability

with nominal X_{cg}. In each such region we investigate how much we can change the centre of gravity without making the system unstable.

3. *Robust nonlinear stability*: Finally, having performed the previous two stages we look for regions in the flight envelope where we can show stability with respect to the nonlinearities and all possible uncertainties, that is mass, calibrated airspeed, Mach number and the position of centre of gravity.

We will show results for the robust linear stability with nominal X_{cg} and a simplified version of the robust nonlinear stability. Why we use a simplified version of the robust nonlinear stability is explained in Sect. 15.3.2. The robust linear stability analysis yielded either very conservative results or resulted in optimisation problems that were computationally too heavy. Thus, we exclude this part of the analysis in the following.

15.3.1 Analysis of Robust Linear Stability for Nominal Centre of Gravity Position

The objective of the proposed analysis is either to show stability of the system over the whole flight envelope, or if this is impossible, localise regions with unclear stability properties. We do this by partitioning the flight envelope with respect to the uncertainties in several stages, as described in Section 15.2.2.

The stability of a system, such as the one in Fig. 15.1, can be checked by computing the induced $\mathcal{H}_2$ gain from the input signals y_r to the output signals y. If this gain is finite the system is stable. For the closed loop system the input signals y_r are missing. However, we have to introduce artificial inputs in order to take every possible initial state into account. This is done in the following way. The closed loop system is described by

$$\begin{aligned} \dot{x} &= Ax + Bw \\ z &= Cx + Dw. \\ w &= \Delta z \end{aligned}$$

We extend this system with extra inputs and outputs and get

$$\begin{aligned} \dot{x} &= Ax + Bw + B_{red}y_r \\ z &= Cx + Dw \\ y &= Cx + Dw \\ w &= \Delta z \end{aligned}$$

where B_{red} is a column compressed version of B, that is the columns of B_{red} form a minimal basis for the column space of B. Now it is possible to find a y_r that models the contribution of any initial state to the signals y and z. In our applications the number of columns in B often is most often greater than the number of states. Hence, we can usually save considerable computational effort by compressing the columns and use B_{red} instead of adding a copy of B to model the influence of initial states.

In the following analysis `iqc_diag` is used to describe the constant real uncertainties in the model. The results will be conservative but the computational time per partition will be much shorter than if we use the other IQCs listed in Sect. 15.2.1. The choice yields a reasonable trade-off between the number of partitions needed and the computational complexity per partition. To get an idea of how conservative this choice is we first consider a simplified problem. We ignore the nonlinearities and assume that only the Mach number, M, and calibrated air speed, V_{cas}, are uncertain. The mass, m, and centre of gravity, X_{cg} are set to their nominal values. First the analysis is performed with the stability region being the left half plane. Then, to allow slowly divergent poles, we let the stability region be the half plane to the left of $\frac{\ln 2}{6}$.

In Fig. 15.2, Fig. 15.3, Fig. 15.5 and Fig. 15.6 the considered flight envelope is the area between the two straight lines. Points in the flight envelope that are stable with respect to the considered stability region are called *clearable* and points that are proven to be stable with respect to the considered stability region are called *cleared*. The cleared partitions are marked with dark rectangles. Points that corresponds to unstable modes are marked with crosses. Clearable, but not cleared points are marked with dots.

In Fig. 15.2 and Fig. 15.3 we use the left half plane as the stability region and consider two partitioning schemes with different coarseness. In Fig. 15.2 we use the split ratios 6, 3, and 2. That is, the parameter space is first split into six partitions, if necessary these partitions are in turn split into three partitions and so on. In Fig. 15.3 we use the split ratios 4, 4, 2 and 2. By performing the analysis using the finer partition the size of the cleared region is increased, but not very much.

In order to get a better feeling of the level of conservativeness of the approach we do another test. Fig. 15.4 shows how far into the left half plane the eigenvalues of the system are located. As before a cross corresponds to an unstable mode. A plus corresponds to a mode where the eigenvalue closest to the imaginary axis has a real part between -0.002 and 0. A star corresponds to a mode where the eigenvalue closest to the imaginary axis has a real part between -0.004 and -0.002. Finally, the points marked by dots correspond to modes with all eigenvalues having a real part less than -0.004. The points in this latter category correspond fairly well to the partitions that can be cleared with the IQCs and partitioning scheme used. To clear the small area in the the lower left corner of Fig. 15.4 a slightly finer partition than the one used has to be performed. Hence, this area does not show up in Fig. 15.2 and Fig. 15.3. The conclusion is that the results are indeed conservative but not very much so.

Fig. 15.5 and Fig. 15.6 illustrate the results of the analysis when we allow slowly divergent modes. A cross corresponds to an unstable mode, that is

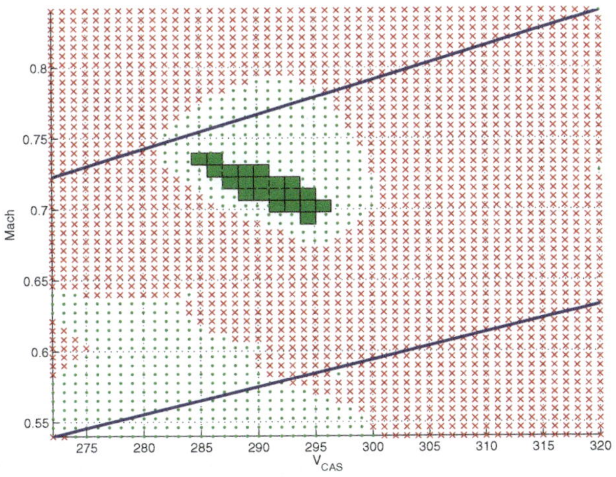

Fig. 15.2 This figure presents the outcome of the analysis for the original model. The analysis for has been performed in 3 stages of partitioning with split ratios of 6, 3 and 2. Cleared partitions are marked with dark rectangles.

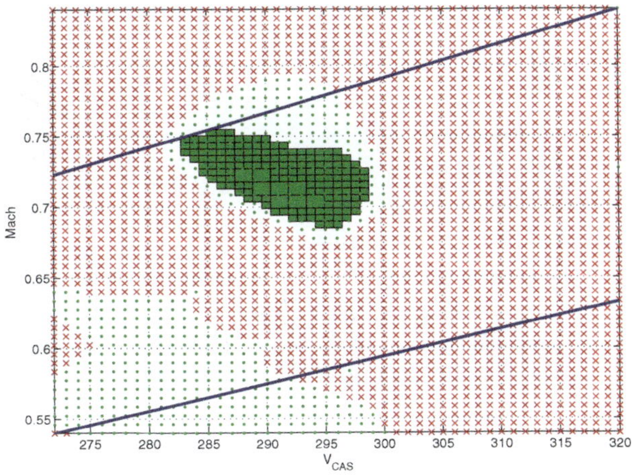

Fig. 15.3 This figure presents the outcome of the analysis for the original model. The analysis for has been performed in 4 stages of partitioning with split ratios of 4, 4, 2 and 2. Cleared partitions are marked with dark rectangles.

where at least one eigenvalue is to the right of $\frac{\ln 2}{6}$. A dot corresponds to a clearable mode, that is where all eigenvalues are to the left of$\frac{\ln 2}{6}$. Cleared partitions are marked by a dark rectangle. We use two different partitioning schemes one with split ratios 6, 3, and 2 and one with split ratios 8, 4, 2, and 2. As can be seen from the figure we have no problem clearing the whole flight

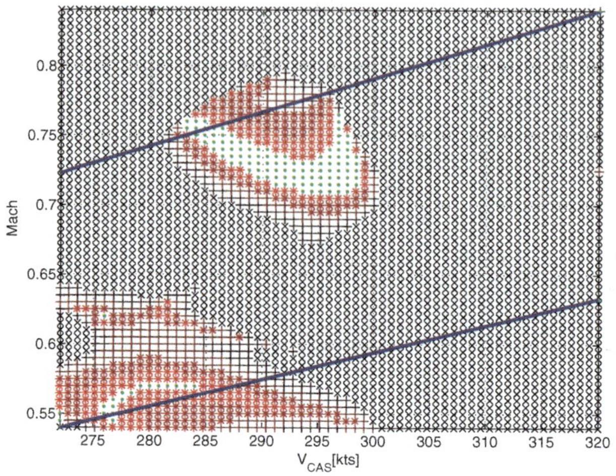

Fig. 15.4 This figure presents the pole mapping of the original model.

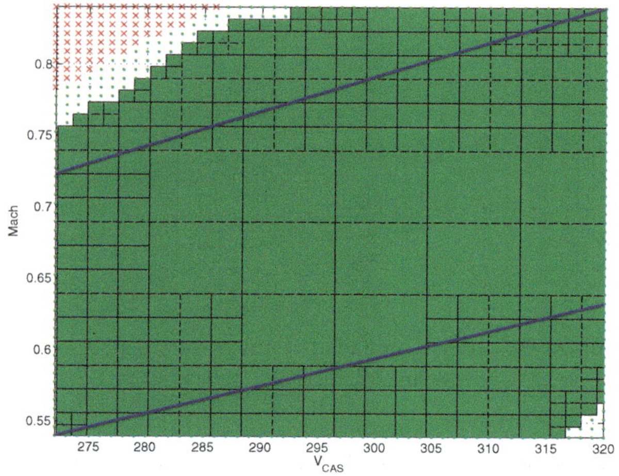

Fig. 15.5 This figure illustrates the partitioning in 3 stages for the model with shifted poles. The split ratios for the different stages are 6,3,2.

envelope for this case. Note, that even with the finer partition we have a small region which we cannot clear even though we know that the all eigenvalues are to the left of $\frac{\ln 2}{6}$.

The model of the nonlinear rigid aircraft consists of sixteen LFR models valid in different regions of the flight envelope. In Fig. 15.7 to Fig. 15.14 the

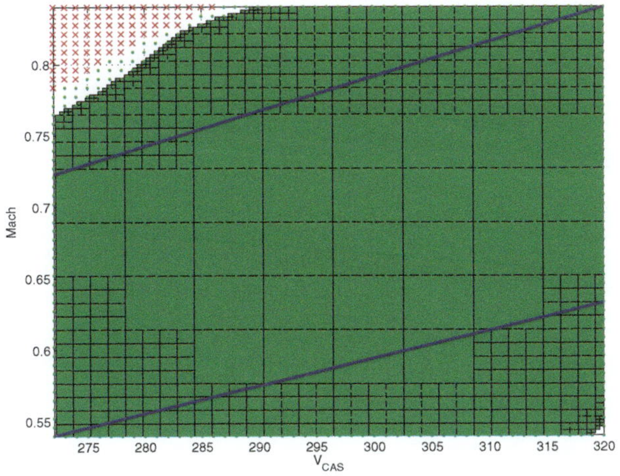

Fig. 15.6 This figure illustrates the partitioning in 4 stages for the model with shifted poles. The split ratios for the different stages are 8,4,2,2.

Table 15.1 Provided LFT models descriptions.

LFT No.	Mass	V_{cas}	Mach	X_{CG}	Total Δ block size	No. States
1	21	37	48	38	144	14
3	35	45	40	24	144	14
14	35	48	32	23	138	14
16	35	48	32	23	138	14

results of the nominal stability analysis for models number 1, 3, 14 and 16 are presented. The dimensions of the uncertainty blocks for the considered models are presented in Table 15.1. The partitioning is conducted in three steps and in order to satisfy the required accuracy, split ratios of 5, 2 and 2 are chosen for the consecutive steps. The time elapsed for performing the 3 stages of partitioning is 339, 277, 348 and 141 hours, for each respective model. Fig. 15.7 and Fig. 15.8 show the cleared regions in the first and second partitioning step for model 1 and 3, respectively. Fig. 15.9 shows the cleared regions in the third partitioning step and Fig. 15.10 displays regions where we cannot prove nor disprove stability for models 1 and 3. Fig. 15.11 and Fig. 15.12 show the cleared regions in the first and second partitioning step for model 14 and 16. Fig. 15.13 shows the cleared regions in the third partitioning step and Fig. 15.14 shows the regions where we cannot prove nor disprove stability. In both cases we consider model 14 and model 16.

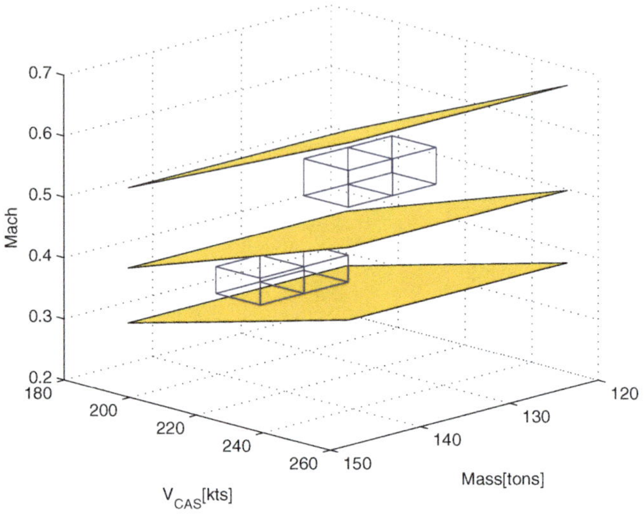

Fig. 15.7 This figure presents the cleared regions within the first stage of the partitioning for models 1 and 3. The light planes represent the validity boundaries for each of the models (the one on the top model 3 and the one on the bottom model 1.). Outer planes also represent the boundaries of the flight envelope.

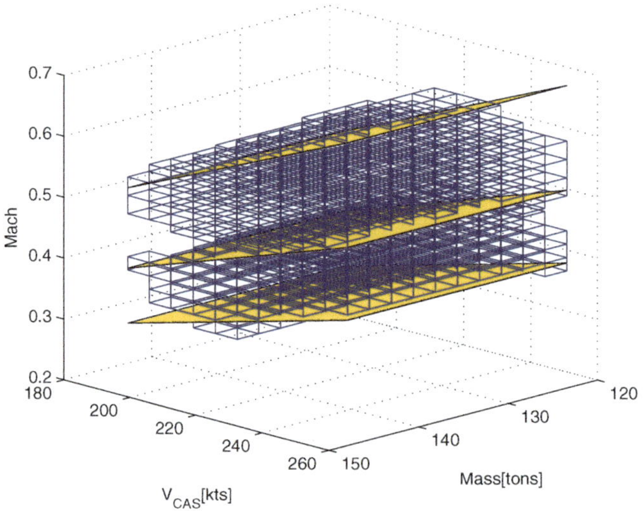

Fig. 15.8 This figure presents the cleared regions within the second stage of the partitioning for models 1 and 3. The light planes represent the validity boundaries for each of the models (the one on the top model 3 and the one on the bottom model 1.). Outer planes also represent the boundaries of the flight envelope.

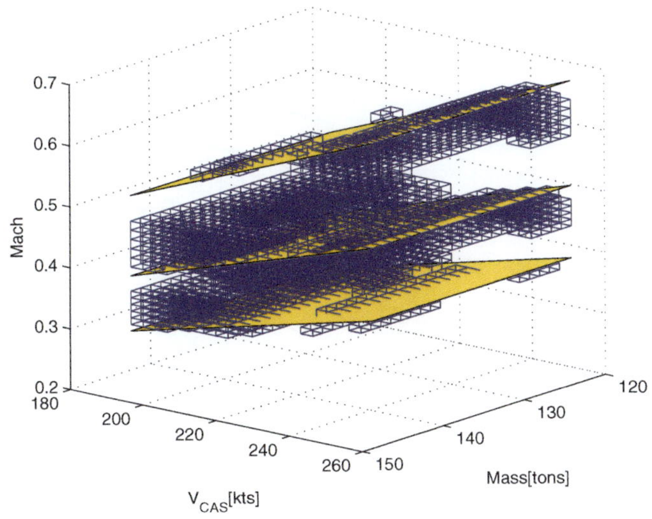

Fig. 15.9 This figure presents the cleared regions within the third stage of the partitioning for models 1 and 3. The light planes represent the validity boundaries for each of the models (the one on the top model 3 and the one on the bottom model 1.). Outer planes also represent the boundaries of the flight envelope.

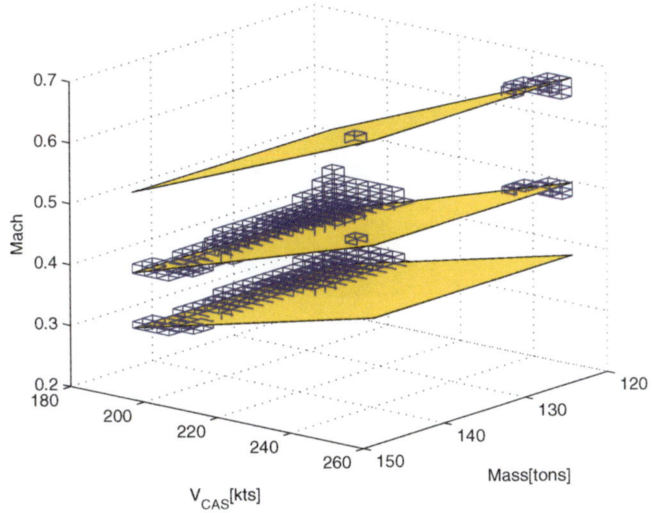

Fig. 15.10 This figure presents the regions with unclear stability properties after the end of the partitioning procedure, for models 1 and 3. The light planes represent the validity boundaries for each of the models (the one on the top model 3 and the one on the bottom model 1.). Outer planes also represent the boundaries of the flight envelope.

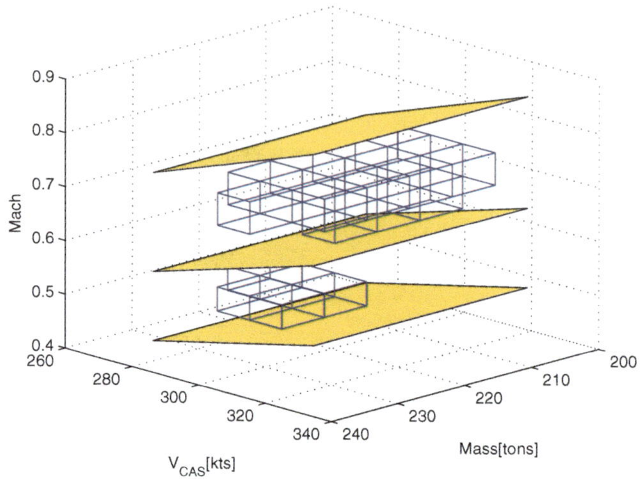

Fig. 15.11 This figure presents the cleared regions within the first stage of the partitioning for models 14 and 16. The light planes represent the validity boundaries for each of the models (the one on the top model 16 and the one on the bottom model 14.). Outer planes also represent the boundaries of the flight envelope.

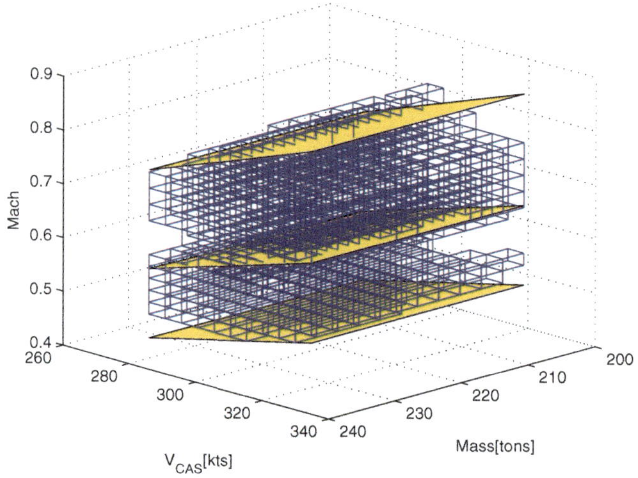

Fig. 15.12 This figure presents the cleared regions within the second stage of the partitioning for models 14 and 16. The light planes represent the validity boundaries for each of the models (the one on the top model 16 and the one on the bottom model 14.). Outer planes also represent the boundaries of the flight envelope.

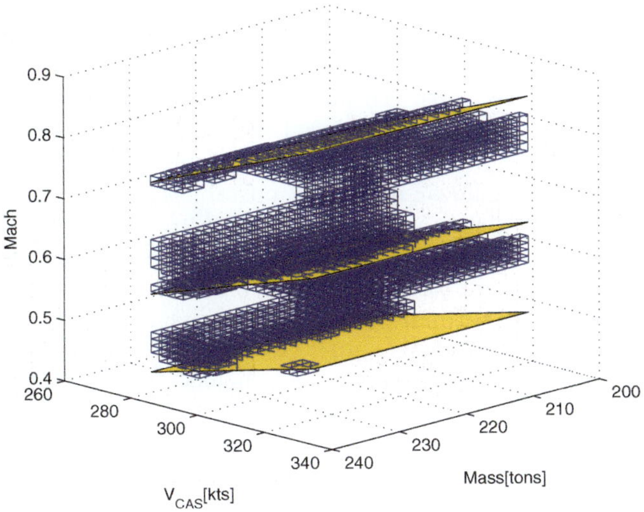

Fig. 15.13 This figure presents the cleared regions within the third stage of the partitioning for models 14 and 16. The light planes represent the validity boundaries for each of the models (the one on the top model 14 and the one on the bottom model 16.). Outer planes also represent the boundaries of the flight envelope.

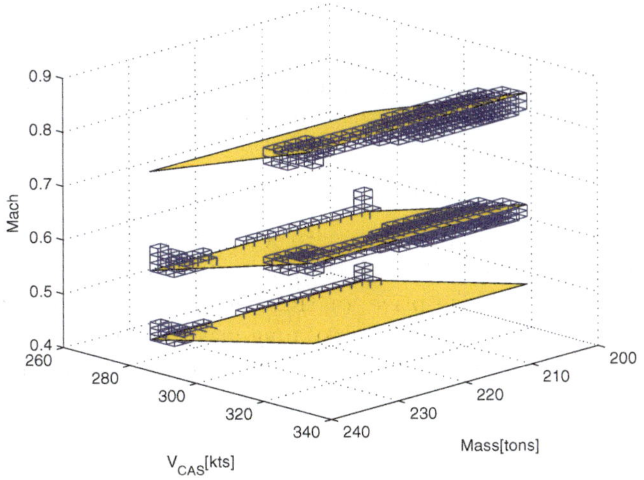

Fig. 15.14 This figure presents the regions with unclear stability properties after the end of the partitioning procedure (right),for models 14 and 16. The light planes represent the validity boundaries for each of the models (the one on the top model 14 and the one on the bottom model 16.). Outer planes also represent the boundaries of the flight envelope.

15.3.2 Analysis of Robust Nonlinear Stability

In this section we present the results from the nonlinear analysis. The procedure is the same as for the robust linear stability with nominal X_{cg}. During preliminary experiments with the models considered it was discovered that if one of the nonlinearities was replaced with a constant gain only slightly larger than 0.7 the resulting closed loop system will be unstable. This does not mean that the nonlinear system is unstable but we cannot use existing IQCs to prove stability. Instead we choose to make some experiments to investigate how conservative the IQCs for sector restricted and slope restricted nonlinearities are for this system considering the little information they utilise. To get an estimate of the level of conservativeness we perform the analysis using the nominal model, i.e. the uncertain parameters are set to their nominal values, and include the nonlinearities. All models considered have 14 states and 8 nonlinearities. We then look for the maximum sector or slope for which we can prove stability.

The analysis using the sector description of the nonlinearities, `iqc_sector`, results in bounds on β very close to zero. The analysis is not very computationally demanding but the results are very conservative. The analysis performed using slope based IQCs leads to less conservative results but as the resulting optimisation problem is high dimensional and the analysis is very computationally demanding. The results for the IQC for slope restricted nonlinearities is presented in Table 15.2. As expected, the results are a little bit

Table 15.2 Estimated maximum allowed slope for description of nonlinearities.

LFT No.	Time Elapsed [min]	Maximum Allowed Slope
1	18	0.6718
2	17	0.6950
3	30	0.6950
4	17	0.7050
5	50	0.6818
6	18	0.6718
7	16	0.6818
8	17	0.6950
9	24	0.6443
10	19	0.6605
11	19	0.6605
12	14	0.6818
13	20	0.6443
14	19	0.6605
15	16	0.6605
16	20	0.6818
Total	6 Hours	

conservative. None of the existing IQCs yield a description of the deadzone nonlinearity that is informative enough. However, it is an active field of research how to introduce less conservative methods from other frameworks, such as Lyapunov based methods, to the IQC framework and provide more accurate descriptions for simple nonlinearities like deadzones and saturations. Hopefully, there will be future results.

15.4 Loads and Comfort Criterion

In this section we present and discuss analysis results for the comfort criterion defined for the unpiloted integral aircraft model. The baseline solution for the comfort criterion is described in Chapter 2 and how to reformulate the problem as an $\mathcal{H}_2$ performance analysis problem and tackle it via IQC techniques is shown in Chapter 10. As this criterion is a performance criterion we want to use the best possible IQCs. Hence, the number of choices we have to make is much more limited than for the stability analysis of the nonlinear rigid aircraft model described earlier.

We consider the LFR models described in Chapter 4 for analysis of the longitudinal linear dynamics of the aircraft. As the analysis is very computationally demanding we only consider models where the fullness ratio of the fuel load for the centre tank (CT) is uncertain. In particular, we consider the following models:

- `OL_lon15_C` : open-loop model with uncertain parameter CT, valid in the domain $[0, 15]$ rad/sec.
- `OL_lon50_C` : open-loop model with uncertain parameter CT, valid in the domain $[0, 50]$ rad/sec.
- `CL_lon15_CXcg` : closed-loop model with uncertain parameter CT and X_{cg} set to a constant value, valid in the domain $[0, 15]$ rad/sec.

The number of state variables n and the size of the Δ block d, that is, the number of repetitions of uncertain parameter CT for these models, are reported in Table 15.3.

Table 15.3 Number of states n and size of Δ block d for the integral LFR models developed for the comfort criterion

Model	n	d
`OL_lon15_C`	12	14
`OL_lon50_C`	30	26
`Cl_lon15_CXcg`	22	18

15.4.1 Computational Times

As has been mentioned above the robust $\mathcal{H}_2$ technique is quite computationally demanding. The average computational times experienced for a single robust $\mathcal{H}_2$ norm computation, i.e.: one criterion, one point in the fuselage and one partition of the uncertainty domain, are the following:

- `OL_lon15_C` : 20 minutes on an Intel XEON 5150 processor with 4 GB RAM
- `CL_lon15_CXcg` : 40 minutes on the same machine as above.
- `OL_lon50_C` : 8 hours on a Intel dual core processor with 16 GB RAM.

In contrast, the baseline solution routine is very fast, as it does not require the solution of any optimisation problem. It will only take a few minutes to compute all the criteria for all points in the fuselage. However, this is not surprising, as we have chosen only 21 grid points and a single uncertain parameter.

15.4.2 Level of Conservatism

To get an idea of the conservativeness of the analysis, the method has been tested on the model `OL_lon15_C`. The values of the robust $\mathcal{H}_2$ norm for all the criteria, all points in the fuselage and all four channels, i.e. $[J_{feet};\ J_{seat};\ J_{back};\ J]$, are shown in Fig. 15.15. The corresponding values provided by the baseline solution, applied to the LFR models, are reported in Fig. 15.16.

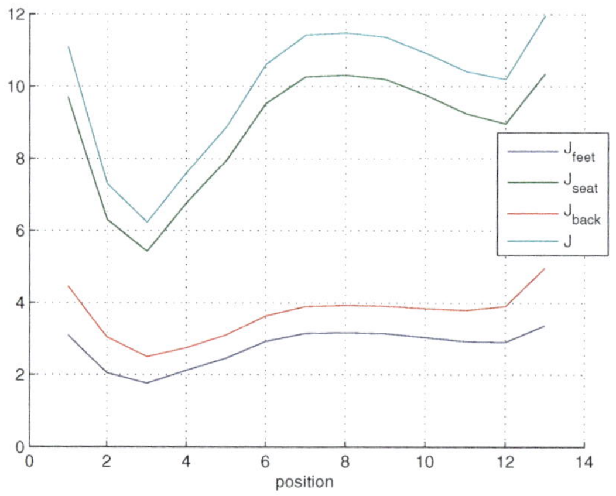

Fig. 15.15 Robust $\mathcal{H}_2$ analysis for model `OL_lon15_C`.

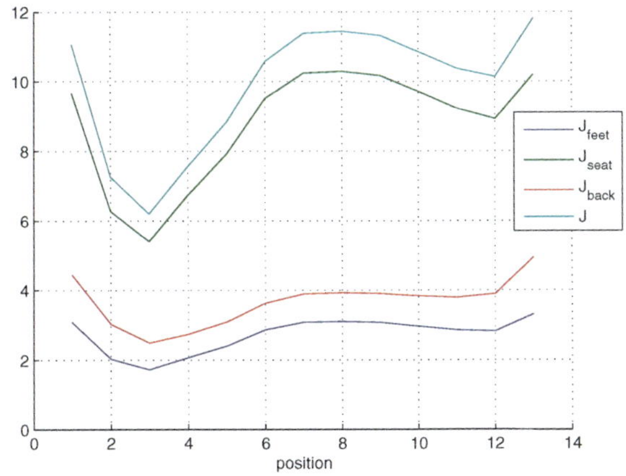

Fig. 15.16 Baseline solution for model `OL_lon15_C`.

It can be observed that the two figures are almost indistinguishable. This suggests that the proposed technique for robust $\mathcal{H}_2$ norm computation is not significantly conservative for the considered LFR model. Partly this has to do with the fact that it is a single input single output system. The lower bound provided by the baseline solution is quite tight. This is due to the fact that the worst-case values for the considered model are attained at the extremes of the uncertainty interval and that the gridding with respect to CT always includes those points. The comparison between the baseline solution for the LFR models and that computed for the original AIRBUS models, may be useful to validate the LFR models. In fact, significant discrepancies are expected for the models whose domain of validity is $[0, 15]$ rad/sec, while more accurate values of the comfort criterion should be obtained on the models valid in the $[0, 50]$ rad/sec range. This is due to the fact that the baseline solution is based on numerical integration on the finite frequency domain $[0.062, 62.8]$ rad/sec. This comparison may represent a valuable guideline for a fine tuning of the LFR models generated for the comfort clearance.

It is worth noting that the values of the comfort index returned by the baseline solution are always smaller than those obtained via the robust $\mathcal{H}_2$ analysis for two reasons: (i) the considered robust $\mathcal{H}_2$ technique is conservative in general; (ii) the baseline solution considers a finite grid of CT values and a finite frequency range, and hence it gives a lower bound to the true criterion; on the contrary, the robust $\mathcal{H}_2$ approach considers the continuum of all possible CT values and all the frequencies, thus providing an upper bound to the criterion.

15.5 Conclusions and Future Research

In this chapter we have presented the results from the IQC based analysis of the nonlinear rigid aircraft model and the integral aircraft model. We have performed stability analysis on the nonlinear rigid aircraft model and considered the comfort and loads criterion for the integral aircraft model. Some parts of the stability analysis performed has been omitted here since it, with the choices of IQCs made et cetera, either yielded very conservative results or were computationally too complex. Also the robust nonlinear analysis had to be somewhat simplified due to problems with destabilising constant gains lying in the same sector as the deadzone nonlinearities considered.

The models developed in the project are very computationally demanding to perform IQC analysis on. As was mentioned in the introduction existing structure exploiting methods for solving KYP-SDPs cannot handle the structure we have in the problems at hand. Hence, it would be very interesting to further investigate if it is possible to find new structure exploiting algorithms that can cope with multipliers with a complex structure and systems with many more inputs than states. Outer approximation methods for IQC analysis which work in the frequency domain are not applicable either as the number of parameters needed for the multipliers is too large. However, there is perhaps possible to develop new methods inspired by interesting recent work done in μ-analysis.

Very often we also encounter numerical issues in complex IQC analysis problems. While years of research has been spent on how to solve Riccati equations in a good way surprisingly little has been done for KYP-SDPs. KYP-SDPs can solve the same problems as can be solved using Riccati equations but also more general ones. Hence, improving numerical properties for this class of semidefinite programs is, in our opinion, a very important area of future research.

References

1. Megretski, A., Rantzer, A.: System analysis via integral quadratic constraints. IEEE Transactions on Automatic Control 42(6), 819–830 (1997)
2. Jönsson, U.: Lecture Notes on Integral Quadratic Constraint. Royal Institute of Technology, Stockholm (2001), `http://www.math.kth.se/~ulfj/5B5744/Lecturenotes.ps`
3. Doyle, J.: Analysis of feedback systems with structured uncertainties. IEE Proceedings, Part D: Control Theory and Applications 129(6), 242–250 (1982)
4. Safonov, M.G.: Stability margins of diagonally perturbed multivariable feedback systems. IEE Proceedings, Part D: Control Theory and Applications 129(6), 251–256 (1982)
5. Roos, C., Biannic, J.M.: On the numerical solution of LMIs derived from the KYP lemma. In: Proceedings of the 5th IFAC Symposium on Robust Control Design (2006)

6. Parrilo, P.: On the numerical solution of LMIs derived from the KYP lemma. In: Proceedings of the 38th IEEE Conference on Decision and Control, Phoenix, Arizona, USA, vol. 3, pp. 2334–2338 (1999)
7. Parrilo, P.: Outer approximation algorithms for KYP-based LMIs. In: Proceedings of the American Control Conference, Arlington, VA, USA, vol. 4, pp. 3025–3028 (June 2001)
8. Kao, C.Y., Megretski, A., Jönsson, U.T.: An algorithm for solving optimization problems involving special frequency dependent LMIs. In: Proceedings of the 2000 American Control Conference, Chicago, Illinois, USA, vol. 1, pp. 307–311 (2000)
9. Kao, C.Y., Megretski, A., Jönsson, U.T.: A cutting plane algorithm for robustness analysis of time-varying systems. IEEE Journal on Automatic Control 46(4), 579–592 (2001)
10. Kao, C.Y., Megretski, A., Jönsson, U.T.: Specialized fast algorithms for IQC feasibility and optimization problems. Automatica 40(2), 239–252 (2004)
11. Wallin, R., Kao, C.Y., Hansson, A.: A cutting plane method for solving KYP-SDPs. Automatica 44(2), 418–429 (2008)
12. Vandenberghe, L., Balakrishnan, V.R., Wallin, R., Hansson, A., Roh, T.: Interior-point algorithms for semidefinite programming problems derived from the KYP lemma. In: Positive Polynomials in Control. LNCIS. Springer, Heidelberg (2005)
13. Liu, Z., Vandenberghe, L.: Low-rank structure in semidefinite programs derived from the KYP lemma. In: Proceedings of the 46th IEEE Conference on Decision and Control, New Orleans, Louisiana, USA, pp. 5652–5659 (2007)
14. Wallin, R., Hansson, A., Johansson, J.H.: A Structure Exploiting Preprocessor for Semidefinite Programs Derived From the Kalman-Yakubovich-Popov Lemma. IEEE Journal on Automatic Control 54(4), 697–704 (2009)
15. Jönsson, U., Kao, C.Y., Megretski, A., Rantzer, A.: A guide to IQC-beta: A Matlab toolbox for robust stability and performance analysis (2004)
16. Fan, M.K.H., Tits, A.L., Doyle, J.C.: Robustness in the presence of mixed parametric uncertainty and unmodeled dynamics. IEEE Transactions on Automatic Control 36(1), 25–38 (1991)
17. Young, P.M.: Robustness with parametric and dynamic uncertainty. PhD thesis, CA Tech. Inst. (1993)
18. Zames, G., Falb, P.L.: Stability conditions for systems with monotone and slope-restricted nonlinearities. SIAM Journal of Control 6(1), 89–108 (1968)

Chapter 16
Application of Nonlinear Programming Methods for Determination of Worst-Case Pilot Inputs

Daniel Skoogh and Fredrik Berefelt

Abstract. This chapter describes the results obtained by applying the optimisation methods presented in Chapter 11 for searching of worst-case manoeuvres. The search for worst-case pilot inputs is performed for a single flight condition and nominal model parameters by using both local and global optimisation algorithms over a parameter space consisting of suitably parameterized pilot inputs.

16.1 Introduction

In this chapter we describe the results obtained by applying the optimisation methods described in Chapter 11 to the worst-case pilot input problem. For all the numerical results we have used the Airbus simulation model described in [1]. All computational results have been obtained using MATLAB Version R2007b. For the optimizations we used dedicated MATLAB implementations of all local and global methods.

In Sect. 16.2 we describe the results obtained from the application of the three local methods: quasi-Newton, cyclic coordinate descent, and pattern search. The performed tests illustrate that local search methods, when properly initialized, are able to determine useful worst-case pilot inputs by employing an acceptable number of function evaluations. We also performed comparisons with similar general purpose tools available in popular optimization software. It is worth mentioning that our implementations are competitive with these tools, in both their timing performances as well as in the achieved accuracy.

Daniel Skoogh · Fredrik Berefelt
Swedish Research Defence Agency (FOI), System Technology,
SE-16490 Stockholm, Sweden
e-mail: daniel.skoogh@foi.se, fredrik.berefelt@foi.se

A. Varga et al. (Eds.): Opti. Based Clearance of Flight Control Laws, LNCIS 416, pp. 299–315.
springerlink.com

In Sect. 16.3 we describe results from application of the following global methods: differential evolution, genetic algorithm, evolution strategies, and dividing rectangles. As expected, the global methods are generally more expensive to use than local methods in terms of necessary number of function evaluations. However, the performed tests showed that, with acceptable costs, it is relatively easy to determine worst-case pilot inputs which violates the controller protection laws (if such inputs exist).

16.2 Application of Local Methods

In this section we describe the results obtained by applying the *Quasi-Newton* method (QN) in combination with the *Cyclic Coordinate Descent* method (CCD), and *Pattern Search* method (PS) to solve the worst-case pilot input determination problem. Since local methods typically converges to local optima, we performed the worst-case search for several randomly generated initial parameter points.

For all local methods we employed the same clearance criteria, the maximum value of the angle of attack in a given time interval $[0, T]$ for a given pilot input. The objective function to be maximized is defined as

$$f(\boldsymbol{x}) = \max_{t\in[0,T]} \alpha(t, \boldsymbol{x}) =: \alpha_{\max}(\boldsymbol{x}) \tag{16.1}$$

where $\alpha(t, \boldsymbol{x})$ indicates the dependence of the angle of attack of time and a parameter vector $\boldsymbol{x}$ which consists of the parameters used for the parametrization of the pilot signal. We will refer throughout this chapter to $t_{\max}$ as the time instant when the maximum angle of attack $\alpha_{\max}$ occurs. From practical considerations, we used a value of T of 150 seconds and 11 intervals of 10 seconds lengths were used to define 10 time grid values for each pilot input. Thus, for each the pitch and roll pilot signal parameters, a total of 20 decision variables have been used. The amplitudes of all pilot signals were constrained to lie in the interval $[-16°, 16°]$. It follows, that the resulting mathematical optimization problem has only simple bound constraints on the optimization variables.

Each parameter set determines the shapes of two pilot inputs (the longitudinal and lateral sticks). The duration of each pilot inputs encompasses 150 seconds, of which the first 10 seconds and the last 30 seconds use null amplitude signals. On the remaining 10 time intervals, linear interpolation is employed between two successive time grid values. Each function evaluation use a simulation run from $t = 0$ to $t = 150$, where the in the starting flight condition corresponds to a cruise flight at altitude of 31834 feet with a speed of 236 knots. Further, the considered aircraft weight was 230000 kg and the center of gravity position was at 0.37%.

16.2.1 The Quasi-Newton Method

For all tests, we employed a centered finite difference scheme to estimate the gradient, with a step size of $h = 0.1°$. For details regarding the choice of the step size, see [2].

In Fig. 16.1, the longitudinal and lateral pilot signals are shown together with the corresponding angle of attack $\alpha(t, \boldsymbol{x}_{\text{start}})$, where $\boldsymbol{x}_{\text{start}}$ is a typical randomly generated starting point for the QN iterations. The dashed lines in Fig. 16.1 correspond to the time instant of the last pilot inputs grid points that affects the maximum angle of attack peak for small changes of the amplitude. In Fig. 16.2, the maximum angle of attack α_{max} is plotted against the iteration number. For this example, the QN method converged after 9 iterations and needed 362 objective function evaluations to determine the local optimiser $\boldsymbol{x}_{\text{max}}$. The corresponding optimum is $f(\boldsymbol{x}_{\text{max}}) = 12.7002°$ and occurs at time instant $t_{\text{max}} = 28.7167s$. Of the total number of 362 objective function evaluations, 124 were used for gradient estimation and 238 for line search. A number of 236 objective function evaluations have been avoided due to the special structure of the problem induced by causality considerations, as discussed in Chapter 11 and [2].

After the convergence of the QN method, we run the CCD method with a step size $h = 0.1°$ used to define the search grid directions. As initialization point of the CCD method we used the termination point of the QN method. For this specific case, the CCD method terminated at the first iteration after

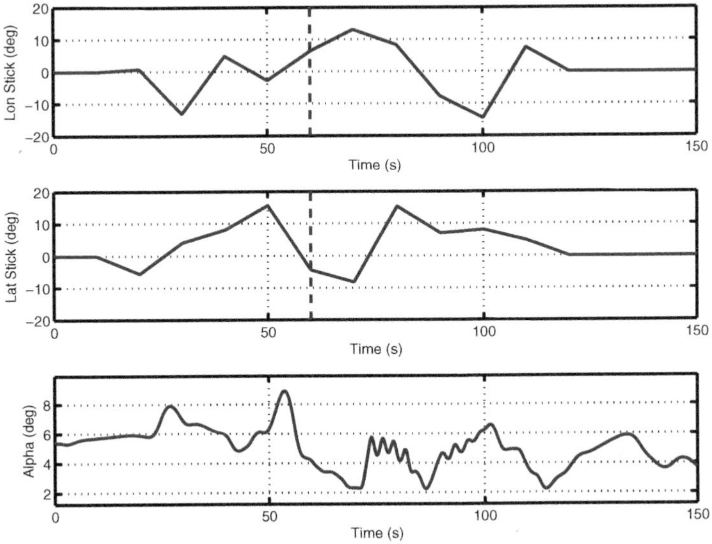

Fig. 16.1 The longitudinal and lateral pilot signals are shown together with the angle of attack $\alpha(t, \boldsymbol{x}_{start})$ corresponding to the starting point of the quasi-Newton iterations.

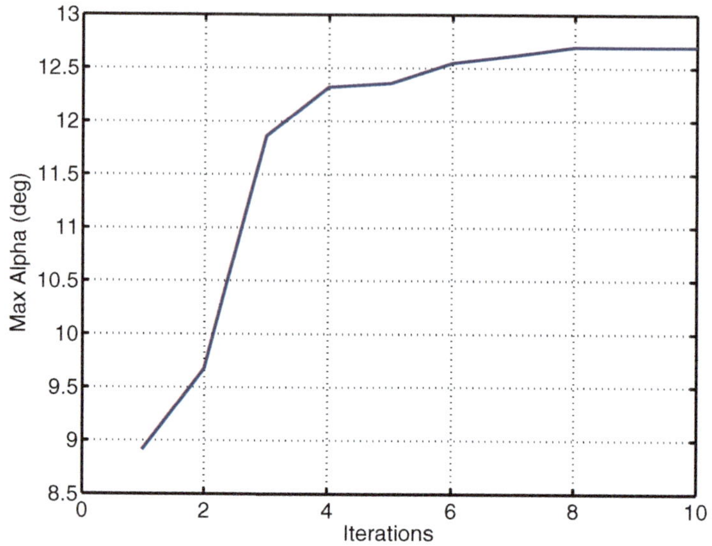

Fig. 16.2 The maximum angle of attack is shown with respect to the number of iterations for the quasi-Newton method.

performing a complete search cycle. In this way, the solution computed by the QN method was confirmed as a local optimiser in one iteration of the CCD method. The advantages of using of using the CCD method in conjunction with QN method is discussed in Chapter 11.

In Fig. 16.3, the longitudinal and lateral pilot signals are shown together with the angle of attack $\alpha(t, \boldsymbol{x}_{\max})$ corresponding to the termination point of the QN iterations. The dashed lines in Fig. 16.3 correspond to the time instant of the last interpolation point that will affect the maximum angle of attack peak for small changes of amplitude.

As a comparison, the objective function (16.4) used for the global search in Section 16.3 was evaluated to $-4.5593°$, i.e. a violation of the $\alpha_{\max}$ protection law by $4.5593°$.

Timing Results

We used both the parallel and the serial version of the QN method, which are described in detail in [3]. To run the QN method in parallel, we used the MATLAB Distributed Computing Toolbox [4][1]. For all computations we haved used a computer with two Xeon 2.66GHz processors, 4 cores on each processor, under Linux.

The QN method converged in 9 iterations and performed 362 objective function evaluations. Using parallel computations on 4 cores, the QN method

[1] The toolbox has now changed the name to: Parallel Computing Toolbox.

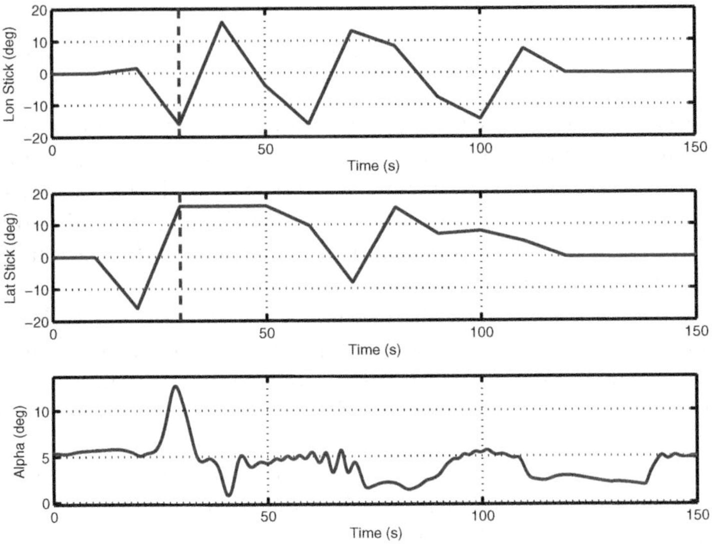

Fig. 16.3 The longitudinal and lateral pilot signals are shown together with the angle of attack $\alpha(t, \boldsymbol{x}_{\max})$ corresponding to the termination point of the quasi-Newton iterations.

needed 2992 seconds. For the same computation performed in serial mode, the QN method needed 7721 seconds. The speedup ratio using 4 cores[2] was thus 2.58, which should be compared to the speedup 3.24 obtained for Monte Carlo simulations. The reasons why the maximum speedup ratio of 4 can usually not be achieved are the lack of full parallelization in the employed code and the shared memory architecture of the employed machine.

Evaluation and Comparison

In what follows, we present more details of the performed experiments with our implementation of the QN method. To increase the robustness of local search, we used 7 randomly generated initial points on which the QN method was applied. After convergence, the CCD method, with a step length of $0.1°$ in each input parameter, was used to refine the local optimum. For comparison, we performed the same experiments with two solvers available in popular optimization packages: the function `fmincon` from the MATLAB Optimization Toolbox [6] and the Fortran code SNOPT [5] via the MATLAB based interface TOMLAB [7]. Both solvers implement the Sequential Quadratic Programming (SQP) method, but also provide options for selecting BFGS quasi-Newton updates for approximating the Hessian of the Lagrangian.

[2] We used 2 cores on each processor. The maximum speedup was obtained using this configuration.

For the QN method, the termination criteria was construced by using the objective function values at iterates $k+1$ and $k-1$ as follows

$$|f(\boldsymbol{x}_{k+1}) - f(\boldsymbol{x}_{k-1})| < \epsilon \tag{16.2}$$

The tolerance ϵ was set to be 10^{-4}. Further, for SNOPT/TOMLAB, the major optimality tolarance was 10^{-6}, which specifies the final accuracy of the dual variables $\boldsymbol{\pi}$ involving the reduced gradient. Finaly, for `fmincon`, the termination tolerance of the objective function value `TolFun` was 10^{-6} and the termination tolerance of the decision variable `TolX` was 10^{-6}.

The Tables 16.1, 16.2, and 16.3 summarize the results for our implementation of the QN method (denoted by QN), `fmincon` and SNOPT/TOMLAB, respectively. Here, NIT denotes the *number of performed iterations* and NFE, the *number of performed function evaluations*. By studying the tables, we see that the objective function value at the QN termination point was much closer to the objective function value of the CCD method than for `fmincon` and SNOPT/TOMLAB, for most starting points. Note also that the maximum difference between the objective function value at the termination point of `fmincon` and CCD was 2.4. The corresponding value for SNOPT/TOMLAB was 0.76, while for the QN it was smaller than 0.1. Overall on this problem, our implementation of the QN method was considerably more robust than both `fmincon` and SNOPT/TOMLAB.

On average, our implementation of the QN method used more function evaluations than `fmincon`, but considerably fewer than SNOPT/TOMLAB. Also, `fmincon` failed for the third initialization point due to a simulation model crash.

This evaluation also shows that for the worst-case pilot input optimisation problem involving noisy function evaluations, it makes good sense to supplement a local search (as performed with the QN method) with a less noise sensitive method like CCD that operates on a grid of points. An alternative to CCD could also be the pattern search method which we will discuss next.

Table 16.1 Evaluation of the implemented QN method.

Starting Point No.	1	2	3	4	5	6	7
NIT for QN	6	9	13	10	27	19	10
NFE for QN	187	362	1060	443	2279	902	968
$\alpha_{\max}$	7.52	12.70	10.74	12.96	15.70	13.02	10.25
$t_{\max}$	16.08	28.72	142.96	39.65	124.04	42.31	98.79
NIT for CCD	1	1	4	3	5	3	5
NFE for CCD	3	5	223	45	335	101	218
$\lvert\alpha_{\max}^{QN} - \alpha_{\max}^{CCD}\rvert$	0	0	0.0280	0.0233	0.0431	0.0962	0.0365

Table 16.2 Evaluation of the `fmincon` software.

Starting Point No.	1	2	3	4	5	6	7
NIT for `fmincon`	4	3	–	21	2	23	36
NFE for `fmincon`	129	105	–	605	63	668	1022
$\alpha_{\max}$	7.52	10.26	–	10.47	9.60	12.02	11.46
$t_{\max}$	16.08	29.62	–	30.74	29.43	43.41	40.50
NIT for CCD	1	1	–	13	8	2	2
NFE for CCD	4	5	–	299	152	48	46
$\lvert\alpha_{\max}^{fmincon} - \alpha_{\max}^{CCD}\rvert$	0	0	–	2.4031	0.2313	0.4662	0

Table 16.3 Evaluation of the TOMLAB/SNOPT software.

Starting Point No.	1	2	3	4	5	6	7
NIT for SNOPT	8	25	38	33	34	14	31
NFE for SNOPT	2790	6152	8940	11113	9678	5004	9637
$\alpha_{\max}$	7.52	13.54	13.15	10.81	14.40	11.31	10.00
$t_{\max}$	16.08	50.68	121.36	41.60	116.64	42.21	99.51
NIT for CCD	1	3	9	4	10	4	2
NFE for CCD	3	93	639	125	835	130	153
$\lvert\alpha_{\max}^{snopt} - \alpha_{\max}^{CCD}\rvert$	0	0.0021	0.0295	0.7672	0.1048	0.2217	0.1715

16.2.2 Pattern Search

The Pattern Search (PS) method was applied to the same test problem and starting points as the QN method. The initial step size was $\Delta_0 = 3.2°$ for each coordinate direction. For a given shrink factor θ, the step sizes used by PS are

$$\Delta = \Delta_0 \theta^M, \tag{16.3}$$

where M is the number of times the step size has been reduced. As termination criterion, we used the smallest allowed relative step size satisfying $\theta^M \geq 0.005$, which corresponds to an absolute step size of 0.016°. An interesting feature is that by using this termination criteria, PS finds a local optimiser and the corresponding optimum on a grid of points with the grid size equal to the last used step length. This makes the termination of PS more robust than of QN, and therefore there was no need to supplement a search by PS with another algorithm.

In Table 16.4, we show the results from the run with PS for a shrink factor of $\theta = 0.5$, for which the fastest convergence results have been obtained. When comparing these results with those for QN in Table 16.1, we see that, on average, PS found higher $\alpha_{\max}$ than the QN method, although the latter found the highest $\alpha_{\max}$ of all methods.

A potential drawback of PS is that the method may converge towards different local optima depending on the ordering of the decision variables. This may cause difficulties, for example, when PS is used in a hybrid global method. In this case, some coordinate directions may be systematically overlooked.

Table 16.4 Evaluation of Pattern Search

Starting Point No.	1	2	3	4	5	6	7
NIT for PS	6	8	13	24	11	9	6
NFE for PS	527	524	839	1225	769	602	686
$\alpha_{\max}$	7.52	13.00	12.20	14.82	15.09	12.93	11.48
$t_{\max}$	16.07	29.19	68.21	40.15	120.18	42.47	98.20

16.3 Application of Global Methods

In this section we present results for worst-case pilot input determination obtained using the global methods *Differential Evolution* (DE), *Evolution Strategies* (ES), *Genetic Algorithm* (GA), and *Dividing Rectangles* (DIRECT). We refer to Chapter 11 for a review of the algorithms and a description of their option parameters.

For the evaluation of global methods, we used an alternative formulation of the worst-case pilot input determination problem, by minimizing the following objective function defined as the "distance" to the peripheral envelope

$$f(\boldsymbol{x}) = \min_{t\in[0,T]} (\alpha_{prot}(t,\boldsymbol{x}) - \alpha(t,\boldsymbol{x})) \tag{16.4}$$

where $\alpha_{prot}(t,\boldsymbol{x})$ is the upper (protection) bound for the angle of attack provided by the control algorithm. Both $\alpha_{prot}(t,\boldsymbol{x})$ and $\alpha(t,\boldsymbol{x})$ depend of the parameter vector $\boldsymbol{x}$ which consists of the parameters used for the parametrization of the pilot signal. This "distance" defined above takes on a negative value when the angle of attack exceeds the upper bound, so minimising the envelope distance means implicitly maximising the angle of attack. The flight condition used for the initialization of the pilot maneuver was: altitude of 20000 ft and speed of 220 kts. The considered aircraft weight was 230000 kg and the center of gravity position was at 0.37%. The auto thrust engagement was on. The search space included only longitudinal and lateral stick inputs, both defined as constant values on 6 time intervals of 2 seconds length (thus $T = 12$). The effective pilot signals to be determined start after 2 seconds and ends at 14 seconds. The total simulation time was 15 seconds, and the pilot inputs are zero outside the active interval [2, 14].

The search space of the optimisation is spanned by the 12 amplitude values, lying in the interval $[-16°, 16°]$. The present setup is comparable to the worst-case pilot input computation performed in [8]. This reference also used step sequences for longitudinal and lateral inputs, but the values of the two sequences were coupled to each other, so there were only 5 decision variables taking only 4 possible discrete values (binary coded GA was employed).

It has been surprisingly easy to find stick inputs that bring the aircraft outside the peripheral envelope defined in [9]. This is why, in all our tests (except DIRECT), we started from randomly generated initial populations which already contained pilot inputs leading to the protection violation.

We did not perform a systematic comparison between the different global methods. To do this, it would be necessary to collect statistics from a great number of runs, which in the present case would demand large computational resources. We have done just an informal investigation of a limited number of sample computations, trying different values of the option parameters. By letting the computations last for a large number of function evaluations, we have tried to expose the long term behavior of the algorithms (several of the runs presented below have actually lasted longer than indicated in the convergence plots).

16.3.1 Differential Evolution

The upper left panel in Fig. 16.4 shows the convergence plot in a run employing DE. The chosen population size was 50 and the values of the two main parameters, the scaling factor F and crossover probability p_{cross}, were 0.5 and 0.9, respectively (see [10]). The plot shows the envelope distance as a function of the accumulated number of objective function evaluations. A negative value implies an angle of attack exceeding the upper bound of the peripheral envelope. Each objective function evaluation involves the simulation of the closed-loop aircraft model to compute the envelope distance within a time horizon of 15 seconds. Each dot in the figure represents the least distance found in a single iteration of the algorithm. Every iteration performs 50 new objective function evaluations.

The envelope distance falls rapidly in the beginning, going below -20 degrees after 750 function evaluations. After 3400 function evaluations, it stays close to -23.8 (translating into an angle of attack of 34.4 degrees). The other three panels in Fig. 16.4 show the longitudinal and lateral stick inputs and the resulting angle of attack versus time of the final worst-case found. The time axis has been extended to 40 seconds in order to reveal what happens after the stick input has become quiescent. As one might expect, the angle of attack returns quickly to values in between the upper and lower bounds (represented by dashed red lines).

Fig. 16.5 shows another DE run with population size $= 100$, $F = 0.5$ and $p_{cross} = 0.8$. The convergence is here much slower. It takes 10200 function evaluations to arrive below -20 and very little progress is made between 5800 and 10200 function evaluations, where it could have been tempting to regard the computation as being converged. But after 10200 function

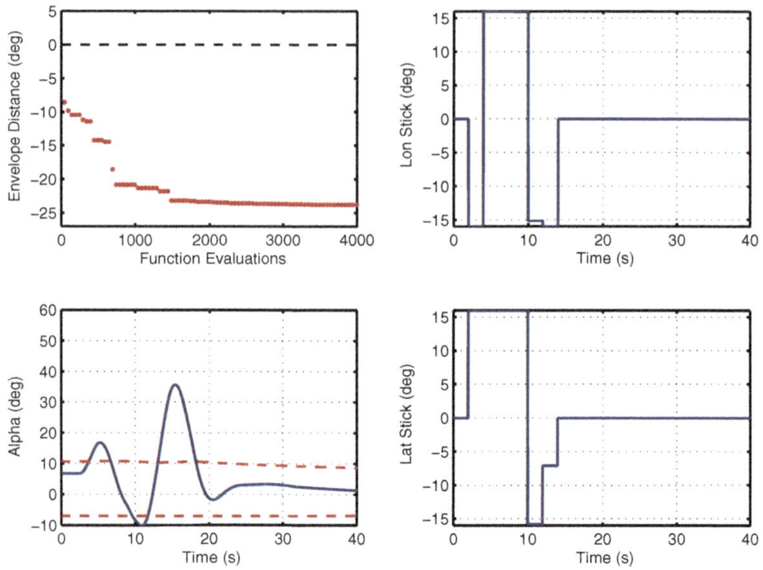

Fig. 16.4 DE: population size = 50, F = 0.5, p_{cross} = 0.9. Convergence plot (upper left), angle of attack (lower left), longitudinal stick input (upper right) and lateral stick input (lower right) of the final worst-case found.

evaluations, the progress is resumed, reaching finally a criterion value of -26.5 after 23000 function evaluations (the corresponding angle of attack is 37.1 degrees). Fig. 16.5 shows also the final worst-case longitudinal/lateral stick inputs and angle of attack versus time.

We want to point out the following observations. The longitudinal input sequence is quite similar to the previous case apart from the first step, while the lateral input is, more or less, the mirror image of the sequence in Fig. 16.4. This makes sense, since the fitness landscape should be symmetric under change of sign of the lateral stick input, and this run just happened to find (approximately) the symmetric partner of the solution found in the previous run. However, the plot of angle of attack versus time in Fig. 16.5 is very different from the one in Fig. 16.4, showing that there is something in the simulation model breaking the "left-right symmetry". In the present case, the angle of attack makes a huge excursion reaching almost 60.5 degrees at t = 22 seconds before it returns inside the bounds at t = 36 (n.b. that the time horizon for the worst-case search is just 15 seconds).

Further, by *increasing* the amplitude value of the *first* step in the *longitudinal* input by a tiny amount ($\sim$ 0.004 degrees), the angle of attack versus time in Fig. 16.5 flips into the shape shown in Fig. 16.4 (the lateral input staying fixed). We have no explanation for this behavior.

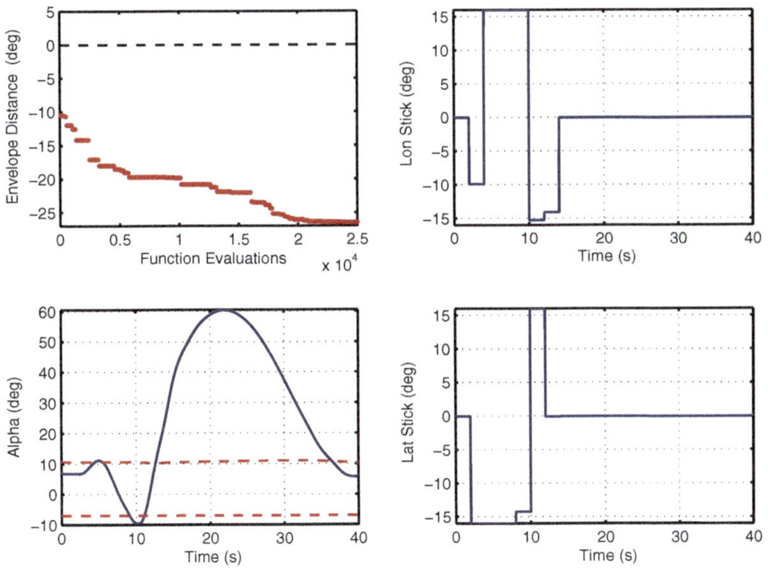

Fig. 16.5 DE: population size = 100, $F = 0.5$, $p_{cross} = 0.8$

A third example run of DE is shown in Fig. 16.6 (population size = 50, $F = 0.5$ and $p_{cross} = 0.8$). It reaches a solution very close to the case in Fig. 16.5 after only 3500 objective function evaluations. Again, changing sign on the lateral stick input turns the angle of attack versus time back into the form of Fig. 16.4.

If this exercise of changing sign on the lateral input is performed on some more "benign" stick sequence, the angle of attack versus time stays constant as it should. In other words, the global optimisers seem to have proved their usefulness in finding an anomalous stick input for the present test problem.

16.3.2 Evolution Strategies

For the application of ES, there are several option parameters to be set. It was not intended to pursue exhaustive investigations for an optimal setting of these parameters for the problem in question. The used values adhere to standard recommendations found in the literature. For the problem at hand, it was quite obvious that a non-elitist selection involving μ parents and λ children (denoted with (μ, λ)-ES in the ES terminology) does not properly work. Therefore, we used all $\mu + \lambda$ parents and children for the selection (denoted with $(\mu + \lambda)$-ES). Non-isotropic mutation was always employed.

Fig. 16.7 gives an example of running $(\mu+\lambda)$-ES with $\mu = 30$, $\lambda = 210$ (also with several other parameters/options described in [10] set to: *initScale* = 1,

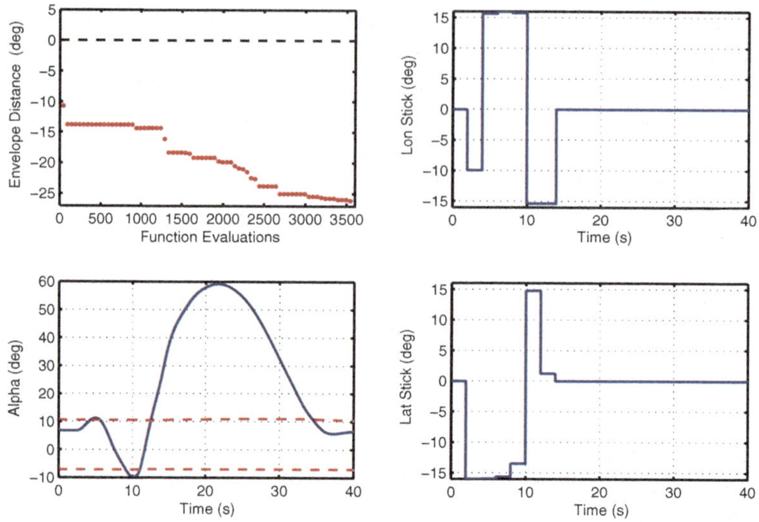

Fig. 16.6 DE: population size = 50, $F = 0.5$, $p_{cross} = 0.8$.

$tauC = 1.5$, using recombination scheme "B"). The envelope distance goes below -20 degrees after 1900 function evaluations, below -25 after 8000 function evaluations, and settles down at -26.5 around 21000 function evaluations. The final worst-case solution is almost identical to the one in Fig. 16.5. We have also seen examples of ES converging to the mirror solution in Fig. 16.4.

Occasionally, we observed that the ES gets caught in what appears to be a local optimum at an envelope distance near -20 degrees. An example is shown in Fig. 16.8 (with parameters/options set to: $\mu = 10$, $\lambda = 100$, $initScale = 3.3$, $tauC = 1$, recombination scheme "A" with $\rho = 2$). The DE computation in Fig. 16.5 seems to be visiting precisely this local optimum during the plateau between 5800 and 10200 objective function evaluations.

16.3.3 Genetic Algorithm

As the last member of the evolutionary algorithms, an example run of GA is given in Fig. 16.9. The population size was 50 and the number of elite members was 5. The computation clearly got stuck at -14 degrees, then suddenly took a jump and moved towards -19.2 degrees. The worst-case solution found resembles the one in Fig. 16.8, but it has not quite reached the local optimum of the latter. The convergence in this example is not good, the balance between the five different variation operators being far from ideal. Instead of using standard GA operators, it would be interesting to construct special crossover and mutation operators that better explored the structure inherent in the problem, for instance the temporal order of the input steps

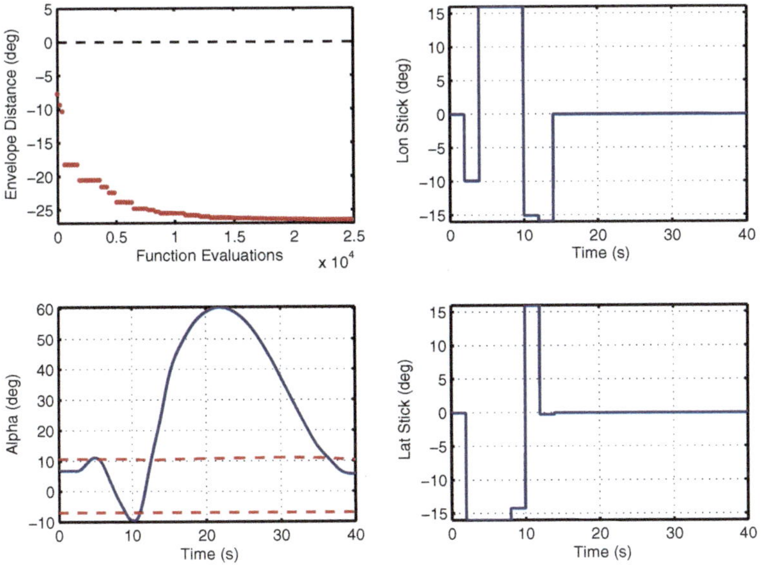

Fig. 16.7 ES: $\mu = 30$, $\lambda = 210$, $initScale = 1$, $tauC = 1.5$

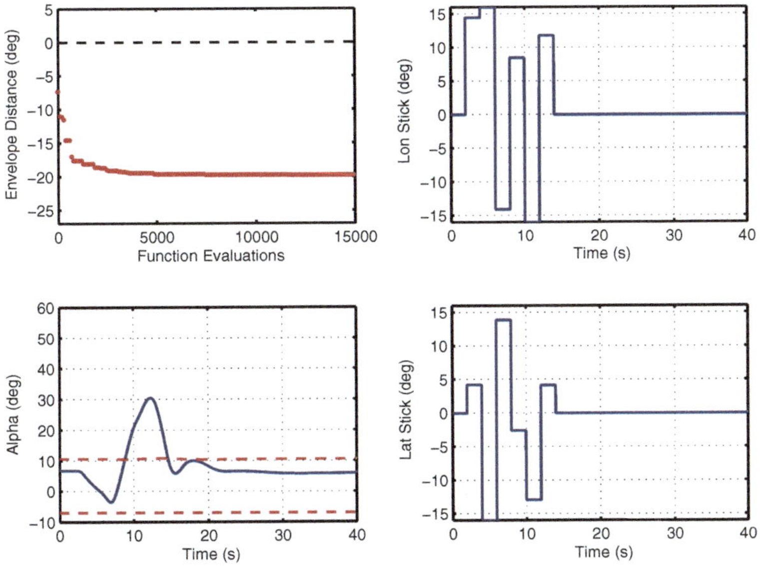

Fig. 16.8 ES: $\mu = 10$, $\lambda = 100$, $initScale = 3.3$, $tauC = 1$.

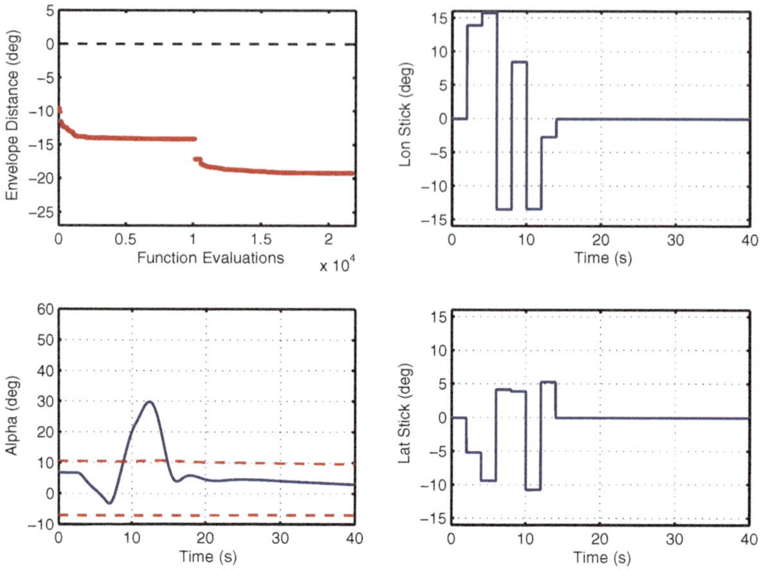

Fig. 16.9 GA: population size = 50, number of elite = 5.

(as earlier steps maybe more decisive for the result than later steps) or the fact that the worst-case seems to lie close to the search space boundary (or even close to its *corners*, as is clearly seen in Fig. 16.4 and Fig. 16.5 and which might be expected also from a physical point of view).

16.3.4 Dividing Rectangles

The result of running the dividing rectangles method (DIRECT) for 26000 function evaluations is shown in Fig. 16.10. If one compares this with the solution obtained already after 2555 evaluations, as shown in Fig. 16.11, it is clear that the only thing the algorithm accomplished during the last 23000 evaluations was to move some of the steps closer to the extreme values ± 16. This just confirms the well known fact that DIRECT has great difficulties finding optima located on the search space boundary.

But a closer look reveals that at least some steps of the input sequences in Fig. 16.11 are in fact not very far from the values of the (presumably) "global optimum" in Fig. 16.4. An obvious idea is then to try a local method and see if it can move the differing steps into place. We did run pattern search with different parameter settings, but without success. Some of the steps that differed would have to be moved across almost the whole feasible interval, something that our implementation of pattern search did not manage to do. It would be interesting to devise a method that *could* refine, in an effective

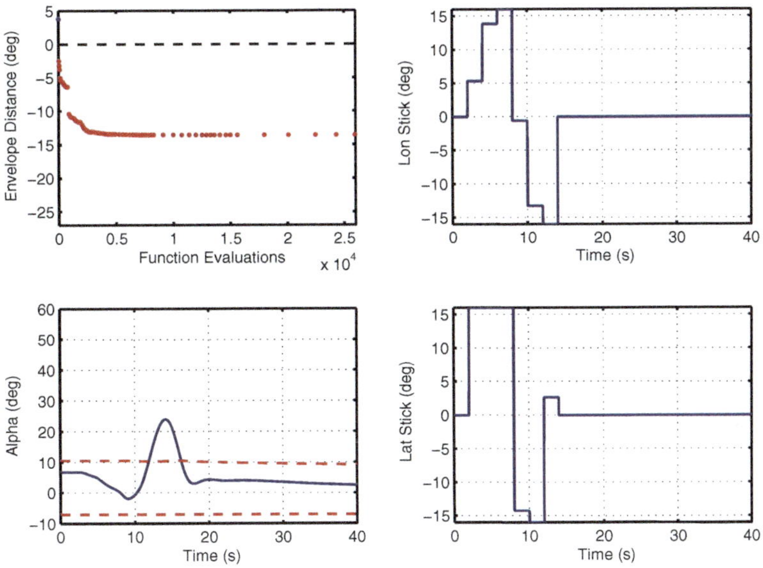

Fig. 16.10 DIRECT, the result after 26000 function evaluations.

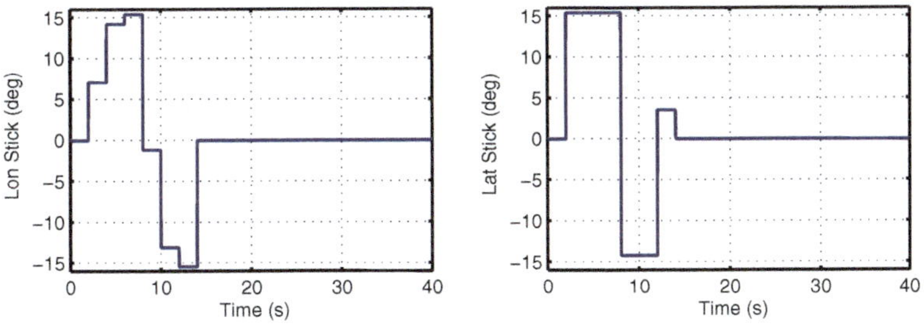

Fig. 16.11 The stick inputs found by DIRECT after 2555 functions evaluations

manner, a coarse search made by DIRECT. (Of course, with hindsight, one expects that the most effective method for the present problem would be to search for an approximate solution only in the corners of the search space boundary.)

16.4 Conclusions

Regarding the local methods, it is difficult to give a definitive answer to the question which of the applied local search methods (PS or QN with CCD) is better suited to address the solution of the worst-case pilot input determination problem. The main difficulties of solving this problem are the relatively high costs of function evaluations and noisy nature of optimization problem. Both these aspects result from the need to perform numerical simulations in each function evaluation. Taking advantage of parallel computations, could substantially alleviate the computational costs for both QN and PS.

The PS method appears to be more robust on the considered problem than the QN method. Recall that for a shrink factor of 0.5, PS performs on average approximately as good as, or even better than QN. However, just as for CCD, there is no proof that the termination point of PS is a sufficient good approximation of a local optimum on the noisy test problem.

The QN method tends to be more efficient in producing good approximations of the worst-case by using gradient information. Unfortunately, noisy function evaluations strongly limits the good guaranteed final convergence. A major cost with QN on the considered problem is the line search. The ratio of the number of objective function evaluations with line search and with gradient estimation will probably get smaller for higher dimensions.

Regarding the global methods, it is hard to declare any of the evolutionary algorithms to be better than the others for the present problem. DE converged more quickly in a couple of cases, but may be this was just by chance. DE is in any case the easiest method to use, having only two option parameters to decide apart from the population size. ES has more parameters and its performance seems to depend quite a bit on their values. It is doubtful whether the self-adaptive mechanism of ES really make any difference in the present problem. Perhaps the objective function is too "awkward" (the fitness landscape may be very irregular indeed). GA in the real-valued representation should be investigated further, especially in regard to the possibility of using variation operators that better explore the search space boundary.

References

1. Puyou, G.: Models delivery: Part 1 - Nonlinear model. COFCLUO Technical Report D1.1.2, Airbus France (2007)
2. Skoogh, D., Berefelt, F.: Final report WP2.5. COFCLUO Technical Report D2.5.5, FOI (January 2010)
3. Skoogh, D.: Final software - Part 2: Local methods for search of worst case manoeuvre, Version 2.0. COFCLUO Technical Report D2.5.4, FOI (January 2010)
4. Mathworks. Distributed Computing Toolbox, Version 3, User's Guide (2007)
5. Gill, P.E., Murray, W., Saunders, A.: User's Guide for SNOPT Version 6, A Fortran Package for Large-Scale Nonlinear Programming (2002)
6. Mathworks. Optimization Toolbox, Version 3, User's Guide (2007)

7. Holmström, K., Göran, A.O., Edvall, M.M.: User's Guide for TOMLAB (2006)
8. Menon, P.P., Bates, D.G., Postlethwaite, I.: Optimisation-based flight control law clearance. In: Bates, D., Hagström, M. (eds.) Nonlinear Analysis and Synthesis Techniques for Aircraft Control. LNCIS, vol. 365, pp. 259–300. Springer, Berlin (2007)
9. Puyou, G.: Selected clearance problems. Part 1 - Nonlinear model. COFCLUO Technical Report D1.1.1, Airbus France (2007)
10. Berefelt, F.: Final software - Part 1: COFOPT, a worst case maneuver search package version 2.1. COFCLUO Technical Report D2.5.4, FOI (January 2010)

Part V
Industrial Evaluation

Chapter 17
Evaluation of Clearance Techniques in an Industrial Context

Guilhem Puyou, Rafael Fernandes de Oliveira, and Adrien Berard

Abstract. This chapter focuses on the application of the methods developed in previous chapters to a common benchmark to compare them (when relevant) regarding their performance and evaluate the easiness of their applicability. The most promising techniques have been further tested in a real industrial context, with more complex requirements to fulfill and using a certified aircraft model as reference. Based on this application an integrated validation strategy is proposed to comply with the industrial validation needs.

17.1 Overview of Basic Methods

In this section we review the clearance methods based on the information provided in previous chapters and analyse them from an user/engineering point of view. For each clearance technique, we point out the main advantages, summarize the best practices as suggested by the authors and briefly explain the level of complexity involved in their use.

17.1.1 Enhanced μ-Analysis Techniques

Criteria and Results

The main appeal of the methods described in Chapter 7, based on the μ-upper bound computation, is their ability to provide guaranteed robust stability margins. In the COFCLUO project, these methods have been used to assess the eigenvalue criteria for the "Nonlinear" and "Integral" benchmarks and the stability margin criteria for the "Integral" benchmark. For the eigenvalue stability analysis, the default practice assumes that eigenvalues should stay

Guilhem Puyou · Rafael Fernandes de Oliveira · Adrien Berard
AIRBUS, Stability and Control Department, Toulouse, France
e-mail: {guilhem.puyou,adrien.berard}@airbus.com, rafael@rafael.aero

A. Varga et al. (Eds.): Opti. Based Clearance of Flight Control Laws, LNCIS 416, pp. 319–357.
springerlink.com

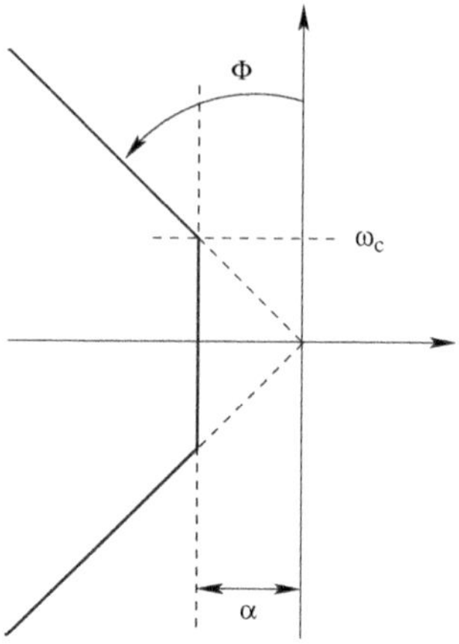

Fig. 17.1 Sector definition

in the left half-plane of the complex space, but the users can define their own sector-based stability domain as in Fig. 17.1, characterized by the following parameters: degree of stability α, minimum damping $\sin\Phi$ and maximum natural frequency ω_c. To define the stability margin requirements, the user should choose values for the maximum phase margin, or the minimum and maximum gain offsets for the gain margin.

Let $c(\delta)$ be a clearance criterion expressing a certain stability requirement. This function depends on the parameter vector $\delta \in \Theta$, where Θ denotes the admissible set of parameter values. A parameter combination δ is *cleared* if $c(\delta) \in C$, where C is the set of admissible values of c. Thus, the set

$$\mathcal{D} = \{\delta \in \Theta \mid c(\delta) \in C\} \tag{17.1}$$

defines the *clearable domain* of parameter values.

The μ-analysis based method provides an estimate $\widetilde{\mathcal{D}}$ of the domain $\mathcal{D}$, such that $\widetilde{\mathcal{D}} \subset \mathcal{D}$. Typical approximations of $\mathcal{D}$ rely on fine gridding of the parameter domain and determine a union of cleared subdomains denoted by $\mathcal{D}^\star$. A precision indicator can be computed as the ratio of the domain $\widetilde{\mathcal{D}}$ hypercube volume and the volume of the approximated domain $\mathcal{D}^\star$ (usually consisting of many small hypercubes). Obviously, this kind of information is useful in a preliminary phase when setting up the methods and calibrating them, to have confidence (or not) in the results. Its use in an extensive way is not planned.

Parameters Tuning and Ease of Use

The user can tune several input parameters of the analysis function as, for example, the sensitivities and tolerances to compute the μ upper bound, the employed algorithm (LMI-based or not), etc. For all these parameters, carefully chosen default values can be used, so that, for practical use, the only parameter to be tuned by the user is the minimum allowed length of the investigated hypercube edge in the Θ domain. By relying on the default settings, the user has therefore only to choose the criterion to be analyzed, define the uncertain parameters δ and the corresponding search the domain Θ through the LFR object definition, and set the minimum length for parametric search space subdivision. Therefore it is rather simple to use the μ-analysis for stability clearance purposes, even without an expert knowledge of this technique.

17.1.2 Convex Relaxations

Criteria and Results

The Lyapunov-based methods described in Chapter 9 are able to deliver a guaranteed robustness margin (e.g., for the left half complex plane condition or shifted imaginary axis condition, both involving to check that each model eigenvalue λ satisfies $\Re(\lambda) \leq \alpha$, $\alpha \in \mathbb{R}$) for either LTI or LTV models. In the present case both have been studied to check aeroelastic stability for the "Integral" benchmark and the un-piloted stability criterion in the presence of saturations for the "Nonlinear" benchmark.

In all cases, the method provides an estimate $\widetilde{\mathcal{D}}$ of the clearable domain $\mathcal{D}$ defined in (17.1), such that $\widetilde{\mathcal{D}} \subset \mathcal{D}$. The same kind of precision indicator as the one proposed for the enhanced μ-analysis techniques has been set up.

Parameters Tuning and Ease of Use

A GUI has been developed to interface with the available stability analysis methods. In addition to the model to be analysed, the user can choose among several relaxation methods (FD, DS, WB), the kind of multipliers (parameter dependent, constant full, constant diagonal), the parametric region to be investigated (default or user defined), the type of tilling approach (progressive or adaptive), the structure of the Lyapunov function (common, affine parameter-dependent, multi-affine parameter dependent) and the number of domain partitions (`NbPart`). As the degrees of freedom available are high, a compromise should be found between precision (increased either by more complex dependencies or higher number of partitions) and computational time. Although no clear best strategy has been highlighted in the application chapter, a default strategy has been proposed (progressive tiling, FD method, simple multipliers (full), affine parameter-dependent Lyapunov function), so that the only remaining decision is the choice of the model (LFR object) that includes the parameter range definition and the number of partitions. It can be noticed that, compared to the enhanced μ-analysis technique, `NbPart` corresponds to the number of domain partitions that are authorized, so that the equivalent minimum subdivision length is equal to $1/2^{NbPart}$.

17.1.3 IQC-Based Analysis Techniques

Criteria and Results

The IQC-based methods described in Chapter 10 can be used to check robust stability, stability margin or robust $\mathcal{H}_2$ performance (i.e., a guaranteed level of $\mathcal{H}_2$-norm of a given transfer function on the whole parametric domain). Within the project, they have been used to check un-piloted stability and eigenvalue criteria, as well as the aeroelastic comfort ($\mathcal{H}_2$) criterion for the "Integral" benchmark. The results are quite different depending on the analysis performed. For stability criteria, the method provides an approximation of the clearable domain $\mathcal{D}$ defined in (17.1). For the comfort criterion, the method provides a worst-case value $c_{max} = \max_{\delta \in \Theta} c(\delta)$ and worst-case argument $\delta_{max} = \arg\max_{\delta \in \Theta} c(\delta)$ as well.

Parameters Tuning and Ease of Use

The software is based on a set of scripts to deal with stability issues. In the same way as for μ-analysis and Lyapunov-based methods, inputs are the LFR and the number of subdivisions allowed for the partition strategy.

For the comfort criterion, it is possible to choose the comfort channel (feet, seat, back, all) and the set of points of interest along the fuselage. The number of partitions of the uncertainty region is also available. Since the comfort analysis is based on the IQC theory, additional degrees of freedom can be used in order to reduce the conservatism of the results, namely, by the choice of position of poles of the dynamic multiplier (10, by default) and its degree (2, by default).

17.1.4 Worst-Case Search-Based Methods

Criteria and Results

The worst-case search-based methods described in Chapters 8 and 11 apply nonlinear programming techniques to minimize/maximize suitably defined clearance criteria to determine worst-case combinations of uncertain parameters. These methods can be in principle applied to all formulated clearance criteria in the clearance benchmark. However, the applications of these methods primarily focussed on time domain criteria involving expensive simulation runs, as manoeuvrability, flight domain protection, or piloted stability criteria in Chapter 13 and on the determination of worst-case pilot inputs in Chapter 16. Each clearance task has been formulated as a minimization problem of a mathematical function $c(\delta)$ depending on uncertain parameters δ over a set of admissible values Θ. The results are the worst-case value found and its associated input parameters

$$c_{min} = \min_{\delta \in \Theta} c(\delta), \qquad \delta_{min} = \arg\min_{\delta \in \Theta} c(\delta)$$

The main appeal of the worst-case search-based methods is the lack of conservativeness via an exhaustive search of the admissible parameter domain Θ. Also, the computed worst-case parameter vector δ_{min} provides useful information to further improvements of control laws. Moreover, provided δ_{min} is a global minimum, then the clearance is achieved if $c_{min} \in C$ (the set of admissible values of c).

Of course, the difficulty to find the global optimum will increase with the complexity of the c function and therefore the choice of an adequate optimisation method will greatly depend on the problem itself. The authors of Chapters 13 and 16 have therefore decided to include several optimisation algorithms in their software and test them on the benchmark with both global (stochastic) and local (gradient-based and gradient-free) methods.

Parameters Tuning and Ease of Use

For the analysis in Chapter 13, the first step is to define a model to serve for the evaluation of the objective function c. This is achieved in the current application by using an augmented aircraft simulation model based on the provided Simulink model. Then optimisation parameters δ can be specified through a dedicated GUI, where the user can activate the parameters for the worst-case search, and define nominal values and simple bounds for them. Additional flexibility is available to experienced users, by allowing to add their own parameters to the list. The setup of the optimization problem for the worst-case pilot determination problem in Chapter 16 relies on special parameterizations of the pilot inputs.

Several optimisation methods and several parameters for each optimisation method are available. According to the tests performed in Chapters 13 and 16, it appears that the evolutionary strategy (ES) is one of the best methods or at least the one which statistically provides the highest probability to get the global optimum for the validation problem. A default tuning of internal algorithm parameters is provided to maximize algorithm efficiency for the current benchmark. The two main parameters are the initial population size and the tolerance to stop the algorithm. The population size should be proportional to twice/three times the number of parameters to be tuned, but this ratio can be increased if run-time is not a constraint, as this will directly influence the ability to find worst cases. Genetic algorithm (GA2) developped in Chapter 11 seems also to be efficient and will be selected for application.

17.2 Methods, Criteria and Models

17.2.1 Techniques Mapping

Fig. 17.2 and Fig. 17.3 show the class of techniques developed and their use on the two benchmarks: the rigid nonlinear benchmark (`NLin`) and the integral benchmark (`Lin`). Two classes of techniques have been employed. The first class is called *LFR-based techniques* and uses simplified closed-loop aircraft models (known as LFT models or LFRs) to perform the robustness analysis. The second class is called *worst-case search-based techniques* and applies nonlinear programming methods to minimise suitably defined clearance criteria to determine worst-case parameter combinations. The underlying models should serve for criteria evaluation purposes and thus can be general purpose nonlinear simulation models or any kind of linear model approximations (including LPV models).

The results delivered by the two method classes have a complementary character. The LFR-based methods are able to deliver a guaranteed stability (and sometimes performance) region, whose estimation can be slightly conservative (but nevertheless guaranteed) depending on the used method and its tuning. These are called "sufficient techniques" in Chapter 18. The potential difficulties with these methods is the lack of full coverage of the admissible parameter domain and the hight costs involved by the generation and validation of low order approximations as LFRs. Moreover, these methods are usually not suited to produce worst-case parameter combinations, but only bounds where the worst-case is supposed to lie.

On the other hand, the worst-case search-based techniques, called "necessary techniques" in Chapter 18, explicitly determine worst-case parameter combinations together with the worst-case criteria values. Thus, if the employed optimization method provides also certificates for the resulting global optima, these results can be used to assess the clearance. Moreover, a single violation of clearance requirements allows to completely deny the clearance. Besides the full coverage of the admissible parameter domain, the computed worst-case parameter vector often provides useful information to further improvements of control laws. However, in general, there is no guarantee that the worst case found by a particular method is a global optimum, and obtaining certificates for global optimimum can involve significant additional efforts. Moreover, even if a worst-case value violates the clearance requirements, the range of values where such violations occur can not be directly computed.

17.2.2 LFR Review

While worst-case search-based techniques can in principle use any kind of input/output model (such as a black box), this is not the case for the methods which use an LFR object to represent the model with uncertainties to be

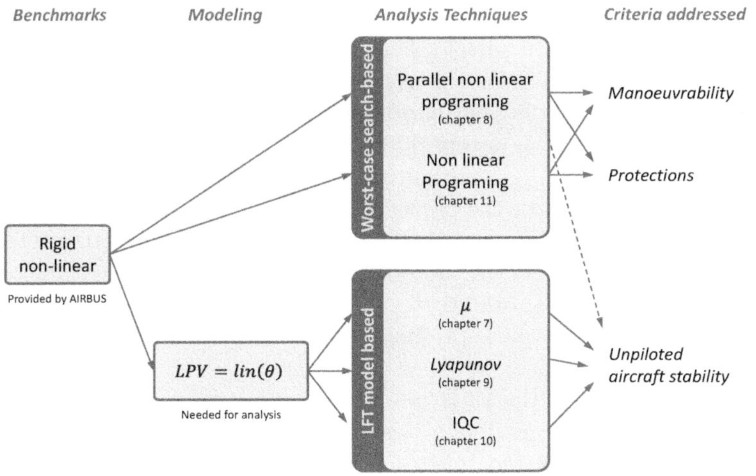

Fig. 17.2 Analysis applied to the rigid nonlinear benchmark clearance

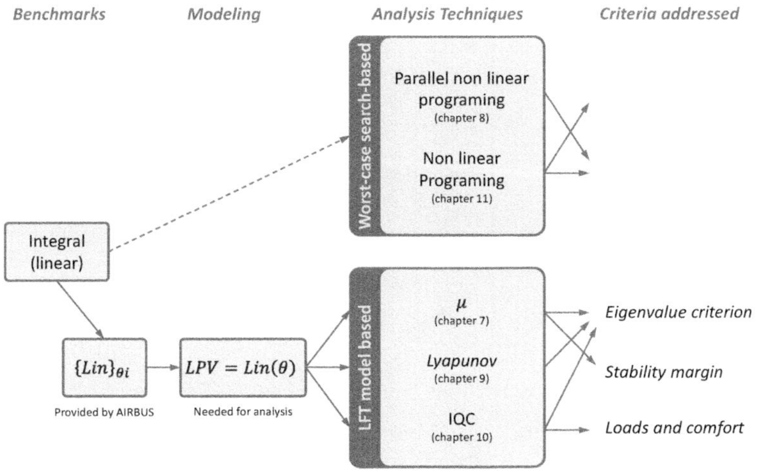

Fig. 17.3 Analysis applied to the integral benchmark clearance

analysed. Hence, for LFR-based techniques, a modeling step is mandatory before the analysis can start. The final result and its relevance to the original problem will then greatly depend on the quality of the model itself (see further details in section 17.3.5). Therefore we discuss here the representativeness of the generated LFR with respect to the original nonlinear model.

Nonlinear Benchmark

To cover the whole flight envelope and the whole range of parameter variations, the flight domain for the rigid nonlinear benchmark has been partitioned into mutually disjunct subdomains on which accurate LPV-models have been generated with acceptable complexity. We restrict our study to a single LFR covering a single subdomain of the flight envelope. Since stability problems usually occur in the extreme Mach and speed conditions, we have selected the LPV-model number 16. In order to make it easier to interpret the results, we also restricted the analysis to only two varying parameters: the Mach number M and calibrated speed V_{cas}. The mass m and center of gravity position X_{cg} have been kept constant at their nominal values: $m = 217t$ and $X_{cg} = 29\%$.

A first check of how the LFR model approximate the nonlinear has been performed using time domain analysis. For this, the effects of small pilot input movements on the stick (δq) on some system outputs (q and θ) have been compared for the original nonlinear model response, the linearized model response and the selected LFR. The time responses shown in Fig. 17.4 are representative of the accuracy of the model in the whole domain. The presence of errors in the static gains reflects the difficulty of generating an LFR to be representative of all the side-stick deflections, due to the nonlinearities that exists in the control laws and that depends on the stick input value itself.

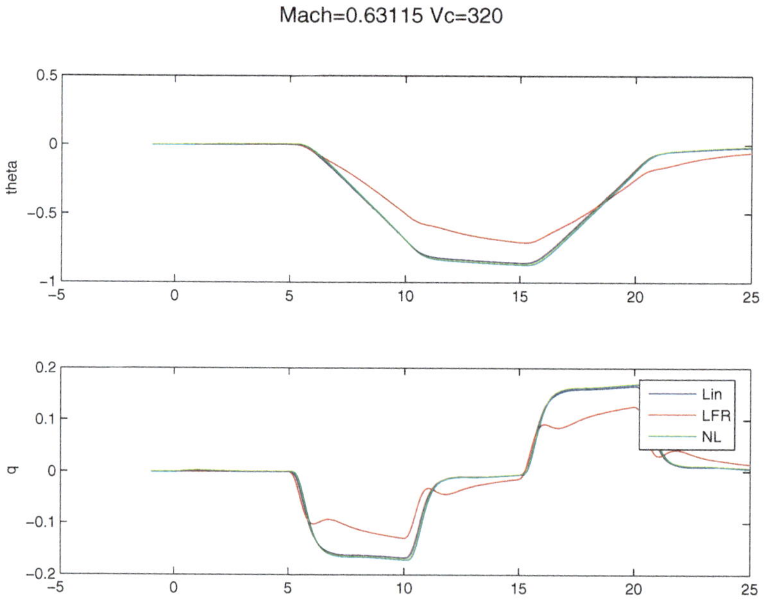

Fig. 17.4 Time domain analysis of the LFRs for longitudinal stick input

The time domain comparisons have been supplemented with analysis which is closely connected to the clearance problem at hand. In our case, the LFR is used exclusively to check the closed-loop system stability over the flight domain. Therefore, the initial goal was to evaluate the errors in the eigenvalues characterizing different handling qualities modes (e.g., short period, phugoid, Dutch roll, etc.) However, since this kind of analysis is really difficult to implement, we refrained only to map the degree of stability of the closed-loop models, defined as the maximum real part of the closed-loop system eigenvalues. In Fig. 17.5 we present the contour plots for the degree of stability for linear model resulted from linearizations of the original nonlinear model and the generated LPV model.

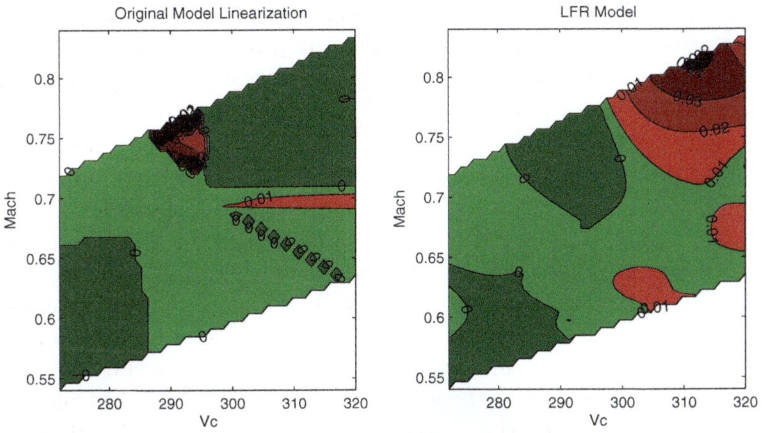

Fig. 17.5 Degrees of stability of LFR vs. linearized models

A difference can be seen in the strict stability domain frontiers (degree of stability equal to zero), but if we observe the results more precisely, we can see that the differences in the stability degree values are small, with a maximum value of 0.05.

Integral Benchmark

The LFR for the integral flexible aircraft benchmark has been generated using a set of linearized state-space models provided for a number of mass/flight point configurations shown in Table 17.1. The approach used in Chapter 4 to generate the LFR of the open-loop model ensures that the resulting LFR is roughly exact on the set of original linear models. To analyse the resulting LFR, accuracy tests have been performed using extra values of V_{cas}. The comparisons presented in this paragraph concern the aircraft model with no *payload* (PL), 50% of *outer tank* (OT) and 50% of *central tank* (CT) for a Mach number of $M = 0.91$. For the generation of the LFR, three speed cases

Table 17.1 Linear stability analysis grid

Variable	Values
V_{cas}	V_1, V_2, V_3
Mach	0.7, 0.86, 0.91
PL	0%, 100%
OT	0%, 50%, 100%
CT	0%, 50%, 100%

were provided ($V_1 = 275$ kts, $V_2 = 320$ kts and $V_3 = 365$ kts). For the sake of comparison, additional flight points are considered at every 15 kts.

Comparisons of eigenvalues and frequency domain analysis results have been performed for the longitudinal open loop system only. Fig. 17.6 and Fig. 17.7 show a comparison between the frequency analysis using LFR and the original model.

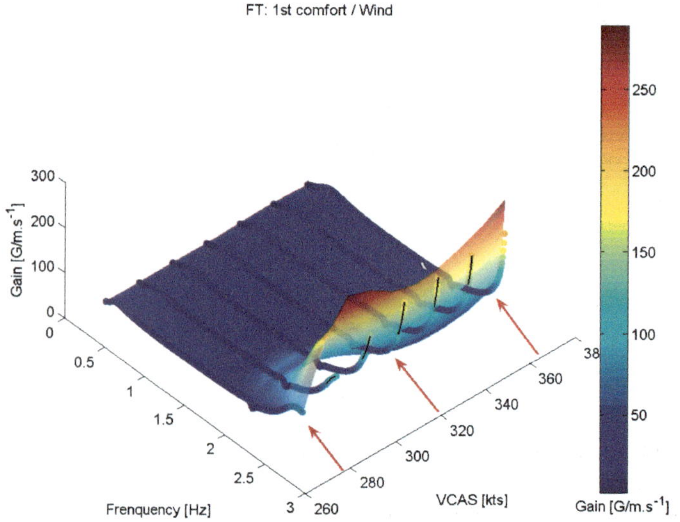

Fig. 17.6 Comparison of the frequency response gains between the generated LFR and original linear models for frequencies up to 2.4 Hz.

Fig. 17.6 shows the variation with the calibrated speed V_{cas} of the frequency response gain of the open-loop transfer function between the up-wind input and the vertical load factor n_z output (at the front of the cabin) for frequencies up to 2.4Hz. This transfer function underlies the definition of the comfort criterion (see Chapter 2). The coloured semi-transparent surface is the gain of the transfer function of the generated LFR, while the coloured

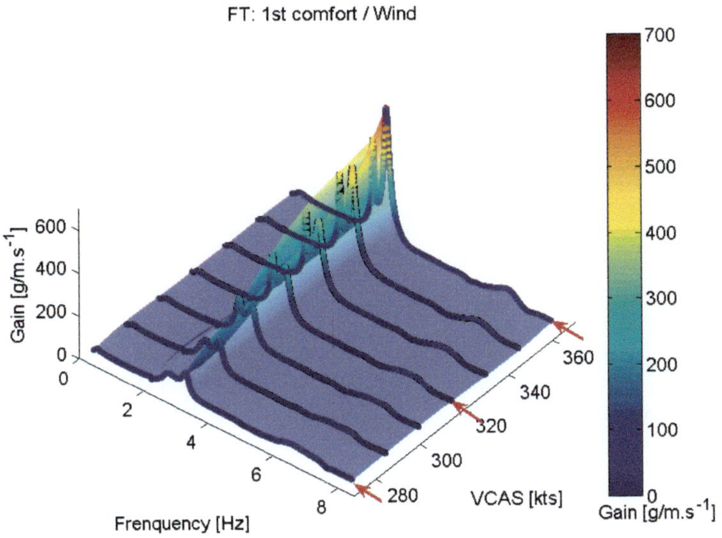

Fig. 17.7 Comparison of the frequency response gains between the generated LFR and original linear models for frequencies up to 8 Hz.

lines are the gain of the transfer function obtained from the linearization of the original model. The red arrows highlight the speeds whose corresponding model was used for the creation of the LFR (V_1=275 kts, V_2=320 kts and V_3=365 kts). As expected, the matching between the LFR and the original model frequency gains is very good at the speeds V_1, V_2 and V_3. The matching is also good in intermediary points between V_2 and V_3. But the LFR seems to be over-evaluating gains for speeds between V_1 and V_2.

Fig. 17.7 shows a comparison of the gain of transfer function between up-wind and the vertical load factor n_z up to a frequency of 8 Hz. As previously observed, the matching is very good for the speeds used for the generation of the LFR (red arrows). Some discrepancies appear for frequencies around 2.5-3 Hz (as already illustrated in the previous figure), but the overall matching is good. The discrepancies found are due to errors in the LFR mode tracking (for speed variations).

Fig. 17.8 illustrates mode tracking when V_{cas} varies. Most of the modes are well approximated by the LFR, but some errors appear in the mode tracking (circled in blue and red). It appears in the area circled in blue that the LFR swapped two modes, as it fails to capture the folding of one of them. This swapping results in some discrepancies between the frequency responses of the models, but these are negligible since the two affected modes make only a small contribution to the response of the aircraft to wind. The LFR mode tracking errors are due to the fact that models for only three speeds were used to generate the LFR.

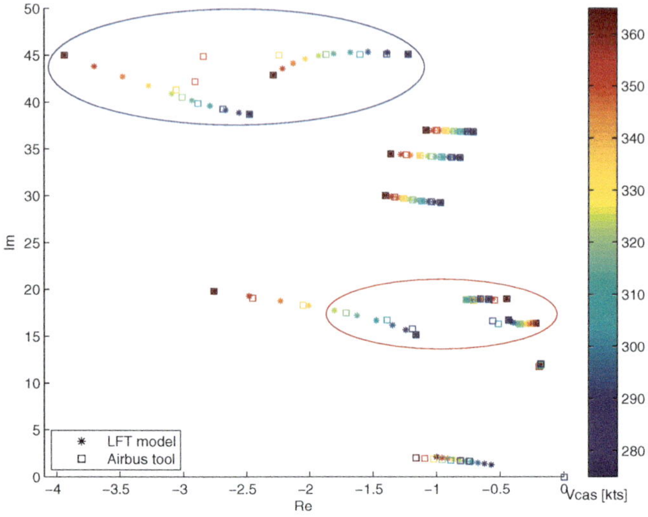

Fig. 17.8 Speed dependency of the real and imaginary parts of the eigenvalues of the state matrix of LFR and original linear model

17.3 Straightforward Applications of LFR Based Methods

All LFR-based methods described in Sect. 17.1 have been used to solve the rigid benchmark stability clearance problem for the LFR number 16. The results have been compared with those obtained by a baseline solution which corresponds to the current industrial practice. The objective of our study was to highlight the improvements provided by new methods over the baseline solution and underline potential drawbacks.

17.3.1 Baseline Solution

In the baseline solution we addressed the linear stability problem over the whole flight domain using a gridding based approach. The analysis consists in linearizing the closed-loop nonlinear aircraft model on a discrete set of flight conditions and parameter values and analysing the closed-loop stability by comparing the maximum real part of the linearized closed-loop system eigenvalues with a given stability degree α.

17.3.2 Enhanced μ-Analysis Techniques

Analysis Run

We ran the stability analysis to delimit the stability region for different stability degree values of α. We tested both the classical stability degree of $\alpha = 0$ and the value $\alpha = \log 2/6$ resulting from the un-piloted aircraft stability requirement taking into account the admissible doubling time of 6s. For both values, we tested two minimum subdivision length (`led`) values: `led` $= 0.1$ and `led` $= 0.01$, for which the analysis results are presented in Fig. 17.9 and Fig. 17.10, respectively. The green squares represent the guaranteed stability domains determined by the analysis function. The results of the gridding approach employed in the baseline solution were also plotted: the red crosses represent unstable points found, and green dots are the stable ones. Gray areas are out of the LFR validity domain. It can be seen in Fig. 17.9 and Fig. 17.10 that the most complex stability domain shape is obtained for α =0. In this case (`led`=0.1), analysis is only able to find a stability domain that corresponds to 58% of the real one. There is therefore a clear interest in using a smaller subdivision length. For `led` $= 0.01$, the method is able to estimate the stability domain within a 5% error, at the cost of an increased computational time.

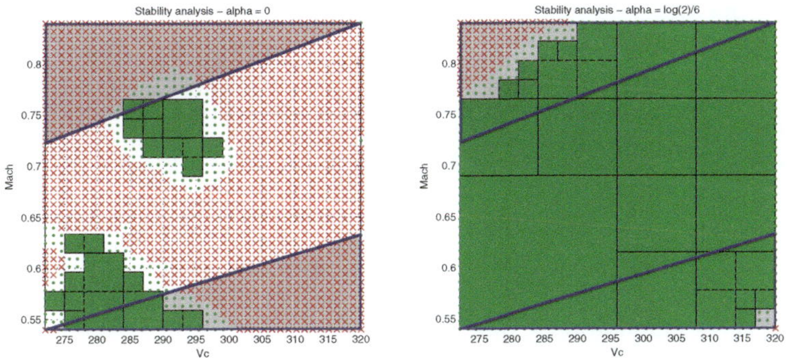

Fig. 17.9 Stability analysis for two α values and `led` $= 0.1$

A sensitivity study was carried out for $\alpha = 0$ to highlight the trade-off that needs to be done between computational time and precision (Fig. 17.11). For a given minimum subdivision length value, we plotted both the precision of the stability domain estimation (called precision indicator) and the time taken to perform this analysis. Of course, the more the subdivision length is reduced, the closer we are to the true stability domain but, unfortunately, longer computational time is necessary. It is difficult to extract default values from this example because the results greatly depend on the shape of the stability domain that we are trying to estimate. If we go back to the first

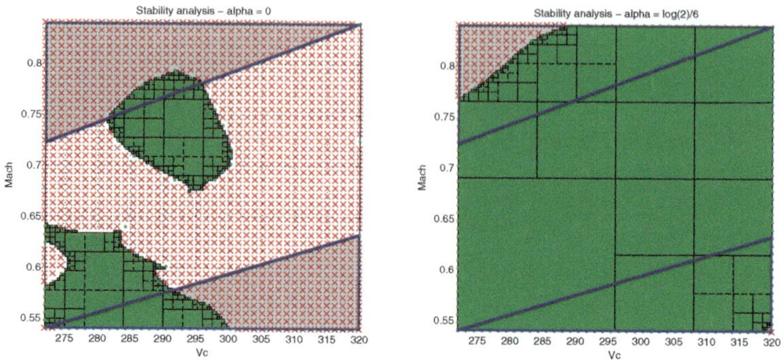

Fig. 17.10 Stability analysis for two α values and led $= 0.01$

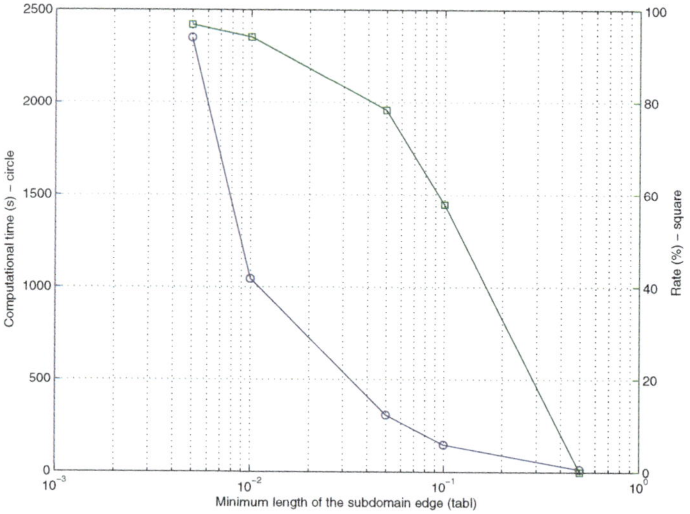

Fig. 17.11 Computational time and achieved coverage of stability domain for $\alpha = 0$.

case (led $= 0.1$), although the estimation error was 42% (precision indicator $= 58\%$) for $\alpha = 0$, it was only 1% for the $\alpha = \log 2/6$ case. It is therefore recommended to start with a middle value (led $= 0.1$ is a good starting point) and then increase precision (i.e., decrease led parameter value) if the accuracy of the result is not sufficient.

Comparison with the Baseline Solution

In the baseline solution, the employed grid is really coarse, so an unstable region can easily be missed (see the case for $\alpha = 0$). In this sense, the μ-based technique seems more powerful. On the other hand, this kind of analysis is more time consuming, so it can be argued that refining the grid would have delivered the same results by identifying the instability regions better. It is interesting to note that, in this case, the the μ-analysis based approach and the baseline solution with a fine gridding (with a step length corresponding to the minimum subdivision length `led`) are equivalent in terms of computational burden. Still, the μ-analysis results are more useful, since they guarantee stability on compact regions instead of proving stability in discrete points.

Conclusion

This method provides a real improvement in terms of computational time (in comparison with existing algorithms). It can handle high order LFR objects (i.e., with high number of states and high order parametric blocks). Moreover, the sub-domain division strategy enables the user to determine complicated stability domain shapes (instead of square domains used in the basic version of μ-analysis). Nevertheless, a trade-off between the resulted accuracy and computational time is necessary. However, the method usually remains fast for a less than 5% error result.

An enhancement provided by the method is to present a stability margin analysis, which is still widely used in the industrial context. Of course results should be treated with care, as their quality depends on LFR accuracy.

17.3.3 Convex Relaxations

Analysis Run

Using the same example as in the previous section, we selected two stability degree values ($\alpha = 0$ and $\alpha = \log 2/6$) and set `NbPart=4`, which is equivalent in the previous μ-analysis method to `led`=0.1. Results are shown in Fig. 17.12, where green squares are guaranteed stability regions, and red squares are those for which at least one unstable eigenvalue has been found. It should be noted that this does not necessarily mean that the entire red square is unstable.

Here is a summary of the achieved results for $\alpha = \log 2/6$ and `NbPart=4`. There were 49 convex optimization problems solved, which totally required 12.1 hours. Thus, about 15 minutes were necessary to solve each optimization problem. In comparison, the time required for the gridding approach was significantly shorter, being only 7 seconds. A positive feature of this method is that the analysis covered the whole parameter region, of which 93.75% is cleared and 6.25% is not cleared. The (red marked) not cleared squares, contain at least one point where the clearance requirement is not fulfilled.

For this analysis, the computational time is quite significant. We measured the computation times for two precision settings: `NbPart=4` and `NbPart=6`. As

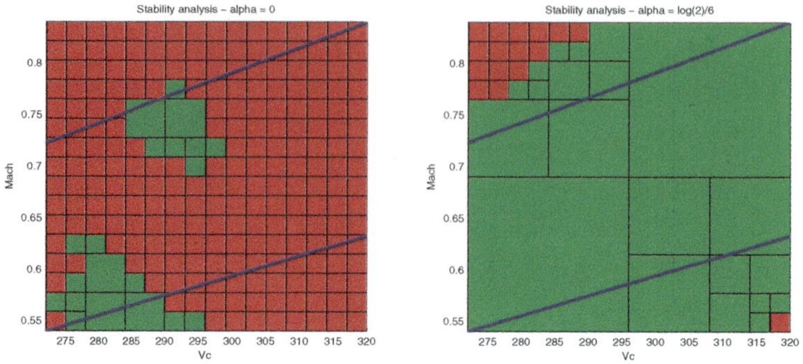

Fig. 17.12 Stability analysis for two α values and `NbPart=4`

Table 17.2 Computational times (hours)

Stability degree	$\alpha = 0$	$\alpha = \log 2/6$
`NbPart=4 (led=0.1)`	16	12.1
`NbPart=6 (led=0.01)`	–	36.4

it clearly appears from Table 17.2, that desired precision should be carefully chosen to avoid excessive computational times.

Comparison with the Baseline Solution

The method produce solid results, with a 100% coverage of the parameter region. This strongly contrasts with the poor coverage of the baseline solution in only discrete points. The price for this higher reliability is a tremendous computational effort, which is orders of magnitudes higher than for the baseline solution. The improvements in reliability of the clearance results involve therefore high costs and need to be carefully pondered.

Conclusion

The default strategy provided by the method can give good accuracy. Since it can provide guaranteed stability regions whose shape is more complex than an hypercube, the tiling strategy makes the result more powerful. Nevertheless, although an effort to make the analysis as quick as possible (and not too conservative) has been provided, the computational burden is still high. We do not report here the different tests we performed on the method and multiplier choices, but nevertheless they did not seem to reduce the required time while maintaining accuracy. Since the LFR object size has a huge impact on computational time, one way of improvement could be to include for the tightest region a LFR reduction step to take advantage of the reduced parametric variation intervals. Although some direct LFR reduction methods can be really fast, this added step is not obviously instantaneous and a compromise should be found at the end.

17.3.4 IQC-Based Analysis Techniques

Analysis Run

To benchmark the method and make a fair comparison with previous techniques, we implemented the stability analysis restricted to the (Mach,V_{cas}) domain with α =0. The results are plotted in Fig. 17.13. We apply the same colour coding as for the μ-analysis technique (see subsection 17.3.2). The computational time of 8500 seconds (i.e., about 2.4 hours) is still acceptable, but the accuracy of the results is poor. Another test was therefore performed on a relaxed stability requirement ($\alpha = \log 2/6$) which provided slightly better results, but with an increased amount of time (about 5.5 hours).

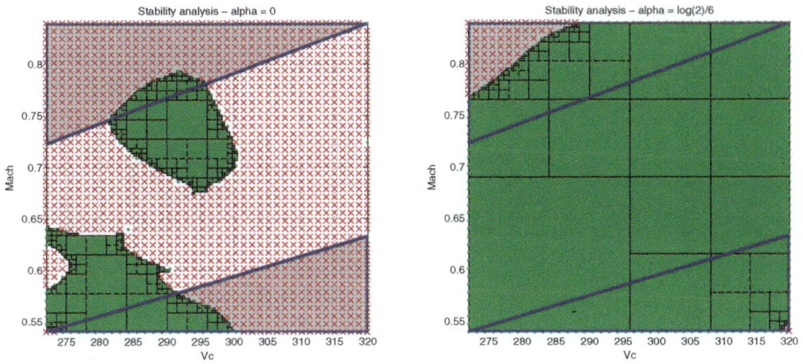

Fig. 17.13 Stability analysis on the (Mach,V_{cas}) domain

Conclusion

In our opinion, these new IQC-based analysis methods are not yet sufficiently mature and should be investigated further. First, computational time is still high, and there is room for a more efficient numerical approach (see the μ-analysis example). Moreover, IQC-based techniques should be carefully considered in solving problems which are more complex than the classical trim point linear stability analysis.

17.3.5 Discussion

All three LFR-based methods are able to solve the stability clearance problem and determine the guaranteed stability domain with a good accuracy, but the methods strongly differ in terms of required computational effort. On the same problem and for the same accuracy requirement (i.e., minimum subdivision length), the computational time varies from 440s to solve the problem with the μ-analysis techniques, 8506s with the IQC-based method and 57622s with the Lyapunov-based approach. Although the use of Lyapunov- and IQC-based techniques may not be optimally adapted to this specific problem (i.e., configuration of parametric dependencies), these methods seem to be not well-suited for a wide use. The μ-analysis technique seems to be the most promising for an immediate integration in an industrial clearance process. For the sake of comparison, we performed the stability a analysis over the full parameter range of V_{cas}, M, m and X_{cg} using both the enhanced μ-analysis and the gridding-based approach (for the baseline solution) on a fine grid. For a minimum subdivision length of 0.0625 it took about 32 hours to analyse the LFR model 16 with the μ-analysis technique. Within the same time it was possible to perform the analysis of $68^4 \approx 21.4$ millions individual parameter points, so that the achieved grid minimum subdivision length corresponds to $1/68 \approx 0.0147$. In this sense, the precision of results is four times higher for the gridding approach. On the other hand, the μ-analysis technique provides more useful results by guaranteeing stability of whole continuous regions rather than the stability of a finite set of points for a gridding-based approach. Therefore, for the baseline solution the stability in intermediary points is only assumed, but without any guarantees. However, for practical use we can therefore consider that both methods have similar performance.

Independently of performance results, these methods are still based on the use of a suitable LFR, which provides an approximation of the nonlinear closed-loop system. Therefore, from the clearance point of view, both the LFR generation and the LFR-based analysis should be considered as part of the clearance process. In Fig. 17.14 we present a comparison of the clearance processes based on a gridding approach (used for the baseline solution) and on LFR-based methods. If we only focus on the clearance results for the determination of the stability domain, we already knew that LFR-based methods are more powerful, since they provide a continuous stability domain S_c, whereas the baseline solution provides only a finite set of points S_d. Then, theoretically the following inclusions must hold: $S_c \subset \mathcal{D}$ and $S_d \subset S_c$. However, there are still two sources of difficulties. The first one regards the conservatism of the analysis method, quantified via the parameter $\bar{\varepsilon}$, that may make $S_d \not\subset S_c$, the second on is the error in the LFR approximation ε, that may make $S_c \not\subset \mathcal{D}$. We saw that conservatism of the method is easy to monitor using the minimum subdivision length. However, the error in the LFR approximation error is an issue that has not been addressed by the partners and especially a way to include it in the analysis result is missing.

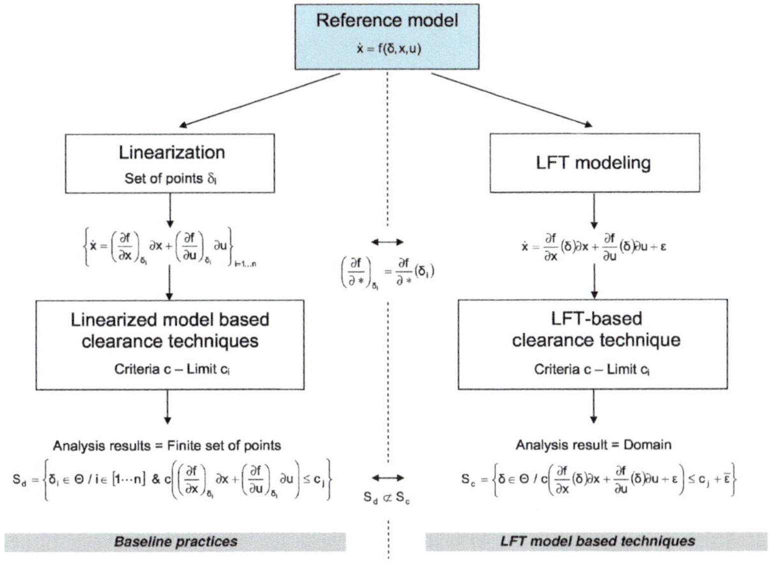

Fig. 17.14 Clearance process - Gridding-based vs. LFR-based approaches

In Chapter 12 the authors propose a way to address this problem, but it has not yet been applied to any concrete clearance problem.

The LFR generation is usually a time-consuming activity, so if we consider both the modeling and analysis as a whole, the LFR-based approach does not seem to be a ready to apply and competitive method compared to the gridding-based approach used in the baseline solution. This could however change in the future if the LFR generation process will be automated and if the LFRs are not only used for the clearance purposes, but also for control laws design.

17.4 Straightforward Applications of Worst-Case Search-Based Methods

LFR-based methods can be used when dealing with clearance problems which can be formulated as linear systems robustness analysis problems. A typical application is stability analysis around constant trim points. However, to handle more complex validation problems, there is a need for high fidelity non-linear models which are used for the evaluation of clearance criteria via simulations of the augmented aircraft model.

A typical example is the validation of aircraft flight domain protections, where it has to be assessed that with any combination of pilot inputs and initial flight conditions, the aircraft remains within the flight envelope boundaries. Such a problem can be addressed by the worst-case search-based methods, which will be studied in this section.

As benchmark model, we use the Simulink model of the augmented aircaft presented in Fig. 2.4 of Chapter 2. For the flight domain validation, two basic problems were selected for comparisons purposes. The first one is related to the aircraft performing a fixed manoeuvre and finding the worst flight conditions for it. The second problem is related to finding the (worst-case) pilot input that causes the most critical response of the aircraft. Both problems aim at validating the maximum angle of attack protection efficiency.

17.4.1 Baseline Solution

Part of the validation process currently used by the industry relies on gridding the input domain, or performing Monte-Carlo simulations in order to assess the behaviour of a system. The flight domain validation is usually performed using a gridding approach in conjunction with a fixed flight manoeuvre, while the determination of worst-case pilot input is handled using Monte-Carlo simulations. In what follows, we describe the baseline solutions for each of these problems.

Fixed Manoeuvre Analysis Using Gridding

The main use of the gridding approach is to prove that the aircraft is able to perform a set of basic manoeuvres without problems. The manoeuvre studied here is shown in Fig. 17.15 and is related to the maximum angle of attack protection, associated with the roll angle protection. It is currently performed during the validation phase of flight control laws, and its outputs are checked for no overshoot of the α_{max} and that the roll angle is limited to at most 45°. This manoeuvre is currently checked for a grid of 54 flight points defined by Table 17.3.

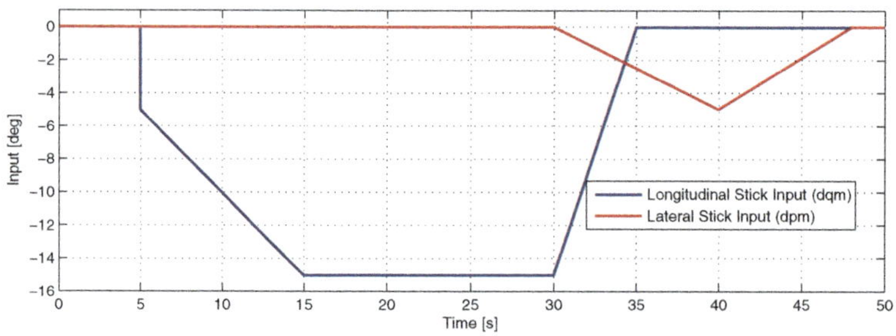

Fig. 17.15 Fixed manoeuvre for angle of attack protection

Table 17.3 Flight domain grid for fixed manoeuvre at altitude 35000 [ft]

Mass	CG Position	Mach						
$0.5 \cdot (MLW + MZFW)$	Forward	0.7	0.75	0.78	0.83	0.85	0.87	0.89
$0.5 \cdot (MLW + MZFW)$	Median	0.7	0.75	0.78	0.83	0.85	0.87	0.89
$0.5 \cdot (MLW + MZFW)$	Aft	0.7	0.75	0.78	0.83	0.85	0.87	0.89
$0.5 \cdot (MLW + MTOW)$	Forward		0.75	0.78	0.83	0.85	0.87	0.89
$0.5 \cdot (MLW + MTOW)$	Median		0.75	0.78	0.83	0.85	0.87	0.89
$0.5 \cdot (MLW + MTOW)$	Aft		0.75	0.78	0.83	0.85	0.87	0.89
$MTOW$	Forward			0.78	0.83	0.85	0.87	0.89
$MTOW$	Median			0.78	0.83	0.85	0.87	0.89
$MTOW$	Aft			0.78	0.83	0.85	0.87	0.89

Worst-Case Pilot Input Determination Using Monte-Carlo Method

To analyse the flight control laws protection in the whole operating domain, a Monte-Carlo method is currently used. Random pilot inputs are applied at random flight points, and protections of the flight envelope are checked. For the tests performed on the Simulink model, a simple parameterization of the pilot inputs was used, with two points in a fixed time step for both

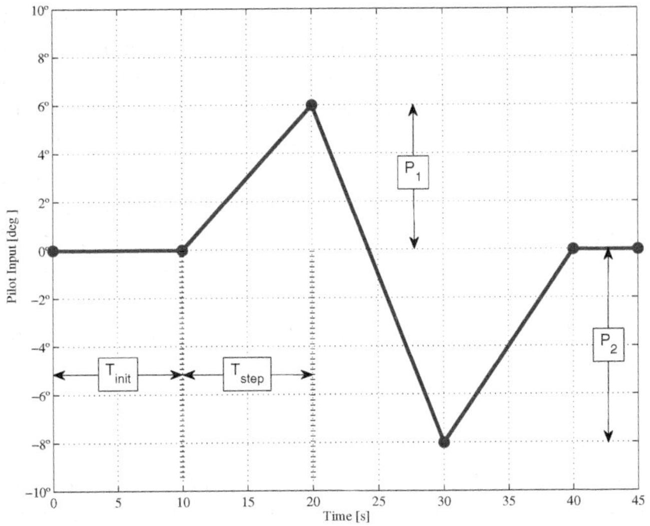

Fig. 17.16 Random pilot input

longitudinal and lateral sticks, as shown in Fig. 17.16. The time-step was taken at 10 seconds, and points were in the range of full stick deflection.

One main advantage of the Monte-Carlo method is repeatability of results (random points are created always using the same seed), so that, when a new version of the flight control laws is developed, the engineer can compare results, looking for improvements and especially for performance degradations. Another advantage is that it is possible to validate several different criteria at the same time, thus reducing the number of runs required to check different aspects of the flight control laws to a single Monte Carlo analysis.

17.4.2 Comparisons with the Baseline Solution

We applied the worst-case search based method to both benchmark problems. The general idea of the optimization based worst-case search is shown in Fig. 17.17. Here, the determination of the worst-case pilot input and worst-case parameter combination is described. However, the case a fixed manoeuvre is also covered, by keeping the same pilot inputs during the optimization run, while varying only the flight parameters. For optimization purposes we only employed the evolutionary strategy (ES) and genetic algorithm (GA) to solve the clearance benchmark problems. In what follows we discuss the main findings of our study.

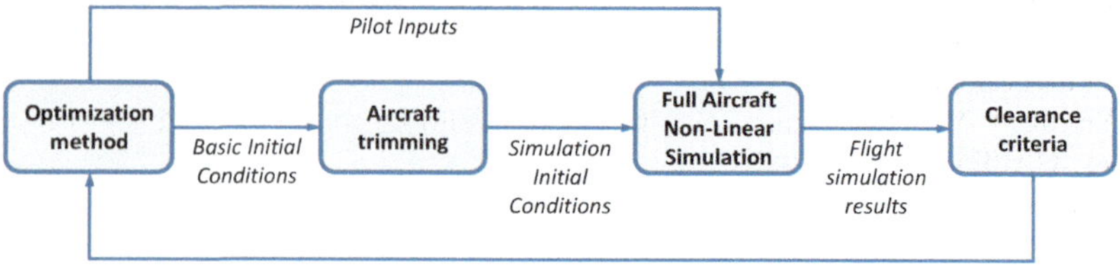

Fig. 17.17 Overview of simulation-based validation

Fixed Manoeuvre - Gridding vs. Optimisation

The gridding approach (using the flight points defined in Table 17.3) is shown to be very limited in comparison to optimisation. It performs a set of 54 simulations, and can find only one worst case with an overshoot of 0.9°. Optimisation was able to identify, using the same manoeuvre and altitude-speed limits, worst cases with overshoot of over 2.0°, as can be seen in Fig. 17.18[1].

---

[1] In Fig. 17.18, since we have formulated the validation problem as a minimization, overshoot is negative. Thus a 2.0° overshoot of the maximum angle of attack value corresponds to a $-2.0°$ value in the third column.

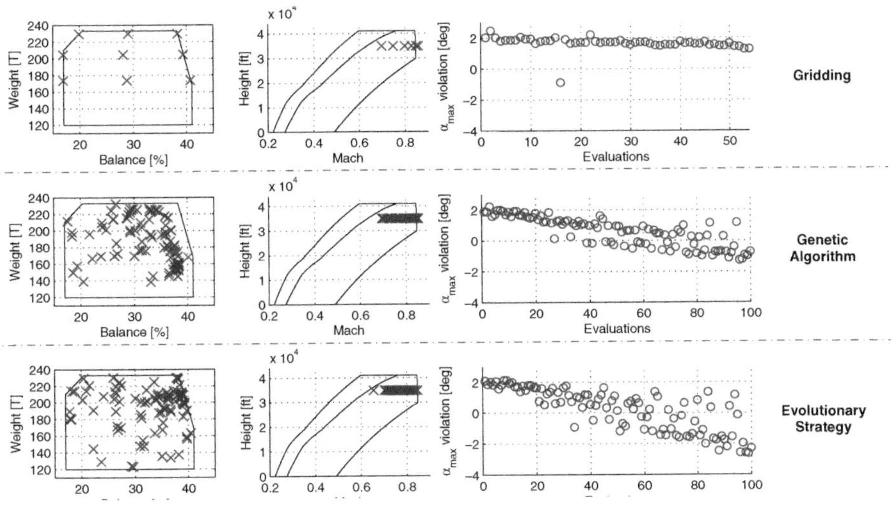

Fig. 17.18 Comparison of results - Gridding vs. optimisation

Worst-Case Pilot Input - Monte-Carlo Method vs. Optimisation

It was found, during comparison of the methods, that one of the main weaknesses of the Monte-Carlo method is that it does not explore the flight domain after a worse case is found. It only reports that there is something wrong but without searching around it for worst points.

On the other hand, optimisation was consistently able to gather worst cases and improve them. The cost of this is that it will take more time to explore the whole flight domain than with the Monte-Carlo method, as the population size is usually small due to run-time constraints, and more than one run is required. Another advantage of the Monte-Carlo method is that many criteria can be checked at the same time. Multi-criteria optimisation is investigated later in section 17.5. Nevertheless, results obtained through optimisation show its real value. Violations found were consistently worse than those found by the Monte-Carlo method (able to find only one α_{max} overshoot of 1.3°), while optimisation could find overshoots up to 8°.

Comparison of the Methods

On the basis of the results reported in Fig. 17.18 and Fig. 17.19, it appears that for these clearance problems the evolutionary strategy is more efficient than the genetic algorithm to find the worst case within a limited number of runs.

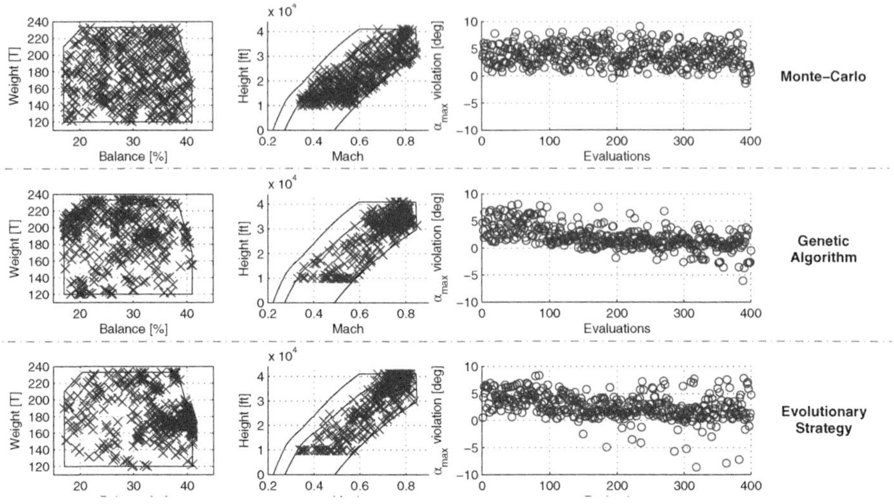

Fig. 17.19 Comparison of results - Monte-Carlo method vs. optimisation

17.5 Industrial Application of Worst-Case Search-Based Methods

As worst-case search-based methods seem to provide promising results and can be adapted to any simulation tool, we decided to implement them in an industrial context.

17.5.1 Coupling of Optimisation and Simulation Model

In the context of the COFCLUO project, a Simulink-based long-range aircraft model was used as a way to represent the industrial application. To implement internally the methods developed for COFCLUO, the Simulink model was replaced with a certified flight loop model that is capable of simulating many AIRBUS aircraft, with different configurations, engines, systems, and flight control law versions, among several other options. The model is a stand-alone application that receives a specially crafted *xml* file representing the simulation inputs (including operating point, pilot inputs for controls and engines) and required outputs. The simulation is run on a dedicated server, and the outcome is processed to calculate each criterion. The whole process is shown in Fig. 17.20.

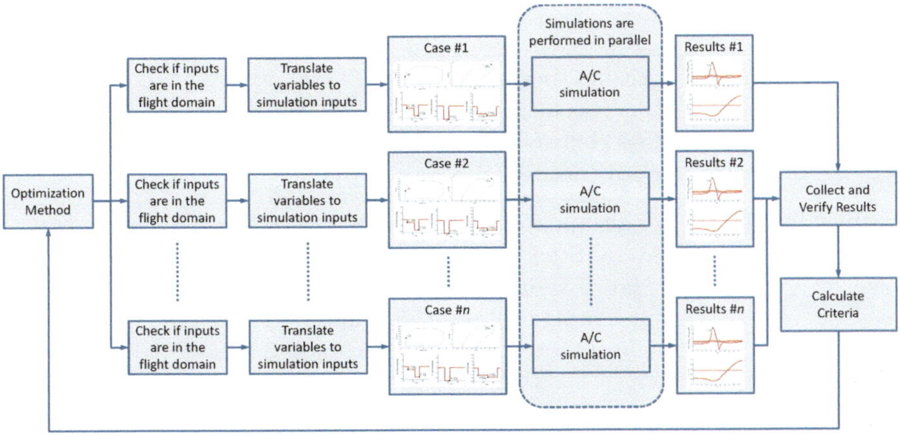

Fig. 17.20 Overview of the optimisation process

17.5.2 Benchmark: Worst-Case Pilot Input

Objective

Current practice within AIRBUS is mainly based on extensive use of simulation on either deterministic or randomly generated grids of points (see section 17.4.1). To test the usefulness of optimisation in a real validation of flight control laws, it is necessary to know if the novel method is better than the current practice (at least maximizing the probability of finding worst cases), so a comparison between the Monte-Carlo method and optimisation was made.

In the case of validation using random pilot inputs, the flight points (weight, position of CG, initial speed and altitude) and pilot inputs (lateral and longitudinal side-stick, pedal, throttle and air-brakes – Fig. 17.21) are taken randomly. The aircraft slats/flaps configuration is taken as clean.

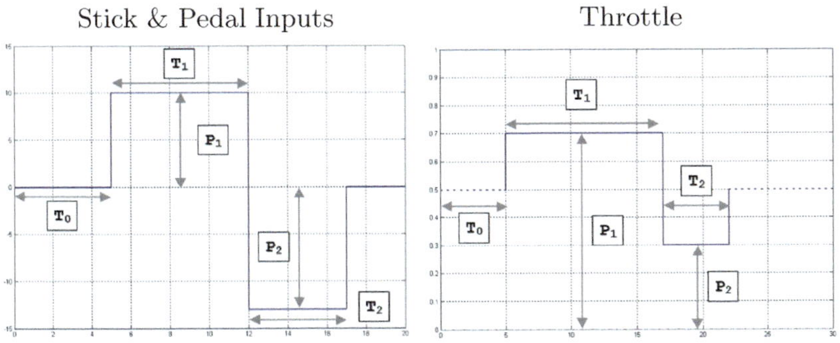

Fig. 17.21 Definition of a random pilot input

Table 17.4 Parameters settings and ranges for the random pilot input test

Flight Domain:	Altitude (between 10 and 43 kft) $\%_{CG}$ (between 0 and 100% of available range) Mass (between OEW and MTOW) Speed (between V_{ls} and V_{MO})
Pilot Inputs:	Longitudinal Stick (2 values, 3 timesteps) Lateral Stick (1 value, 3 timesteps) Pedal (2 values, 3 timesteps) Left Throttle (2 positive values, 3 timesteps) Right Throttle (2 positive values, 3 timesteps)

The settings of search parameters and their variation ranges are given in the Table 17.4. The total number of parameters amounts to 27.

The currently used Monte-Carlo method checks for violations in many criteria, and while this could be implemented in a multi-objective setting, it was decided that a comparison with only two time domain criteria would be done first, both for time constraints and not to have unexpected influence on the outcome. The protections for $\alpha < \alpha_{max}$ and $V_{cas} < V_D$ were selected, as they require, at least different flight conditions ($\alpha < \alpha_{max}$ usually being a low-speed problem, and $V_{cas} < V_D$ the opposite). The criteria are defined as

$$c_{\alpha_{max}} = \min_t \left[\alpha_{max}(t) + \alpha_{offset}(t) - \alpha(t)\right] \tag{17.2}$$

$$c_{V_{cas}<V_D} = \min_t \left[-\frac{V_{cas}(t)}{V_D(t)}\right] \tag{17.3}$$

where α_{max} and V_D are respectively the maximum angle of attack and maximum dive speed that the aircraft is allowed to reach, as calculated by the aircraft flight computers during simulation. There is an allowed offset (α_{offset}) of at most 2° for the α_{max} protection, that is a function of aircraft indicated speed. Sample violations are shown in Fig. 17.22.

Baseline Method - Monte-Carlo Method

The current validation of flight control laws is done in two separate parts, one using a gridding approach and the other the Monte-Carlo method. The purpose of the gridding approach is to check, in fixed flight points, how the control laws are performing, while the Monte-Carlo method is used to ensure the completeness of the protections. For this, random pilot inputs are applied at random flight conditions, and the simulation results are then checked for any misbehaviour and the probability of violations.

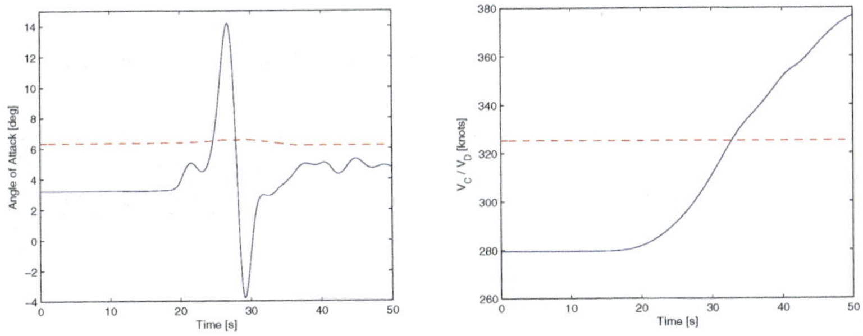

Fig. 17.22 Selected protections: $\alpha < \alpha_{max}$ (left) and $V_{cas} < V_D$ (right), red line represents protection boundaries

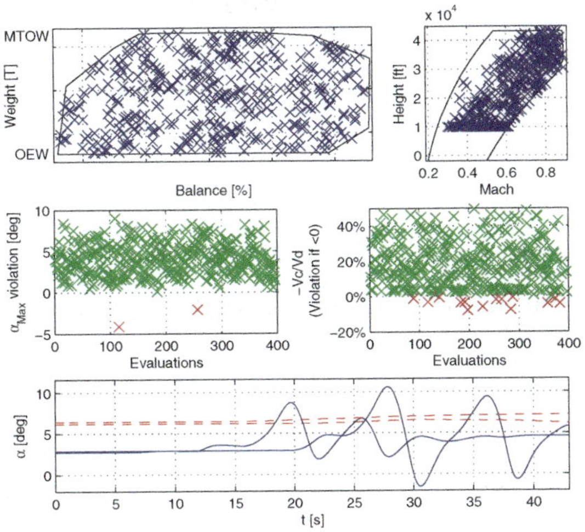

Fig. 17.23 Monte-Carlo simulation results for the random pilot input test

A typical Monte-Carlo simulation run with 400 points is presented in Fig. 17.23, where it can be seen that the method is only capable of finding some small violations of 5° for the α_{max} protection (-5° on the "α_{max} violation" plot) and at most 7% for speed protection (-7% on the "-V_{cas}/V_D" plot).

COFCLUO Approach

Different criteria can also be dealt with either by separate single-objective optimisations or by multi-objective optimisations.

By focusing on a single criterion, the single-objective optimisation method should be better (i.e. more efficient) at finding worst case values for each criterion. Nevertheless, Pareto-based multi-objective optimisation methods are supposed to provide a set of optimal solutions corresponding to different weighted linear combinations of the selected criteria. Hence single-objective is a special case of multi-objective which should not just focus on solutions that violate all criteria at the same time (Fig. 17.24). One of the main advantages, compared to single-objective optimisation, is to share information generated during simulation results for different criteria.

Therefore, multi-objective optimisation using the concept of Pareto optimum was tested. After simulations are performed and results analysed for violations, these violations are ranked, so that the non-dominated worst case solutions have a higher probability of being selected during optimisation (Fig. 17.24).

Both single-objective and multi-objective optimisations were run. Results showed, at least on our application, a clear saving of time without any lack of optimality when using the Pareto approach. It will therefore be the baseline approach for subsequent optimisation.

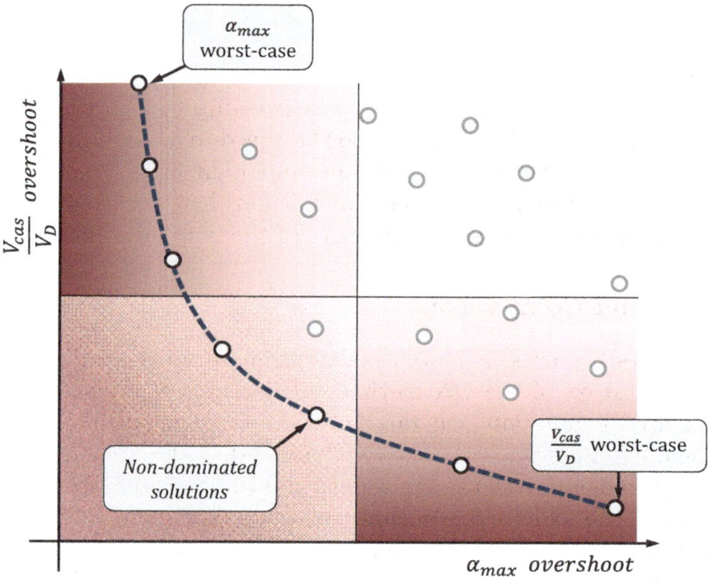

Fig. 17.24 Pareto frontier for $\alpha < \alpha_{max}$ and $V_{cas} < V_D$ criteria

Comparison of Results

Pareto frontiers for Monte-Carlo method and for optimisation are shown in Fig. 17.25. It can be seen that Monte-Carlo method was not capable of finding significant worst cases while optimisation was able to find and refine them. One interesting point is that, although initially the two criteria are supposed to be antinomic (one corresponds to high speed and the other to low speed), optimisation was able to find points where both criteria were violated at the

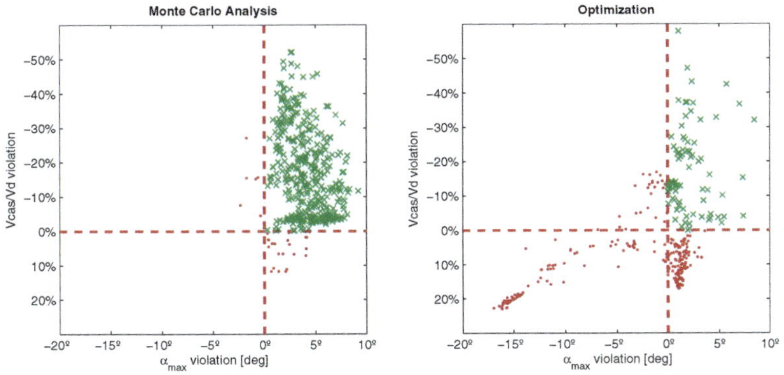

Fig. 17.25 Comparison of Pareto frontiers

same time. This may be an isolated case but it demonstrates the interest of using multi-objective optimization anyway: either a method is able to find worst cases for several criteria in the same area and then you can save time, either worst cases are decoupled and then Pareto front will grow in several directions at the same time.

Conclusion

The use of optimisation methods for validation of flight control laws has proven to be interesting, but it is also important for the engineer to have a way to first check if changes in flight control laws result in improvements or regressions and then evaluate the importance of the worst cases with respect to the ones which fulfil the specifications. Simply replacing the Monte-Carlo method by optimisation is not ideal, as Monte-Carlo method has its usefulness for comparing different versions of the flight control laws, as inputs are created by a pseudo-random number generator that can be initialized with the same seed. In this sense, the ideal outcome we have found is to complement the Monte-Carlo method with optimisation, in a way that optimisation will start from the previous results found by Monte-Carlo simulations. This actually improves optimisation, as it is better to select the initial points from a more comprehensive population than restricting it to a small set.

17.5.3 Enhancements

This section presents enhancements of the original method proposed in the previous chapters.

Selection of Optimisation Methods

Previous studies by [1, 2, 3, 4] have shown that stochastic methods, such as genetic algorithms (GA) have exceptional robustness for the type of problem studied here, but generally require many evaluations before convergence. Initial studies within COFCLUO found another evolutionary method, called Evolutionary Strategy (ES) that yields better results in the clearance challenge, but as different methods became available, a new comparison of methods was made using the validation scheme proposed. Besides the GA method, with two implementations denoted GA1 (from the MATLAB software) and GA2 (from 11), and the ES method (from 8), also a new method developed internally called *Variability Preservation Search* (VIPS) (see reference [7]) has been tested using single criterion optimisation, to solve the $\alpha < \alpha_{max}$ protection issue. The initial population was created with 100 points, and was kept the same for all runs.

Each method was allowed to run for 400 simulations, and 20 runs were made to compare the ability to converge to a worst case. From the results in Fig. 17.26 and summarized in Table 17.5, it can be seen that both GA2 and VIPS consistently found violations, with VIPS being able to find worst cases more often (in the current context where the number of runs is limited). The ES method was also able to find worst cases in every run, with a minimal violation of 3.4° , but was unable to improve them beyond a certain point. The cumulative probability of finding worst cases is presented in Fig. 17.27.

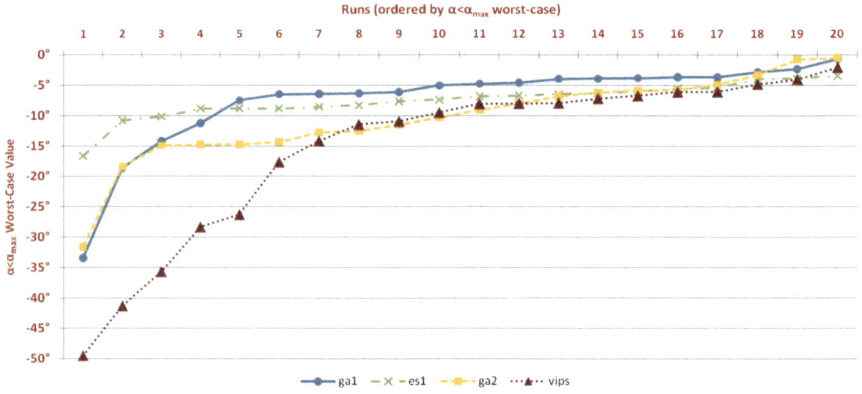

Fig. 17.26 Results for $\alpha < \alpha_{max}$, different optimisation methods starting with same initial population

As explained above, the capability of dealing with more than one criterion at the same time is also required. Multi-objective optimisations are generally performed by stochastic methods, and many techniques for this have already been studied, such as fitness-sharing, Pareto domination and non-dominated Pareto ranking [8]. Here, only the concept of non-dominated Pareto fronts, as presented by [9], was used to assess the performance of multi-objective optimisation methods. Both $\alpha < \alpha_{max}$ and $V_{cas} < V_D$ protections are now checked. The methods tested were: the baseline Monte-Carlo method, NSGA2

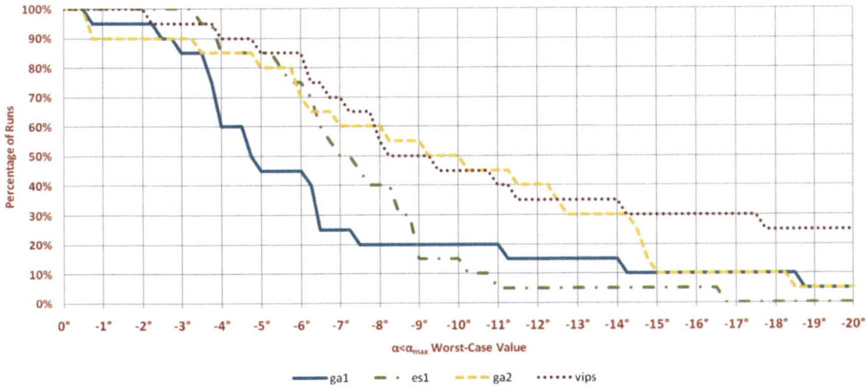

Fig. 17.27 Distribution of worst-cases for $\alpha < \alpha_{max}$ protection, different optimisation methods starting with same initial population

Table 17.5 Summary of results for different optimisation methods for $\alpha < \alpha_{max}$ violation

Method	Maximum	Minimum	Average	Std. Deviation
GA1	$-33.4°$	$-0.7°$	$-7.4°$	$7.4°$
ES	$-16.7°$	$-3.4°$	$-7.5°$	$2.9°$
GA2	$-31.6°$	$-0.5°$	$-10.3°$	$7.1°$
VIPS	$-49.6°$	$-2.1°$	$-15.3°$	$13.6°$

Table 17.6 Summary of multi-objective results for $\alpha < \alpha_{max}$ and $V_{cas} < V_D$ violations

Method	Worst	Minimum	Average	Worst	Minimum	Average
NSGA2	$-9.9°$	$1.4°$	$-0.4°$	-11.7%	3.78%	-2.7%
Monte-Carlo	$-8.6°$	$-1.7°$	$-3.1°$	-17.2%	-6.4%	-11.1%
GA2	$-17.9°$	$-2.4°$	$-8.6°$	-26.1%	-12.2%	-21.2%
VIPS	$-54.2°$	$-3.8°$	$-11.6°$	-38.7%	-11.9%	-18.7%

(from [9]), GA2 and VIPS. The setup for this test consisted of 20 runs with 400 evaluations each, starting with an initial population of 20 random points (keeping the same between methods during each run).

It can be noted from Fig. 17.28 and Fig. 17.29 that NSGA2 performed worse than Monte-Carlo method. This shows the necessity of tuning optimisation methods to the highly non-linear COFCLUO problem (as also explained by [3,10]). Between the other two GA2 and VIPS, both have notable

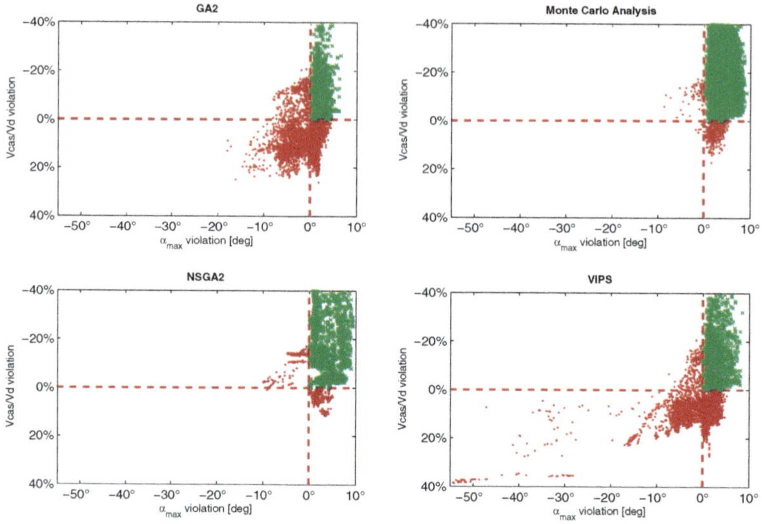

Fig. 17.28 All violations found for each criteria, 20 runs for each method

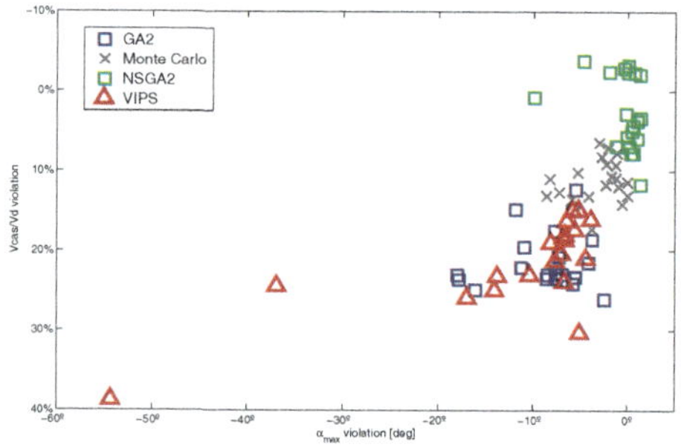

Fig. 17.29 Final worst-case value for each run

results, finding worst-cases in 100% of the runs, with an average violation of $-8.6°/-21.2\%$ for GA2, and $-11.6°/-18.7\%$ for VIPS.

Effects of Parallelisation

Generate-and-test methods, such as genetic algorithms, evolutionary strategy and Monte-Carlo method, are well suited to parallelisation due to their nature. As cases are independent of each other, multiple simulations can be run at the same time, thus reducing the total elapsed time. Currently, parallelisation is made using a simulation server that takes input data from optimisation, performs full aircraft simulation, and then returns the results. A study of parallelisation speedup was made, and results are shown in Fig. 17.30.

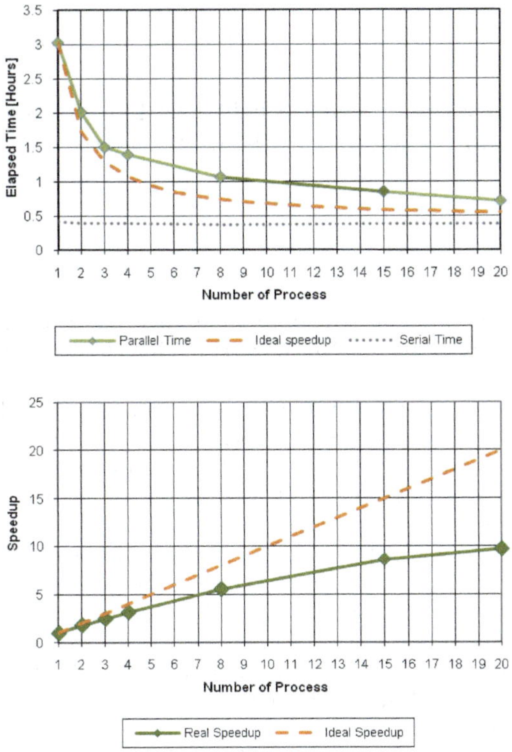

Fig. 17.30 Optimisation runtime and speedup, 500 simulations

Although it is possible to significantly reduce the computational time required, there is still a huge gap between the ideal and achieved speedup. This is related to some time-consuming operations that are performed in a serial manner, such as reading and plotting results. Therefore, further improvements can be made to reduce this time, making this validation tool more efficient.

Local Single-Objective Optimisation

One problem found in global methods is that, although they are able to find worst-cases with good probability, they sometimes have limited convergence performance, as investigated by [5, 6]. Unfortunately we are both limiting the number of runs and using multi-objective methods with a wide searching area. In order to improve results, a local single-objective approach was tested (in addition to multi-objective optimisation), using the pattern search (PS) method. Use of a global single-objective method before a local one would have been wiser, but results were not better when limiting the evaluation number. The pattern search method takes previously found worst cases, and

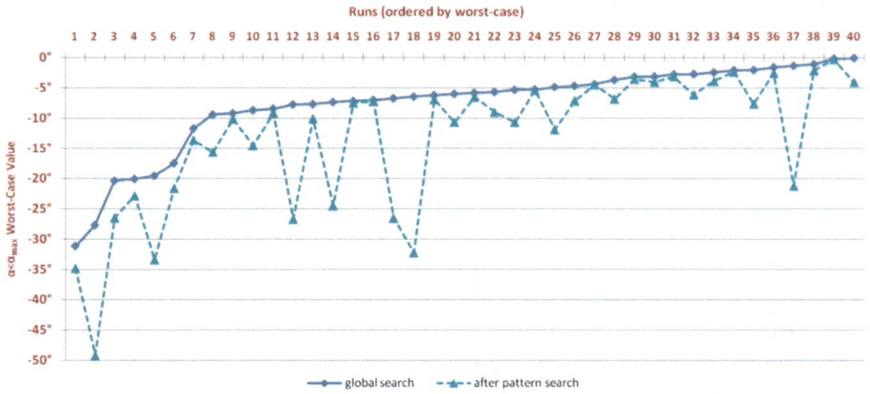

Fig. 17.31 Improvements obtained by pattern search, results for $\alpha < \alpha_{max}$

uses them as the initial point for a pattern search optimisation. Basically, each input variable is incremented monotonically by a small δ, and when a better point is found, the method starts to move in that direction, until improvement stops. Then it starts again with this new point, reducing δ as necessary until convergence.

The use of pattern search after optimisation was tested with 40 runs using $\alpha < \alpha_{max}$ protection, and was found to be worthwhile. Results in Fig. 17.31, before and after pattern search, show improvements for most of the worst cases tested, with an average improvement of 5.5°.

A typical result of hybrid optimisation is shown in Fig. 17.32. The method started with a previously found worst case, running 50 simulations (a small number to test its usefulness when many worst cases will have to be tested). A significant improvement of the worst-case was found, increasing the α_{max} overshoot from $-4.53°$ to $-8.85°$, by making small changes in four key variables, as detailed in Table 17.7 and Fig. 17.34. From the 27 parameters that define the simulation, it was possible to identify 4 of them that improved results.

There is still a high computational cost of using local methods, as due to their nature, they are hard to parallelise compared to global stochastic methods. Therefore, it is suggested that local optimisation should be restricted to some selected worst cases, after previous runs of a global method were able to find them, but got stuck in improving results. A possible way to parallelise this is to perform several pattern search runs at the same time, using different worst cases found.

Sensitivity to Inputs

From an engineering point of view, it is of great interest to know what caused a bad response from the flight control laws in each worst case. Currently, this

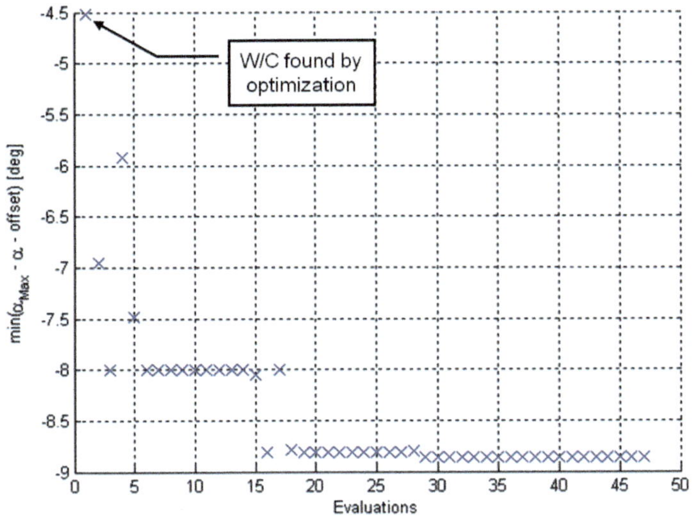

Fig. 17.32 Improvement in worst case using pattern search

Table 17.7 Changes in four worst-case parameter values[a]

Parameter[b]	Start Point	After pattern search
Mass [% of MTOW]	68.30%	67.17%
$pc_{manche}^{T_0}$	5.48°	5.86°
$pc_{manche}^{P_1}$	9.66°	10.19°
$throttle_{left}^{T_0}$	11.94°	11.56°

[a] All other parameter values are kept unmodified.
[b] Definition of parameters (see Fig. 17.21):
$pc_{manche}^{T_0}$ is the lateral stick position during initial step time T_0;
$pc_{manche}^{P_1}$ is the first step amplitude variation of lateral stick, P_1;
$throttle_{left}^{T_0}$ is the left throttle position during initial step time T_0.

is done by hand, trying to identify relevant parameters in the simulation. A sensitivity analysis is implemented by changing each input variable by a configurable amount and analysing results to see which parameter has most influence. It requires only twice the number of parameters function evaluations to be performed, and is suitable for parallelisation, so the cost of performing a sensitivity analysis is low, taking, with the current implementation, less than 20 minutes to run. Fig. 17.33 illustrates the basic principle of sensitivity analysis for the worst-case solution.

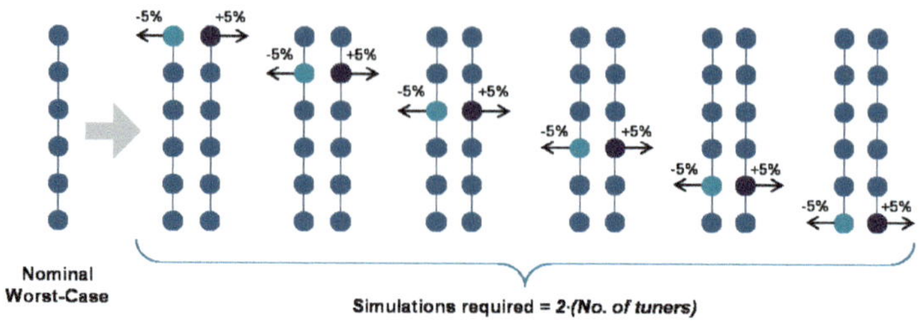

Fig. 17.33 Sensitivity analysis

Fig. 17.34 shows a sensitivity report before and after local optimisation, providing information on which input parameters are influencing the results. It can be seen that pattern search correctly identified the parameters $Mass$, $pc_{manche}^{T_0}$ and $pc_{manche}^{P_1}$ as the ones capable of improving the worst case, and there was still room for improvement, especially in the parameter $pc_{manche}^{P_1}$, but the algorithm was unable to find it with the small number of function evaluations defined.

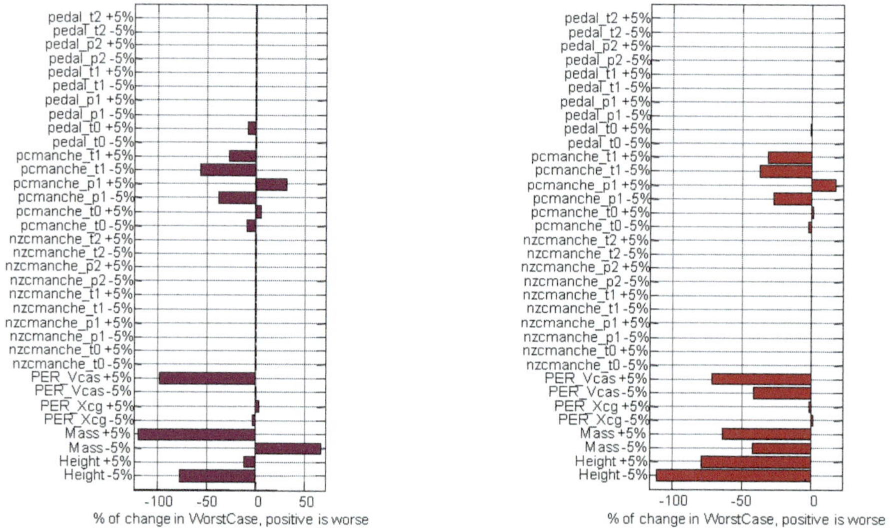

Fig. 17.34 Sensitivity report before (left) and after pattern search (right) - Some parameters are not shown

17.5.4 Proposed Validation Workflow

As for the clearance problem, it is important for the engineers to have, not only for certification but during the design phase, a way to check if flight control laws are performing adequately in the main part of the operating domain. Moreover, it is interesting to compare different versions, to verify improvements and regressions. Monte-Carlo method is useful for this, as it can be initialized with the same seed between runs. But Monte-Carlo method alone does not provide the necessary insight compared to the results obtained with optimisation. In this sense, the ideal outcome is to complement Monte-Carlo simulations with optimisation, as described in Fig. 17.35. The proposed validation scheme starts with performing Monte-Carlo simulations, and then, different multi-objective optimisation runs are performed. Selected worst cases are then used as initial points for a hybrid-optimisation using the classical pattern search method. Finally, sensitivity analysis is used to give a comprehensive view of what is causing the unwanted response, and to guide the engineer in understanding and solving the problem.

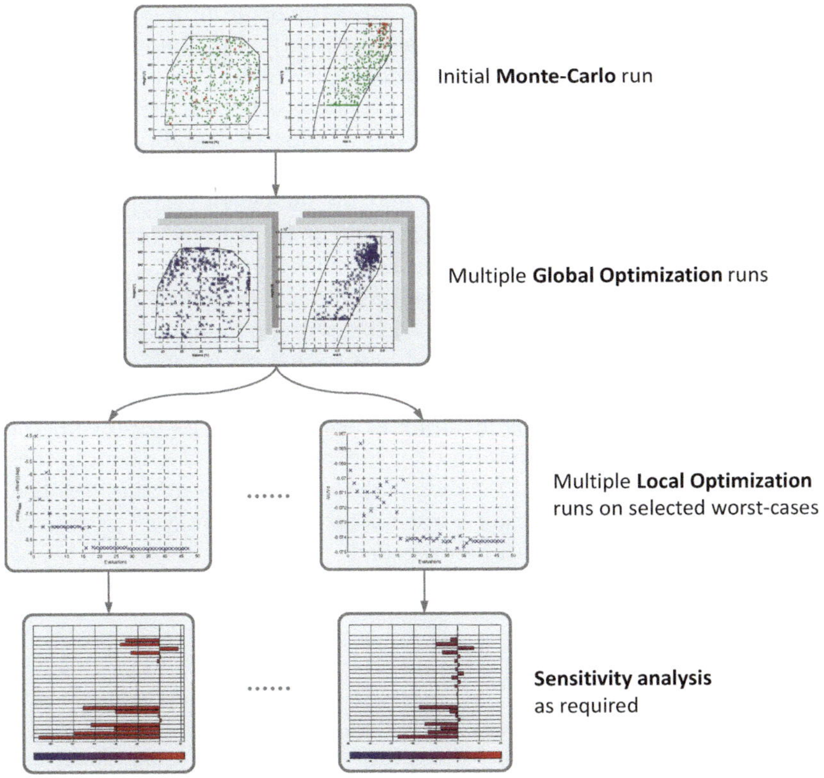

Fig. 17.35 Validation workflow

17.6 Conclusion

LFR-based analysis methods are interesting to be used in the first design validation loop, to perform linear analyses such as stability checks or robustness margins for uncertainties. The computational burden seems to be no longer an issue, and results are more powerful than current baseline gridding-based solution. But an efficient modeling process is still needed, which is something that may be missing today (even after this project). Two main issues are still open. The first is the automation of the aircraft LFR generation process, but this is mainly a matter of developing the required tools. The second is probably more difficult to handle, since it is related to the way to integrate modeling errors in the final result provided by the analysis. A preliminary approach has been initiated in the paper, but it relies on a gridding based approach. Whatever solution that could be found to these two open points, it appears that a generalized use of LFRs for controller design and analysis should be built up to make modeling time investment profitable, focusing on the open loop aircraft model only. Regarding analysis techniques, we have mainly tried to solve an LPV stability problem. We should think about usefulness of techniques (Lyapunov and IQC) to solve other problems and even address more complex ones, such as the piloted aircraft stability issue.

Worst-case search methods, coupling optimisation and simulation on the full nonlinear model, appear to be really mature and easy to implement in an industrial context. The results are quite different from those obtained with the gridding or Monte-Carlo based approach. There is no longer the idea of a domain, only worst-case values and arguments are provided. However the methods provided are usually better suited (i.e. with an higher probability) to find worst cases than Monte-Carlo-based approaches with the same amount of simulations. Optimisation is therefore no longer a competitor, but should be used as a complement to current approaches. A mixed use in a complete workflow has been set up to respond to validation needs. Although these methods are powerful, experience shows that they do not stop the user from keeping a critical view on the results. Special attention should be paid to the model validity domain which can be exceeded.

References

1. Fielding, C., Varga, A., Bennani, S., Selier, M. (eds.): Advanced Techniques for Clearance of Flight Control Laws. LNCIS, vol. 283. Springer, Berlin (2002)
2. Forssell, L.S., Hyden, A.: Flight control system validation using global nonlinear optimisation algorithms. In: Proc. of the European Control Conference, Cambridge, United Kingdom (2003)
3. Menon, P.P., Bates, D.G., Postlethwaite, I.: Optimisation-based flight control law clearance. In: Bates, D., Hagström, M. (eds.) Nonlinear Analysis and Synthesis Techniques for Aircraft Control. LNCIS, vol. 365, pp. 259–300. Springer, Berlin (2007)

4. Menon, P., Bates, D., Postlethwaite, I.: Computation of worst-case pilot inputs for clearance of flight control laws. In: Proc. of the 16 th IFAC World Congress, Prague, Czech Repulic (2005)
5. Menon, P., Bates, D., Postlethwaite, I.: Hybrid Optimisation Schemes for the Clearance of Flight Control Laws. In: Proc. of the 16 th IFAC World Congress, Prague, Czech Repulic (2005)
6. Menon, P.P., Kim, J., Bates, D.G., Postlethwaite, I.: Clearance of nonlinear flight control laws using hybrid evolutionary optimisation. IEEE Transactions on Evolutionary Computation 10(6), 689–699 (2006)
7. Oliveira, R.F., Puyou, G.: Clearance of Flight Control Laws using Multi-objective optimisation. In: Proc. of 2010 International Conference on Genetic and Evolutionary Methods, Las Vegas, USA (2010)
8. Smith, J.E., Eiben, A.E.: Introduction to Evolutionary Computing. Springer, Berlin (2003)
9. Deb, K., Pratap, A., Agarwal, S., Meyarivan, T.: A fast and elitist multi-objective genetic algorithm: NSGA-II. IEEE Transactions on Evolutionary Computation 6(2), 182–197 (2002)
10. Ferreres, G., Puyou, G.: Flight control law design for a flexible aircraft: Limits of performance. Journal of Guidance Control and Dynamics 29(4), 870–878 (2006)

Chapter 18
Concluding Remarks and Industrial Perspective

Anders Hansson and Philippe Menard

Abstract. In this chapter the achievements of the COFCLUO project are summarized. The possible exploitation of the results is discussed. Finally, directions for future research are given.

18.1 Summary of Achievements

The optimization techniques for *Clearance of Flight Control Laws* (CFCL) developed within the COFCLUO project have been tested on two realistic and relevant benchmarks with a very positive outcome.

The first benchmark considers CFCL for a rigid aircraft with a non-linear model of the closed-loop flight control system. Two major issues have driven this choice: performance assessment in a nonlinear framework, and validation of requirements on the whole flight domain considering a wide class of pilot inputs and wind perturbations. The second benchmark considers CFCL for a flexible aircraft with a linear model, for which the bending modes are very close to the rigid body modes.

For implementing different clearance criteria for a range of optimization-based approaches, different types of parametric models were needed to be employed. A non-linear dynamics aircraft model with explicit parametric dependencies has been developed together with appropriate flight control laws to be cleared. Also a collection of so-called integral linear models depending on relevant parameters have been provided to model flexible aircraft configurations. A criteria library has been defined and implemented starting from the

Anders Hansson
Division of Automatic Control, Linköping University, SE-581 83 Linköping, Sweden
e-mail: `hansson@isy.liu.se`

Philippe Menard
AIRBUS OPERATION SAS, 316 Route de Bayonne, Toulouse, France
e-mail: `philippe.menard@airbus.com`

A. Varga et al. (Eds.): Opti. Based Clearance of Flight Control Laws, LNCIS 416, pp. 359–365.
springerlink.com

specifications of the benchmark problem for both the integral linearized as well as the non-linear closed-loop aircraft models. The availability of reliable, accurate and fast trimming and linearization tools was essential for generating linearized models. Such tools have been developed and served for efficient and accurate criteria evaluations in optimization-based worst-case search. The trimming and linearization tools have been also used to obtain parameter dependent linear models, the so-called *linear parameter varying* (LPV) models, which can be alternatively described using *linear fractional transformation* (LFT) based representation of system matrices. The LFT-based models for the closed-loop aircraft models, both nonlinear and integral models, have been generated to serve for analysis purposes. The LPV-modelling and LFT-generation activities have been pursued practically during the whole project period by improving successively the quality of approximations, developing new LPV-approximation methods and generating LFT-models of lower complexity.

The clearance techniques can be grouped in two different categories:

1. Sufficient techniques based on solving convex optimization problems and using LFT models of the aircraft.
2. Necessary techniques based on solving nonlinear optimization problems and using standard nonlinear differential equation models of the aircraft.

In the first category, LFT models have to be developed and then convex optimization problems are solved. In the case when the analysis method delivers a positive answer, it is for sure known that the whole region of the flight envelope and the whole region of uncertain parameters considered are cleared. However, if the method delivers a negative answer, nothing is known, i.e. it could be the case that the region considered is safe, but the method was not able to provide that answer. Therefore the methods in this category are conservative, i.e. they are so-called sufficient techniques for CFCL. Also the methods might provide the wrong answer in the case the LFT models do not approximate the nonlinear differential equations of the aircraft accurately enough.

In the second category, no LFT models need to be developed. In the case a method finds a violation of a clearance criteria, it is for sure known that there is a point in the flight envelope which is not cleared. In the case when the optimization algorithm is able to find the global optimum, at least one unsafe point in a region that has unsafe points will be found. However, in such a case presumably a whole neighborhood of points will contain only unsafe points. This is the reason why this category of methods are called necessary techniques for CFCL. However, solving nonlinear optimization problems is difficult, and often only local optima are found. Because of this, it may happen that these methods fail to find unsafe points, although such point may exist. Hence, in general, if the method cannot find an unsafe point, usually nothing is known, i.e. it could be the case that the regions considered are not safe because the method has missed an unsafe point, or the region is safe because there are no unsafe points.

Because of what has been said above the two different categories of methods are complementary to one another. In case a method in category 1 cannot say anything for sure, it could be that a method in category 2 may detect an unsafe point. Also in case a method in category 2 cannot say anything for sure, it could be the case that a method in category 1 can clear the whole region under consideration. Of course it can also be the case that methods from both categories employed together cannot say anything for sure. Notice that it should never be possible for a method in category 1 to clear a region at the same time as a method in category 2 finds an unsafe point, unless the LFT models do not approximate the nonlinear differential equations accurately enough. This fact can be used to validate or invalidate the LFT models.

In comparison with the Monte Carlo simulation based approach currently used in the industry for both robustness assessment and performance certification, the sufficient techniques are promising, though in some cases too slow. The usefulness of the sufficient techniques relies on accurate enough approximations of the nonlinear differential equations with LFTs. However, the more accurate the approximations are, the larger the dimension of the LFT models are, resulting in longer computational times for the sufficient techniques for clearance. Hence there is trade-off between accuracy in the results and the time it takes to obtain the results. It should be stressed, that if the LFTs are accurate, then the sufficient techniques can guarantee that a whole region is cleared and not only the generated points as is the case with Monte Carlo simulation.

The necessary techniques are usually able to find cases that are worse than those obtained with Monte Carlo simulations when using the same number of simulations. However, they are not able to determine compact domains of parameter values containing only safe (cleared) and unsafe (not cleared) points. The resulting worst cases are only isolated points, which can be used to complement the results of Monte Carlo analysis techniques in guiding the designers in their understanding of what may lead to these worst cases.

To summarize, the project has resulted in new modelling techniques as well as new optimization based clearance techniques. Usefulness of LFTs has been demonstrated as well as the potential and maturity of global optimization. Huge efforts have been made to make large-scale computations feasible by exploiting problem structure and inherent computational parallelisability features. Easy to use software, like GUIs, global tuners, and guidelines have been developed to assist the users. It has been demonstrated on an industrial benchmark that the new techniques are very promising, and it is the intention of AIRBUS to use some of these new techniques in their development process.

18.2 Industrial Perspective

Industry has made a substantial step forward thanks to the COFCLUO project and a part of the delivered methods will be used in a development context within AIRBUS. An A350 toolbox is already prototyped for use by the control design teams to secure their flight controls performance and robustness, combining COFCLUO derived methods with Monte-Carlo analysis. The resulting mixed validation strategy is being compared with usual industrial practises in terms of worst case discovery capabilities, and productivity.

Beyond validation plans used internally for securing robustness of flight control systems, AIRBUS could also propose to airworthiness authorities that they include these kinds of methods in their recommended clearance process in addition to the current recommendations. Some of the COFCLUO deliverables will be developed into production quality clearance tools. These tools may either be sold or licensed, or used in-house or for consulting services. The results from the project are useful not only for clearance of flight control laws for civil aircraft but also for military aircraft. Many of the results obtained are general and can be adapted for clearance of control laws for other vehicles, such as unmanned aerial vehicles, cars and trucks. Flight clearance for unmanned aerial vehicles is expected to call for different approaches and pose even more challenging issues. For the car industry, one application of optimization-based clearance of control laws could be to improve the reliability of existing systems, such as vehicle stability control and traction control. Another application in future control systems development is automatic obstacle avoidance. The results obtained can also be used in the validation of many other different types of systems, and thus the results will strengthen the ability of European industry to validate safety-critical systems in general.

18.3 Future Research

Several important challenges remain for future research. A few areas that are of key interest for future research are highlighted below.

18.3.1 Influence of Internal Controller Structure

Flight control systems are basically non-linear and time-varying systems. They include many non-differentiable nonlinearities (e.g., position and rate limiters, multi dimensional interpolation based gain-scheduling) and non-smooth behaviour as control law switching. The latter happens, for example, when the controller is switching from normal laws to protection laws as the aircraft gets closer to its flight envelop border. For these kinds of controllers the objective functions used for clearance are often not differentiable, and hence global optimization methods using only objective function values are the main candidates for an optimization-based clearance approach. Since these methods need typically an extremely large number of function evaluations to converge, it is essential to be able to evaluate the objective functions in an efficient way. It is believed that in the future the aircraft manufacturer should invest a sustained effort in the design of control architectures and the implementation of control laws to be better tailored to fit the efficiency requirements of the clearance approach.

18.3.2 Identification of Critical Parameter Domains

The workload for guidance and control design and validation steadily increased over time, and therefore the complexity of the validation process increased as well. Being more and more time consuming, the validation phase calls for enhanced methods to guide engineers in their worst case search strategies. Even if sufficient techniques are sometimes conservative, they have the advantage, that in the case they are successful, the search for critical cases can be eliminated or at least drastically reduced. On the other hand, necessary methods able to continuously explore the parameter space are of key interest, if they could also provide some information about the nature of critical scenarios or determine whole compact domains of critical operation points, instead of providing only isolated worst cases. For the industrial clearance, it is desirable that CFCL techniques can assist designers in finding features of low robustness regions and families of critical scenarios. That is why the clearance approach based on Monte-Carlo techniques is still a valuable analysis tool in the industry. A merging of the two clearance methodologies would be of highest interest to enhance the current industrial clearance practice.

18.3.3 Low Computational Burden Challenge

The computational burden associated with the optimization-based clearance is an important cost factor for the future acceptance of new verification and validation strategies in the industry. The speed of numerical computations has dramatically increased for a long period of time, partly due to increase of processors performance and partly due to development of more sophisticated algorithms. However, for the last 5 years processor performance has not increased significantly. To compensate for this speed stagnation, multi-core and multi-processor computers have seen an increased use. In addition to this, computational clusters and grids have emerged as another way to speed up computations. These new parallel architectures offer various new opportunities to reduce the computational burden of the clearance process, but also pose new challenges on the development of new analysis algorithms able to exploit the inherent parallelism in many analysis tasks.

18.3.4 Clearance of Systems with Failures

The CFCL addressed in this book covers only the case of failure free aircraft. However, the CFCL is also mandatory for the analysis of failure recovery cases implemented in the flight control laws. The multitude of failure scenarios, encompassing structural failures of aircraft, actuators and sensor failures, as well as their combinations, leads to a complicated clearance procedure with combinatorial complexity. To cover all flight conditions, failure combinations and recovery procedures, the analysis involves mixed discrete and continuous event scenarios which pose new challenges for the development of automatic clearance procedures. One way to address these clearance problems, is to formulate them as *mixed integer nonlinear programming* (MINLP) problems involving optimization of both discrete and continuous variables. Finding worst cases and critical scenarios requires the use of special methods like branch-and-bound, outer-approximation, or cutting plane, which represents extensions of continuous nonlinear programming techniques to problems with discrete variables.

18.3.5 Human Pilot Modelling Challenge

The CFCL process requires a complete closed-loop flight control system model to simulate various flight scenarios operated by the pilot in manual mode. For this purpose, usually a so-called pilot model is employed, whose primary task is to generate the required control demands. The complexity of pilot models differs from case to case, for example, different models can be used for take-off and for landing, to mention only two typical situations. More complex models are necessary to execute complex manoeuvres, as for example, those necessary to evaluate protection laws violation criteria (see Chapter 13) or recovery manoeuvres from failures cases. Therefore, the development of human pilot models with realistic biomechanical features is an important research challenge for the CFCL.

18.3.6 Other Challenges

The COFLCUO Project has clearly illustrated various challenges which the optimization-based CFCL faces. Here we only present a short list, to recall them once again.

Developing accurate low-order LFT models. Sufficient methods fully rely on these models, and therefore the generation of LPV/LFT models still requires sustained research in both theoretical and computational areas. Exploiting analytical information is one way to arrive to better models.

Efficient solution of very large-scale semi-definite programs. This is required by the application of sufficient techniques for clearance. The nowadays required excessive times are not acceptable in many cases.

Merging optimization and Monte-Carlo analysis-based techniques. The integration of different clearance techniques serves for a better coverage, increased performance and overall reliability.

Extension of sufficient methods to nonlinear simulation criteria. Addressing time-varying and/or transient conditions have been addressed so far only with necessary techniques. To provide clearance guarantees, extension of sufficient techniques to the analysis of nonlinear simulation criteria is desirable.

Embedding of CFCL into the design cycle. The optimization-based CFCL provides many useful information (e.g., worst-case parameter combinations) which can serve to enhance the performance of the FCS or even to redesign the flight control laws. The definition of a systematic design cycle which embeds the CFCL process would potentially contribute to reduced global costs for controller tuning and assessment.

Part VI
Surveys on Flight Control Clearance

Chapter 19
Assessment of Aircraft Flight Controllers Using Nonlinear Robustness Analysis Techniques

Peter Seiler, Gary J. Balas, and Andrew K. Packard

Abstract. The current practice to validate flight control laws relies on applying linear analysis tools to assess the closed loop stability and performance characteristics about many trim conditions. Nonlinear simulations are used to provide further confidence in the linear analyses and also to uncover dynamic characteristics, e.g. limit cycles, which are not revealed by the linear analysis. This chapter reviews nonlinear analysis techniques which can be applied to systems described by polynomial dynamic equations. The proposed approach is to approximate the aircraft dynamics using polynomial models. Nonlinear analyses can then be solved using sum-of-squares optimization techniques. The applicability of these methods is demonstrated with nonlinear analyses of an F/A-18 aircraft and NASA's Generic Transport Model aircraft. These nonlinear analysis techniques can fill the gap between linear analysis and nonlinear simulations and hence used to provide additional confidence in the flight control law performance.

19.1 Introduction

The current practice to validate flight control laws relies on applying linear analysis tools to assess the closed loop stability and performance characteristics about many trim conditions. Nonlinear simulations are then used to provide further confidence in the linear analyses and also to uncover dynamic characteristics, e.g. limit cycles, which are not revealed by the linear

Peter Seiler · Gary J. Balas
Aerospace Engineering & Mechanics, University of Minnesota,
Minneapolis, MN, 55455, USA
e-mail: {seiler,balas}@aem.umn.edu

Andrew K. Packard
Mechanical Engineering, University of California, Berkeley, CA, 94720, USA
e-mail: apackard@berkeley.edu

A. Varga et al. (Eds.): Opti. Based Clearance of Flight Control Laws, LNCIS 416, pp. 369–397.
springerlink.com

analysis. This approach is well-suited for validation of current commercial and military aircraft. However, there are drawbacks of this approach. First, the process is rather time-consuming and requires many well-trained control and simulation engineers. Second, most adaptive control laws lead to nonlinear, time-varying closed loop dynamics. Thus the current practice is not applicable to validating systems with adaptive control laws. There is a need for analytical tools to assess the performance of nonlinear feedback systems.

This chapter reviews an approach to reformulate nonlinear analysis problems into a form which can be solved using available software tools. The approach is applicable to nonlinear systems described by polynomial dynamics and it relies on connections between sums of squares (SOS) polynomials and positive semidefinite matrices. A polynomial p is a sum of squares if it can be expressed as $p = \sum_{i=1}^{m} f_i^2$. This connection was made in the work by Parrilo [1, 2] and has led to research on computational tools for estimating regions of attraction, reachability sets, input-output gains, and robustness with respect to uncertainty. The reader is referred to [3, 4, 5, 6, 7, 8, 9, 10, 11, 12, 13, 14, 15, 16, 17, 18, 19, 20] and the references contained therein. There are two key ideas in this approach. First, sufficient conditions for many nonlinear analysis problems can be formulated as set containment conditions involving either a Lyapunov function or a storage function. Second, the set containment conditions can be reformulated as polynomial non-negativity conditions using a generalized version of the S-procedure [21]. This approach will be described in more detail in the remainder of the chapter.

These nonlinear analysis techniques can fill the gap between linear analysis and nonlinear simulations. Linearized analysis is only valid over an infinitesimally small neighborhood of the equilibrium point/null input. The proposed approach provides an improvement over linearized analysis in that the results are valid over a provable region of the state/input space [20]. Moreover, the nonlinear analysis tools can complement the linear analysis tools and nonlinear simulations to provide additional confidence in the flight control law performance.

The remainder of the chapter has the following outline. The next section provides a brief review of background material including SOS polynomials, their connections to positive semidefinite matrices, and SOS programming problems. Sect. 19.3 describes the formulation of several nonlinear analysis problems in terms of optimizations with SOS constraints. This section also provides a discussion of the computational approaches to solve these problems. In Sect. 19.4 the proposed approach is applied to compare the performance of two F/A-18 control laws in their ability to suppress a loss-of-control motion known as the Falling Leaf Mode. The tools are also used to compute reachable set estimates for NASA's Generic Transport Model. Finally, conclusions are given in Sect. 19.5.

19.2 Sum of Squares Optimization

This section provides a brief review of computational methods for sum-of-squares polynomial optimizations. A polynomial p is a *sum of squares* (SOS) if there exist polynomials $\{f_i\}_{i=1}^m$ such that $p = \sum_{i=1}^m f_i^2$. As a simple example, $p = x^2 - 4xy + 7y^2$ is a sum of squares since $p = f_1^2 + f_2^2$ where $f_1 = (x - 2y)^2$ and $f_2 = 3y^2$. This section first presents the notation and background material. Next connections between semidefinite matrices and SOS polynomials are described. Finally the software available to solve SOS optimization problems is discussed.

19.2.1 Background

19.2.1.1 Polynomial Notation

$\mathbb{R}[x]$ denotes the set of all polynomials in variables $\{x_1, \ldots, x_n\}$ with real coefficients. $\mathbb{N}$ denotes the set of nonnegative integers, $\{0, 1, \ldots\}$, and $\mathbb{N}^n$ is the set of n-dimensional vectors with entries in $\mathbb{N}$. For $\alpha \in \mathbb{N}^n$, a monomial in variables $\{x_1, \ldots, x_n\}$ is given by $x^\alpha \doteq x_1^{\alpha_1} x_2^{\alpha_2} \cdots x_n^{\alpha_n}$. The degree of a monomial is defined as $\deg x^\alpha \doteq \sum_{i=1}^n \alpha_i$. In this notation a polynomial in $\mathbb{R}[x]$ is simply a finite linear combination of monomials:

$$p \doteq \sum_{\alpha \in \mathcal{A}} c_\alpha x^\alpha = \sum_{\alpha \in \mathcal{A}} c_\alpha x_1^{\alpha_1} x_2^{\alpha_2} \cdots x_n^{\alpha_n}$$

where $c_\alpha \in \mathbb{R}$ and $\mathcal{A}$ is a finite collection of vectors in $\mathbb{N}^n$. Using the definition of deg for a monomial, the degree of p is defined as $\deg p \doteq \max_{\alpha \in \mathcal{A}} \; [\deg x^\alpha]$.

A polynomial p is a *sum of squares* (SOS) if there exist polynomials $\{f_i\}_{i=1}^m$ such that $p = \sum_{i=1}^m f_i^2$. The set of SOS polynomials is a subset of $\mathbb{R}[x]$ and is denoted as $\Sigma[x]$. Note that if p is a sum of squares then $p(x) \geq 0$ $\forall x \in \mathbb{R}^n$. Thus $p \in \Sigma[x]$ is a sufficient condition for a polynomial to be globally non-negative. The converse is not true, i.e. non-negative polynomials are not necessarily SOS polynomials. This is related to one of the problems posed by Hilbert in 1900 [22].

19.2.1.2 Semidefinite Programming

This brief review of semidefinite programming (SDP) is based on a survey by Vandenberghe and Boyd [23] and a monograph by Boyd, et al. [21]. A symmetric matrix $F \in \mathbb{R}^{n \times n}$ is positive semidefinite if $x^T F x \geq 0$ for all $x \in \mathbb{R}^n$. Positive semidefinite matrices are denoted by $F \succeq 0$. A semidefinite program is an optimization problem of the following form:

$$\begin{array}{ll} \min_\lambda & c^T \lambda \\ \text{subject to: } F_0 + \sum_{k=1}^r \lambda_k F_k \succeq 0 & \end{array} \tag{19.1}$$

The symmetric matrices $F_0, \ldots, F_r \in \mathbb{R}^{n\times n}$ and the vector $c \in \mathbb{R}^r$ are given data. The vector $\lambda \in \mathbb{R}^r$ is the decision variable and the constraint, $F_0 + \sum_{k=1}^r \lambda_k F_k \succeq 0$, is called a linear matrix inequality. Equation (19.1) is referred to as the primal problem. The dual associated with this primal problem is:

$$\begin{array}{rl} \max_Z & -\mathbf{Tr}\left[F_0 Z\right] \\ \text{subject to:} & \mathbf{Tr}\left[F_k Z\right] = c_k \; k = 1, \ldots, r \\ & Z \succeq 0 \end{array} \tag{19.2}$$

where $Z = Z^T \in \mathbb{R}^{n\times n}$ is the decision variable for the dual problem. $\mathbf{Tr}\left[\cdot\right]$ denotes the trace of a matrix. This dual problem can be recast in the form of Equation (19.1) and thus it is also a semidefinite program. While the primal and dual forms may look restrictive, these formulations are quite versatile and SDPs find applications in many problems of interest. Moreover, SDPs are convex and quality software exists to solve these problems. In particular, SeDuMi [24, 25] is a freely available MATLAB toolbox that simultaneously solves the primal and/or dual forms of a semidefinite program.

In some cases, the only goal is to find a decision variable that satisfies the linear matrix inequality constraint. These are semidefinite programming feasibility problems. The following is an example:

$$\text{Find } \lambda_1, \ldots, \lambda_r \in \mathbb{R} \text{ such that } F_0 + \sum_{k=1}^r \lambda_k F_k \succeq 0 \tag{19.3}$$

19.2.2 Connections between SOS Polynomials and Semidefinite Matrices

Theorem 19.1 below gives a concrete statement of the connection between sums of squares and positive semidefinite matrices. Two facts that follow from [26] (refer to Theorem 1 and its preceding Lemma) are required:

1. If p is a sum of squares then p must have even degree.
2. If p is degree $2d$ ($d \in \mathbb{N}$) and $p = \sum_{i=1}^m f_i^2$ then $\deg f_i \leq d \; \forall i$.

Next, define z as the column vector of all monomials in variables $\{x_1, \ldots, x_n\}$ of degree $\leq d$:[1]

$$z \doteq \left[1,\; x_1,\; x_2,\; \ldots,\; x_n,\; x_1^2,\; x_1x_2,\; \ldots,\; x_n^2,\; \ldots,\; x_n^d\right]^T \tag{19.4}$$

[1] Any ordering of the monomials can be used to form z. In Equation (19.4), x^α precedes x^β in the definition of z if:

$$\deg x^\alpha < \deg x^\beta \quad \text{or} \quad \deg x^\alpha = \deg x^\beta \text{ and the first nonzero entry of } \alpha - \beta \text{ is } > 0$$

There are $\binom{k+n-1}{k}$ monomials in n variables of degree k. Thus z is a column vector of length $l_z \doteq \sum_{k=0}^{d} \binom{k+n-1}{k} = \binom{n+d}{d}$. If f is a polynomial in n variables with degree $\le d$, then f is a finite linear combination of monomials of degree $\le d$. Consequently, there exists $a \in \mathbb{R}^{l_z}$ such that $f = a^T z$. The proof of the following theorem, introduced as the "Gram Matrix" method by Choi, Lam, and Reznick [27], is included for completeness. This result can be found more recently in [28].

Theorem 19.1. *Suppose $p \in \mathbb{R}[x]$ is a polynomial of degree $2d$ and z is the $l_z \times 1$ vector of monomials defined in Equation (19.4). Then $p \in \Sigma[x]$ if and only if there exists a symmetric matrix $Q \in \mathbb{R}^{l_z \times l_z}$ such that $Q \succeq 0$ and $p = z^T Q z$.*

Proof
($\Rightarrow$) If p is a SOS, then there exists polynomials $\{f_i\}_{i=1}^m$ such that $p = \sum_{i=1}^m f_i^2$. As noted above, $\deg f_i \le d$ for all i. For each f_i there exists a vector $a_i \in \mathbb{R}^{l_z}$ such that $f_i = a_i^T z$. Define the matrix $A \in \mathbb{R}^{l_z \times m}$ whose i^{th} column is a_i and define $Q \doteq AA^T \succeq 0$. Then $p = z^T Q z$.

($\Leftarrow$) Assume there exists $Q = Q^T \in \mathbb{R}^{l_z \times l_z}$ such that $Q \succeq 0$ and $p = z^T Q z$. Define $m \doteq rank(Q)$. There exists a matrix $A \in \mathbb{R}^{l_z \times m}$ such that $Q = AA^T$. Let a_i denote the ith column of A and define the polynomials $f_i \doteq z^T a_i$. Then $p = z^T(AA^T)z = \sum_{i=1}^m f_i^2$. ■

19.2.3 Software for SOS Optimizations

A sum-of-squares program is an optimization problem with a linear cost and SOS constraints on the decision variables [29]:

$$\begin{aligned} &\min_{u \in \mathbb{R}^n} c_1 u_1 + \cdots + c_n u_n \\ &\text{subject to:} \\ &a_{k,0}(x) + a_{k,1}(x)u_1 + \cdots + a_{k,n}(x)u_n \in \Sigma[x] \;\; k = 1, \ldots N_s \end{aligned} \tag{19.5}$$

The polynomials $\{a_{k,j}\}$ are given as part of the optimization data and $u \in \mathbb{R}^n$ are decision variables. In Sect. 19.3 it will be shown that many nonlinear analysis problems can be posed within this optimization framework.

Theorem 19.1 provides the link to convert an SOS program into a semidefinite-programming problem. For example, the constraint $a_{k,0}(x) + a_{k,1}(x)u_1 + \cdots + a_{k,n}(x)u_n \in \Sigma[x]$ can be equivalently written as:

$$a_{k,0}(x) + a_{k,1}(x)u_1 + \cdots + a_{k,n}(x)u_n = z^T Q z \tag{19.6}$$

$$Q \succeq 0 \tag{19.7}$$

Q is a new matrix of decision variables that is introduced when an SOS constraint is converted to an LMI constraint. Equating the coefficients of $z^T Q z$ and $a_{k,0}(x) + a_{k,1}(x)u_1 + \cdots + a_{k,n}(x)u_n$ imposes linear equality constraints on the decision variables u and Q. Thus, Equation (19.6) can be rewritten as a set of linear equality constraints on the decision variables. All SOS constraints in Equation (19.5) can be replaced in this fashion with linear equality constraints and LMI constraints. As a result, the SOS program in Equation (19.5) can be written in the SDP dual form (Equation (19.2)).

While this may appear cumbersome, there is software available to perform the conversion. For example, SOSTOOLS [29], Yalmip [30], and SOSOPT [31] are freely available MATLAB toolboxes for solving SOS optimizations. These packages allow the user to specify the polynomial constraints using a symbolic toolbox. Then they convert the SOS optimization into an SDP which is solved with SeDuMi [25, 24] or another freely available SDP solver. Finally these toolboxes convert the solution of the SDP back to a polynomial solution.

A drawback is that the size of the resulting SDP grows rapidly if the SOS optimization involves polynomials with many variables and/or high degree. For a generic degree $2d$ polynomial p in n variables, the Gram matrix representation involves $l_z := \binom{n+d}{d}$ monomials. An SOS constraint on p is enforced via a positive semidefinite constraint on the $l_z \times l_z$ Gram matrix Q. For example, for a generic degree $2d = 8$ polynomial in $n = 8$ variables, the Gram matrix has dimension $l_z = 495$. The size of this positive semidefinite constraint is at or near the limits of current semidefinite programming solvers. While various techniques can be used to exploit the problem structure [32], this computational growth is a generic trend in SOS optimizations. Some methods which use simulation to ease this computational growth have been developed [16, 17, 18].

19.3 Nonlinear Analysis Tools

Many nonlinear analysis problems can be formulated as sum of squares programming problems. This connection was made in the work by Parrilo [1, 2] and has led to research on computational tools for estimating regions of attraction, reachability sets, input-output gains, and robustness with respect to uncertainty. The reader is referred to [3, 4, 5, 6, 7, 8, 9, 10, 11, 12, 13, 14, 15, 16, 17, 18, 19, 20] and the references contained therein. The key idea is that sufficient conditions for these nonlinear analysis problems can typically be formulated as set containment conditions involving either a Lyapunov function or a storage function. The set containment conditions can be reformulated as polynomial non-negativity conditions using a generalized version of the S-procedure [21]. These problems can then be solved as SOS programs since SOS polynomials are globally non-negative. In this section this approach is described in more detail for region of attraction estimation, $L_2 - L_2$ input-output gain calculation, and estimation of reachability sets with L_2 bounded inputs. Analysis problems with different signal norms and/or with model uncertainty are described in the references given above. Software to perform all analyses described in this section is available at [31].

19.3.1 Region of Attraction Estimation

This section describes the computational method to estimate a region of attraction (ROA). Consider an autonomous nonlinear dynamical system of the form:

$$\dot{x} = f(x), \ x(0) = x_0 \tag{19.8}$$

where $x \in \mathbb{R}^n$ is the state vector and $f : \mathbb{R}^n \to \mathbb{R}^n$ is a multivariable polynomial. Assume that $x = 0$ is a locally asymptotically stable equilibrium point. Formally, the ROA is defined as:

$$\mathcal{R} = \left\{ x_0 \in \mathbb{R}^n : \text{ If } x(0) = x_0 \text{ then } \lim_{t \to \infty} x(t) = 0 \right\} \tag{19.9}$$

Computing the exact ROA for nonlinear dynamical systems is difficult. There has been significant research devoted to estimating invariant subsets of the ROA [1, 3, 4, 15, 33, 34, 35, 36, 37]. The approach taken here is to restrict the search to ellipsoidal approximations of the ROA. Given an $n \times n$ matrix $N = N^T > 0$, define the shape function $p(x) := x^T N x$ and level set $\mathcal{E}_\beta := \{x \in \mathbb{R}^n \ : \ p(x) \leq \beta\}$. $p(x)$ defines the shape of the ellipsoid and β determines the size of the ellipsoid $\mathcal{E}_\beta$. The choice of p is problem dependent and reflects dimensional scaling information as well as the importance of certain directions in the state space. Given the shape function p, the problem is to find the largest ellipsoid $\mathcal{E}_\beta$ contained in the ROA:

$$\begin{aligned} \beta^* &= \max \beta \\ &\text{subject to: } \mathcal{E}_\beta \subset \mathcal{R} \end{aligned} \tag{19.10}$$

Determining the best ellipsoidal approximation to the ROA is still a challenging computational problem. Instead, lower and upper bounds for β^* satisfying $\underline{\beta} \leq \beta^* \leq \bar{\beta}$ are computed. If the lower and upper bounds are close then the largest ellipsoid level set, defined by Equation (19.10), has been approximately computed.

The upper bounds are computed via a search for initial conditions leading to divergent trajectories. If $\lim_{t \to \infty} x(t) = +\infty$ when starting from $x(0) = x_{0,div}$ then $x_{0,div} \notin \mathcal{R}$. If $\bar{\beta}_{div} := p(x_{0,div})$ then $\mathcal{E}_{\bar{\beta}_{div}} \not\subset \mathcal{R}$ which implies $\beta^* \leq \bar{\beta}_{div}$ and $\mathcal{E}_{\beta^*} \subseteq \mathcal{E}_{\bar{\beta}_{div}}$. An exhaustive Monte Carlo search is used to find the tightest possible upper bound on β^*. Specifically, random initial conditions are chosen starting on the boundary of a large ellipsoid: Choose x_0 satisfying $p(x_0) = \beta_{try}$ where β_{try} is sufficiently large that $\beta_{try} \gg \beta^*$. If a divergent trajectory is found, the initial condition is stored and an upper bound on β^* is computed. β_{try} is then decreased by a factor of 0.995 and the search continues until a maximum number of simulations is reached. There is a trade-off involved in choosing the factor 0.995. A smaller factor results in a larger reduction of the upper bound for each divergent trajectory but it typically limits the accuracy of the upper bound. No divergent trajectories can be found when $\beta_{try} < \beta^*$ and this roughly limits the upper bound accuracy to $\beta^*/(\text{factor})$. The value of 0.995 is very close to one and was chosen to obtain an accurate upper bound on β^*. $\bar{\beta}_{MC}$ will denote the smallest upper bound computed with this Monte Carlo search.

The lower bounds are computed using Lyapunov functions and recent results connecting sums-of-squares polynomials to semidefinite programming. Computing these bounds requires the vector field $f(x)$ in Equation (19.8) to be a polynomial function. The computational algorithm is briefly described here and full algorithmic details are provided elsewhere [5,6,7,10,14,16,18,19]. Lemma 19.1 is the main Lyapunov theorem used to compute lower bounds on β^*. This specific lemma is proved by [14] but very similar results are given in textbooks, e.g. by [38].

Lemma 19.1. *If there exists $\gamma > 0$ and a polynomial $V : \mathbb{R}^n \to \mathbb{R}$ such that:*

$$V(0) = 0 \text{ and } V(x) > 0 \; \forall x \neq 0 \tag{19.11}$$

$$\Omega_\gamma := \{x \in \mathbb{R}^n \; : \; V(x) \leq \gamma\} \text{ is bounded.} \tag{19.12}$$

$$\Omega_\gamma \subset \{x \in \mathbb{R}^n \; : \; \nabla V(x) f(x) < 0\} \cup \{0\} \tag{19.13}$$

then for all $x \in \Omega_\gamma$, the solution of Equation (19.8) *exists, satisfies $x(t) \in \Omega_\gamma$ for all $t \geq 0$, and $\Omega_\gamma \subset \mathcal{R}$.*

A function V, satisfying the conditions in Lemma 19.1 is a Lyapunov function and Ω_γ provides an estimate of the region of attraction. If $x = 0$ is asymptotically stable, a linearization can be used to compute a Lyapunov function. Let $A := \left.\frac{\partial f}{\partial x}\right|_{x=0}$ be the linearization of the dynamics about the origin and compute $P > 0$ that solves the Lyapunov equation $A^T P + PA = -I$. $V_{LIN}(x) := x^T P x$ is a quadratic Lyapunov function that satisfies the conditions of Lemma 19.1 for sufficiently small $\gamma > 0$. V_{LIN} can be used to compute a lower bound on β^* by solving two maximizations:

$$\begin{aligned} \gamma^* := & \max \gamma \\ & \text{subject to: } \Omega_\gamma \subset \{x \in \mathbb{R}^n \; : \; \nabla V_{LIN}(x) f(x) < 0\} \end{aligned} \tag{19.14}$$

$$\begin{aligned} \underline{\beta} := & \max \beta \\ & \text{subject to: } \mathcal{E}_\beta \subset \Omega_{\gamma^*} \end{aligned} \tag{19.15}$$

The first maximization finds the largest level set of V_{LIN}, Ω_{γ^*}, such that Lemma 19.1 can be used to verify $\Omega_{\gamma^*} \subseteq \mathcal{R}$. The second maximization finds the largest ellipsoid $\mathcal{E}_\beta$ contain within Ω_{γ^*}.

The set containment constraints can be replaced with a sufficient condition involving non-negative functions [14]. The next Lemma provides this sufficient condition. This lemma is a generalization of the S-procedure which has been frequently applied in control theory [21]. The function s appearing in the Lemma is called a multiplier.

Lemma 19.2. *Define two sets* $A := \{x \in R^n : f_A(x) \geq 0\}$ *and* $B := \{x \in R^n : f_B(x) \geq 0\}$*. If there exists a function* $s(x) \geq 0\ \forall x$ *such that* $f_B(x) - f_A(x)s(x) \geq 0\ \forall x$ *then* $A \subseteq B$.

Proof
Assume there exists a function $s(x) \geq 0\ \forall x$ such that $f_B(x) - f_A(x)s(x) \geq 0$ $\forall x$. Take any $x \in A$. Then $f_B(x) \geq f_A(x)s(x) \geq 0$. Thus x is also in B. ∎

The Positivstellensatz is a result from algebraic geometry that can be used to construct necessary and sufficient algebraic conditions for this set containment condition (See [1,2] and the references contained therein). In addition, more general set containment constraints, e.g. sets with nonstrict inequalities, can be handled. The condition in Lemma 19.2 can be interpreted as a simplification of the most general Positivstellensatz conditions. This simple condition will be used in the remainder of the paper since it typically requires much less computation than the most general Positivstellensatz conditions.

Applying this Lemma to $\mathcal{E}_\beta \subset \Omega_{\gamma^*}$ in Optimization ((19.15)) leads to:

$$\begin{aligned} \underline{\beta} := \max_{\beta,\ s(x)} \beta & \\ \text{subject to: } & s(x) \geq 0\ \forall x \\ & -(\beta - p(x))\, s(x) + (\gamma^* - V_{LIN}(x)) \geq 0\ \forall x \end{aligned} \tag{19.16}$$

The function $s(x)$ is a decision variable of the optimization, i.e. its coefficients are decision variables that are computed as part of the optimization. It is straight-forward to show that the two non-negativity conditions in Optimization ((19.16)) are a sufficient condition for the set containment condition in Optimization ((19.15)). If $s(x)$ is restricted to be a polynomial then both constraints involve the non-negativity of polynomial functions. Restricting a polynomial to be SOS is a sufficient condition for the polynomial to be non-negative. Replacing the non-negativity conditions in Optimization ((19.16)) with SOS constraints leads to an SOS optimization problem:

$$\begin{aligned} \underline{\beta} := \max \beta & \\ \text{subject to: } & s(x) \in \Sigma[x] \\ & -(\beta - p(x))s(x) + (\gamma^* - V_{LIN}(x)) \in \Sigma[x] \end{aligned} \tag{19.17}$$

As described in Sect. 19.2 there is freely available software to solve such SOS optimizations. $\underline{\beta}_{LIN}$ will denote the lower bound obtained from Optimization ((19.17)) using the quadratic Lyapunov function obtained from linearized analysis.

Unfortunately, $\underline{\beta}_{LIN}$ is usually orders of magnitude smaller than the upper bound $\bar{\beta}_{MC}$. Better lower bounds $\underline{\beta}$ can be computed by also optimizing the choice of the Lyapunov function. This leads to an optimization problem that is bilinear in the Lyapunov function and a multiplier function.

Specifically, the SOS constraint that arises due to the set containment condition in Equation (19.14) is:

$$-(\gamma - V)s_2 - (\nabla V \cdot f + l_2) \in \Sigma[x] \tag{19.18}$$

where s_2 is another SOS multiplier function and and $l_2(x) = -\epsilon_2 x^T x$ where ϵ_2 is a small positive constant on the order of 10^{-6}. If both V and s_2 are allowed to vary then the term Vs_2 in this constraint will be bilinear in the unknown coefficients of V and s_2. Thus optimizing over V and the set containment multipliers is not a convex problem due to this bilinearity and hence heuristic solution methods are required. Several methods to compute better Lyapunov functions exist, including V-s iterations [5,6,7,10], bilinear optimization [14], and the use of simulation data [16,18]. The V-s iteration is now described in more detail. The Lyapunov function $V(x)$ in the iteration is initialized with the linearized Lyapunov function V_{LIN}. The iteration also uses functions $l_1(x) = -\epsilon_1 x^T x$ and $l_2(x) = -\epsilon_2 x^T x$ where ϵ_1 and ϵ_2 are small positive constants on the order of 10^{-6}. The V-s iteration algorithm steps are provided below.

1. γ **Step**: Hold V fixed and solve for s_2 and γ^*

$$\gamma^* := \max_{s_2 \in \Sigma[x], \gamma} \gamma \quad \text{s.t.} \; -(\gamma - V)s_2 - (\nabla V \cdot f + l_2) \in \Sigma[x]$$

2. β **Step**: Hold V, γ^* fixed and solve for s_1 and $\underline{\beta}$

$$\underline{\beta} := \max_{s_1 \in \Sigma[x], \beta} \beta \quad \text{s.t.} \; -(\beta - p)s_1 + (\gamma^* - V) \in \Sigma[x]$$

3. V **step**: Hold s_1, s_2, $\underline{\beta}$, γ^* fixed and solve for V satisfying:

$$\begin{aligned} &-(\gamma^* - V)s_2 - (\nabla V \cdot f + l_2) \in \Sigma[x] \\ &-(\underline{\beta} - p)s_1 + (\gamma^* - V) \in \Sigma[x] \\ &V - l_1 \in \Sigma[x], V(0) = 0 \end{aligned}$$

4. Repeat as long as the lower bound $\underline{\beta}$ continues to increase.

Software and additional documentation on the V-s iteration is provided at [31]. The basic idea of the iteration is to avoid the bilinearity in Vs_2 by holding either s_2 or V fixed. Each step of this iteration is a linear SOS optimization that can be solved with available software. In the V-s iteration, the Lyapunov functions are allowed to have polynomial degree greater than two. Increasing the degree of the Lyapunov function will improve the lower bound at the expense of computational complexity. The computational time grows rapidly with the degree of the Lyapunov function. Simulation data can also be used to construct a good initial candidate V for this iteration [17].

The V step requires additional discussion. An interior-point solver is used to find a feasible solution to the LMI feasibility problem in the V step. The Lyapunov function V that is used in the γ and β steps will be feasible for the constraints in the V step. Thus it possible for the solver to simply return the same Lyapunov function that was used in the γ and β steps. While this is possible, it typically happens that the solver returns a different V that allows both γ and β to be increased at the next iteration. An informal justification for this behavior is now given. The constraint $-(\gamma^* - V)s_2 - (\nabla V \cdot f + l_2) \in \Sigma[x]$ is active after the γ step. In the V step an interior point method obtains a new feasible V by computing decision variables that are at the analytic center of the set specified by the LMI constraints. The V step typically returns a feasible V that is "pushed away" from the constraints. Loosely, the new feasible V satisfies $-(\gamma^* - V)s_2 - \left(\nabla V \cdot f + \tilde{l}_2\right) \in \Sigma[x]$. where $\tilde{l}_2 \geq l_2$. $\tilde{l}_2 \geq l_2$ means the next γ step has freedom to increase γ while still satisfying the constraint with l_2. A more formal theory for the behavior of this feasibility step is still an open question.

19.3.2 Input-Output Gains

This section describes a computational method to compute L_2 input-output gains. More details on this problem as well as computing gains with other signal norms can be found in [13, 19, 20]. Consider nonlinear dynamical systems of the form:

$$\begin{aligned} \dot{x} &= f(x,u) \\ y &= h(x) \end{aligned} \tag{19.19}$$

where $x \in \mathbb{R}^{n_x}$ is the state vector, $u \in \mathbb{R}^{n_u}$ is the input, and $y \in \mathbb{R}^{n_y}$ is the output. Assume f is a $n_x \times 1$ polynomial function of x and u such that $f(0,0) = 0$. Also assume that h is an $n_y \times 1$ polynomial function of x such that $h(0) = 0$. Denote this system by $\mathcal{S}$.

Define the L_2 norm of a signal as $\|u\|_2 := \sqrt{\int_0^\infty u^T(t)u(t)dt}$. u is an L_2 signal if this integral is finite. The L_2-L_2 input-output gain of the system is defined as $\|\mathcal{S}\| := \sup_{u \in L_2, \|u\|_2 \neq 0} \frac{\|y\|_2}{\|u\|_2}$. A "local" input-output gain of the system can also be defined as $\|\mathcal{S}\|_R := \sup_{u \in L_2, 0 < \|u\|_2 \leq R} \frac{\|y\|_2}{\|u\|_2}$. For linear systems the magnitude of the output scales proportionally with the magnitude of the input and hence the ratio $\frac{\|y\|_2}{\|u\|_2}$ does not depend on $\|u\|_2$. Thus $\|\mathcal{S}\|_R = \|\mathcal{S}\|$ for all $R > 0$. For a nonlinear system, the local gain depends on the magnitude of the input and hence $\|\mathcal{S}\|_R$ and $\|\mathcal{S}\|$ need not be equal. The class of possible inputs increases with increasing values of R and so $\|\mathcal{S}\|_R$ is a monotonically increasing function of R and $\|\mathcal{S}\|_R \geq \|\mathcal{S}\|$ for all $R > 0$.

Lemma 19.3 provides a sufficient condition for the local L_2-L_2 input-output gain to be less than γ. This specific lemma can be found in [13, 19, 20] but similar results are given in textbooks [38, 39].

Lemma 19.3. *If there exists a $\gamma > 0$ and a continuously differentiable function $V : \mathbb{R}^n \to \mathbb{R}$ such that:*

- $V(0) = 0$ *and* $V(x) \geq 0 \;\; \forall x \in \mathbb{R}^{n_x}$
- $\{(x,u) \in \mathbb{R}^{n_x+n_u} \; : \; V(x) \leq R^2\} \subseteq \{(x,u) \in \mathbb{R}^{n_x+n_u} \; : \; \frac{\partial V}{\partial x} f(x,u) \leq u^T u - \gamma^{-2} y^T y\}$

then $x(0) = 0$ *and* $\|u\|_2 \leq R$ *implies* $\|y\|_2 \leq \gamma \|u\|_2$.

Proof
A sketch of the proof is provided. Assume that $\frac{\partial V}{\partial x} f(x,u) \leq u^T u - \gamma^{-2} y^T y$ holds along the trajectories of the system $\mathcal{S}$ from time 0 to T. Integrating with respect to time yields:

$$V(x(T)) - V(x(0)) \leq \int_0^T \left(u^T u - \gamma^{-2} y^T y \right) dt \tag{19.20}$$

If $x(0) = 0$ and $\|u\|_2 \leq R$ then Equation (19.20) implies that $V(x(T)) \leq \|u\|_2^2 \leq R^2$. Thus the state trajectories satisfy $V(x(T)) \leq R^2 \; \forall T \geq 0$ and it is valid to assume $\frac{\partial V}{\partial x} f(x,u) \leq u^T u - \gamma^{-2} y^T y$ holds along the system trajectories. Moreover, Equation (19.20) implies that $\int_0^T (y^T y) \, dt \leq \gamma^2 \int_0^T (u^T u) \, dt$ since $V(0) = 0$ and $V(x) \geq 0 \; \forall x$. $\|y\|_2 \leq \gamma \|u\|_2$ follows by letting $T \to \infty$. ■

Lemma 19.3 provides a sufficient condition to prove $\|S\|_R \leq \gamma$ in terms of a storage function, V. This lemma involves one non-negativity condition on the storage function and one set containment condition. The generalized S-procedure (Lemma 19.2) can again be used to convert the set containment condition into a function non-negativity constraint. This leads to the following optimization for computing upper bounds on the local L_2-L_2 gain:

$$\gamma^* := \min_{V,s,\gamma} \gamma \tag{19.21}$$

subject to:

$$s(x,u) \in \Sigma[x,u], V(x) \in \Sigma[x], V(0) = 0 \tag{19.22}$$

$$u^T u - \gamma^{-2} h(x)^T h(x) - \nabla V \cdot f(x,u) - s(x,u) \left(R^2 - V(x) \right) \in \Sigma[x,u] \tag{19.23}$$

The constraint in Equation (19.23), if satisfied, ensures that $\{(x,u) \in \mathbb{R}^{n_x+n_u} \; : \; V(x) \leq R^2\} \subseteq \{(x,u) \in \mathbb{R}^{n_x+n_u} \; : \; \frac{\partial V}{\partial x} f(x,u) \leq u^T u - \gamma^{-2} y^T y\}$. Since SOS polynomials are non-negative everywhere this follows by applying the generalized S-procedure in Lemma 19.2. By Lemma 19.3, $\|S\|_R \leq \gamma$ for any γ for which the constraints are valid. γ^* is the smallest upper bound on $\|S\|_R$ which can be found with this sufficient condition.

This optimization problem involves SOS constraints on $s(x,u)$ and $V(x)$ (Equation (19.22)). The coefficients of the polynomials $s(x,u)$ and $V(x)$ are decision variables in the optimization. The constraint in Equation (19.23) is an SOS constraint on a polynomial of x and u. Unfortunately this constraint is bilinear in the decision variables since it involves a term of the form $s(x,u) \cdot V(x)$. This problem is nonconvex in the decision variables and again heuristics are required. This can be solved directly using bilinear matrix inequality solvers [17]. Alternatively, a V-s iteration can be formulated for this input-output gain problem. This iteration can be initialized with V as the quadratic storage function obtained from linear analysis [6]. Simulation data can also be used to construct a good initial candidate V for this iteration [17].

19.3.3 Reachable Sets

Again consider nonlinear dynamical systems, $\mathcal{S}$, in the form of Equation (19.19). The reachable set $\mathcal{G}(\gamma)$ is the set of states that can be reached from $x(0) = 0$ with an input satisfying $||u||_2^2 \leq \gamma$. Formally,

$$\mathcal{G}(\gamma) := \{x_f \in \mathbb{R}^n \ : \ \exists T, \ u(t) \text{ defined on } [0,T] \text{ s.t. } ||u||_2^2 \leq \gamma \text{ and } x(T) = x_f\} \tag{19.24}$$

The reachable set depends on the input energy to the system. Lemma 19.4 provides a sufficient condition for computing an outer bound on $\mathcal{G}(\gamma)$ [14,19]. Similar reachable set results for linear systems can be found in [21].

Lemma 19.4. *If there exists a $\gamma > 0$ and a continuously differentiable function $V : \mathbb{R}^n \to \mathbb{R}$ such that:*

- $V(0) = 0$ *and* $V(x) \geq 0 \ \ \forall x \in \mathbb{R}^{n_x}$
- $\{(x,u) \in \mathbb{R}^{n_x+n_u} \ : \ V(x) \leq \gamma\} \subseteq \{(x,u) \in \mathbb{R}^{n_x+n_u} \ : \ \nabla V \cdot f(x,u) \leq u^T u\}$

then $\mathcal{G}(\gamma) \subseteq \{x \ : \ V \leq \gamma\} := \Omega_\gamma$

As in the previous sections, the generalized S-procedure can be used to convert the set containment constraint into an SOS constraint. Also, an ellipsoidal approximation of $Reach(\gamma)$ is easier to visualize and understand. As in the ROA estimation problem, a shape function $p(x) := x^T N x$ and level set $\mathcal{E}_\beta := \{x \in \mathbb{R}^n \ : \ p(x) \leq \beta\}$ can be introduced. The problem is then to find the smallest ellipsoid $\mathcal{E}_\beta$ that contains $\mathcal{G}(\gamma)$. $\mathcal{E}_\beta$ then provides an outer approximation for the reachable set. This leads to the following optimization problem:

$$\begin{aligned} \beta^* := \min_{V,s,\gamma,\beta} \ & \beta \\ \text{subject to: } & \\ & s(x,u) \in \Sigma[x,u], \ V(x) \in \Sigma[x], \ V(0) = 0 \\ & (\beta - p) - (\gamma - V)s_2 \in \Sigma[x] \\ & -((\nabla V \cdot f - u^T u) + (\gamma - V)s_1) \in \Sigma[x,u] \end{aligned} \tag{19.25}$$

Again this problem is bilinear due to the Vs_1 term in the last constraint. A V-s iteration can be used to compute an ellipsoidal bound $\mathcal{E}_\beta$ on $\mathcal{G}(\gamma)$ for any value of γ.

19.3.4 Summary of Approach

The computational tools for other nonlinear problems (estimating regions of attraction, reachability sets, input-output gains with other signal norms, and robustness with respect to uncertainty) all essentially follow the same steps as used in the previous sections. Specifically, a Lyapunov or storage function type theorem is used to derive a sufficient condition for the nonlinear system to have a particular performance/stability property. Lyapunov and storage functions are naturally restricted to be positive definite and this can be enforced using SOS constraints. Additional conditions can typically be formulated as set containment conditions. These set containment conditions can then be converted into function non-negativity constraints using the generalized S-procedure. Since SOS polynomials are non-negative everywhere, the non-negativity constraints can be relaxed and written as SOS constraints. In many cases this sequence of constraint reformulations leads to either a linear or bilinear SOS programming problem which yields a bound on a particular systems property (e.g. inner approximations to regions of attraction or upper bounds on system gains). Bilinear problems can be solved using one of the methods described above. Simulations or gradient searches can be used to compute dual bounds (e.g. outer approximations to regions of attraction or lower bounds on system gains). For example, lower bounds on the local gain can be computed using a power method derived for a finite horizon optimal control problem [40]. This approach provides an improvement over linearized analysis in that the results are valid over a provable region of the state/input space rather than for an infinitesimally small neighborhood of the equilibrium point/null input. Further details on this statement can be found in [20].

19.4 Examples

This section performs nonlinear analyzes for an F/A-18 and NASA's Generic Transport Model. The software used to perform these analyses is available at [31].

19.4.1 ROA Estimation for an F/A-18

The US Navy F/A-18 A/B/C/D Hornet aircraft with the original baseline flight control law experienced a number of out-of-control flight departures since the early 1980's. Many of these incidents have been described as a falling leaf motion of the aircraft [41]. The falling leaf motion has been studied extensively to investigate the conditions that lead to this behavior. The complex dynamics of the falling leaf motion and lack of flight data from the departure events pose a challenge in studying this motion. An extensive revision of the baseline control law was performed by NAVAIR and Boeing in 2001 to suppress departure phenomenon, improve maneuvering performance and to expand the flight envelope [41]. The revised control law was implemented on the F/A-18 E/F Super Hornet aircraft after successful flight tests. These flight tests included aggressive maneuvers that demonstrated successful suppression of the falling leaf motion by the revised control law. This section uses nonlinear region of attraction estimation to compare the closed-loop stability properties of the baseline and revised control laws. Additional details on this analysis can be found in [42].

The falling leaf motion of an aircraft can be characterized as large, coupled out-of-control oscillations in the roll (p) and yaw (r) direction combined with large fluctuations in angle-of-attack (α) and sideslip (β) [43, 41]. Fig. 19.1 shows the main characteristics of the falling leaf motion [43,41]. This out-of-control mode exhibits periodic in-phase roll and yaw rates with large amplitude fluctuations about small or zero mean. The roll and yaw rate generation is mainly due to the large sideslip oscillation. During large sideslip and angle-of-attack motion, the dihedral effect (roll caused by sideslip) of the aircraft wings becomes extremely large and the directional stability becomes unstable. The like-signs of these two values are responsible for the in-phase motion. The roll rate motion can easily reach up to $\pm 120^\circ/s$, while the yaw rate motion can fluctuate around $\pm 50^\circ/s$. During this motion, the value of angle-of-attack can reach up to $\pm 70^\circ$ with sideslip oscillations between $\pm 40^\circ$ [43]. The required aerodynamic nose-down pitching moment is exceeded by the pitch rate generation due to the inertial coupling of the in-phase roll and yaw rates. The reduction in pitching moment is followed by a reduction in normal force, eventually causing a loss of lift in the aircraft. A distinguishing feature of the falling leaf motion is that α vs. β plot produces a mushroom shape curve as seen in Fig. 19.1. For more details on the falling leaf motion, readers are encouraged to refer to the papers by Jaramillo & Ralston [43] and Heller, David & Holmberg [41].

The F/A-18 Hornet is a high performance, twin engine fighter aircraft built by the McDonnell Douglas (currently known as the 'Boeing') Corporation. Each engine is a General Electric, F404-GE-400 rated at 16,100-lbf of static thrust at sea level. The aircraft features a low sweep trapezoidal wing planform with 400 ft^2 area and twin vertical tails [44]. Table 19.1 lists the aerodynamic reference and physical parameters of the aircraft. The conventional F/A-18 Hornet has five pairs of control surfaces: stabilators, rudders, ailerons, leading edge flaps, and trailing edge flaps. However, only the symmetric stabilator, differential aileron and differential rudder are considered as control effectors for the analysis performed in this section. Longitudinal control or pitch control is provided by the symmetric deflection of the stabilators. Deflection of differential ailerons is used to control the roll or lateral

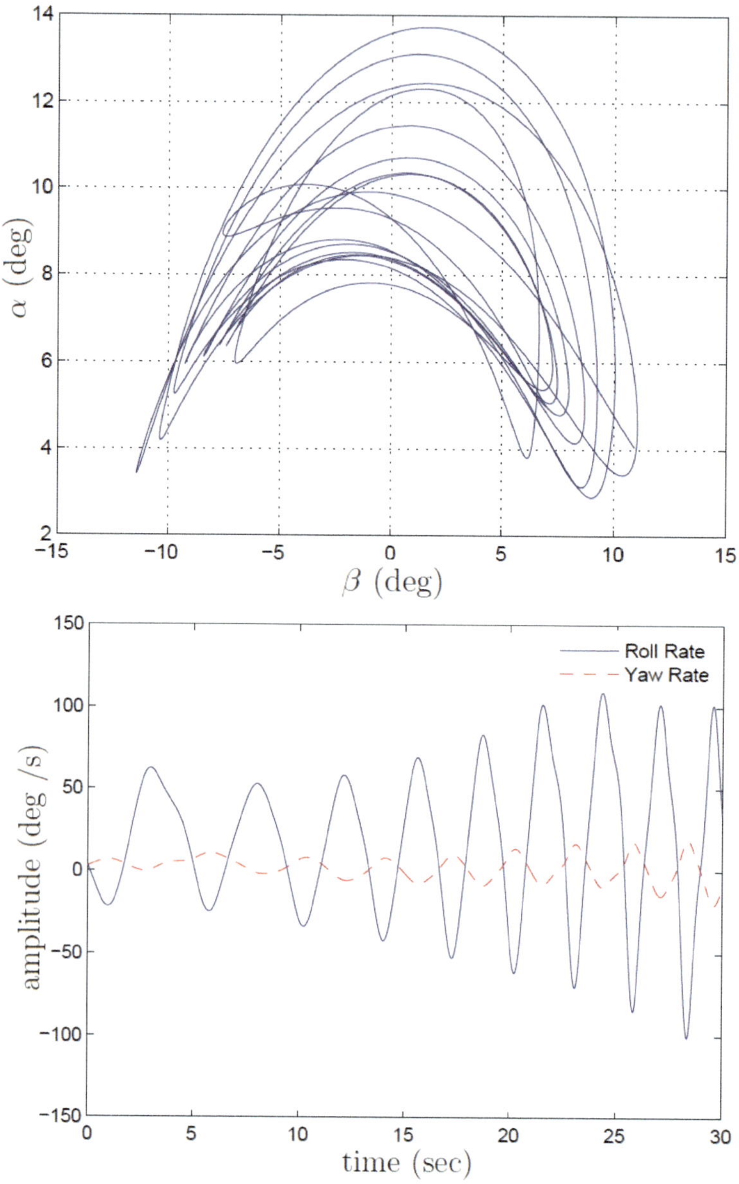

Fig. 19.1 Characteristic behavior of falling leaf motion

direction, while differential deflection of rudders provide directional or yaw control. There is a coupling between roll and yaw dynamics.

The conventional 6DOF aircraft equations of motion are described in Stengel [45], Cook [46], and Napolitano and Spagnuolo [47] are primarily driven by the aerodynamic forces and moments acting on the aircraft. Many flight

Table 19.1 Aircraft parameters

Wing Area, S_{ref}	400 ft^2
Mean Aerodynamic Chord (c)	11.52 ft
Wing Span, b_{ref}	37.42 ft
Weight	33310 lbs
I_{xx}	23000 slug-ft^2
I_{yy}	151293 slug-ft^2
I_{zz}	169945 slug-ft^2
I_{xz}	-2971 slug-ft^2

experiments have been performed to estimate the stability and control derivatives of the F/A-18 High Alpha Research Vehicle (HARV) [48,49,50,51]. The F/A-18 HARV has similar aerodynamic characteristics as the F/A-18 Hornet [52] with the exception of the F/A-18 HARV having thrust vectoring control. Hence, the F/A-18 HARV aerodynamic data are used to construct the aerodynamic coefficient data.

The nonlinear region of attraction analysis requires the aircraft dynamics to be described via a polynomial model. The computational burden of SOS optimization also restricts the model to cubic degree polynomials. Hence, a six state cubic degree polynomial model of the F/A-18 aircraft for roll-coupled maneuvers [53] was constructed for the region of attraction estimation. The polynomial model captures the key characteristics of the full 6 DOF model. This polynomial model is derived based on the characteristics of the falling leaf motion. During the falling-leaf motion, the velocity is usually on the order of 250 ft/s [43]. Hence velocity is assumed to be constant and equal to 250 ft/s in the construction of the 6-state polynomial model. Aggressive maneuvers, like bank turns, are more likely to put the aircraft in the falling leaf motion compared to straight and level flight. Hence, steady bank turn maneuvers with zero climb rate ($\dot{\theta} = 0$) are considered. As a result two other states, pitch angle (θ) and yaw angle (ψ), can be assumed constant in the six state model. Thrust effects in the sideslip direction are also neglected. Small angle approximations are used for the trigonometric terms in the full 6 DOF model to derive a polynomial representation of the aircraft dynamics. Finally, a polynomial least squares fit of the aerodynamic data over a gridded α - β space of $-20° \leq \beta \leq 20°$, and $-10° \leq \alpha \leq 40°$ is performed to obtain the cubic polynomial model. Further details of this polynomial model approximation are provided in [42].

The baseline controller structure for the F/A-18 aircraft closely follows the *Control Augmentation System (CAS)* presented in the report by Buttrill, Arbuckle, and Hoffler [44]. The revised F/A-18 flight control law is described in the papers by Heller, David, & Holmberg [41] and Heller, Niewoehner, & Lawson [54]. The objective of the revised flight control law was to improve the departure resistance characteristics and full recoverability of the

F/A-18 aircraft without sacrificing the maneuverability of the aircraft [41]. The significant change in the revised control law was the additional sideslip (β in rad) and sideslip rate ($\dot{\beta}$ in rad/s) feedback to the aileron actuators. The sideslip feedback plays a key role in increasing the lateral stability in the $30-35°$ range of angle-of-attack. The sideslip rate feedback improves the lateral-directional damping. Hence, sideslip motion is damped even at high angles-of-attack. This feature is key to eliminating the falling leaf mode, which is an aggressive form of in-phase Dutch-roll motion. There are no direct measurements of sideslip and sideslip rate. Therefore, these signals are estimated for feedback. The sideslip and the sideslip rate feedback signals are computed based on already available signals from the sensors and using the kinematics of the aircraft.

The remainder of this section compares the regions of attraction for the baseline and revised control laws. The V-s iteration described in Sect. 19.3.1 is used to compute these ROA estimates. The analysis is performed for the F/A-18 aircraft operating at a steady ($\beta = 0$) bank turn of $\phi = 60^o$. This ROA analysis uses the cubic polynomial models for 60° steady bank turn maneuver. The ordering of the state vector is $x^T := [\beta,\ \mathrm{p},\ \mathrm{r},\ \phi,\ \alpha,\ \mathrm{q},\ x_c]$. The shape matrix for the ellipsoid is chosen to be

$$N := (5)^2 \cdot \mathrm{diag}\,(5^o, 20^o/s, 5^o/s, 45^o, 25^o, 25^o/s, 25^o)^{-2}.$$

This roughly scales each state by the maximum magnitude observed during flight conditions. The factor of $(5)^2$ normalizes the largest entry of the matrix N to be equal to one. The ellipsoid, $x^T N x = \underline{\beta}$, defines the set of initial conditions for which the control law will bring the aircraft back to its trim point. This provides valuable information about the closed-loop stability characteristics. If the aircraft is perturbed due to a wind gust or other upset condition but remains in the ellipsoid then the control law will recover the aircraft and bring it back to trim. In other words the ellipsoid defines a safe flight envelope for the F/A-18. Hence, the ROA provides a measure of how much perturbation the aircraft can tolerate before it becomes unstable. The value of the $\underline{\beta}$ can be thought of as 'nonlinear stability margin'.

As previously mentioned, increasing the degree of the Lyapunov function will improve the lower bound estimate of the ROA. ROA bounds were first computed using the quadratic Lyapunov function from linearized analysis. This method has been proposed for validation of flight control laws [55]. The bound $\underline{\beta}_{LIN} = 8.05 \times 10^{-5}$ was computed for the baseline control law and $\underline{\beta}_{LIN} = 1.91 \times 10^{-4}$ for the revised control. Unfortunately these lower bounds are not particularly useful since they are two to three orders of magnitude smaller than the corresponding upper bounds computed via Monte Carlo search. Next, lower bounds were computed with the V-s iteration using quadratic (degree 2) and quartic (degree 4) Lyapunov functions. The V-s iteration with quadratic Lyapunov functions gives $\underline{\beta}_2 = 3.45 \times 10^{-3}$ for the baseline control law and $\underline{\beta}_2 = 9.43 \times 10^{-3}$ for the revised control law.

These bounds took several minutes to compute. The V-s iteration with quartic Lyapunov functions is $\underline{\beta}_4 = 1.24 \times 10^{-2}$ for the baseline control law and $\underline{\beta}_4 = 2.53 \times 10^{-2}$ for the revised control law. The bounds for the baseline and revised controllers took seven and five hours, respectively, to compute. These bounds are significantly larger than the bounds obtained for the linearized Lyapunov function. A sixth order Lyapunov function would lead to improved lower bounds but with a significant increase in computation time.

The Monte Carlo search, described in Sect. 19.3.1, was used to compute an upper bound on the ROA estimate. A search was performed with 2 million simulations each for the baseline and revised control laws. The baseline control law provides an upper bound of $\bar{\beta}_{MC} = 1.56 \times 10^{-2}$ whereas the revised control law provides an upper bound of $\bar{\beta}_{MC} = 2.95 \times 10^{-2}$. The search also returns an initial condition x_0 on the boundary of the ellipsoid, i.e. $p(x_0) = x_0^T N x_0 = \bar{\beta}_{MC}$, that causes the system to go unstable. Hence, the value of the $\bar{\beta}_{MC}$ provides an upper bound of the ROA for the F/A-18 aircraft. This is complementary information to that provided by the Lyapunov-based lower bounds. The Monte Carlo search returned the following initial condition for the closed system with the baseline control law:

$$x_0 = [-1.1206^o,\ -12.3353^o/s,\ 1.5461^o/s,\ -5.8150^o,\ 28.9786^o,\ 9.9211^o/s,\ 0]^T$$

This initial condition satisfies $p(x_0) = 1.56 \times 10^{-2}$ and the closed-loop system with the baseline control law diverges from this initial condition. Decreasing the initial condition slightly leads to a stable response. For the revised control law the Monte Carlo search returned the following initial condition:

$$x_0 = [0.3276^o,\ -8.0852^o/s,\ 2.8876^o/s,\ -2.1386^o,\ 44.8282^o,\ 9.9829^o/s,\ 0]^T$$

This initial condition satisfies $p(x_0) = 2.95 \times 10^{-2}$ an the closed-loop system with the revised control law diverges from this initial condition. Decreasing the initial condition slightly leads to a stable response.

The lower and upper bounds on β^* can be visualized by plotting slices of the ellipsoidal approximation. Fig. 19.2 and Fig. 19.3 show slices of the inner/outer approximations of the best ellipsoidal ROA approximation for both the baseline and revised control laws. The slices are in the α-β (Fig. 19.2) and p-r (Fig. 19.3) planes. The solid lines show the slices of the inner bounds obtained from quartic Lyapunov analysis. Every initial condition within the solid ellipses will return to the trim condition (marked as a '+'). The dashed lines show the slices of the outer bounds obtained from Monte Carlo analysis. There is at least one initial condition on the outer ellipsoid which leads to a divergent trajectory. The initial condition leading to a divergent trajectory does not necessarily lie on the slice of the ellipsoid shown in the figure. The closeness of the inner and outer ellipsoids means that we have solved, for engineering purposes, the best ROA ellipsoid problem. Recall the aerodynamic coefficients were fitted over a gridded α - β space of $-20° \leq \beta \leq 20°$, and

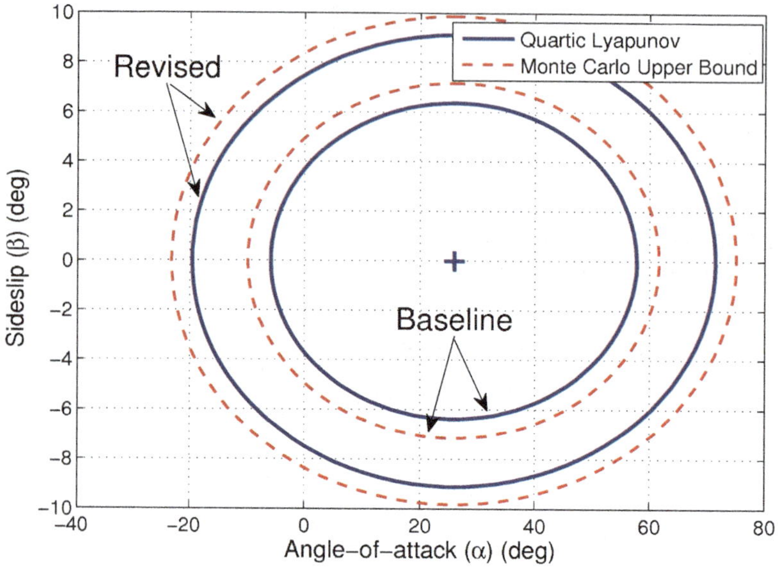

Fig. 19.2 ROA estimates in α-β plane for baseline and revised flight control law around steady 60° bank turn

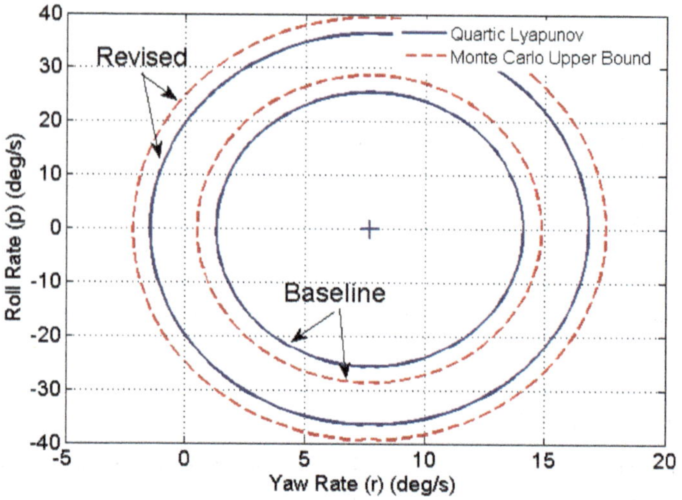

Fig. 19.3 ROA estimates in p-r for baseline and revised flight control law around steady 60° bank turn

$-10° \leq \alpha \leq 40°$. Hence, the model is not valid over the entire region shown in this figure. The lower bounds for the revised controller are larger than the upper bounds for the baseline controller. Thus the ROA for the revised controller is provably larger than the ROA for the baseline controller.

19.4.2 Reachable Set Estimation for NASA's GTM

NASA's Generic Transport Model (GTM) is a remote-controlled 5.5 percent scale commercial aircraft [56,57]. The main GTM aircraft parameters are provided in Table 19.2. NASA constructed a high fidelity 6 degree-of-freedom Simulink model of the GTM with the aerodynamic coefficients described as look-up tables. This section describes the construction of a polynomial model for the longitudinal dynamics of the GTM. This polynomial model is then used to estimate the reachable set for the open-loop longitudinal dynamics. Details on the polynomial modeling are provided in [58]. [59] provides additional motivation for using reachable sets for estimating the safe flight envelope for an aircraft.

Table 19.2 Aircraft and environment parameters

Wing Area, S	5.902 ft^2
Mean Aerodynamic Chord, $\bar{c}$	0.9153 ft
Mass, m	1.542 slugs
Pitch Axis Moment of Inertia, I_{yy}	4.254 slugs-ft^2
Air Density, ρ	0.002375 slugs/ft^3
Gravity Constant, g	32.17 ft/s^2

The longitudinal dynamics of the GTM are described by a standard four-state longitudinal model [60]:

$$\dot{V} = \frac{1}{m}\left(-D - mg\sin(\theta - \alpha) + T_x \cos\alpha + T_z \sin\alpha\right) \tag{19.26}$$

$$\dot{\alpha} = \frac{1}{mV}\left(-L + mg\cos(\theta - \alpha) - T_x \sin\alpha + T_z \cos\alpha\right) + q \tag{19.27}$$

$$\dot{q} = \frac{(M + T_m)}{I_{yy}} \tag{19.28}$$

$$\dot{\theta} = q \tag{19.29}$$

where V is the air speed (ft/s), α is the angle of attack (rad), q is the pitch rate (rad/s) and θ is the pitch angle (rad). The control inputs are the elevator deflection δ_{elev} (deg) and engine throttle δ_{th} (percent).

The drag force D (lbs), lift force L (lbs), and aerodynamic pitching moment M (lb-ft) are given by:

$$D = \bar{q}SC_D(\alpha, \delta_{elev}, \hat{q}) \tag{19.30}$$
$$L = \bar{q}SC_L(\alpha, \delta_{elev}, \hat{q}) \tag{19.31}$$
$$M = \bar{q}S\bar{c}C_m(\alpha, \delta_{elev}, \hat{q}) \tag{19.32}$$

where $\bar{q} := \frac{1}{2}\rho V^2$ is the dynamic pressure (lbs/ft^2) and $\hat{q} := \frac{\bar{c}}{2V}q$ is the normalized pitch rate (unitless). C_D, C_L, and C_m are unitless aerodynamic coefficients computed from look-up tables provided by NASA.

The GTM has one engine on the port side and one on the starboard side of the airframe. Equal thrust settings for both engines is assumed. The thrust from a single engine T (lbs) is a function of the throttle setting δ_{th} (percent). $T(\delta_{th})$ is specified as a ninth-order polynomial in NASA's high fidelity GTM simulation model. T_x (lbs) and T_z (lbs) denote the projection of the total engine thrust along the body x-axis and body-z axis, respectively. T_m (lbs-ft) denotes the pitching moment due to both engines. T_x, T_z and T_m are given by:

$$T_x(\delta_{th}) = n_{ENG}T(\delta_{th})\cos(\epsilon_2)\cos(\epsilon_3) \tag{19.33}$$
$$T_z(\delta_{th}) = n_{ENG}T(\delta_{th})\sin(\epsilon_2)\cos(\epsilon_3) \tag{19.34}$$
$$T_m(\delta_{th}) = r_zT_x(\delta_{th}) - r_xT_z(\delta_{th}) \tag{19.35}$$

$n_{ENG} = 2$ is the number of engines. $\epsilon_2 = 0.0375$ rad and $\epsilon_3 = -0.0294$ rad are angles that specify the rotation from engine axes to the airplane body axes. $r_x = 0.4498$ ft and $r_z = 0.2976$ ft specify the moment arm of the thrust.

The following terms of the longitudinal are approximated by low-order polynomials:

1. Trigonometric functions: $\sin(\alpha)$, $\cos(\alpha)$, $\sin(\theta - \alpha)$, $\cos(\theta - \alpha)$
2. Engine model: $T(\delta_{th})$
3. Rational dependence on speed: $\dfrac{1}{V}$
4. Aerodynamic coefficients: C_D, C_L, C_m

Constructing polynomial approximations for the trigonometric functions, engine model, and rational dependence on speed is relatively straight-forward. The trigonometric functions are approximated by Taylor series expansions: $\sin z \approx z - \frac{1}{6}z^3$ and $\cos z \approx 1 - \frac{1}{2}z^2$ for z in units of radians. For the engine model, a least squares technique is used to approximate the ninth order polynomial function $T(\delta_{th})$ by a third order polynomial. The least squares technique is also used to compute a linear fit to $\dfrac{1}{V}$ over the desired range of interest from 100 ft/s to 200 ft/s. Finally, polynomial least squares fits are computed for the aerodynamic coefficient look-up table provided by NASA. A degree seven polynomial model is obtained after replacing all non-polynomial terms with their polynomial approximations. The polynomial model takes the form:

$$\dot{x} = f(x, u) \tag{19.36}$$

where $x := [V(\text{ft/s}),\ \alpha(\text{rad}),\ q(\text{rad/s}),\ \theta(\text{rad})]$, and $u := [\delta_{elev}(\text{deg}),\ \delta_{th}(\%)]$. The degree seven polynomial model $f(x, u)$ is provided in [58]. The quality of the polynomial approximation was assessed by comparing the trim conditions and simulation responses of the polynomial model and the original model with look-up tables.

The remainder of the section describes the estimation of the reachable set for the open-loop longitudinal dynamics of the GTM. The trim condition for the analysis is:

$$x_{trim} := [150 \text{ ft/s}, 0.047 \text{ rad}, 0 \text{ rad/s}, 0.047 \text{ rad}] \tag{19.37}$$

The throttle input is held at its trim value and the reachable set is computed for elevator inputs around the trim elevator input. The shape function is $p(x) := x^T N x$ where:

$$N := \text{diag}\,(50 \text{ ft/s}, 0.35 \text{ rad}, 0.87 \text{ rad/s}, 0.35 \text{ rad})^{-2} \tag{19.38}$$

Upper bounds $\bar{\beta}$ were computed such that $\mathcal{G}\gamma \subseteq \mathcal{E}_{\bar{\beta}}$. These bounds were computed for many values of γ using the method described in Sect. 19.3.3. A nonlinear optimal control problem can be approximately solved to compute "worst-case" inputs for this reachable set problem [40]. The inputs are worst-case in the sense of maximizing $p(x_f)$ subject to the constraint $\|u\|_2^2 \leq \gamma^2$. This worst-case algorithm provides lower bounds $\underline{\beta}$ for the reachable set problem. These lower bounds provide complementary information to the upper bounds computed using SOS methods. Specifically the lower bounds prove that there is an $x_f \in \mathcal{G}\gamma$ such that $p(x_f) = \underline{\beta}$.

Fig. 19.4 shows the lower and upper bounds computed for the GTM. The upper bounds computed using quadratic and quartic (degree 4) storage functions are shown in squares and diamonds, respectively. Each point on the upper bound curve with quadratic storage functions took several minutes to compute. Only three points were computed for quartic storage functions since each data point on this curve took several hours to compute. Degree six storage functions would lead to improved bounds but with a significant increase in computation time. The lower bound computed using the method in [40] is shown with circles. This curve is quite far from the quartic upper bound. The reachable set for the linearized plant is drawn as a black dashed curve. This is a straight line because scaling the norm of the input scales the distance that can be reached. The optimal input for the linear plant can be computed via an optimal control problem. This worst-case input computed from the linear plant is shown in Fig. 19.5. Simulating the nonlinear system with this worst-case input (scaled to achieve $\|u\|_2^2 = \gamma$) achieved the lower bound shown in squares in Fig. 19.4. This lower bound is very close to the quartic upper bound. Thus the worst-case input computed from the linear plant is also a bad input for the nonlinear plant. It appears that the GTM

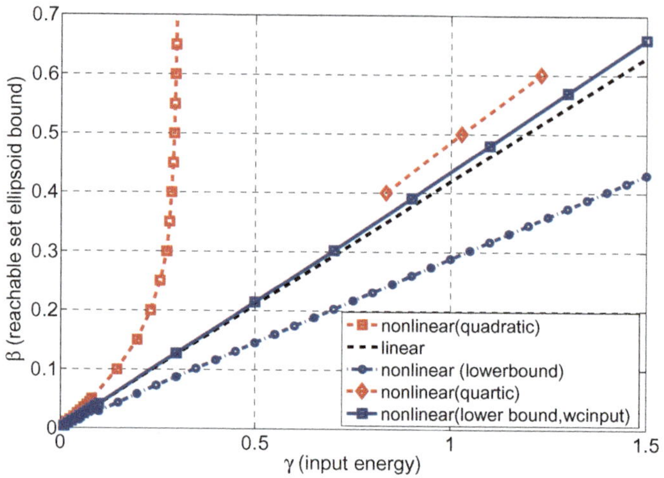

Fig. 19.4 Reachable set bounds

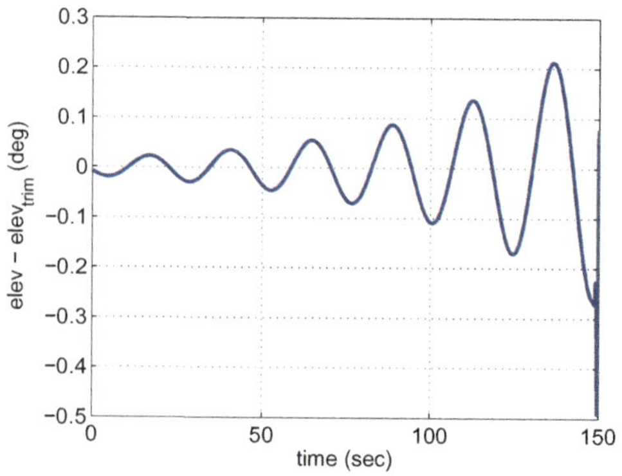

Fig. 19.5 Worst-case input computed from the linear plant

dynamics are well-approximated by the linearization at this flight condition. This statement was further investigated by simulating the full 6DOF GTM model with the worst-case input computed from the linear plant (scaled to have $\|u\|_2^2 = 1$). The 6DOF GTM model includes many additional modeling details including lateral dynamics, actuator dynamics, etc. Surprisingly the response of the full 6DOF GTM model with this input was very similar to the response of the polynomial model with this input. A comparison of the simulation responses is shown in Fig. 19.6. The states (α,θ,q) have units of (rad,rad,rad/s) but are plotted in units of (deg,deg,deg/s) for ease of interpretation. For this input the polynomial model achieved $\underline{\beta} = 0.4803$ and the full GTM achieved $\underline{\beta} = 0.3478$. This nonlinear reachable analysis provides

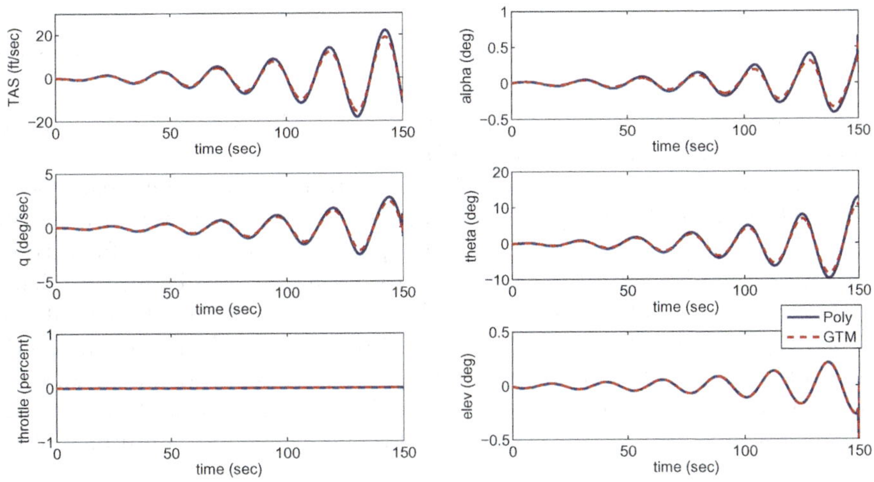

Fig. 19.6 Simulation of polynomial and 6DOF GTM models with worst-case input

confidence that the linear model can be used as a good approximation at this flight condition.

19.5 Conclusion

This chapter described the use of sum-of-squares optimizations for analyzing nonlinear polynomial systems. In particular, optimizations with SOS constraints were formulated for computing region of attraction estimates, bounds on L_2 to L_2 gain, and reachable sets. Many other nonlinear analysis problems can be formulated within this optimization framework. The approach was applied to compare the performance of two F/A-18 control laws in suppressing a loss-of-control motion known as the falling leaf mode. The reachable set for NASA's Generic Transport Model was also estimated. These nonlinear analysis tools can fill the gap between linear analyses, which are valid only for infinitesimally small neighborhoods about an equilibrium, and nonlinear simulations. These tools can be used to provide additional confidence when validating the performance of a flight control law. Significant work remains to be done to reduce the computational cost and enable these techniques to be applied to moderate-sized systems (systems with more than ≈ 8 states).

Acknowledgments

This research was partially supported under the NASA Langley NRA contract NNH077ZEA001N entitled "Analytical Validation Tools for Safety Critical Systems" and the NASA Langley NNX08AC65A contract entitled 'Fault

Diagnosis, Prognosis and Reliable Flight Envelope Assessment." The technical contract monitors are Dr. Christine Belcastro and Dr. Suresh Joshi respectively.

References

1. Parrilo, P.: Structured Semidefinite Programs and Semialgebraic Geometry Methods in Robustness and Optimization. PhD thesis, California Institute of Technology (2000)
2. Parrilo, P.: Semidefinite programming relaxations for semialgebraic problems. Mathematical Programming Ser. B 96(2), 293–320 (2003)
3. Tibken, B.: Estimation of the domain of attraction for polynomial systems via LMIs. In: Proceedings of the IEEE Conference on Decision and Control, pp. 3860–3864 (2000)
4. Hachicho, O., Tibken, B.: Estimating domains of attraction of a class of nonlinear dynamical systems with LMI methods based on the theory of moments. In: Proceedings of the IEEE Conference on Decision and Control, pp. 3150–3155 (2002)
5. Jarvis-Wloszek, Z.: Lyapunov Based Analysis and Controller Synthesis for Polynomial Systems using Sum-of-Squares Optimization. PhD thesis, University of California, Berkeley (2003)
6. Jarvis-Wloszek, Z., Feeley, R., Tan, W., Sun, K., Packard, A.: Some controls applications of sum of squares programming. In: Proceedings of the 42nd IEEE Conference on Decision and Control, vol. 5, pp. 4676–4681 (2003)
7. Tan, W., Packard, A.: Searching for control Lyapunov functions using sums of squares programming. In: 42nd Annual Allerton Conference on Communications, Control and Computing, pp. 210–219 (2004)
8. Chesi, G.: On the estimation of the domain of attraction for uncertain polynomial systems via LMIs. In: Proceedings of the IEEE Conference on Decision and Control, pp. 881–886 (2004)
9. Papachristodoulou, A.: Scalable analysis of nonlinear systems using convex optimization. PhD thesis, California Institute of Technology (2005)
10. Jarvis-Wloszek, Z., Feeley, R., Tan, W., Sun, K., Packard, A.: Controls Applications of Sum of Squares Programming. In: Positive Polynomials in Control. LNCIS, vol. 312, pp. 3–22. Springer, Heidelberg (2005)
11. Chesi, G., Garulli, A., Tesi, A., Vicino, A.: Lmi-based computation of optimal quadratic lyapunov functions for odd polynomial systems. International Journal of Robust and Nonlinear Control 15, 35–49 (2005)
12. Prajna, S.: Optimization-Based Methods for Nonlinear and Hybrid Systems Verification. PhD thesis, California Institute of Technology (2005)
13. Tan, W., Packard, A., Wheeler, T.: Local gain analysis of nonlinear systems. In: Proceedings of the American Control Conference, pp. 92–96 (2006)
14. Tan, W.: Nonlinear Control Analysis and Synthesis using Sum-of-Squares Programming. PhD thesis, University of California, Berkeley (2006)
15. Tibken, B., Fan, Y.: Computing the domain of attraction for polynomial systems via BMI optimization methods. In: Proceedings of the American Control Conference, pp. 117–122 (2006)

16. Topcu, U., Packard, A., Seiler, P., Wheeler, T.: Stability region analysis using simulations and sum-of-squares programming. In: Proceedings of the American Control Conference, pp. 6009–6014 (2007)
17. Tan, W., Topcu, U., Seiler, P., Balas, G., Packard, A.: Simulation-aided reachability and local gain analysis for nonlinear dynamical systems. In: Proceedings of the IEEE Conference on Decision and Control, pp. 4097–4102 (2008)
18. Topcu, U., Packard, A., Seiler, P.: Local stability analysis using simulations and sum-of-squares programming. Automatica 44(10), 2669–2675 (2008)
19. Topcu, U.: Quantitative Local Analysis of Nonlinear Systems. PhD thesis, University of California, Berkeley (2008)
20. Topcu, U., Packard, A.: Linearized analysis versus optimization-based nonlinear analysis for nonlinear systems. Submitted to the 2009 American Control Conference (2009)
21. Boyd, S., El Ghaoui, L., Feron, E., Balakrishnan, V.: Linear Matrix Inequalities in System and Control Theory. Studies in Applied Mathematics, vol. 15. SIAM, Philadelphia (1994)
22. Reznick, B.: Some concrete aspects of Hilberts 17th problem. Contemporary Mathematics 253 (2000)
23. Vandenberghe, L., Boyd, S.: Semidefinite programming. SIAM Review 38(1), 49–95 (1996)
24. Sturm, J.F.: Using SeDuMi 1.02, a MATLAB toolbox for optimization over symmetric cones. Optimization Methods and Software, 625–653 (1999)
25. Sturm, J.: SeDuMi version 1.05 (2001), `http://sedumi.ie.lehigh.edu/`
26. Reznick, B.: Extremal PSD forms with few terms. Duke Mathematical Journal 45(2), 363–374 (1978)
27. Choi, M.D., Lam, T.Y., Reznick, B.: Sums of squares of real polynomials. Proceedings of Symposia in Pure Mathematics 58(2), 103–126 (1995)
28. Powers, V., Wörmann, T.: An algorithm for sums of squares of real polynomials. Journal of Pure and Applied Algebra 127, 99–104 (1998)
29. Prajna, S., Papachristodoulou, A., Seiler, P., Parrilo, P.A.: SOSTOOLS: Sum of squares optimization toolbox for MATLAB (2004)
30. Lofberg, J.: Yalmip: A toolbox for modeling and optimization in MATLAB. In: Proceedings of the CACSD Conference, Taipei, Taiwan (2004)
31. Balas, G.J., Packard, A., Seiler, P., Topcu, U.: Robustness analysis of nonlinear systems (2009), `http://www.aem.umn.edu/~AerospaceControl/`
32. Gatermann, K., Parrilo, P.: Symmetry groups, semidefinite programs, and sums of squares. Journal of Pure and Applied Algebra 192, 95–128 (2004)
33. Vannelli, A., Vidyasagar, M.: Maximal Lyapunov functions and domains of attraction for autonomous nonlinear systems. Automatica 21(1), 69–80 (1985)
34. Hauser, J., Lai, M.C.: Estimating quadratic stability domains by nonsmooth optimization. In: Proceedings of the American Control Conference, pp. 571–576 (1992)
35. Genesio, R., Tartaglia, M., Vicino, A.: On the estimation of asymptotic stability regions: State of the art and new proposals. IEEE Transactions on Automatic Control 30(8), 747–755 (1985)
36. Davison, E.J., Kurak, E.M.: A computational method for determining quadratic Lyapunov functions for nonlinear systems. Automatica 7, 627–636 (1971)

37. Chiang, H.-D., Thorp, J.S.: Stability regions of nonlinear dynamical systems: A constructive methodology. IEEE Transactions on Automatic Control 34(12), 1229–1241 (1989)
38. Vidyasagar, M.: Nonlinear Systems Analysis, 2nd edn. Prentice-Hall, Englewood Cliffs (1993)
39. Khalil, H.K.: Nonlinear Systems, 3rd edn. Prentice-Hall, Englewood Cliffs (2002)
40. Tierno, J., Murray, R., Doyle, J.C., Gregory, I.M.: Numerically efficient robustness analysis of trajectory tracking for nonlinear systems. AIAA Journal of Guidance, Control, and Dynamics 20, 640–647 (1997)
41. Heller, M., David, R., Holmberg, J.: Falling leaf motion suppression in the F/A-18 Hornet with revised flight control software. In: AIAA Aerospace Sciences Meeting, number AIAA-2004-542 (2004)
42. Chakraborty, A., Seiler, P., Balas, G.: Applications of linear and nonlinear robustness analysis techniques to the f/a- 18 flight control laws. In: AIAA Guidance, Navigation, and Control Conference, number AIAA-2009-5670 (2009)
43. Jaramillo, P.T., Ralston, J.N.: Simulation of the F/A-18D falling leaf. In: AIAA Atmospheric Flight Mechanics Conference, pp. 756–766 (1996)
44. Buttrill, S.B., Arbuckle, P.D., Hoffler, K.D.: Simulation model of a twin-tail, high performance airplane. Technical Report NASA TM-107601, NASA (1992)
45. Stengel, R.F.: Flight Dynamics. Princeton University Press, Princeton (2004)
46. Cook, M.V.: Flight Dynamics Principles. Wiley, Chichester (1997)
47. Napolitano, M.R., Spagnuolo, J.M.: Determination of the stability and control derivatives of the NASA F/A-18 HARV using flight data. Technical Report NASA CR-194838, NASA (1993)
48. Marcello, R., Napolitano, A.C.: Paris, and Brad A. Seanor. Estimation of the lateral-directional aerodynamic parameters from flight data for the NASA F/A-18 HARV. In: AIAA Atmospheric Flight Mechanics Conference, number AIAA-96-3420-CP, pp. 479–489 (1996)
49. Lluch, C.D.: Analysis of the out-of-control falling leaf motion using a rotational axis coordinate system. Master's thesis, Virginia Polytechnic Institue and State Unniversity (1998)
50. Marcello, R., Napolitano, A.C.: Paris, and Brad A. Seanor. Estimation of the longitudinal aerodynamic parameters from flight data for the NASA F/A-18 HARV. In: AIAA Atmospheric Flight Mechanics Conference, number AIAA-96-3419-CP, pp. 469–478
51. Iliff, K.W., Wang, K.-S.C.: Extraction of lateral-directional stability and control derivatives for the basic F-18 aircraft at high angles of attack. NASA TM-4786 (1997)
52. Illif, K.W., Wang, K.-S.C.: Retrospective and recent examples of aircraft parameter identification at NASA dryden flight research center. Journal of Aircraft 41(4) (2004)
53. Schy, A.A., Hannah, M.E.: Prediction of jump phenomena in roll-coupled maneuvers of airplanes. Journal of Aircraft 14(4), 375–382 (1977)
54. Heller, M., Niewoehner, R., Lawson, P.K.: High angle of attack control law development and testing for the F/A-18E/F Super Hornet. In: AIAA Guidance, Navigation, and Control Conference, number AIAA-1999-4051, pp. 541–551 (1999)

55. Heller, M., Niewoehner, R., Lawson, P.K.: On the validation of safety critical aircraft systems, part i: An overview of analytical & simulation methods. In: AIAA Guidance, Navigation, and Control Conference, number AIAA 2003-5559 (2003)
56. Cox, D.: The GTM DesignSim v0905 (2009)
57. Murch, A.M., Foster, J.V.: Recent NASA research on aerodynamic modeling of post-stall and spin dynamics of large transport airplanes. In: 45th AIAA Aerospace Sciences Meeting and Exhibit, Reno, Nevada (2007)
58. Chakraborty, A., Seiler, P., Balas, G.: Nonlinear region of attraction analysis for flight control verification and validation. Control Engineering Practice (2010) (submitted for publication)
59. Pandita, R., Seiler, P., Balas, G.: Reachability and region of attraction analysis applied to gtm dynamic flight envelope assessment. In: AIAA Guidance, Navigation, and Control Conference, number AIAA-2009-6258 (2009)
60. Stevens, B.L., Lewis, F.L.: Aircraft Control and Simulaion. John Wiley & Sons, Chichester (1992)

Chapter 20
Validation and Verification (V&V) of Safety-Critical Systems Operating under Off-Nominal Conditions

Christine M. Belcastro

Abstract. *Loss of control* (LOC) remains one of the largest contributors to aircraft fatal accidents worldwide. Aircraft LOC accidents are highly complex in that they can result from numerous causal and contributing factors acting alone or more often in combination. Hence, there is no single intervention strategy to prevent these accidents. Research is underway at the *National Aeronautics and Space Administration* (NASA) in the development of advanced onboard system technologies for preventing or recovering from loss of vehicle control and for assuring safe operation under off-nominal conditions associated with aircraft LOC accidents. The transition of these technologies into the commercial fleet will require their extensive *validation and verification* (V&V) and ultimate certification. The V&V of complex integrated systems poses highly significant technical challenges and is the subject of a parallel research effort at NASA. This chapter summarizes the V&V problem and presents a proposed process that could be applied to complex integrated safety-critical systems developed for preventing aircraft LOC accidents. A summary of recent research accomplishments in this effort is referenced.

20.1 Introduction: Motivation for Off-Nominal Conditions

Aircraft LOC accidents can result from numerous causal and contributing factors that are collectively referred to in this chapter as "off-nominal conditions". "Off-nominal" conditions include adverse conditions occurring onboard the vehicle, such as system failures, external hazards, such as inclement weather, and abnormal flight conditions, such as stall/departure. A more

Christine M. Belcastro
NASA Langley Research Center, Hampton, VA 23681-0001, USA
e-mail: christine.m.belcastro@nasa.gov

A. Varga et al. (Eds.): Opti. Based Clearance of Flight Control Laws, LNCIS 416, pp. 399–419.
springerlink.com

detailed description of off-nominal conditions associated with aircraft LOC accidents is given in Sect. 20.1.1.

Current aircraft autopilot systems are primarily designed for operation under nominal conditions, and sometimes disengage and return control authority to the pilot under off-nominal conditions. Future aircraft control systems will be expected to provide resilience under off-nominal conditions and operate as a component of a larger resilient flight system. Control resilience will need to be designed into future systems to provide the capability to mitigate off-nominal conditions and provide recovery back to a stable operational mode whenever possible. This capability will be developed as part of a holistic approach to reduce aircraft LOC accidents. The broader resilient flight system will include vehicle health management, flight safety management, and reliable crew interface management functions.

V&V becomes much more difficult for safety-critical resilient systems operating under off-nominal conditions. The objectives of this chapter are to address V&V issues associated with future safety-critical resilient flight systems operating under off-nominal conditions and to propose a comprehensive V&V research framework to address these issues. The remainder of Sect. 20.1 describes aircraft loss of control in more detail (Sect. 20.1.1) and presents a future resilient flight system concept (Sect. 20.1.2). Sect. 20.2 defines the V&V problem associated with future resilient flight systems, describes problem complexity and key technical challenges, identifies V&V process requirements, and summarizes a research approach being taken at NASA. Sect. 20.3 presents a comprehensive V&V process that can serve as an initial research framework for addressing future integrated resilient flight systems. Sect. 20.4 briefly discusses the status of this research and references a detailed summary of research accomplishments made at NASA Langley. Sect. 20.5 provides a chapter summary and some concluding remarks. The primary emphasis of this chapter is on the validation component of V&V for advanced flight control systems.

20.1.1 Aircraft LOC

LOC remains one of the largest worldwide contributors to aircraft fatal accidents. For example, a summary of worldwide commercial jet airplane accidents from 2000 through 2009 [1] is shown in Fig. 20.1. As indicated in the figure, *in-flight loss of control* (LOC-I) is the largest accident category for transport aircraft weighing more than 60,000 pounds, and resulted in 20 accidents and 1,848 total fatalities. The data in Fig. 20.1 show the number of fatalities for accident categories defined by the *Commercial Aviation Safety Team* (CAST) and the *International Civil Aviation Organization* (ICAO). A full definition of the CAST/ICAO accident categories is provided in Table 20.1.

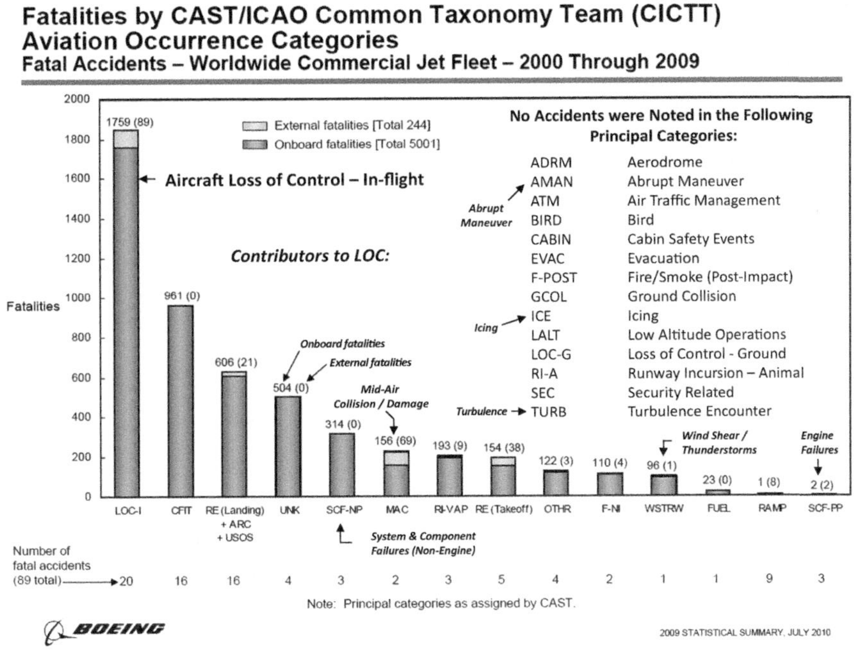

Fig. 20.1 Aircraft Accident Statistics for Worldwide Commercial Jet Fleet, 2000-2009 [1]

Aircraft LOC is a highly complex event. Some contributors to aircraft LOC are denoted in Fig. 20.1. Although some LOC factors noted in Fig. 20.1 were not determined to be primary causal factors of any accidents in this class of vehicles (i.e., over 60,000 lbs.) during the stated time period, in general they have been found to contribute to LOC accidents and are therefore noted for completeness. Causal and contributing factors associated with aircraft LOC can occur individually, but more often occur in various combinations. A detailed analysis of 126 aircraft LOC accidents is presented in [2], in which worst case combinations of LOC accident precursors, i.e., causal and contributing factors, and their time sequences are identified. These factors, or "off-nominal conditions," can be categorized as: adverse conditions occurring onboard the aircraft, including faults, failures, damage, crew error, etc.; external hazards and disturbances, including icing, wind shear, wake vortices, turbulence, terrain and obstacles, other aircraft, etc.; and abnormal flight or upset conditions, including unusual attitudes, stall, stall/departure, etc..

Aircraft LOC clearly involves operation under off-nominal conditions, which motivates the use of the term "off-nominal conditions" to designate the associated causal and contributing factors. LOC accidents occur across all vehicle classes, from small aircraft through large transports, and configuration types, from single to multiple engines, including both jet and propeller.

Table 20.1 CAST/ICAO Accident Categories

AMAN	Abrupt Maneuver
ADRM	Aerodrome
ARC	Abnormal Runway Contact
ATM	Air Traffic Management/Communications, Navigation, Surveillance
CABIN	Cabin Safety Events
CFIT	Controlled Flight into or Toward Terrain
EVAC	Evacuation
F-NI	Fire/Smoke (Non-Impact)
F-POST	Fire/Smoke (Post-Impact)
FUEL	Fuel Related
GCOL	Ground Collision
ICE	Icing
LALT	Low Altitude Operations
LOC-G	Loss of Control – Ground
LOC-I	Loss of Control – In flight
MAC	Midair/Near Midair Collision
OTHR	Other
RAMP	Ground Handling
RE	Runway Excursion
RI-A	Runway Incursion – Animal
RI-VAP	Runway Incursion – Vehicle, Aircraft or Person
SEC	Security Related
SCF-NP	System/Component Failure or Malfunction (Non-Power Plant)
SCF-PP	System/Component Failure or Malfunction (Power Plant)
TURB	Turbulence Encounter
USOS	Undershoot/Overshoot
UNK	Unknown or Undetermined
WSTRW	Wind Shear or Thunderstorm

LOC also occurs across all operational categories, scheduled and unscheduled, and flight phases, including takeoff, cruise, and approach.

Because of the scope and complexity of aircraft LOC events, i.e., accidents and incidents, there is no single intervention strategy for preventing them. Improved crew training and operational procedures for off-nominal conditions can enable improved crew response during LOC events. Advanced onboard systems that provide resilience to off-nominal conditions can enable improved situational awareness and vehicle response under LOC events. A holistic approach for preventing aircraft LOC accidents is presented in the next section.

20.1.2 Future Advanced System Concept

Improved capabilities are needed for off-nominal conditions that enable effective crew training, enhanced situational awareness, and onboard resilience. Underlying technologies to achieve these capabilities have been the subject of research at NASA over the past decade, i.e., since the year 2000, within NASA's *Aviation Safety Program* (AvSP). Core technology areas of research include: 1.) dynamics modeling and simulation for off-nominal conditions; 2.) diagnostics and prognostics for detecting, identifying, and characterizing off-nominal conditions in real time or near real time; 3.) resilient control technologies for mitigation of off-nominal conditions and vehicle recovery; and 4.) crew interface technologies for improved situational awareness and decision support especially under off-nominal conditions.

These core technology areas must be coordinated during both development and operation. V&V technologies must also be developed and applied to these technology areas for an improved understanding of safe and unsafe regions of operation under off-nominal conditions, and for the ultimate certification of these technologies.

An integrated system concept can be developed based on these technologies for preventing aircraft LOC accidents in the future. One such future concept, called the *Aircraft Integrated Resilient Safety Assurance and Failsafe Enhancement* (AIRSAFE) System, is shown in Fig. 20.2.

The shading and block shapes of Fig. 20.2 designate the four core technology areas just discussed. Medium shading represents vehicle health management functions, no shading represents crew interface management

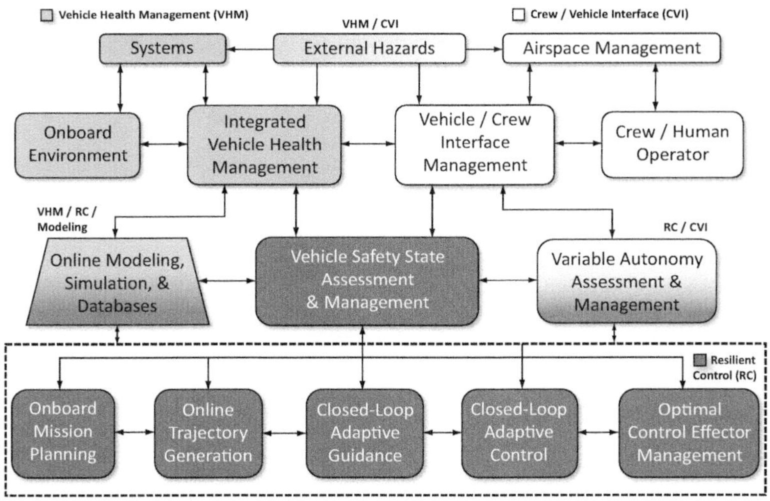

Fig. 20.2 Aircraft Integrated Resilient Safety Assurance and Failsafe Enhancement (AIRSAFE) System Concept

functions, and dark shading represents flight safety management and resilient control functions. The trapezoidal shape represents modeling and simulation functions for off-nominal conditions. Multi-shaded blocks represent shared functions between multiple technology areas. A detailed description of the functional capabilities and interfaces associated with the AIRSAFE System concept is contained in [3, 4].

The V&V of future integrated systems, such as the AIRSAFE System concept of Fig. 20.2, poses numerous technical challenges. In particular, there is no current V&V capability for complex integrated safety-critical systems operating under off-nominal conditions. This problem is the subject of Sect. 20.2.

20.2 V&V Problem

The V&V of integrated safety-critical systems that are designed for operation under off-nominal conditions is a complex problem. The V&V process must ultimately lead to system certification. The *Federal Aviation Administration* (FAA) in the United States and the *Joint Aviation Authorities* (JAA) in Europe have developed extensive and compatible certification specifications. The *Federal Aviation Regulation* (FAR) and *Joint Aviation Regulation* (JAR) Part 25 provides the certification specifications for transport category aircraft, and Sect. 1309 applies to equipment and systems installed onboard aircraft. An excerpt from FAR 25.1309 is provided below, and JAR 25.1309 is nearly identical.

Part 25 AIRWORTHINESS STANDARDS: TRANSPORT CATEGORY AIRPLANES
Sec. **25.1309**: *Equipment, systems, and installations.*

(*a*) *The equipment, systems, and installations whose functioning is required by this subchapter, must be designed to ensure that they perform their intended functions under any foreseeable operating condition.*
(*b*) *The airplane systems and associated components, considered separately and in relation to other systems, must be designed so that –*
  (1) *The occurrence of any failure condition which would prevent the continued safe flight and landing of the airplane is extremely improbable, and*
  (2) *The occurrence of any other failure conditions which would reduce the capability of the airplane or the ability of the crew to cope with adverse operating conditions is improbable.*

(*c*) *Warning information must be provided to alert the crew to unsafe system operating conditions, and to enable them to take appropriate corrective action. Systems, controls, and associated monitoring and warning means must be designed to minimize crew errors which could create additional hazards.*
(*d*) *Compliance with the requirements of paragraph* (*b*) *of this section must be shown by analysis, and where necessary, by appropriate ground, flight, or simulator tests. The analysis must consider –*

(1) *Possible modes of failure, including malfunctions and damage from external sources.*
(2) *The probability of multiple failures and undetected failures.*
(3) *The resulting effects on the airplane and occupants, considering the stage of flight and operating conditions, and*
(4) *The crew warning cues, corrective action required, and the capability of detecting faults.*

The terminology "extremely improbable" in FAR 25.1309 translates to an average probability per flight hour for catastrophic failure conditions of 10^{-9}, and "improbable" failure conditions are those having a probability on the order of 10^{-5} or less per flight hour (but greater than 10^{-9}). The development of a V&V process for demonstration of compliance to FAR/JAR 25.1309 is extremely challenging for complex integrated systems designed for operation under off-nominal conditions, such as the AIRSAFE System concept of Fig. 20.2. In fact, the V&V problem for these systems poses a key technology barrier to their implementation and transition into the fleet. There are currently no comprehensive V&V processes for certifying advanced safety-critical control systems, commercial or military, for effective operation under off-nominal conditions, or even for adaptive and potentially non-deterministic systems. The following subsections discuss V&V problem complexity and key technical challenges for the AIRSAFE future system concept, V&V process requirements for meeting those challenges, and a research approach being taken at NASA to address V&V of future safety-critical systems.

20.2.1 V&V Problem Complexity and Technical Challenges

V&V of safety-critical integrated systems operating under off-nominal conditions can be thought of and analyzed as a complex multidimensional problem [5]. V&V problem complexity can be discussed in terms of system complexity, operational complexity, and V&V process complexity.

System complexity arises from integrating vehicle health management functions, resilient control functions, flight safety assessment and prediction functions, and crew interface and variable autonomy functions. Each of these functions is characterized by algorithmic diversity that must be addressed in the V&V process. Vehicle health management involves diagnostic and prognostic algorithms that utilize stochastic decision-based reasoning and extensive information processing and data fusion. Resilient control functions can involve adaptive control algorithms that utilize time-varying parameters and/or hybrid system switching. Flight safety management may involve diagnostic and prognostic reasoning algorithms as well as control theoretic algorithms. Crew interface functions involve displays that are human-factor-based and require information processing, and variable autonomy will require assessment and reasoning algorithms. Onboard modeling functions will involve system identification algorithms and databases. All four core functions are software based and will involve various levels of logic and discrete mathematics-based abstractions and combinations. Subsystem integration will also involve significant software and possible hardware complexity.

The second aspect of V&V complexity arises from operational complexity. Normal operating conditions of the future may extend beyond current-day operational limits. Moreover, safe operation under off-nominal conditions that could lead to LOC events will be a focus of the system design. In particular, operation under abnormal flight conditions, external hazards and disturbances, adverse onboard conditions, and key combinations of these conditions will be a major part of the operational complexity required for future safety-critical systems. Future air transportation systems [6] must also be considered under operational complexity, such as requirements for dense all-weather operations, self separation of aircraft, and mixed capabilities of aircraft operating in the same airspace, including current and future vehicle configurations as well as piloted and autonomous vehicles.

The third aspect of V&V complexity pertains to the V&V process itself. A wide variety of analytical methods will be needed to evaluate stability and performance of various and dissimilar system functions, robustness to adverse and abnormal conditions, and reliability under errors, faults, failures, and damage. Simulation methods will require the development of high-fidelity models that characterize off-nominal conditions and their multidisciplinary effects on the vehicle. The capability for multidisciplinary subsystem integration must also be available in a simulation environment, as well as the inclusion of pilot-in-the-loop effects. Simulation capability must range from desk-top batch operation to hardware/pilot-in-the-loop fixed/motion-based evaluations. Experimental test capability must include ground and flight testing of hardware/software systems, allow for multidisciplinary subsystem integration, and enable realistic emulation of off-nominal conditions. The V&V process must itself be assessed for its predictive capability to effectively infer safe system operation under off-nominal conditions associated with aircraft LOC events that cannot be fully replicated during V&V. The V&V process assessment must be able to quantify a level of confidence in this inference.

Operation under off-nominal conditions over a wide envelope of flight conditions results in a very large operational space with multidisciplinary coupled effects. Due to the huge operational space, there are too many conditions to fully analyze, simulate, and test. While there are numerous technical challenges associated with this problem, some key technical challenges are summarized below.

- Development and Validation of Physics-Based Off-Nominal Conditions and Effects Models
  - Requires modeling of
    - ▷ adverse onboard conditions (e.g., faults, failures, damage)
    - ▷ abnormal flight conditions (e.g., unusual attitudes, stall, stall/departure, other vehicle upset conditions)

    - ▷ external hazards and disturbances (e.g., icing, wind shear, wake vortices, turbulence)
    - ▷ worst-case combinations, as determined from LOC Accident/Incident data
  - Requires data and/or experimental methods for off-nominal conditions, which may not be available or easily obtained
  - Can involve multidisciplinary coupled effects
  - Cannot fully replicate in-flight LOC environment

- V&V of Adaptive Diagnostic, Prognostic, and Control Algorithms Operating under Off-Nominal Conditions

  - Involves a variety of nonlinear mathematical constructs (e.g., inference engines, probabilistic methods, physics-based, neural networks, artificial intelligence, etc.)
  - May involve onboard adaptation that may result in stochastic system behavior
  - Involves fusion and reasoning algorithms for sensor data, information processing, and decisions
  - Requires methods for establishing probabilities of
    - ▷ false alarms and missed detections
    - ▷ incorrect identifications and decisions
    - ▷ loss of stability, recoverability, and control
  - Requires methods and metrics for establishing off-nominal condition coverage, reliability, and accuracy for diverse algorithms and multiple objectives
  - Requires integrated multi-disciplinary system assessment methods
    - ▷ performance assessment
    - ▷ error propagation and effects assessment
    - ▷ inter-operability effectiveness assessment

- System Verification and Safety Assurance

  - Involves large-scale complex interconnected software systems
  - Involves potentially fault tolerant and reconfigurable hardware
  - May involve adaptive and reasoning algorithms with stochastic behavior
  - Requires verification methods for a complex system of systems

- V&V Predictive Capability Assessment

  - Requires methods to demonstrate compliance to certification standards for an extensive set of off-nominal conditions and their combinations that cannot be fully replicated
  - Requires methods for determining and quantifying level of confidence in V&V process and results for demonstrating compliance

These technical challenges can be utilized in defining V&V process requirements, as presented in Sect. 20.2.2.

20.2.2 V&V Process Requirements

In carrying out V&V of complex integrated safety-critical systems operating under off-nominal conditions, it is necessary to expose system weaknesses and vulnerabilities, and to be able to identify safe and unsafe operational conditions, regions, and their boundaries. This is a key point. It is not sufficient, for example, to demonstrate that a system appears to work in a few selected flight regimes or under a small subset of off-nominal conditions. In fact, it is necessary to define a comprehensive integrated V&V process for these systems, and to utilize this process as a research framework to identify gaps in current V&V capabilities. Moreover, it is critical to define a V&V process that effectively and efficiently utilizes analysis, simulation, and experimental testing to assist in exposing system deficiencies and limitations over a very large operational space. The V&V process must clearly demonstrate compliance to certification specifications, such as FAR/JAR 25.1309, and quantify a level of confidence in this compliance.

Key components of the V&V process include algorithm validation, system verification, and V&V predictive capability assessment. Each of these V&V components requires the development of methods, tools, and testbeds to perform analysis, simulation/ground testing, and flight testing. Moreover, each method, tool, and testbed must be developed to assess system operation under off-nominal conditions associated with aircraft LOC accidents in order to reduce or prevent them in the future. V&V metrics must be defined for the diverse set of algorithms associated with the subsystems and integrated system, and new methods, tools, and testbeds developed as needed to assess these metrics. Based on an analysis of the V&V problem, the V&V process requirements for future systems designed for operation under off-nominal conditions, such as the AIRSAFE System concept, can be defined as depicted in Fig. 20.3. This figure shows V&V process components, methods, and some example algorithm validation metrics that are required for AIRSAFE subsystem and integrated system technologies. The core V&V methods of analysis, simulation/ground testing, and flight testing are applicable to each of the core V&V components and take on different meanings for each. Metrics must be developed for assessment of each core component using the appropriate methods. Although Fig. 20.3 shows some example metrics for algorithm validation, and illustrates that these are dependent on the algorithm type, metrics are needed for each core V&V component.

System validation is a confirmation that the algorithms are performing the intended function under all possible operating conditions. Validation is not merely a demonstration that the system works under the design condition and selected test conditions, but a comprehensive process that involves analytical, simulation/ground testing, and flight testing. The validation subprocess must be capable of identifying potentially problematic regions of operation, and their boundaries, and exposing system limitations - particularly for operation under off-nominal conditions. Fig. 20.3 presents some of the methods and metrics needed for the analysis, simulation/ground testing, and flight testing of algorithms associated with AIRSAFE System technologies. New methods, tools, testbeds, and metrics must be established for algorithms that cannot be thoroughly evaluated using existing methods. For

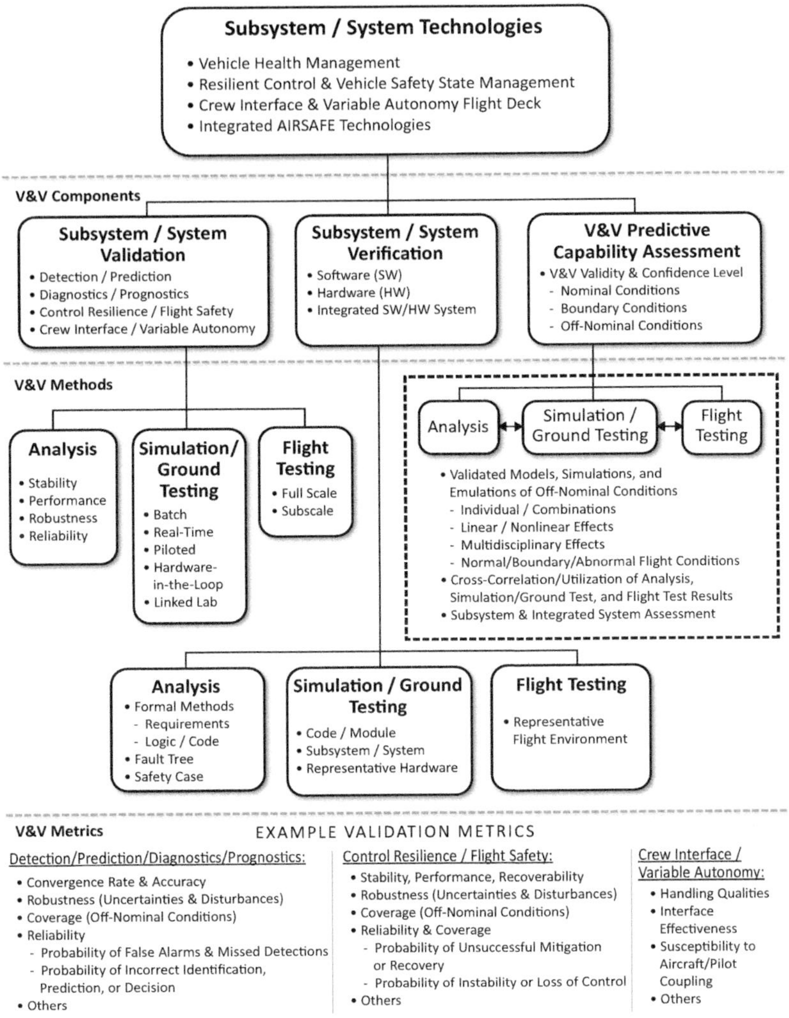

Fig. 20.3 V&V Process Requirements for the AIRSAFE System Concept

example, adaptive control systems may require new methods and metrics for their effective analysis. Moreover, methods and metrics may vary depending on the algorithm being considered. For example, stability of detection and prediction algorithms may imply convergence rate and accuracy rather than the traditional control-theoretic meaning of stability. Performance of diagnostic and prognostic algorithms may be characterized by probabilities associated with correct detection and diagnosis of system faults or failures, whereas performance of control systems may be characterized by tracking capability or evaluation of some other control objective. Robustness for all algorithms must be evaluated relative to uncertainties, including parameter variations and unmodeled system dynamics, and disturbances, including signal and system noise and turbulence. Coverage of off-nominal conditions must also be clearly defined and evaluated for effectiveness in dealing with these conditions. Examples of reliability metrics are given in the figure for detection/prediction and control theoretic algorithms. Crew interface and variable autonomy algorithms must be evaluated for handling qualities and interface effectiveness, and *aircraft-pilot coupling* (APC), or *pilot-induced oscillation* (PIO), susceptibility under off-nominal conditions. Moreover, real-time partitioning effectiveness between the human and automation must be evaluated under off-nominal and emergency conditions. Simulation and ground testing includes traditional batch, real-time, piloted, and hardware-in-the-loop methods, as well as a linked lab capability for the integration and evaluation of multidisciplinary technologies. Flight testing includes traditional full-scale testing to evaluate pilot/system interactions, as well as sub-scale testing to evaluate algorithm effectiveness and dynamics models under off-nominal conditions that are too risky for full-scale testing.

Verification of the system is a confirmation that the validated algorithms have been correctly implemented in software and hardware. This is also a nontrivial task. Formal methods are utilized for analytically verifying with proofs that the system requirements are fully defined and met by the implementation. Fault-tree and safety case analyses of the system implementation must also be performed. Testing of code is performed at various levels of system build-up, including evaluation of the code on representative or actual hardware to be fielded. Flight testing also requires the use of representative avionics hardware systems and flight environments under nominal and off-nominal conditions. Although none are given in Fig. 20.3, verification metrics must be clearly defined and evaluated.

V&V predictive capability assessment is an evaluation of the validity and a level of confidence that can be placed in the V&V process and its results for operation under nominal and off-nominal conditions. The need for this evaluation arises from the inability to fully evaluate these technologies under actual LOC conditions. A detailed disclosure is required of model, simulation, and emulation validity for the off-nominal conditions being considered in the V&V, as well as interactions that have been neglected and assumptions that have been made during design. Cross-correlations should be utilized between analytical, simulation and ground test, and flight test results in order to corroborate the results and promote efficiency in covering the very large space of operational and off-nominal conditions being evaluated. The level of confidence in the V&V process and results must be established for subsystem technologies as well as the fully integrated system. This includes an evaluation of error propagation effects across subsystems, and an evaluation of integrated system effectiveness in mitigating off-nominal conditions. Metrics for performing this evaluation are also needed.

20.2.3 Research Approach

An approach taken at NASA for addressing V&V has been in the development of metrics, methods, software tools, and testbeds that facilitate the evaluation of safety-critical systems operating under off-nominal conditions. A high-level V&V concept was developed which integrates analytical, simulation, and experimental methods. Analytical methods must be developed, with theoretical extensions where needed, as well as user-friendly software tools to assess algorithm stability, performance, robustness, and reliability under off-nominal conditions. Simulation methods must be developed to facilitate Monte Carlo analysis and piloted evaluations under off-nominal conditions. In addition, advanced high-fidelity databases, models, and simulation enhancements must be developed to characterize off-nominal conditions and their impacts on vehicle dynamics and control. Experimental testbeds must be developed to facilitate testing under off-nominal conditions in ground-based laboratory tests as well as in-flight tests. The full integrated V&V process must also be demonstrated, evaluated, and refined using realistic LOC test scenarios, subsystems, and systems. The following sections present a V&V research framework developed at NASA Langley and a brief summary of recent accomplishments in this research.

20.3 V&V Process and Research Framework

Based on the V&V process requirements of Fig. 20.3, a detailed V&V process can be developed for complex integrated resilient systems, such as the AIRSAFE System concept of Fig. 20.2. A high-level overview of the integrated V&V process is presented in Fig. 20.4. The shading of the blocks correlates to core AIRSAFE subsystem functions depicted in Fig. 20.2 – that is, dark gray correlates to resilient control functions, light gray represents health management functions, and white is associated with crew interface functions. Multi-shaded boxes in Fig. 20.4 represent evaluation of the associated integrated subsystem functions. Analysis, simulation, and experimental V&V components are organized in the V&V process of Fig. 20.4 moving from left to right, and system evaluation becomes more highly integrated moving to the center and to the right. Also as indicated in Fig. 20.4, results from the V&V process are utilized as an iterative process for refining the algorithm design of each subsystem. The remainder of this section will present a more detailed description of the control-related components of the V&V process, including methods and interfaces. This is depicted in Fig. 20.4 by the dotted box around the lower two rows of the process. Reference [5] provides a detailed description of the entire process.

A set of recommended V&V methods for resilient control system functions is presented in Fig. 20.5 and Fig. 20.6, which depict analysis and simulation methods and simulation and experimental methods, respectively. For process

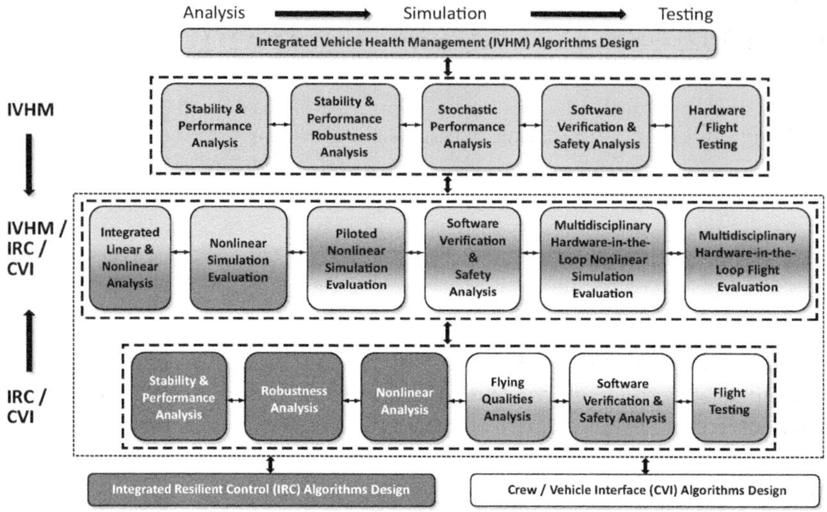

Fig. 20.4 V&V Process Overview

continuity, the right-most blocks of Fig. 20.5 are repeated as the left-most blocks of Fig. 20.6. The methods listed in each block include those that are currently well understood and available as software tools, as well as some that are in need of further research. Moreover, additional methods can be identified and added to each block. In this way, new methods and tools can be identified.

The "Stability and Performance Analysis" block in the lower left of Fig. 20.5 includes standard stability and performance linear analysis methods, including: eigenvalue and eigenvector analysis, transient and steady-state response, and controllability/observability analysis. These methods are well understood for standard linear time-invariant systems, but are not as well understood for hybrid and adaptive systems. Failure and damage coverage must also be considered relative to stability and performance implications.

The "Robustness Analysis" block includes standard μ-Analysis methods (see Chap. 7) as well as nonlinear extensions (see Chap. 19) for analyzing stability and performance robustness to uncertainties. Uncertainty modeling methods that generate a *Linear Fractional Representation* (LFR) of the uncertain system must be utilized for characterizing linear and nonlinear parameter variations (see Chaps. 3 – 6) and unmodeled dynamics. Robustness methods that enable the evaluation of hybrid systems switching effects, adaptive systems, stochastic uncertainties, and time-delay effects must also be considered, as well as robustness and worst case analysis for fault/failure/damage conditions and external disturbances.

The "Nonlinear Analysis" block of Fig. 20.5 includes bifurcation analysis of nonlinear dynamic and controlled systems, controllability and observability in a nonlinear sense, such as degree of controllability and observability as a

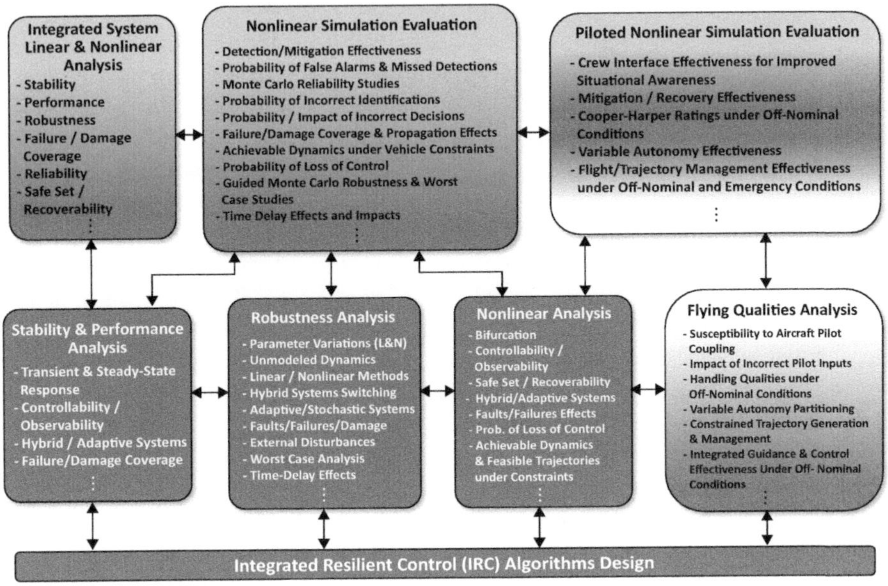

Fig. 20.5 V&V Process for Resilient Control Functions – Analysis and Simulation Methods

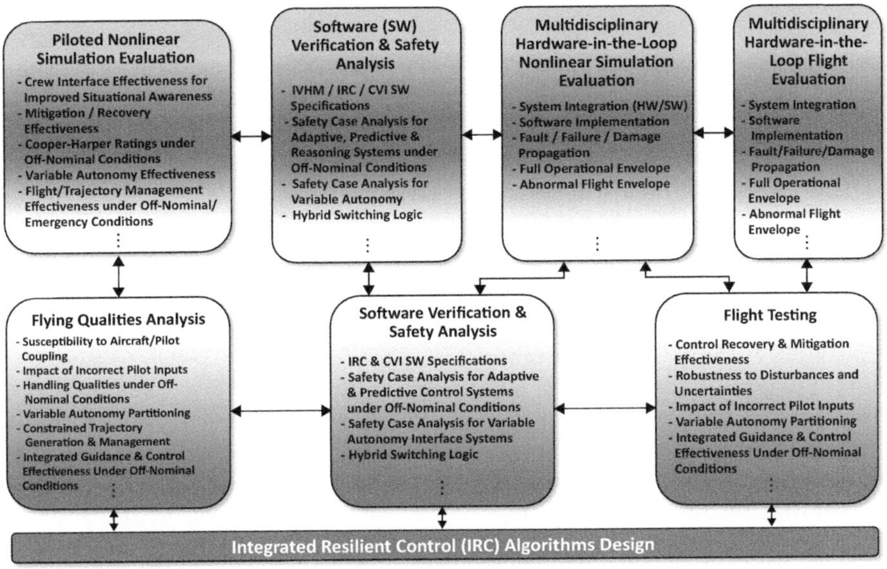

Fig. 20.6 V&V Process for Resilient Control Functions – Simulation and Experimental Methods

function of the changing parameters, and safe set and recoverability analysis. Safe set and recoverability analysis enables the determination of safe operating regions within which recovery to stable trim points can be achieved, as well as the identification of boundaries to unsafe regions from which recovery may not be guaranteed or even possible. Nonlinear analysis of hybrid and adaptive systems, fault and failure effects, and achievable dynamics of constrained or impaired vehicles must also be considered. A method for analytically determining the Probability of LOC in a nonlinear sense must also be developed.

These analysis methods must then be applied to the integrated health management system, including failure detection and identification functions for critical control components, and resilient control system, including failure mitigation functions, as indicated by the "Integrated System Linear and Nonlinear Analysis" block.

The "Flying Qualities Analysis" block evaluates resilient control system effectiveness relative to a pilot being in the loop, and may integrate pilot models and/or crew interface functions. This analysis includes methods to assess susceptibility to PIO, impact of inappropriate pilot inputs, handling qualities under off-nominal conditions, effectiveness of variable autonomy partitioning between automatic control resilience functions and human-involved control, effectiveness of trajectory generation and management under vehicle impairment or damage, and integrated guidance and control effectiveness under off-nominal conditions.

Nonlinear simulation evaluations are performed to assess: the effectiveness of the detection and mitigation algorithms and their integration; the probability and impact of false alarms, missed detections, incorrect identifications, and incorrect decisions; failure/damage coverage and propagation effects; achievable dynamics under vehicle failures or damage; and time delay effects associated with failure detection, identification, and mitigation. Guided Monte Carlo studies, guided by analysis results to further explore potentially problematic operational regions, can be utilized to assess these and other reliability metrics, robustness under uncertainties, and worst-case combinations of flight and impairment conditions. Nonlinear simulations are used in evaluating the vehicle health management and resilient control subsystems individually and in combination. The crew interface subsystem is assessed in piloted simulation evaluations individually and as part of the integrated system to evaluate: crew interface effectiveness in improving situational awareness under off-nominal conditions; mitigation and recovery effectiveness, including variable levels of autonomy; handling qualities under off-nominal conditions, using Cooper-Harper metrics and extensions; variable autonomy interface effectiveness; and flight/trajectory management under off-nominal and emergency conditions.

Fig. 20.6 shows the progression to subsystem and integrated system evaluations that involve the software/hardware implementations. Formal verification and safety case analysis methods are utilized to assess system requirements and specifications, implementation integrity of adaptive and predictive/reasoning systems under off-nominal conditions, hybrid switching logic, and the variable autonomy interface. Various levels of system integration and implementation are evaluated through laboratory tests and flight tests, using both full-scale and sub-scale vehicles. Ground and flight test methods are utilized to assess system integration, software implementation, fault/failure/damage mitigation effectiveness, and upset recovery effectiveness under off-nominal conditions throughout and beyond the normal flight envelope. Robustness to uncertainties, reliability and coverage, variable autonomy interface effectiveness, and impacts of inappropriate crew responses are also assessed. Sub-scale vehicle flight tests are utilized for high-risk conditions that would not be feasible in a manned vehicle, and full-scale flight tests are performed to evaluate the *crew/vehicle interfaces* (CVI) in flight while using the appropriate timescale.

The V&V process depicted in Fig. 20.5 and Fig. 20.6 is integrated across the various methods, with information being exchanged between each block. Information exchange is indicated with double-headed arrows. Reference [5] provides a detailed description of information exchange throughout the process. As an example, consider a subset of the process shown in Fig. 20.5 and depicted below in Fig. 20.7.

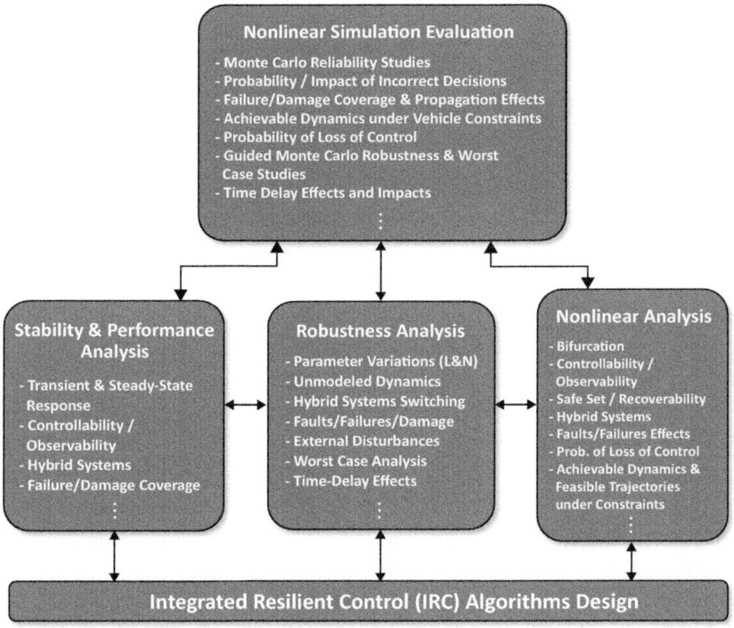

Fig. 20.7 V&V Subprocess for Resilient Control Functions – Analysis and Simulation Methods

This figure contains analysis and simulation methods that are applied to the resilient control functions. To illustrate information exchange between subprocess components, consider a subset of these methods as presented in Fig. 20.8.

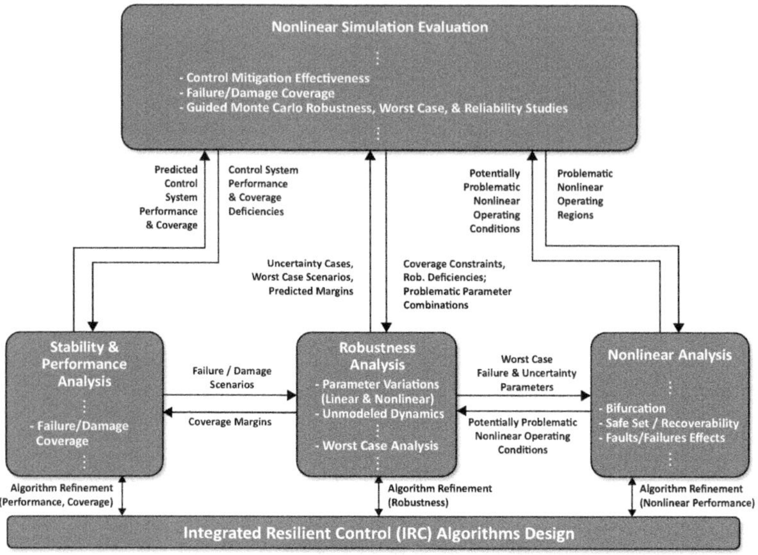

Fig. 20.8 Example V&V Subprocess Interfaces for Resilient Control Functions – Analysis and Simulation Methods

Starting with the lower left block of Fig. 20.8, failure/damage scenarios are evaluated in the "Stability and Performance Analysis" block based on the failure and damage profiles being mitigated in the resilient control design. The stability and performance analysis results define the effective coverage of these failure/damage scenarios. This information can be provided for use in the "Robustness Analysis" block to generate parametric and non-parametric uncertainty models, and for performing a worst case analysis. Using robustness analysis techniques, failure/damage coverage margins can be generated as well as worst case failure, damage, and uncertainty combinations. These results can be utilized by the nonlinear analysis tools, such as bifurcation, safe set and recoverability, and failure effect analyses, to identify potentially problematic nonlinear operating regions. The nonlinear analysis results can be utilized in re-evaluating robustness in these regions. Analysis results related to stability and performance, such as failure/damage coverage predictions, robustness, including uncertainties, worst case scenarios, and predicted margins, and nonlinear properties, such as potentially problematic operating conditions, are utilized, corroborated, or disputed during nonlinear simulation evaluations. Simulation results are then utilized by the analysis components during re-evaluation. The analysis and simulation results are also utilized as part of an iterative design process. Each evaluation method provides a basis for improved system design, as depicted in Fig. 20.8. The subsequent analysis and simulation results might then be utilized to generate test scenarios for use in piloted simulation evaluations (not shown in Fig. 20.8). It is conjectured that the use of analytical, simulation, and experimental results in a coordinated manner will provide a means to effectively and efficiently identify problematic flight conditions, off-nominal conditions, uncertainties, and combinations of these without having to perform exhaustive testing.

Recent NASA research that pertains to the V&V process thus described is briefly summarized and referenced in Sect. 20.4.

20.4 V&V Research Status and Recent Accomplishments

Significant resources and effort have been invested by NASA in addressing the V&V of future advanced safety-critical systems. For the last decade, this work has largely been planned and funded by the system research projects focused on vehicle health management, flight-critical system design, and resilient control technology development under the NASA *Aviation Safety Program* (AvSP). This research has resulted in the development of analytical methods and software tools, simulation-based methods, and experimental testbeds for the validation of safety-critical systems operating under off-nominal conditions related to aircraft loss of control [7, 8]. These results are summarized in [9]. Software verification methods and tools were also developed under this research effort, and a new effort under the AvSP is currently being planned to focus on the V&V of software-intensive systems [10]. This new effort will develop V&V methods that can be applied to the Next Generation Air Transportation System.

20.5 Summary and Concluding Remarks

Aircraft loss of control is a significant contributor to accidents and fatalities, resulting in the highest number of fatalities among the worldwide commercial jet fleet. It is also the most complex accident category, resulting from numerous causal and contributing factors that occur individually or more often combine to result in a loss of control accident or incident. These factors are off-nominal conditions that occur onboard the aircraft, as external disturbances, or as abnormal flight conditions. To address aircraft loss of control, NASA is developing onboard systems technologies to: prevent and detect faults, failures, and damage through the development of vehicle health management technologies; provide improved situational awareness to the crew through the development of advanced flight deck technologies; and to provide the capability to mitigate off-nominal conditions through the development of resilient aircraft control technologies. A future technology concept, called the AIRSAFE System, for integrating these technologies and providing onboard flight safety assurance is envisioned. These technologies are being developed for safety-critical operation under off-nominal conditions.

The V&V of safety-critical systems operating under off-nominal conditions poses significant technical challenges. This chapter has provided an analysis of this V&V problem, and has described a research approach being taken at NASA to address it. High-level V&V process requirements were defined, which integrate analytical, simulation, and experimental methods, software tools, and testbeds. A detailed V&V process was defined for application to the AIRSAFE System concept, and a detailed description was provided of the methods and some example interfaces involved in the controls-related components. Research progress at NASA in the development of analytical, simulation, and experimental methods was briefly discussed and referenced.

Acknowledgement and Dedication. The V&V research process, methods and software tools, and the AIRSAFE System concept presented in this chapter were developed in collaboration with Dr. Celeste M. Belcastro of NASA Langley Research Center, who lost her selfless and courageous battle with cancer and passed from this life on August 22, 2008. This chapter and continued work in this area are dedicated to her memory.

References

1. Statistical Summary of Commercial Jet Airplane Accidents, Worldwide Operations, 1959-2009. Boeing Commercial Airplanes (July 2010), `http://www.boeing.com/news/techissues/pdf/statsum.pdf`
2. Belcastro, C.M., Foster, J.V.: Aircraft Loss-of-Control Accident Analysis. In: AIAA Guidance, Navigation and Control Conference, Toronto, Canada (2010)
3. Belcastro, C.M., Belcastro, C.M.: Future Research Directions for the Development of Integrated Resilient Flight Systems to Prevent Aircraft Loss-of-Control Accidents, Part I: System Technologies. NASA TM (under final preparation)
4. Belcastro, C.M., Jacobson, S.: Future Integrated Systems Concept for Preventing Aircraft Loss-of-Control Accidents. In: AIAA Guidance, Navigation and Control Conference, Toronto, August 2-5 (2010)
5. Belcastro, C.M., Belcastro, C.M.: Future Research Directions for the Development of Integrated Resilient Flight Systems to Prevent Aircraft Loss-of-Control Accidents, Part II: Validation and Verification. NASA TM (under preparation)
6. Joint Planning and Development Office, Concept of Operations for the Next Generation Air Transportation System, Version 3 (October 2009), `http://www.jpdo.gov/library.asp`
7. Belcastro, C.M., Belcastro, C.M.: On the Validation of Safety Critical Aircraft Systems, Part I: An Overview of Analytical & Simulation Methods. In: AIAA Guidance, Navigation and Control Conference, Austin, Texas, USA (2003)

8. Belcastro, C.M., Belcastro, C.M.: On the Validation of Safety Critical Aircraft Systems, Part II: An Overview of Experimental Methods. In: AIAA Guidance, Navigation and Control Conference, Austin, Texas, USA (2003)
9. Belcastro, C.M.: Validation and Verification of Future Integrated Safety-Critical Systems Operating under Off-Nominal Conditions. In: AIAA Guidance, Navigation and Control Conference, Toronto, Canada (2010)
10. Validation and Verification for Flight-Critical Systems Assessment of Critical Research Activities. NASA Aeronautics Research Mission Directorate, Aviation Safety Program, November 25 (2009)

Chapter 21
Clearance of Flight Control Laws for Carefree Handling of Advanced Fighter Aircraft

Robert Stich

Abstract. Clearance of fighter aircraft poses additional challenges compared to civilian transports due to the high agility and performance of the vehicles, their larger flight envelope, different mission types and higher instability. The carefree handling concept of modern fighters, which is a mandatory requirement for the flight control laws to reduce pilot work load, further increases the challenges for the assessment task. Carefree handling means, that the flight control system allows any pilot inputs while it takes care of the structural and aerodynamic limits of the aircraft and simultaneously maximizes aircraft performance. The carefree clearance covers however only manoeuvres that are meaningful for the respective configuration and phase of flight. The required analysis techniques to provide such a clearance are explained.

21.1 Introduction

This chapter explains how a carefree handling clearance is performed for the Eurofighter Typhoon at EADS Defense and Securities Business Unit Military Air Systems (MAS). An aircraft with a carefree handling *flight control system* (FCS) responds to whatever demands and manoeuvres the pilot performs taking into account the structural and aerodynamic limits whilst simultaneously maximizing performance. This poses unique challenges for the clearance process including a dedicated set of manoeuvre sequences. The assessment is based on nonlinear simulation, linear stability analysis, manned simulation and flight test.

During the nonlinear simulation based assessment, carefree handling manoeuvres are simulated for different aircraft configurations (mass, CoG,

Robert Stich
Cassidian, Air Systems, Flight Control,
(former EADS Defense & Security, Military Air Systems)
Rechliner Straße, 85077 Manching, Germany
e-mail: `Robert.Stich@cassidian.com`

A. Varga et al. (Eds.): Opti. Based Clearance of Flight Control Laws, LNCIS 416, pp. 421–442.
springerlink.com

stores) and envelope points as well as for many different uncertainty combinations. The resulting time histories are automatically evaluated with respect to a given set of criteria and 'worst-case' configurations and uncertainties are selected and used for further robustness analysis.

The extremal values of the angle-of-attack and vertical-load-factor obtained during the nonlinear simulation, and the known margins of the flight envelope are utilized to define the aircraft operational boundaries for the linear stability analysis. The complete linear clearance process is highly automated and is described in this chapter.

Detailed handling qualities evaluations are performed in manned simulations, based mainly on the results of the nonlinear assessment. Results of the nonlinear simulations are further used to derive limits for monitoring critical parameters during flight testing. After an envelope expansion flight test campaign and an aerodynamic model modification the nonlinear simulation is further utilised for the validation of the aerodynamic model via dedicated post-flight analysis.

21.2 Plant Description and Clearance Process

For control laws design and subsequent clearance a good understanding of the controlled system is essential. This is especially true for an extremely complex high performance fighter type aircraft. Thus, a short description of the Eurofighter Typhoon will be given. Working in a multi-national/multi-company environment causes some unique concurrent engineering challenges which will be described with respect to the Eurofighter clearance process.

Fig. 21.1 Eurofighter Typhoon aircraft

21.2.1 Aircraft Description

The Eurofighter Typhoon is a delta-canard-configuration fourth generation fighter aircraft (see Fig. 21.1). In the sub- and transonic speed regimes the aircraft is aerodynamically unstable in the pitch axis for superior performance and agility. It is propelled by twin EJ200 engines with a thrust-to-weight-ratio larger than one. Its top speed is Mach 2 and the load factor ranges from -3g to +9g. The aircraft features swing role capability, i.e. changing in-flight from the air-to-surface to the air-to-air role according to the attached stores. For more information about the aircraft see [1].

The Eurofighter FCS is a full-authority quadruplex digital fly-by-wire system providing high agility and excellent handling qualities. Primary control surfaces are canard, trailing edge flaps and rudder. Airbrake, leading edge slats and intake cowl flaps are used as secondary control surfaces.

During failure-free flight the FCS provides carefree handling characteristics to minimise pilot workload and improve flight safety. The Eurofighter carefree handling concept will be described in more detail in section 21.3. In addition to carefree handling some emergency features have been included into the FCS design to ensure maximum safety of aircraft operation.

If the aircraft decelerates towards its minimum airspeed the auto-low-speed recovery function takes over and recovers the aircraft at a safe speed before control is handed over to the pilot again. A disorientation recovery function can be activated by the pilot in case of spatial disorientation.

21.2.2 Control Laws Clearance Process

For the Eurofighter clearance a model-based approach is used. This means, that the clearance is achieved based on validated models of the aircraft and its subsystems including uncertainties. Model validation and updating is done via flight testing and other system tests.

The system design responsibility for the Eurofighter flight control laws is with Military Air Systems. Thus the 'FCS Joint Team' (FJT) which performs the flight-control-system design and coordinates the clearance efforts is managed by Military Air Systems and is situated at its Manching plant. It is made up of engineers from all core *Eurofighter Partner Companies* (EPCs) whose participants are given in Table 21.1. The FJT defines the assessment-rules [2], [3], splits the assessment tasks between the EPCs and coordinates the results. A schematic description of this process is given in Fig. 21.2.

For a clearance it is essential, that every partner team uses the same models in the same way. In a multi-partner environment this requires an intensive use of checkcases. Beside utilising the same aircraft models, different partner companies use different tool sets based on their experience. This ensures additional safety because it avoids a common tool error at all EPCs.

Table 21.1 List of core Eurofighter-Partner-Companies

Country	Company
Germany	European Aeronautic Defence and Space Company - Defense and Security, Military Air Systems (EADS-D)
Italy	Alenia Aeronautica (ALN)
Spain	European Aeronautic Defence and Space Company - Casa (EADS-C)
United Kingdom	BAE Systems (BAES)

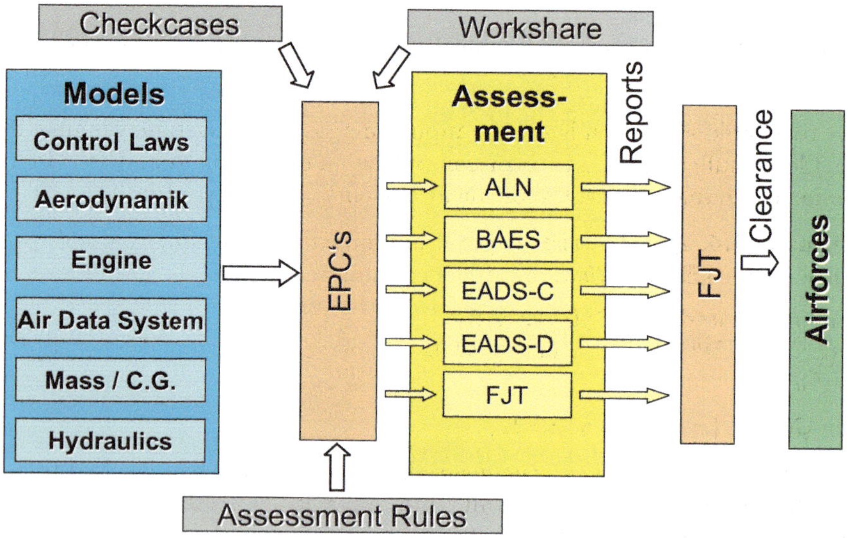

Fig. 21.2 Eurofighter control laws clearance process

21.3 Carefree Handling

Carefree handling is an outstanding feature of the Eurofighter. Protection against departure, spin, overstress etc. is provided by the FCS. This allows the pilot to concentrate on his mission and not on flying the aircraft. However the pilot is only allowed to do meaningful manoeuvres for the actual phase of flight, i.e. carefree does not mean careless. A full carefree handling clearance is only provided for failure-free flight, every other flight phase is not carefree.

For the Eurofighter, failure-free flight is defined such that, either no warning is provided to the pilot, or the failure consequences are so minor, that a failure-free clearance is given nevertheless. This means for instance, that an aircraft with a known first failure of the air data system is still considered to be failure-free.

21.3.1 Carefree Handling Definition

Carefree handling as it is agreed with the Eurofighter customers has the following meaning: *Free use of flight controls during manoeuvres which are meaningful from an operational point of view for air-to-surface attack and air-to-air combat.* This implies for example, that a manoeuvre that is meaningful in the air-to-air role may not make sense for air-to-surface missions and thus will not be part of the carefree handling clearance for an air-to-surface aircraft. This distinction is done to avoid unnecessary conservatism and manoeuvre authority restriction.

Carefree handling rules allow up to full and highly aggressive cross-cockpit stick-inputs including stick-reversals. Pedal inputs may be used in addition. Manoeuvres may be linked in any order and include continuous rolling. They can be performed at any throttle setting between idle and max-reheat.

For clearance purposes an extensive evaluation has been performed by the FJT together with the customer airforces and the infinite number of possible manoeuvres has been reduced to a fixed set as described in [2]. These manoeuvres are categorised according to their likelihood into three categories (frequent, less frequent and rare). The manoeuvre categories are further associated with the tolerance concept (see subsection 21.4.3 for more details). Depending on the aircraft role (e.g., air-to-surface) only a certain subset of the manoeuvres has to be cleared. The given carefree handling manoeuvre set of [2] may be modified during the assessment for worst case identification.

21.3.2 Manoeuvres Category Definition

For design and clearance purposes manoeuvres are divided into three categories according to [2] labeled M1, M2 and M3. M1 are full carefree manoeuvres, M2 are relaxed carefree manoeuvres and M3 are non-carefree manoeuvres.

The most demanding manoeuvre category is M1 (**carefree**). Manoeuvres in this category are only applicable for failure-free flight and have no limitations for the pilot within the carefree philosophy described in subsection 21.3.1. A prerequisite for a M1 clearance is a valid M3 (linear) clearance. A M1 clearance is performed via nonlinear and manned simulation using the carefree handling manoeuvres described in subsection 21.3.1.

If a M1 clearance is not possible due to the aircraft responses found in the assessment, a M2 clearance with relaxed manoeuvres compared to the M1 definition is performed. Just like M1 this category is only applicable for failure-free flight. The pilot inputs are typical for general day-to-day flying including take-off and landing and cover rapid inputs up to full-forward/full-back stick. Maximum roll-rate is limited and only one roll is allowed. By definition this category is not carefree but envelope protection is still provided. The M2 clearance requires an existing (linear) M3 clearance to start with and is performed using nonlinear analysis and manned simulation with the appropriate manoeuvres.

The lowest category M3 is also called 'symmetric and gentle flying' and is not carefree. This type of manoeuvre is the most basic one and must be cleared for every possible flight condition including failure-free flight and all system-failure cases. If a M3 clearance is not possible in a region, then flying there will be prohibited and this typically will trigger changes within the FCS. Control inputs up to full-forward/full-back stick are possible if they are done slowly and gently. Roll rate and bank angle are further limited compared to the M2 category for these manoeuvres.

For final control law delivery a carefree handling M1 clearance is required throughout the service envelope. With customer approval some areas may be limited to M2 manoeuvres. For flight test purposes, excluding carefree handling, a M3 manoeuvre clearance may be sufficient.

21.4 Assessment Model

The assessment is based on the aircraft equations of motion together with models of the aircraft and its systems, the atmosphere and pilot-inputs. A combination of the different configuration parameters and uncertainties is called a 'configuration' and is the baseline of the assessment.

The assessment model is calculated for the six-degrees-of-freedom equations-of-motion of a rigid body. Elastic effects are considered as part of the aerodynamic dataset. For the linear evaluation and part of the nonlinear-simulations the speed is set constant, i.e. the model is reduced to five degrees-of-freedom. This is done to get a better understanding of the local effects throughout the flight envelope.

21.4.1 Aircraft Model

The mass model describes mass, CoG, inertias etc. of the aircraft to be assessed. All of these variables are highly dependent not only on fuel quantity, but also on the aircraft stores, i.e. external tanks, missiles and other weapons. Stores may have a larger effect on mass/CoG as fuel. Due to fuel slosh the CoG can vary significantly during maneuvering and these extreme CoG positions (pitch and lateral) must be considered during assessment.

Aerodynamic forces and moments depend on the usual parameters, like Mach number (M), dynamic pressure ($\bar{q}$), angle-of-attack (AoA), angle-of-sideslip (AoS), control surface position etc. The presence of external stores is a major difference between civil and military aircraft. They have a strong effect on aerodynamic forces and moments and may exceed the influence of different fuel loadings. The effect of elastic deformations on the aircraft is modelled within the aerodynamic data set via specific factors. For linear analysis the aerodynamic coefficients are numerically linearized to calculate the associated derivatives. Uncertainty factors are defined for linear derivatives and moment offsets.

Since stores have a significant effect on mass/CoG and aerodynamic forces and moments, a dedicated numbering system for configuration control is used. If a new store is not covered by the mass/CoG range of existing stores, or has unique aerodynamic features, the control laws must be adapted and the assessment process has to be repeated.

The engine model is of less importance since there is only negligible influence on short term dynamics. The constant speed assumption further enables a good identification of problem areas. For the part of the manoeuvre clearance which is done with 6 degree-of-freedom, as for example decelerations, a model including engine dynamic effects is available, which covers the complete thrust range from idle via max-dry to max-reheat. Compared to civil engines the bypass-ratio is small and the engine reacts relatively fast. The afterburner is a unique feature of military aircraft.

The model of the flight-control-system contains the complete nonlinear control laws and sensor systems. The most important sensor models represent the air-data-system and the inertial-measurement-unit. The fuel/stores model calculates aircraft mass and CoG and provides these data for the FCS. The models cover the failure-free state and failure cases. All of these subsystems and associated errors are of interest for both the linear and nonlinear assessment. Furthermore models of the actuator and hydraulic systems are available. They cover the specific system dynamics and behaviour and also deterioration effects and system failures.

For nonlinear simulations further models describe atmospheric disturbances like gusts and turbulence. Additional models, like e.g. landing-gear, head-up- or head-down-displays, are required during manned simulation exercises.

21.4.2 Aircraft Configuration

The assessment process is based on aircraft configurations. They describe a specific set of input parameters, i.e. mass/CoG, stores, air-data-system settings, fuel system failure, aerodynamic uncertainty, manoeuvre type etc. and get a unique name. All subsequent tasks like trim, linear/nonlinear analysis and postprocessing always refer to these configuration names.

Nonlinear simulations utilise lateral CoG from fuel slosh whereas the linear assessment is usually performed for a symmetric aircraft. This is done since differences of the linear system stability margins between a symmetric and a moderate asymmetric aircraft are minor. If the lateral CoG offset results from asymmetric stores and thus is larger, the linear assessment also includes this effect. For manned simulation the same configurations as for the nonlinear simulations are used.

For an assessment task more than hundred configurations will be used for the linear evaluation. Due to the large number of manoeuvres required for a carefree handling clearance several hundred nonlinear configurations are assessed.

21.4.3 Tolerance Concept

The term *tolerances*, as used throughout this chapter, is synonym for all kinds of model uncertainties (e.g., uncertain parameters in the aerodynamics, fuel stores) and manoeuvre categories and is the basis for addressing robustness checks.

The aircraft aerodynamic model as well as the subsystem models contain various uncertainties which are partly known and partly unknown. The known uncertainties are modelled as uncertainty models and their effects must be observed during the assessment, as described below. The unknown uncertainties are characteristics of the aircraft which are not precisely known, and therefore are not covered by corresponding uncertainty models. Suitable clearance margins are used to take into consideration the effects of these uncertainties.

For all systems such as aerodynamics, air data and fuel/stores-system model uncertainties have been defined. Each model uncertainty exists in a nominal and toleranced version, with the latter one being larger and less likely. Note that nominal performance does not necessarily imply a failure-free system e.g. an air data system with a single failure defines the nominal ADS performance. As described in subsection 21.3.1 carefree handling manoeuvres are also divided into nominal and toleranced categories with respect to their probability in the air-to-air/air-to-surface role. Thus for nonlinear simulations manoeuvre selection adds an additional uncertainty factor.

A tolerance concept on how to combine these uncertainties has been defined. The most basic check is a combination of all models/manoeuvres in nominal condition. If for these basic calculations any clearance requirement is violated, so-called 'noncompliance' checks are performed. They are done to check the robustness of the aircraft via several combinations of nominal and toleranced uncertainties. If the noncompliance calculations exhibit sufficient margins and cliff edge effects are absent, the corresponding limit violation can be cleared. The prescribed evaluation procedure for noncompliance calculations combines the different probabilities of each uncertainty to obtain at least the minimal global value required according to the general safety regulations.

21.4.4 Assessment Tools

For trim, nonlinear simulations and various post-processing tasks in house developed Fortran programs are utilized at MAS. Trim results (aerodynamic model and FCS gains) are linearized by Fortran programs for linear-evaluations. The resulting data are then used to build a linear model in MATRIXx Systembuild from which the linear results (frequency responses, eigenvalues etc.) are derived using MATRIXx tools. Other Eurofighter partner companies use different assessment tool sets including MATLAB.

21.5 Assessment Methods

For a carefree handling clearance several separate assessments take place. Before any analysis can be performed, it is required to trim the aircraft over the desired flight envelope on a grid of values for altitude, Mach number and AoA/Nz. The trim results give a first insight into the aircraft capabilities. Then linear analysis and nonlinear simulations are performed for selected trim results. For linear assessment a fine AoA/Nz grid is used. Neutral, full-forward- and full-back-stick initial conditions are utilized for the nonlinear simulations. Additional handling qualities evaluations and clearance checks are done in manned simulation. The ultimate validation is done during flight test. All of these assessment methods will be described in more detail in the following sections.

Most of the assessment time (about 65%) is spent on nonlinear assessment, wile the linear assessment accounts for about 30% of the time. The amount of time dedicated for manned simulation is below 5%. Flight tests for FCS assessment have been very rare during the last years due to model maturity.

21.5.1 Trim Analysis

Trim analysis is the most basic assessment method used. It provides a first insight in limiting factors like extreme surface deflections or maximum stationary vertical load factor capability. In addition it is the foundation for the subsequent linear analysis and nonlinear simulations which require trimmed points as initial conditions.

For trimming based on very generic inputs like altitude, Mach number, trim-type and associated parameter-range a dedicated trim program has been developed. For angle-of-attack (AoA), vertical-load-factor (Nz), or pitch-stick trim, either fixed values or a parameter-range from a minimum to a maximum value in fixed intervals can be selected. The results of these trim runs are stored in files. They can be used for a much quicker re-trim run, if some parameters of minor influence on the trim results e.g., aerodynamic uncertainty type, have been changed. After trim solutions have been found, the aircraft is linearized around those points if required. Approximately 4,500 trim points are calculated for each configuration in a typical envelope.

Envelope regions where trim is not possible for a given configuration can provide an insight into insufficient control power if surfaces reach their trim limit. On the other side, the amount of remaining surface deflection from the trim values to the limits is an indicator of maximum control power available for manoeuvres.

The low speed envelope boundary is reached, if the lift obtained for full-back-stick equals the weight of the configuration, i.e. level horizontal flight can be sustained. The envelope below this minimum speed is of less operational relevance. The region where maximum angle-of-attack and load factor occur simultaneously in the trim results defines the corner speed.

21.5.2 Linear Analysis

Linear analysis around a trim point is used for stability assessment. Since aerodynamic uncertainties are only valid in the vicinity of a reference (trim) point they are best used together with the linearized aircraft model. Envelope regions with stability problems can be easily identified and described and the gain and phase margin calculated.

Classical gain and phase margin evaluations are used to assess system robustness. Gain and phase margins are defined to provide a safety offset and to cover uncertainties in the aircraft models. Based on the requirements of the MILSPEC [4] several robustness boundaries have been defined for gain- and phase margin (see [3], [4]). They are referred to as 'diamonds' due to their shape. An example is shown in Fig. 21.3. Note that, in accordance with the tolerance concept different stability margins are defined for average (nominal) and toleranced configurations. For robustness checks, reduced (inner) stability margins for gain and phase are defined for both cases in Fig. 21.3.

Frequency responses and eigenvalues for longitudinal and lateral subsystems are assessed. A program automatically evaluates if the required gain and phase margins are violated and records the margin reserve or the level of instability. In addition to the Nichols plots, the associated eigenvalues are calculated as well. Linear handling qualities criteria, like e.g. phase rate criterion, absolute and relative amplitude, attitude dropback, bandwidth frequency, flight path time delay and roll mode time constant, are also evaluated. More details about these criteria can be found in MILSPEC [5].

To avoid unnecessary calculations for configurations/envelope areas with stable aircraft responses a three step approach is used as described in Fig. 21.4. For critical case selection in a first step several 'test'-configurations are trimmed and analysed throughout the envelope. This is done in narrow AoA/Nz steps from the lower to the upper target. To add margin for dynamic overshoots a certain AoA/Nz range beyond the targets is additionally trimmed. The 'test'-configurations are using several worst case assumptions, which may not exist for the real aircraft, e.g. heaviest configuration together with most aft center-of-gravity. Several thousand Nichols plots are calculated and compared to the average outer diamond of Fig. 21.3.

In the second step several hundred realistic configurations (stores, mass, center-of-gravity, air-data-system-error, aerodynamic-uncertainty combinations) are calculated in an AoA/Nz range around the problem areas found for the 'test'-configurations described above. The linear analysis results created in this step are once again automatically evaluated for gain and phase margin (toleranced outer boundary of Fig. 21.3), eigenvalues and other criteria. If one of these requirements is not passed, the configuration and flight condition is recorded as a violation.

If the angle-of-attack/vertical-load-factor value of a violation is a certain margin beyond the extremal AoA/Nz value found during nonlinear carefree handling maneuvers it is marked as outside the assessment envelope and can be ignored. Local aerodynamic discontinuities and/or control power losses can result in instabilities, insufficient stability margins, unstable eigenvalues etc. If these violations occur within the envelope and are localised over a very limited AoA/Nz/AoS range, they are acceptable and thus can be cleared. However it has to be checked in nonlinear/manned simulations, that these local effects do not harm handling qualities. The 'outside' and 'local' detection is performed automatically by software.

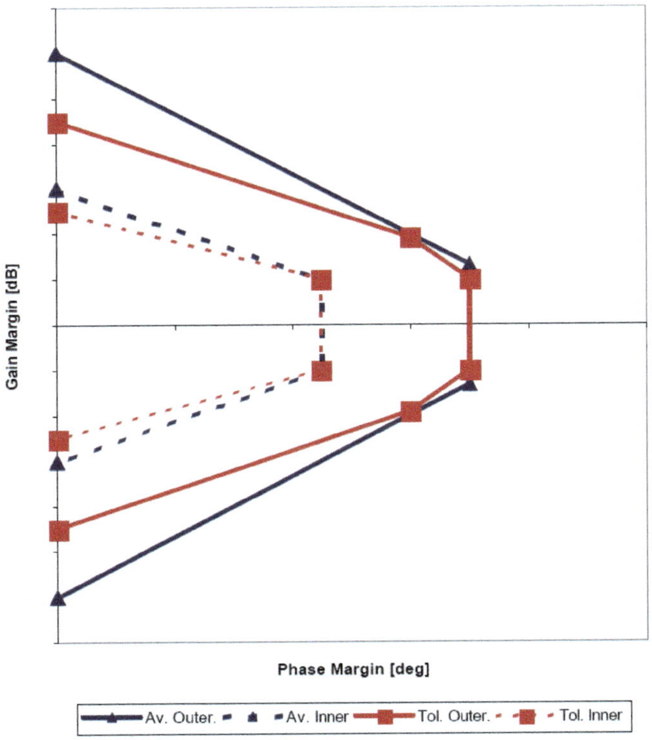

Fig. 21.3 Stability margin requirements for Nichols plots from [3]

For those violations that are neither outside the envelope nor local, non-compliance work is done in the third step. For this purpose configurations with several combined uncertainties or enlarged single uncertainties in accordance with the tolerance concept described in subsection 21.4.3 are created to check for cliff-edge effects. They are calculated for an AoA/Nz range around the remaining violations from the previous assessment step. The linear results are again automatically evaluated for stability margins against the toleranced inner boundary of Fig. 21.3, eigenvalues and other criteria. If the requirements are passed this proves the robustness of the linear system stability to off-nominal conditions. In this case the violation is acceptable and thus can be cleared.

For the remaining violations a manual assessment task is required. This includes a manual check of the outside envelope and local in AoA/Nz/AoS criteria, further variation of the affected uncertainties for additional robustness checks etc. If a violation occurs in a region where nonlinear control law features exist, the linear model may not be an accurate representation of the FCS. In this case a violation can be assessed by using the nonlinear simulation and adding phase and gain shifts to the actuator signals. The amount of the gain or phase shift in this case represents the linear requirement. If the time history response indicates a stable characteristic, or in the extreme case, a neutrally damped oscillation the linear violation is acceptable.

If a violation is not acceptable despite all of the efforts described above, a limitation for M3 maneuvering must be issued, which automatically prohibits M2 and M1 (carefree handling) flying. In this case a repair of the flight control laws is required to obtain a carefree handling clearance.

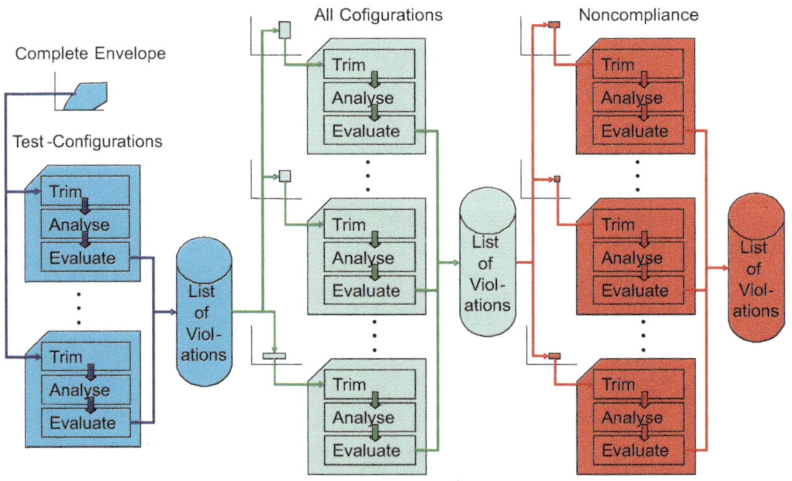

Fig. 21.4 Flow chart of linear assessment process

21.5.3 Nonlinear Simulation for Carefree Handling

Nonlinear simulation is utilised for two reasons. The first is to assess the performance of the nominal aircraft/system/manoeuvre combination according to the tolerance concept (see subsection 21.4.3). The second one, which makes up the bulk of the assessment task, is to demonstrate overall closed loop stability, or check the departure risk during carefree manoeuvres. Uncertainties on models and manoeuvres are applied in accordance with the tolerance concept.

The nonlinear simulation includes all kinds of nonlinear effects from aerodynamic characteristics, nonlinear control law features, cross axis and inertia coupling, engine performance, actuator performance and degradation due to position, rate and acceleration limitations, atmospheric disturbances etc. Nonlinear simulation thus improves the understanding of the aircraft subsystems and their interaction. The result of a nonlinear simulation is a time history, i.e. a recording of several parameters with respect to time.

The time histories created during the carefree handling manoeuvres are automatically evaluated with respect to several criteria, some of which are depicted in Fig. 21.5. The most important check is if the aircraft departs during a manoeuvre. This is done via monitoring, amongst others, the extremal angle-of-attack and angle-of-sideslip values. If a departure is found, additional robustness checks via noncompliance work are required and a limitation may be issued, triggering repairs in the FCS. In addition to the departure check criterion, extremal values reached during the manoeuvres are recorded for various parameters like angle-of-attack, angle-of-sideslip, load-factors etc. For parameters limited by the FCS like angle-of-attack and vertical-load-factor, exceedance of the target values is recorded. For parameters for which a FCS command signal exists, the difference between the command and the actual value is evaluated. Duration of control surface position saturation, rate saturation and actuator hydraulic load saturation (i.e. hinge moment exceeds actuator force) is monitored. For dedicated manoeuvres it is checked automatically that pitch/lateral oscillations can not be excited.

Two methods are used to enter pilot inputs into the nonlinear simulation. The first uses timed commands for pitch- and lateral-stick, pedal, throttle and airbrake. In some cases, the duration of a manoeuvre (e.g., a 360° roll) may significantly vary with aircraft inertias and flight condition. For these cases many (slightly different) pilot input specification files would be required to assess the prescribed carefree handling manoeuvres of subsection 21.3.1. To simplify the manoeuvre set-up, a second method to define pilot inputs has been developed, consisting of a simple event triggered pilot model to determine the required control inputs. This kind of manoeuvre definition works for all aircraft configurations and initial conditions and proved to be very effective for carefree handling calculations.

To obtain a maximum number of parameter variations within a limited number of calculations without compromising safety margins, a phased approach is used as described in Fig. 21.6. Before a carefree handling clearance can be started, trim results for the manoeuvre initial conditions must be created. Since these initial conditions are common for all intended manoeuvres, a dedicated trim condition database is created. Trim results are calculated for a fine grid of angle-of-attack/vertical-load-factor values to provide flexibility for the initial conditions.

Based on the trim results, the required maneuvers are performed at full-forward, full-back and neutral stick initial conditions. The basic calculations are done with constant speed, i.e. with 5 degrees-of-freedom, to better localise problem areas. In addition, for a large part of the envelope the engines provide enough thrust for constant speed even at full back stick. After all manoeuvre sets are calculated, the time histories are evaluated automatically as described above and the results combined in a database. Then for each altitude and Mach number point the time history with the worst response, e.g the

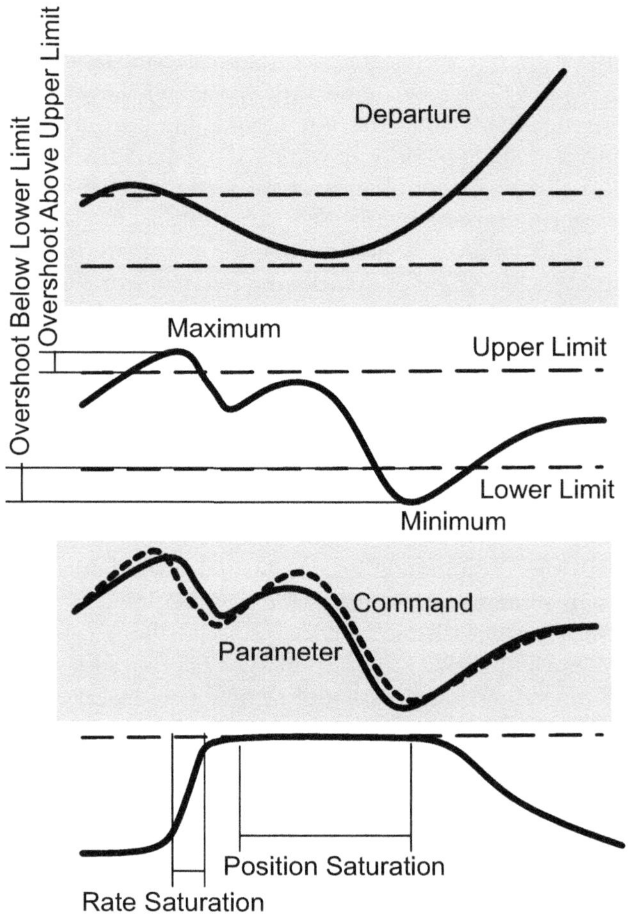

Fig. 21.5 Evaluation criteria for nonlinear simulation

highest overshoot, longest time in saturation etc., is selected out of all the assessed configurations, manoeuvres and initial stick positions. These time histories are marked as 'worst cases'. To further increase the variation of the assessment parameters the 'worst case' manoeuvres are modified with respect to the following criteria to obtain the so called 'additional' manoeuvres:

- Mirrored stick inputs
- Manoeuvre with airbrake extended
- Manoeuvre with half the original control inputs
- Manoeuvre with 6 degrees-of-freedom, i.e. speed is variable and thrust can be varied

After these 'additional' manoeuvres have been calculated, the new time histories are automatically evaluated with respect to the same criteria as described above and the results added to the database file. The extremal angle-of-attack and vertical-load-factor values found in this database for each altitude and

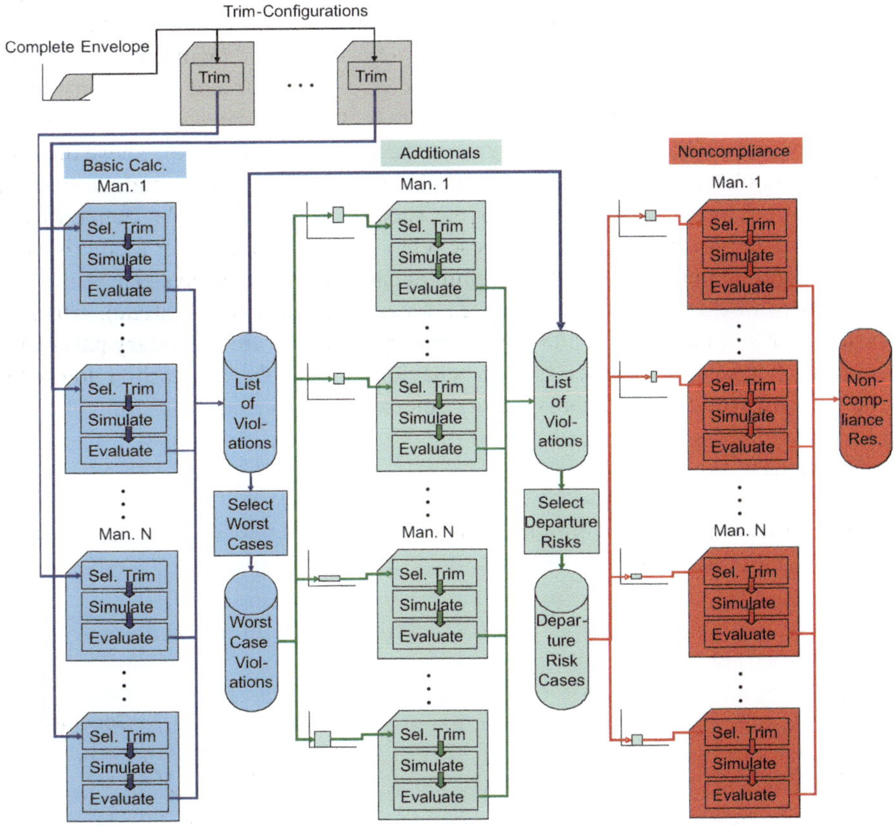

Fig. 21.6 Flow chart of nonlinear assessment process

Mach number point are used to determine the AoA/Nz target values up to which the linear analysis must be performed.

The results of the evaluations of all time histories for the nonlinear assessment criteria must be compared with the associated requirements of [2]. If a requirement is violated this can be accepted, if it only occurs for certain cases which can be excluded via procedure/briefing note, like e.g. problems with extended airbrake. Another possibility to accept a violation is, if the probability of its occurrence is below a certain margin. If an assessment criterion indicates a deterioration of handling qualities, manned simulations are required. The problem can be cleared, if the test pilots classifies the handling qualities as acceptable, otherwise a briefing note or a limitation will be required.

All time histories that violate a criterion that affects closed loop stability, e.g. control surface position or rate saturation, are marked for noncompliance work in the database. For these cases dedicated evaluations are required to prove the existence of sufficient stability margins (robustness). The simulations are carried out assuming reasonable atmospheric disturbances or increased model errors according to the tolerance concept. Time histories marked for noncompliance work are re-calculated with the following modifications:

- Increased air-data-system uncertainties
- Several aerodynamic uncertainties
- Atmospheric disturbances
- Control power variation

After these calculations are done, the new time histories are once again automatically evaluated with respect to the assessment criteria. Manual simulations with fine tuned inputs to destabilize the aircraft are performed in addition. If the time histories do not show a departure tendency, the aircraft behaviour is acceptable despite the original criterion violation. It may however be necessary to check remaining handling qualities issues in manned simulation.

In addition dedicated 6 degrees-of-freedom manoeuvres like decelerations are calculated and evaluated in a similar way as described above with respect to the assessment criteria. This is done to cover acceleration/deceleration dependent effects in the aerodynamics, engine model and FCS.

If no departure tendency can be detected, the handling qualities criteria have been satisfied and a M3 (linear) clearance is available, the most important requirements for a carefree handling (M1) clearance are met. Remaining handling qualities issues after the nonlinear simulation and highly interactive maneuvers must be assessed in manned simulation before a M1 clearance can finally be issued. If a M1 clearance is not obtainable, the clearance process is repeated with less demanding manoeuvres according to the M2 definition with the aim of providing at least a M2 clearance. This may be acceptable for example for flight test purposes or certain flight phases.

21.5.4 Nonlinear Simulation for Dedicated Flight Test Manoeuvres

For aerodynamic data gathering flight tests, a defined excitation of the control surfaces is desired. Due to the interaction with the FCS, specific test inputs are required. For this purpose the Eurofighter control laws offer a unique 'Frequency and Bias Inputs' (FBI) functionality. Please see Fig. 21.7 for a schematic description. If necessary FBI manoeuvres must be cleared in addition to the standard carefree handling (M1) manoeuvres.

After being triggered by the pilot a frequency input adds a sinusoidal signal to the selected nominal control surface signal to enforce a clean excitation. A bias input can be used to modify the trim position of a control surface, e.g. the canard, or to shift target values like angle-of-attack limit or roll-rate limit. The bias part allows for a controlled stepwise approach for envelope expansion. The nonlinear assessment determines the extremal FBI signals that can be used safely during flight test with a process similar to the one described for the carefree handling clearance.

21.5.5 Manned Simulation

Manned simulation uses the same aircraft and environment models as the nonlinear simulation. The main difference is pilot participation via the cockpit and visual system which is attached to the simulation model. This pilot-in-the-loop possibility is essential for some handling qualities evaluations and for checking problem areas found in nonlinear simulations for relevance.

The simulator department at Military Air Systems Manching plant offers several facilities ranging from a one channel visual system to the 'dome' with an almost 360° field of view (see Fig. 21.8). Using the 'dome' greatly improves the level of realism and thus it is preferred for most manned simulation tasks.

Manned simulation is used for manoeuvres with high pilot involvement, like e.g. tracking tasks, air-to-air-refuelling, taxiing, takeoff and landing. These kind of manoeuvres are hard to model in the nonlinear simulation due to the required realistic pilot model.

If the linear and/or nonlinear assessment results in any flight limitations, they must be checked in manned simulation for their applicability. Problem areas found during the nonlinear assessment must be checked in manned simulation for their severity and whether the pilot is able to sustain the flight condition long enough for the problem to develop. Manual recovery from system failures and other transient manoeuvres are also best evaluated in manned simulation.

Handling qualities are further assessed in the simulator. In this case the use of the 'dome' simulation is mandatory. It may be necessary to obtain several pilot opinions, including service pilots, for a consolidated statement. Despite the sophisticated 'dome' environment some handling qualities deficiencies identified in flight test may be difficult to reproduce.

21.5.6 Flight Test

Flight test is the ultimate assessment method. It is however the most risky and expensive part of the assessment and thus preferably all significant shortcomings of the aircraft should be ironed out before. Flight test is further the only kind of assessment that takes place in the real world environment including pilot, atmospheric disturbances, load factors etc.

Flight tests are performed for a variety of reasons, like e.g. model validation, system tests, stores integration, man-machine interface tests, aerodynamic data gathering and envelope expansion. Most of the flight testing is covered by the carefree handling clearance. Aerodynamic data gathering especially with envelope expansion requires some additional support for increased safety.

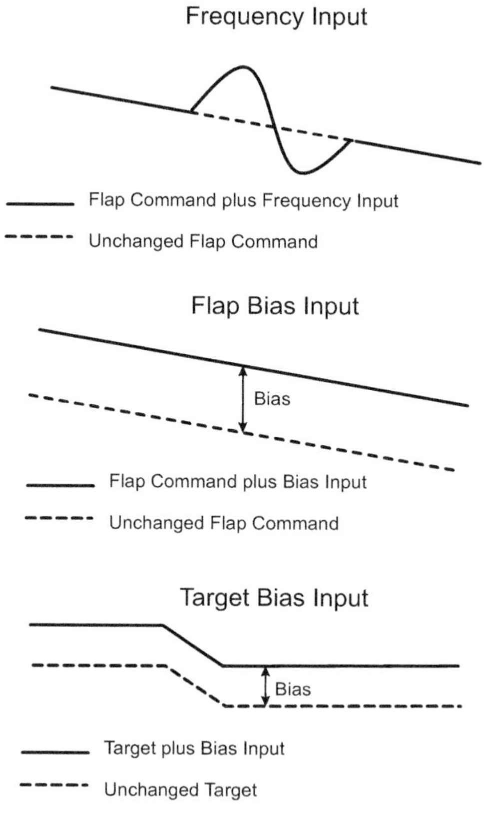

Fig. 21.7 Frequency and bias input schematic

Fig. 21.8 'Dome' simulation facility

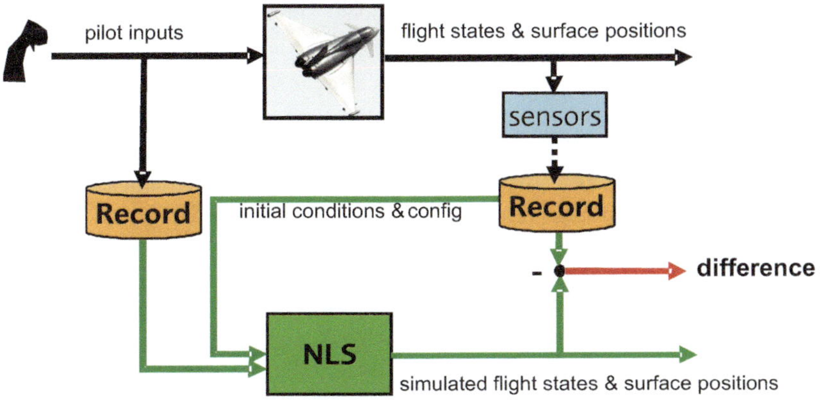

Fig. 21.9 Diagram of post-flight analysis process

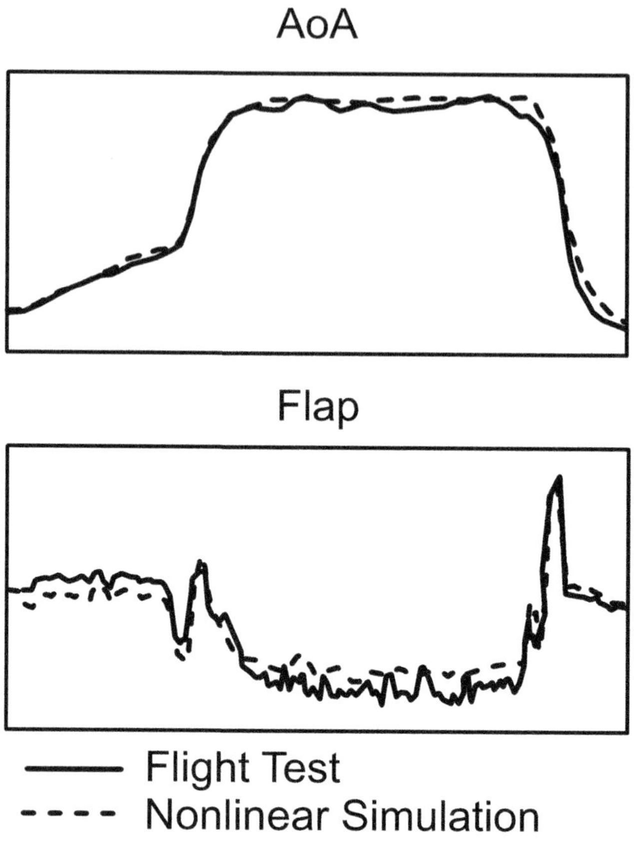

Fig. 21.10 Comparison of post-flight nonlinear simulation results with flight test data

During aerodynamic data gathering and especially envelope expansion flight testing, critical parameters like angle-of-attack, surface deflections etc. are monitored versus the corresponding values obtained during the nonlinear simulation. If a significant discrepancy occurs, this indicates that the models used for the clearance may deviate from reality and the flight test has to be aborted for further evaluation.

Correlation of flight test data and nonlinear simulation is generally very good, especially in the envelope regions where all the aircraft models (aerodynamic, engine, sensors etc.) are validated. This is of course only true for smooth atmospheric conditions. During model validation, e.g. during aerodynamic data gathering for envelope expansion, larger differences may occur. In this case the relevant models have to be fine tuned.

21.5.7 Post-flight Analysis for Aerodynamic Dataset Validation

After aerodynamic data gathering flight tests, the recorded flight test data can be used to validate the aerodynamic model. In this process the same models used to derive the clearance, i.e. to 'predict' the aircraft response, are utilized to calculate or 're-predict' a specific flight test manoeuvre during post-flight analysis and compare it with the actual flight test data.

The analysis process is depicted in Fig. 21.9. During a manoeuvre, pilot inputs, control surface positions and sensor data are recorded. After the flight the recorded pilot inputs, configuration information (mass, CoG, inertias, stores etc.) and initial conditions (altitude, speed, Mach number, angle-of-attack, sideslip-angle, etc.) are used as inputs to the nonlinear simulation to re-produce the manoeuvre. The resulting flight states and control surface positions are then compared with those recorded during the test flight. If the aerodynamic dataset is correct, there should be only minor differences between the calculated and measured values. The type and size of the deviations can give valuable information about how to improve the aerodynamic model. Unfortunately atmospheric disturbances encountered during flight test can cause significant differences and make this task more complex and difficult. An example with the results of a post-flight analysis is shown in Fig. 21.10.

21.6 Conclusion

Providing a clearance for an advanced fighter aircraft is a complex and very challenging task due to the high agility and performance of the vehicles, their larger flight envelope, demanding mission types and high instability. The carefree handling requirement of the FCS further increases the complexity of the clearance task.

For these reasons a typical carefree handling clearance for Eurofighter Typhoon requires huge amounts of stability analysis calculations (typically 700,000 per task) and nonlinear simulations (typically 600,000 per task). Considering all assessment tasks to be done for a flight control laws or aerodynamic model version (more than one task done by each EPC) approximately 12 million linear and 10 million nonlinear assessment runs are performed. To keep those data volumes manageable a high degree of automation is required for creating, calculating and evaluating the data throughout the assessment chain from basic calculations up to the robustness checks, both for linear and nonlinear assessment. At the end only the 'tip of the iceberg' can be evaluated manually. Without this layered approach the number of data to be created would be higher by approximately a factor of 15 to 20.

To ensure safe performance of flight test manoeuvres associated with FCS clearance or aerodynamic data gathering, assessment results can be used to define limits against which flight test data are monitored. This is done in real time for several critical parameters. Aerodynamic model changes based on flight test results are validated by comparing the measured flight test parameters with the ones calculated during post-flight analysis.

To further improve the quality of the assessment process for the Eurofighter Typhoon the FJT has started the implementation of a stochastic clearance process using Monte Carlo simulations. This is done in addition to the established EPC clearance processes. The aim is to cover exotic parameter combinations that might be overlooked otherwise.

Currently unmanned vehicles are becoming of increased importance in the military world. Clearance of these aircrafts will cause additional challenges for the assessment process due to e.g., increased reliability demands and clearance of sense and avoid systems, albeit some other factors may be relaxed e.g., the flight envelope, compared to manned fighter aircraft.

References

1. Eurofighter Jagdflugzeug GmbH, `http://www.eurofighter.com`
2. Sortino M (2009) EFA Flight Mechanics Design and Clearance Requirements, EADS, CD-J-462-F-0003(9)
3. Stenglin P.: EF2000 Flight Control Laws Design and Clearance Requirements, EADS, R-J-462-F-0001(3) (2003)
4. N.N., Flight Control Systems - Design, Installation and Test of Piloted Aircraft, MIL-DTL-9490E (2008)
5. N.N., Military Standard Flying Qualities of Piloted Aircraft, MIL-STD-1797A (2004)

Appendix A
Nomenclature

List of Standard Symbols

Symbol	Definition
(A, B, C, D)	Quadruple of system state-space realization matrices (state, input, output, direct feedthrough)
$\mathcal{B}_{\boldsymbol{\Delta}}$	Ball of structured perturbations
c	Aerodynamic chord (m)
$\mathbb{C}$	Field of complex numbers or complex plane
C_x	Coefficient of axial force
C_y	Coefficient of side force
C_z	Coefficient of normal force
C_l	Coefficient of rolling moment
C_m	Coefficient of pitching moment
C_n	Coefficient of yawing moment
$C_{z,\alpha_{max}}$	Maximum lift coefficient value
δC_x	Relative uncertainty in C_x
δC_y	Relative uncertainty in C_y
δC_z	Relative uncertainty in C_z
$\mathrm{diag}(A_1, \ldots, A_k)$	Block-diagonal matrix with diagonal blocks A_1, ..., A_k
$\mathcal{F}_l(M, \Delta)$	Lower linear fractional transformation matrix
$\mathcal{F}_u(M, \Delta)$	Upper linear fractional transformation matrix
g	Acceleration due to gravity (m/s^2)
h	Altitude (feet)
$G(s)$	Transfer-function matrix with coefficients in $\mathbb{R}$ or $\mathbb{C}$
$\mathcal{H}_2$	Space of square integrable analytic functions on the right half-plane
$\mathcal{H}_\infty$	Space of essentially bounded analytic functions on the right half-plane
$\Im(\cdot)$	Imaginary part of a complex quantity
L	Wingspan (m)
$\mathcal{L}_2$	Space of square integrable functions
m	Aircraft total mass (t)
m_p	Estimated aircraft total mass (t)
M	Mach number

Symbol	Definition
$\overline{M}$	Normalized Mach number
M_D	Design diving Mach number
M_{MO}	Maximum operating Mach number
$\mathbb{N}$	Set of natural numbers
$\mathcal{N}(\mu, \sigma)$	normal distribution with mean μ and variance σ
n_y	Load factor along y-axis (g)
n_z	Load factor along z-axis (g)
δn_z	Load factor correction along z-axis (g)
$\delta n_{z,c}$	Demanded load factor along z-axis (g)
p	Body-axis roll rate (deg/s)
δp_m	Stick lateral deflection angle (deg)
δp_c	Demanded roll angle (deg)
q	Body-axis pitch rate (deg/s)
δq_c	Demanded pitch deflection angle (deg)
δq_m	Stick longitudinal deflection angle (deg)
r	Body-axis yaw rate (deg/s)
δr_c	Demanded yaw angle (deg)
δr_m	Pedal deflection angle (deg)
$\mathbb{R}$	Field of real numbers
$\mathbb{R}_-$	Set of negative real numbers
$\mathbb{R}_+$	Set of positive real numbers
$\Re(\cdot)$	Real part of a complex quantity
s	Complex variable in Laplace operator
S	Wing planform area (m^2)
t	Time (s)
trace(A)	Trace of matrix A
V_{cas}	Calibrated air speed
δV_{cas}	Uncertainty in V_{cas}
ΔV_{cas}	Normalized calibrated air speed
V_D	Design diving speed
V_{ls}	Minimum selectable speed
V_{MO}	Maximum operating speed
V_{s1g}	Stall speed
V_{tas}	True air speed
X_{cg}	Centre of gravity position along x-axis
$X_{cg,p}$	Estimated centre of gravity position along x-axis
ΔX_{cg}	Normalized centre of gravity position along x-axis

List of Greek Symbols

Symbol	Definition
α	Angle of attack (deg)
α_{max}	Maximum value of the angle of attack (deg)
β	Angle of sideslip (deg)
β_c	Demanded sideslip (deg)
β_{est}	Estimated sideslip (deg)
χ	Yaw angle (deg)
γ	Flight path angle (deg)
δ	Vector of system parameters
Δ	LFT uncertainty matrix
$\boldsymbol{\Delta}$	Set of allowable structured perturbations
δ_{CT}	Amount of fuel in the center tank
δ_{OT}	Amount of fuel in the outer tank
δ_{PL}	Embarked payload
$\delta_{A_{il}}$	Inner left aileron deflection (deg)
$\delta_{A_{il,c}}$	Demanded inner left aileron deflection (deg)
$\delta_{A_{ir}}$	Inner right aileron deflection (deg)
$\delta_{A_{ir,c}}$	Demanded inner right aileron deflection (deg)
$\delta_{A_{ol}}$	Outer left aileron deflection (deg)
$\delta_{A_{ol,c}}$	Demanded outer left aileron deflection (deg)
$\delta_{A_{or}}$	Outer right aileron deflection (deg)
$\delta_{A_{or,c}}$	Demanded outer right aileron deflection (deg)
$\delta_{S_{l,2,3}}$	Left spoilers 2 and 3 deflections (deg)
$\delta_{S_{l,2,3,c}}$	Demanded left spoilers 2 and 3 deflections (deg)
$\delta_{S_{r,2,3}}$	Right spoilers 2 and 3 deflections (deg)
$\delta_{S_{r,2,3,c}}$	Demanded right spoilers 2 and 3 deflections (deg)
$\delta_{S_{l,4,5}}$	Left spoilers 4 and 5 deflections (deg)
$\delta_{S_{l,4,5,c}}$	Demanded left spoilers 4 and 5 deflections (deg)
$\delta_{S_{r,4,5}}$	Right spoilers 4 and 5 deflections (deg)
$\delta_{S_{r,4,5,c}}$	Demanded right spoilers 4 and 5 deflections (deg)
δ_{THS}	Trimmable horizontal stabilizer deflection (deg)
$\delta_{THS,c}$	Demanded trimmable horizontal stabilizer deflection (deg)
δ_E	Elevator deflection (deg)
$\delta_{E,c}$	Demanded elevator deflection (deg)
δ_R	Rudder deflection (deg)
$\delta_{R,c}$	Demanded rudder deflection (deg)
μ	Structured singular value
$\mu_{\boldsymbol{\Delta}}(M)$	Structured singular value of M with respect to $\boldsymbol{\Delta}$
ϕ	Bank angle (deg)
ϕ_c	Demanded bank angle (deg)
λ_j	Generic j-th eigenvalue
Π	Thrust value

Symbol	Definition
θ	Pitch angle (deg)
ρ	Density of air (kg/m^3)
ω	Frequency (rad/sec)
ψ	Heading (deg)

Index

Lecture Notes in Control and Information Sciences

Edited by M. Thoma, F. Allgöwer, M. Morari

Further volumes of this series can be found on our homepage:
springer.com

Vol. 416: Varga, A.; Hansson, A.;
Puyou, G. (Eds.):
Optimization Based Clearance of Flight Control Laws
451 p. 2012 [978-3-642-22626-7]

Vol. 412: Fridman, L.; Moreno, J.;
Iriarte R. (Eds.):
Sliding Modes after the First Decade of the 21st Century
595 p. 2011 [978-3-642-22163-7]

Vol. 411: Kaczorek, T.;
Selected Problems of Fractional Systems Theory
344 p. 2011 [978-3-642-20501-9]

Vol. 410: Bourlès, H.; Marinescu, B.;
Linear Time-Varying Systems
637 p. 2011 [978-3-642-19726-0]

Vol. 409: Xia, Y.; Fu, M.; Liu, G.-P.;
Analysis and Synthesis of Networked Control Systems
198 p. 2011 [978-3-642-17924-2]

Vol. 408: Richter, J.H.;
Reconfigurable Control of Nonlinear Dynamical Systems
291 p. 2011 [978-3-642-17627-2]

Vol. 407: Lévine, J.; Müllhaupt, P.:
Advances in the Theory of Control, Signals and Systems with Physical Modeling
380 p. 2010 [978-3-642-16134-6]

Vol. 406: Bemporad, A.; Heemels, M.;
Johansson, M.:
Networked Control Systems
appro. 371 p. 2010 [978-0-85729-032-8]

Vol. 405: Stefanovic, M.; Safonov, M.G.:
Safe Adaptive Control
appro. 153 p. 2010 [978-1-84996-452-4]

Vol. 404: Giri, F.; Bai, E.-W. (Eds.):
Block-oriented Nonlinear System Identification
425 p. 2010 [978-1-84996-512-5]

Vol. 403: Tóth, R.;
Modeling and Identification of Linear Parameter-Varying Systems
319 p. 2010 [978-3-642-13811-9]

Vol. 402: del Re, L.; Allgöwer, F.;
Glielmo, L.; Guardiola, C.;
Kolmanovsky, I. (Eds.):
Automotive Model Predictive Control
284 p. 2010 [978-1-84996-070-0]

Vol. 401: Chesi, G.; Hashimoto, K. (Eds.):
Visual Servoing via Advanced Numerical Methods
393 p. 2010 [978-1-84996-088-5]

Vol. 400: Tomás-Rodríguez, M.;
Banks, S.P.:
Linear, Time-varying Approximations to Nonlinear Dynamical Systems
298 p. 2010 [978-1-84996-100-4]

Vol. 399: Edwards, C.; Lombaerts, T.;
Smaili, H. (Eds.):
Fault Tolerant Flight Control
appro. 350 p. 2010 [978-3-642-11689-6]

Vol. 398: Hara, S.; Ohta, Y.;
Willems, J.C.; Hisaya, F. (Eds.):
Perspectives in Mathematical System Theory, Control, and Signal Processing
appro. 370 p. 2010 [978-3-540-93917-7]

Vol. 397: Yang, H.; Jiang, B.;
Cocquempot, V.:
Fault Tolerant Control Design for Hybrid Systems
191 p. 2010 [978-3-642-10680-4]

Vol. 396: Kozlowski, K. (Ed.):
Robot Motion and Control 2009
475 p. 2009 [978-1-84882-984-8]

Vol. 395: Talebi, H.A.; Abdollahi, F.;
Patel, R.V.; Khorasani, K.:
Neural Network-Based State Estimation of Nonlinear Systems
appro. 175 p. 2010 [978-1-4419-1437-8]

Vol. 394: Pipeleers, G.; Demeulenaere, B.;
Swevers, J.:
Optimal Linear Controller Design for Periodic Inputs
177 p. 2009 [978-1-84882-974-9]